The

IACUC

HANDBOOK

Second Edition

The

IACUC

HANDBOOK

Second Edition

Edited by

Jerald Silverman
Mark A. Suckow
Sreekant Murthy

CRC Press
Taylor & Francis Group
Boca Raton   London   New York

CRC Press is an imprint of the
Taylor & Francis Group, an informa business

CRC Press
Taylor & Francis Group
6000 Broken Sound Parkway NW, Suite 300
Boca Raton, FL 33487-2742

International Standard Book Number-10: 0-8493-4010-1 (Hardcover)
International Standard Book Number-13: 978-0-8493-4010-9 (Hardcover)

Library of Congress Cataloging-in-Publication Data

The IACUC handbook / edited by Jerald Silverman, Mark A. Suckow, and Sreekant Murthy.
    p. cm.
"A CRC title."
Includes bibliographical references and index.
ISBN-13: 978-0-8493-4010-9 (alk. paper)
ISBN-10: 0-8493-4010-1 (alk. paper)
    1. Animal welfare. 2. Laboratory animals. 3. National Institutes of Health (U.S.). Institutional Animal Care and Use Committee. I. Silverman, Jerald. II. Suckow, Mark A. III. Murthy, Sreekant. IV. National Institutes of Health (U.S.). Institutional Animal Care and Use Committee

HV4708.I23 2007
179'.3--dc22
    2006024571

Visit the Taylor & Francis Web site at
http://www.taylorandfrancis.com

and the CRC Press Web site at
http://www.crcpress.com

Preface to the Second Edition

Since the publication of the first edition of this book there have been few changes to those federal laws, regulations, and policies that affect IACUCs, but there have been numerous written opinions and clarifications on how to interpret them. Add to this the continuing experience (and often frustration) of IACUCs with the day-to-day implementation of federal policies and regulations, and the need for a second edition of *The IACUC Handbook* became apparent.

Among the changes affecting IACUCs that occurred since the first edition of this book, the Farm Security and Rural Investment Act of 2002 amended the definition of an *animal* in the AWA. Before the amendment, the AWA excluded *all* birds; now, the AWA excludes only those birds bred for research, while still excluding rats of the genus *Rattus*, and mice of the genus *Mus*, bred for research. The regulations of the AWA were also changed to reflect the new definition of an *animal*. At the time of publication, the USDA was still developing standards for the regulation of those birds now included in the AWA. The PHS Policy definition of an animal has not changed.

Also in 2002, the NIH allowed institutions with Animal Welfare Assurances to submit verification of IACUC approval for animal use procedures on a grant application subsequent to peer review, but prior to award. Previously, IACUC approval was required earlier on in the grant review process. At the time of the change some IACUCs were of the opinion that this would significantly decrease their burden, but subsequent experience suggests that the modification has had little practical effect on the operations of the majority of IACUCs.

Another interesting change, leading to an entire new chapter in this edition, has been the apparent increase in the number of organizations that have seen the need to initiate some form of postapproval monitoring (PAM) of activities on IACUC protocols. Whereas some institutions have considered PAM to be a policing action to assure regulatory compliance, Chapter 30 of this edition emphasizes the collaborative effort needed to make PAM beneficial to the institution, IACUC, and animals.

In 2001 the American Veterinary Medical Association published updated guidelines for the euthanasia of animals. Those guidelines have been incorporated into this edition as needed. The definition of a *field study* under the AWA regulations was changed; however, this was more of a rewording than a change that has had any significant effect on animal use procedures.

This second edition of *The IACUC Handbook* not only incorporates new information on existing laws, regulations, policies and "how to" advice, but also features a greatly expanded survey of IACUC practices from institutions around the nation. In some chapters we have permitted the repetition of items that are similar to those found in other chapters. These have been cross-referenced to provide a broader perspective of expert opinions. We hope all of this will be of value to you.

This new edition has contributions from numerous authors and contributors, all of whom gave freely of their time and expertise to provide our readers with the most up-to-date information possible. Our sincerest thanks go to all of them. We also thank Dr. John Miller of the Association for Assessment and Accreditation of Laboratory Animal Care International (AAALAC), who kindly reviewed all aspects of the text that referred to AAALAC.

The IACUC has a unique place in the oversight of animal care and use activities for most organizations in the U.S. It is a self-regulatory committee, representing its institution to the federal government. Equally importantly, it helps represent to the public the basic standards of animal care and concerns for animal welfare to which researchers, institutions, and laboratory animal science professionals subscribe. Yet, if self-regulation is to be continued, then we must go the extra mile and strive to reach even higher standards of animal care and use. For that to occur, institutions and their IACUCs must be leaders, not followers.

Jerald Silverman, D.V.M.
Mark A. Suckow, D.V.M.
Sreekant Murthy, Ph.D.

Preface to the First Edition

The concept of regulatory oversight as a means to facilitate the welfare of animals used in research has existed for approximately 120 years, beginning with the 1876 Cruelty to Animals Act in England. In the U.S., somewhat disjointed local efforts were undertaken to ensure animal welfare. These efforts met with varied degrees of success.

The Health Research Extension Act of 1985 directed the National Institutes of Health to construct guidelines and recommendations for development of institutional committees that would oversee the use of vertebrate animals in teaching, testing, and research on a local level. The U.S. Department of Agriculture was quick to follow with similar requirements. In this way, the Institutional Animal Care and Use Committee, or IACUC, was born.

The 15 years since the inception of requirements for establishing the IACUC has seen continued evolution of the issues addressed by IACUCs and the means by which review of activities is conducted. For example, the explosion of transgenic technology has presented IACUCs with new challenges in terms of ensuring animal health and welfare. There also has been ongoing development of alternative methods which either replace animal use or significantly reduce animal numbers or distress. This has led investigators to invest deliberate time and effort into the search for alternatives and, whenever possible, to incorporate alternative methods into their research.

Central to the mission of the IACUC is training. The IACUC must ensure that investigators, research technicians, animal care personnel, and others are adequately trained to conduct research with animals in a proper and humane manner. However, it is equally critical that the IACUC itself be adequately trained and knowledgeable with regard to successful interpretation and implementation of regulations and guidelines for proper animal care and treatment. This mission must be accomplished without impeding the scientific enterprise.

The intent of this book is to address those questions and problems often confronting IACUCs and give accurate and succinct answers to them. *The IACUC Handbook* is divided into distinct chapters. For most chapters, the information is provided in "question and answer" format, with the authors offering their personal opinions against a background of regulatory information. We have arbitrarily included, under the regulatory heading, pertinent sections from the Animal Welfare Act and its regulations, the Health Research Extension Act of 1985, the Public Health Service Policy on Humane Care and Use of Laboratory Animals, the 1996 edition of the *Guide for the Care and Use of Laboratory Animals*, and selected publications intended to clarify the Animal Welfare Act regulations and the Public Health Service Policy. Many authors also have provided results from informal surveys as examples of the approaches institutions have taken to solve complex issues. These surveys are not scientific; however, they serve to provide the reader with additional opinions.

This book could not have been written without the chapter authors, chapter contributors, and the many laboratory animal scientists who provided information for the informal surveys. We thank all of you. We would like to give special recognition and thanks to Dr. W. Ron DeHaven and the staff of APHIS/AC, Dr. Nelson Garnett and the staff of NIH/OPRR, and, in particular, Dr. Carol Wigglesworth of NIH/OPRR. APHIS/AC and NIH/OPRR judiciously reviewed every chapter, providing invaluable advice and corrections. Although we have attempted to make the information provided in this book as accurate as possible, we recognize that errors may occur. To that end, we encourage you, the reader, to provide feedback to any of the editors and chapter authors, and to suggest additional questions that might be addressed in future editions of this book.

There can be little question that the role and impact of the Institutional Animal Care and Use Committee will continue to grow in scope and complexity. We hope that this book will provide a foundation — a solid footing — for those attempting to understand the many and varied responsibilities of this committee.

Jerald Silverman, D.V.M.
Mark A. Suckow, D.V.M.
Sreekant Murthy, Ph.D.

Introduction to the First Edition

It has been approximately 13 and 8 years, respectively, since the implementation of the current PHS Policy and Part 2 of the USDA regulations, which first defined the federal requirements for Institutional Animal Care and Use Committees (IACUCs) to oversee certain specific animal care and use program functions. Since then, the IACUC has evolved as the premier instrument of animal welfare oversight within the majority of biomedical research institutions in the U.S. In the interim, the biomedical research community has gained a wealth of experience based, in many cases, on trial and error and on anecdotal information shared through informal communications and other means. Many excellent resources have been developed, but until now, few have provided specific opinion on "best practices" at the facility level.

The format of this book is unique. Answers to a series of practical everyday questions are addressed from the perspective of the applicable regulatory language, from the opinions of knowledgeable and experienced professionals in the field, and in the form of responses to informal surveys on selected institutional policies and practices.

The sections labeled "regulatory" have been reviewed by staff at NIH/OPRR and APHIS/AC for consistency with the Public Health Service Policy on Humane Care and Use of Laboratory Animals (PHS Policy) and the USDA Animal Welfare Act Regulations (AWAR). Every effort has been made to apply correct interpretations in the context of the specific issues being discussed. However, in this highly nuanced field, readers are cautioned not to apply these interpretations out of context or to extend them beyond their intended meaning. Accordingly, such interpretations should not be seen as formulating new federal regulations or policies, and readers are advised to refer to source documents and direct consultations with NIH/OPRR and APHIS/AC when in doubt.

The information provided in the sections labeled "survey" gives an interesting view of how several other institutions may be addressing certain issues, but a note of caution is advised here also. Some responses may appear to indicate deviations from federal policy and regulation. Because variables, such as the type of institution, the species involved, and the type of oversight that is applicable, are not always specified, some answers may reflect differences in applicability of rules or understanding of questions, rather than noncompliance.

For the experienced and sophisticated reader, this book provides a wealth of useful information and taps into the collective experience and wisdom of the various authors. Both NIH/OPRR and APHIS/AC commend the authors and editors for their outstanding efforts and for moving the biomedical research community forward in its formulation of best practices and commonly accepted professional guidance in this complex arena.

Nelson Garnett, D.V.M
W. Ron DeHaven, D.V.M.

Some Comments about the Surveys

The surveys reported in this text should be considered informal reports of how different IACUCs around the nation approach some of the questions presented in the chapters. They should not be considered scientifically valid in terms of methodology. The intent of the surveys is to provide the reader with a general sense of how various IACUCs approach day-to-day concerns, in particular when issues that arise are not clearly defined by federal laws, regulations, or policies.

Using a database of NIH/OLAW-accredited institutions, approximately 1,100 questionnaires were mailed to IACUC Chairpersons throughout the U.S. The packet contained instructions and a stamped, self-addressed return envelope. Instructions included a statement that a basic understanding of IACUC operations was assumed and common, although not necessarily required procedures (such as including rats and mice in the APHIS/AC pain and distress classification system) might be used in the survey questions. Neither signatures nor other identifying information was requested. All raw data were collected by one of the editors (J.S.) and sent to the appropriate chapter authors. Approximately 200 questionnaires were returned, and of those approximately 170 provided usable data. IACUC Chairpersons were given the option of responding to the entire (and quite long) questionnaire form or responding to one half of the form (randomly chosen by the editor). The majority responded to the entire form. As not all chairpersons responded to all questions (even when they chose to complete the entire form) and because some responses were not usable, the number of respondents to different questions in the same chapter may vary.

Unless otherwise noted, in each chapter the number of respondents is represented by the denominator and the actual number of responses is represented by the numerator (in some questions more than one response to a question was possible). In a small number of survey responses the totals do not add up to the sum of the parts as a consequence of number rounding. Six initial questions were asked of all respondents.

1. What is the nature of your organization?

- Primarily academic respondents   90/171
- Primarily pharmaceutical (with drug discovery)   14/171
- Primarily private nonpharmaceutical research or testing   36/171
- Primarily laboratory animal breeding   5/171
- Primarily governmental research or testing   21/171
- Primarily zoological or other exhibitor, with associated research   0/171
- Other   5/171

Note: "Primarily private nonpharmaceutical research or testing" included contract research organizations.

2. Does your organization currently have any level of AAALAC accreditation?

- Yes   111/175
- No   64/175

3. What is the approximate size of your organization (all locations, all job categories, including nonresearch)?

- About 1–100 full-time-equivalent (FTE) employees   35/176
- About 101–500 FTE employees   39/176
- About 501–1000 FTE employees   17/176
- About 1001–5000 FTE employees   42/176
- Over 5000   43/176

4. Is your organization legally required to comply with the Animal Welfare Act for all or some of its animal-related work?

- Yes   170/177
- No   7/177

5. Is your organization legally required to comply with the PHS Policy for all or some of its animal-related work?

- Yes   167/172
- No   5/172

We also asked one question about the personal perception of the individual completing the questionnaire about his or her own IACUC. The responses were largely as anticipated.

6. We would like a general sense of your opinion of your own IACUC. Therefore, do you personally consider your IACUC to be

- Not very diligent in its efforts to assure the welfare of animals   0/178
- Somewhat diligent, but probably should do a better job of assuring the welfare of research animals   9/178
- Diligent. We strike a good balance between research needs and the welfare of animals   85/178
- Very diligent. We tend to put the welfare of the animal first, but never forget they are there for research   82/178
- Too diligent. I think the committee has gone overboard with animal welfare and forgotten about research   0/178
- Other   2/178

The Editors

Jerald Silverman, D.V.M., received degrees in vertebrate zoology and veterinary medicine from Cornell University, and a degree in nonprofit organization management from the New School for Social Research. His background includes private veterinary practice and laboratory animal practice in private and academic research institutions. Since 2002 he has been Director of the Department of Animal Medicine and Professor of Pathology at the University of Massachusetts Medical School, and Adjunct Professor of Environmental and Population Health at Tufts University, Cummings School of Veterinary Medicine.

Dr. Silverman is a diplomate of the American College of Laboratory Animal Medicine and past president of the American Society of Laboratory Animal Practitioners. He has published over 60 research and clinical papers and book chapters and has coordinated the Protocol Review column for *Lab Animal* since 1988. He has long been an advocate for the advancement of the moral stature of animals, particularly those used in biomedical research. He enjoys teaching, nature photography, and watching his daughter grow up.

Mark Suckow, D.V.M., is Director of the Freimann Life Science Center at the University of Notre Dame, Notre Dame, Indiana. He received his D.V.M. from the University of Wisconsin in 1987, and completed a post-doctoral residency program in laboratory animal medicine at the University of Michigan in 1990. He is a diplomate of the American College of Laboratory Animal Medicine.

Dr. Suckow has published more than 80 scientific papers, books, and book chapters. He was co-editor of the first edition of *The IACUC Handbook* (CRC Press, 2000). He has presented seminars and papers at a variety of meetings on diverse topics related to research, professional ethics, and mentorship.

Dr. Suckow was honored as the 1996 Young Investigator of the Year by the American Association for Laboratory Animal Science. In 1998, he received the Excellence in Research Award from the American Society of Laboratory Animal Practitioners. He is the 2006 President of the American Association for Laboratory Animal Science.

Sreekant Murthy, Ph.D., is Professor of Medicine and Vice Provost for Research Compliance at Drexel University, Philadelphia. He received his doctoral degree from the Philadelphia College of Pharmacy and Science in 1973. He completed his Postdoctoral Research Assistantship at the General Clinical Research Center at Temple University Hospital.

Dr. Murthy has published more than 90 scientific papers and more than 150 abstracts and short papers in peer-reviewed journals and books. He has made almost 200 invited presentations in diverse scientific fields and topics on the responsible and ethical conduct of research. He has served on the editorial boards of several journals and he is a peer reviewer for grants and papers for multiple scientific journals.

Chapter Authors

David R. Archer, Ph.D.   Assistant Professor, Department of Pediatrics, Emory University School of Medicine, Atlanta, Georgia

Ron E. Banks, D.V.M., Dipl. A.C.V.P.M., Dipl. A.C.L.A.M.   Director, Office of Animal Welfare Assurance, Duke University, Durham, North Carolina

Kathryn A. L. Bayne, M.S., Ph.D., D.V.M., Dipl. A.C.L.A.M., C.A.A.B.   Senior Director and Director of Pacific Rim Activities Association for Assessment and Accreditation of Laboratory Animal Care International, Waikoloa, Hawaii

Marcy Brown, B.S., M.A.   Director, Animal Care and Use Committee, University of California, Berkeley, California

Larry Carbone, D.V.M., Ph.D.   Senior Veterinarian, Laboratory Animal Resource Center, University of California San Francisco, San Francisco, California

Marilyn J. Chimes, D.V.M., Dipl. A.C.L.A.M., J.D.   Attorney, Jenner & Block LLP, Chicago, Illinois

Susan Stein Cook, D.V.M., M.S., Dipl. A.C.L.A.M.   Consultant in Laboratory Animal Medicine, Williamston, Michigan; Veterinarian, Central Michigan University, Mount Pleasant, Michigan

Stephen K. Curtis, D.V.M., Dipl. A.C.L.A.M.   Associate Director, Alamogordo Primate Facility (NIH), Holloman AFB, New Mexico

Peggy J. Danneman, V.M.D., M.S., Dipl. A.C.L.A.M.   Senior Director, Laboratory Animal Health Services, The Jackson Laboratory, Bar Harbor, Maine

Bernard J. Doerning, D.V.M., Dipl. A.C.L.A.M.   Attending Veterinarian, Senior Scientist, Central Product Safety, The Procter & Gamble Company, Cincinnati, Ohio

Kathryn A. Donohue, J.D.   Associate General Counsel, Drexel University, Philadelphia, Pennsylvania

Laure Bachich Ergin, J.D.   Associate General Counsel and Managing Attorney, Drexel University, Philadelphia, Pennsylvania

Diane J. Gaertner, D.V.M., Dipl. A.C.L.A.M.   Director, University Laboratory Animal Resources, and Professor, Department of Pathobiology, School of Veterinary Medicine, University of Pennsylvania, Philadelphia, Pennsylvania

Cynthia S. Gillett, D.V.M., Dipl. A.C.L.A.M.   Director, Research Animal Resources, University of Minnesota, Minneapolis, Minnesota

Ed J. Gracely, Ph.D. Associate Professor, Department of Family, Community, and Preventive Medicine, Drexel University College of Medicine, Philadelphia, Pennsylvania

Michael J. Huerkamp, D.V.M., Dipl. A.C.L.A.M. Director, Division of Animal Resources, and Associate Professor, Pathology and Laboratory Medicine, Emory University, Atlanta, Georgia

Karen M. Janes, B.S. Associate Director, Office of Research and Sponsored Programs, and Director, Research Integrity and Compliance, Rutgers, the State University of New Jersey, New Brunswick, New Jersey

Alicia Z. Karas, M.S., D.V.M., Dipl. A.C.V.A. Assistant Professor, Department of Clinical Sciences, Cummings School of Veterinary Medicine at Tufts University, North Grafton, Massachusetts

Kathy Laber, D.V.M., M.S., Dipl. A.C.L.A.M. Professor and Vice Chair, Department of Comparative Medicine, Medical University of South Carolina, Charleston, South Carolina; Director, Veterinary Medical Unit, Ralph H. Johnson Veterans Administration Medical Center, Charleston, South Carolina

Neil S. Lipman, V.M.D., Dipl. A.C.L.A.M. Director, Research Animal Resource Center, Memorial Sloan-Kettering Cancer Center and the Weill Medical College of Cornell University; Professor of Veterinary Medicine, Laboratory Medicine and Pathology, Weill Medical College of Cornell University; Laboratory Member, Sloan-Kettering Institute, New York, New York

Michael D. Mann, Ph.D. Professor Emeritus of Cellular and Integrative Physiology, University of Nebraska Medical Center, Omaha, Nebraska

Kathleen D. Moody, V.M.D., M.S., Dipl. A.C.L.A.M. Consultant in Laboratory Animal Medicine, Newtown, Connecticut

Sreekant Murthy, Ph.D. Professor of Medicine, and Vice Provost for Research Compliance, Drexel University, Philadelphia, Pennsylvania

Christian E. Newcomer, V.M.D., M.S., Dipl. A.C.L.A.M. Associate Provost for Animal Research and Resources, and Associate Professor, Department of Molecular and Comparative Pathobiology, Johns Hopkins University, Baltimore, Maryland

Gwenn S.F. Oki, M.P.H. Director, Research Subjects Protection, City of Hope National Medical Center and the Beckman Research Institute, Duarte, California

Scott E. Perkins, V.M.D., M.P.H., Dipl. A.C.L.A.M. Director, Division of Laboratory Animal Medicine, Tufts University and Tufts-New England Medical Center, Boston, Massachusetts; Assistant Professor, Department of Environmental and Population Health, Tufts University School of Veterinary Medicine, North Grafton, Massachusetts

Ernest D. Prentice, Ph.D. Associate Vice Chancellor for Academic Affairs, University of Nebraska Medical Center, Omaha, Nebraska

Lester L. Rolf, Jr., Ph.D., D.V.M. Chief, Laboratory Animal Surgical Services, University Laboratory Animal Resources, University of Pennsylvania, Philadelphia, Pennsylvania

Harry Rozmiarek, D.V.M., M.S., Ph.D., Dipl. A.C.L.A.M. Director, Laboratory Animal Medicine, Fox Chase Cancer Center; Professor Emeritus, Laboratory Animal Medicine, University of Pennsylvania, Philadelphia, Pennsylvania

Howard G. Rush, D.V.M., M.S., Dipl. A.C.L.A.M. Associate Professor and Director, Unit for Laboratory Animal Medicine, University of Michigan Medical School, Ann Arbor, Michigan

Jerald Silverman, D.V.M., M.P.S., Dipl. A.C.L.A.M. Director, Department of Animal Medicine, and Professor, Department of Pathology, University of Massachusetts Medical School, Worcester, Massachusetts

Alison Smith, D.V.M., Dipl. A.C.L.A.M. Associate Professor, Department of Comparative Medicine, and Clinical Director, Division of Laboratory Animal Resources, Medical University of South Carolina, Charleston, South Carolina

Harold F. Stills, Jr., D.V.M., Dipl. A.C.L.A.M. Director, Laboratory Animal Resources, and Professor of Pathology, Boonshoft School of Medicine, Wright State University, Dayton, Ohio

Mark A. Suckow, D.V.M., Dipl. A.C.L.A.M. Director, Freimann Life Science Center, and Associate Research Professor, Department of Biological Sciences, University of Notre Dame, Notre Dame, Indiana

Robin Lyn Trundy, M.S., R.B.P., C.B.S.P. Biological Safety Officer, University of Tennessee, Knoxville, Tennessee

Richard C. Van Sluyters, O.D., Ph.D. Associate Dean for Student Affairs, School of Optometry, and Chair, Animal Care and Use Committee, University of California, Berkeley, Berkeley, California

Patricia A. Ward, B.S., M.P.A., L.A.T.G. Director, Research Regulatory Affairs, Office of Research and Graduate Studies, University of Michigan Medical School, Ann Arbor, Michigan

Chapter Contributors

Robert Ceru   Michigan State University, East Lansing, Michigan

Kristin Erickson   Michigan State University, East Lansing, Michigan

Kathleen L. Smiler, D.V.M., Dipl. A.C.L.A.M.   Consultant, Lakeville, Michigan

Contents

1

Origins of the IACUC

Harry Rozmiarek

Introduction

The need to provide regulatory oversight to assure animal welfare in the research laboratory was recognized in Victorian England over 120 years ago and led to the 1876 Cruelty to Animals Act. Modifications of this law that exist today include "Home Office" oversight and registration of individuals who conduct research procedures using animals. In the U.S., the earliest national law addressing animal welfare was the 28-hour law enacted in 1887. This law primarily governs animals being transported for market and does not specifically address animals in the research laboratory.

The New York Anticruelty Bill of 1866 and the formation of Humane Societies in New York (1866), Pennsylvania (1867), Massachusetts (1868), and Washington, D.C. (1870), addressed the use of animals in research, as did the formation of the American Antivivisection Society in Jenkintown, PA, in 1883. Nevertheless, none of these actions had national authority or scope. The first official law addressing the care and use of laboratory animals in the U.S. was the Laboratory Animal Welfare Act of 1966. An amendment in 1970 changed the name to the Animal Welfare Act (AWA), by which it is known today.

Background

Prior to the 20th century the responsibility for animals used in research in the U.S. was placed directly in the hands of the research investigator. The quality of animal care and animal welfare varied tremendously among research laboratories. Laboratories even within the same school or institution had different animal care policies and standards of care. Animal care was frequently the responsibility of a diener, who provided food, basic care, and much of the animal manipulation. Basic nutrition and sanitation were often inadequate and no environmental or housing standards were available. In many instances, the animal handling staff meant well but was not adequately trained or qualified. In others, animal welfare and even adequate care were of low priority. Rodents were commonly fed cereal grains and dogs and cats received table scraps. Backyard and part-time breeders provided rodents of variable genetic background, and the animals harbored a variety of parasites, pathogenic bacterial agents, and viruses. Long-term studies with rodents were impossible, as animals suffered from malnutrition and clinical disease and accurate research data were difficult to obtain. Dogs and cats were obtained individually, often provided in crates or gunnysacks and obtained from variable sources. Breeders of good-quality, standardized animals for research did not exist.

Central Management and Organization of Laboratory Animal Care

One of the first major research institutions in the U.S. that provided central management and care for research animals directed by a highly trained and qualified individual was the University of Chicago, where in 1945 Dr. Nathan Brewer was hired to direct a centrally managed program of laboratory animal care. Dr. Brewer had provided animal facility management at the University of Chicago as early as 1930 but had since received formal training as a veterinarian and obtained a Ph.D. in physiology. A number of other institutions in the country were following similar paths as the need for better and more standardized animal care and welfare was recognized. However, it was not until the 1940s that a number of professionals with a major interest in laboratory animal care began meeting in the Chicago area; this was undoubtedly stimulated and encouraged by Dr. Brewer. Among these was Dr. W.T.S. Thorpe, director of the newly organized Laboratory Aids Branch at the National Institutes of Health (NIH), which eventually developed into the National Center for Research Resources. In 1950 these meetings led to the formal establishment of the Animal Care Panel, which in 1967 became the American Association for Laboratory Animal Science (AALAS). Dr. Brewer served as the first national president of this organization and with others stimulated many other activities and organizations instrumental in shaping animal care and use policies and laws in the U.S.[1]

Officially established and conducting annual conferences on animal care, the Animal Care Panel appointed a committee in 1961, headed by Dr. Bennett Cohen, charged with providing animal care and use guidelines for research facilities. Their product was the publication in 1963 of the first edition of the *Guide for Laboratory Animal Facilities and Care* (later titled *Guide for the Care and Use of Laboratory Animals*). Future editions of this publication were supported by the NIH and guided by the ILAR. The National Academy Press, under the auspices of the National Research Council, published the seventh and current edition of the *Guide* in 1996.[2] This single document serves as the "bible" for laboratory animal care and use policies and guidelines in the U.S. It is excerpted from and referenced in all major guidelines and regulations and has been translated into and is being published in at least 11 other languages. Over 500,000 copies have been distributed throughout the world. The *Guide* expects every institution to have an Institutional Animal Care and Use Committee (IACUC) "to oversee and evaluate the institution's animal program, procedures, and facilities to ensure that they are consistent with the recommendations in the *Guide*" and other similar documents.

AAALAC

In 1963, the AFAB saw a need to evaluate the standard of animal care and use that was developed in its new *Guide*. This was accomplished in 1964 when two-person teams from the AFAB visited 26 major institutions in the U.S. that conducted research with animals and evaluated their policies and programs using the standards in the new *Guide*. This board soon saw the need to function independently and in 1965 incorporated in the state of Illinois as the American Association of Laboratory Animal Care (AAALAC). The first two institutions accredited by AAALAC were the University of Louisville in Kentucky and Howard University in Washington, D.C. This independent accrediting agency changed its name in 1996 to the Association for Assessment and Accreditation of Laboratory Animal Care International. Still using the *Guide* as its primary reference document, AAALAC now has over 650 accredited institutions in the U.S. and more than 50 in 27 foreign countries under its umbrella of accreditation. The AAALAC program has always expected a mechanism of regular review and assurance (an IACUC or similar committee) as part of a good animal care and use program.

Animal Welfare Act

Prior to 1960 the U.S. did not have a national law addressing laboratory animals. Local humane societies were actively promoting protection for pets and sometimes for farm animals as well. Concurrently, the scientific community was improving the quality of animal care and well-being in the research laboratory. The increasing need for research dogs and cats was partially fulfilled by animal dealers who obtained these animals in various ways and sold them to research laboratories. A series of articles and news reports on animal neglect and abuse at the hands of animal dealers culminated in a major article in *Life* magazine in 1966. This was accompanied by pictures showing animal neglect. It suggested a need for regulation and a system of enforcement. Catalyzed in part by this article, the Laboratory Animal Welfare Act was passed by Congress in 1966 (Public Law 89–544), and for the first time in the U.S. there were legal standards for laboratory animal care and use. The U.S. Department of Agriculture (USDA) was named the responsible agency for implementing and enforcing this new law and it promptly began providing regulations.[3] Although the act covered all laboratory animals, it was enacted primarily to protect dogs and cats and to counteract the business of pets being stolen by dealers and then sold for research purposes. Research laboratories and dealers were now required to register their facilities or be licensed and undergo inspection by USDA personnel, who made out reports and issued citations for noncompliance. These early inspections did not extend into the research laboratory, where animal care and use remained under the direction of the research investigator. Public groups concerned about the acquisition and welfare of animals destined for research use continued their activities, which ranged from local and national humane societies concerned about animal welfare and well-being to radical animal rights groups opposed to the use of animals for any reason. The activity of such groups seemed to escalate and become more vocal in the early 1980s. This activity peaked in a series of illegal break-ins and vandalism by a terrorist group and moved to the forefront of public opinion soon after two incidents involving "animal cruelty" and insensitivity in two well-known research institutions.[4,5] This climate raised public concern and visibility of animals in research and served as a catalyst for amendments and clarifications of guidelines and regulations providing for animal welfare and animal well-being.

A series of legislative bills and meetings in the early to mid-1980s, spearheaded by Senator Bob Dole, George Brown, John Melcher, and others, stimulated interest in including a requirement for an IACUC as part of the USDA regulations. Public Law 99-198, Food Security Act of 1985, Subtitle F – Animal Welfare, also called the Improved Standards for Laboratory Animals Act, was enacted December 23, 1985. This clarified what was meant by *humane care* by mentioning specifics such as sanitation, housing, and ventilation.

It directed the Secretary of Agriculture to establish regulations to provide exercise for dogs and an adequate physical environment to promote the psychological well-being of non-human primates. It also specified that pain and distress must be minimized in experimental procedures and that alternatives to such procedures be considered by the principal investigator (PI). The act also defined practices that were considered to be painful and stated that no animal can be used in more than one major operative experiment from which it recovered (with listed exceptions). Significantly, it established the IACUC with a description of its roles, composition, and responsibilities to the Animal and Plant Health Inspection Service (APHIS). It also included the formation of an information service at the National Agricultural Library to assist those regulated by the act in training employees, searching for ways to reduce or replace animal use, preventing the unintended duplication of research; it also provided information on how to decrease pain or distress. The final section of the amended act explained the penalties for release of trade secrets by regulators and the regulated community. In a *Federal Register* publication of August 31, 1989, the USDA published final rules for Parts 1 and 2 of the Animal Welfare Act Regulations (AWAR), as required by the 1985 Animal Welfare Act (AWA) amendment, and, effective October 30, 1989, the USDA for the first time required that each registered research institution appoint an IACUC of no fewer than three members, which "serves as the agent of the research facility that ensures that the facility is in full compliance with the Act." The final rules for 9 CFR Part 3 (Standards) were published on February 15, 1991, and became effective on March 18, 1991. The USDA still regularly inspects registered institutions but relies heavily on the IACUC to ensure that institutional practices are in good order.

A number of amendments to the AWA have led to regulations that now include animal transportation, marine mammals, and animals in the research laboratory. However, the AWA still excludes common laboratory rats (*Rattus norvegicus*) and mice (*Mus musculus*), birds bred specifically for research, and farm animals used in production agriculture. The most probable reasons for these exclusions are insufficient funds and staff. As nearly 90% of laboratory animals are rats and mice, their inclusion would have a significant impact on the workload of the USDA inspection staff.

U.S. Public Health Service Policy

The first U.S. Public Health Service (PHS) Policy on Humane Care and Use of Laboratory Animals went into effect in 1973, was revised in 1979, and was revised again in 1986.[6] All PHS policies on this subject evolved from an NIH policy published in 1971. That Policy referenced several NIH and PHS statements on appropriate care and humane treatment of laboratory animals, among them the *Guide*. It introduced the animal care committee as a means of local assurance and included all live vertebrate animals. Each revision placed more specific responsibilities on that committee.

The 1971 NIH Policy stated that institutions or organizations using warm-blooded animals in research or teaching supported by NIH grants, awards, or contracts would "assure the NIH that they will evaluate their animal facilities in regard to the maintenance of acceptable standards for the care, use, and treatment of such animals." The institution could either show that it was accredited by a recognized professional laboratory animal accrediting body or44 that it had established its own committee to carry out that assurance function. The minimal number of committee members was not stated, but at least one member had to be a doctor of veterinary medicine. Guidelines to be followed included the *NIH Guide*, all applicable portions of the AWA, and an appended set of guidelines known as the *Principles for the Use of Laboratory Animals*. The committee was required to inspect the animal facilities of the institution at least once a year and report its findings and recommendations to responsible officials of the institution. Records of activities and recommendations were required and were to be available for inspection by NIH representatives. Under this Policy, institutions accredited by AAALAC were not required by NIH to have an evaluation committee.

The first PHS Policy replaced the NIH Policy on July 1, 1973, and continued to accept AAALAC accreditation in lieu of an institutional committee. The minimal number of committee members was now set at three, and unaccredited or partially accredited institutions were required to have a committee. Only when institutions used a "significant number of animals" was a veterinarian required on the committee. If the number of animals used in activities supported by the Department of Health, Education, and Welfare (DHEW) was not "significant," a veterinarian was not required, but one of the committee members must then be "a scientist with demonstrated expertise in the care and use of laboratory animals." If such a person was not available, a veterinarian employed on a consultant basis was an acceptable alternative. A "significant number of animals" was not defined, and final determination was made by the DHEW from animal inventory information provided as part of the institution's assurance statement. The Policy for institutional review of applications and proposals stated, "Grantee and contractor institutions are encouraged to review their applications and proposals in the light of the pertinent provisions of the Animal Welfare Act, the standards set by the Institute of Laboratory Animal Resources (the NIH Guide), and the DHEW Principles for the Use of Laboratory Animals, and to familiarize their staff with these provisions, standards, and principles." However, there was no requirement under this Policy that institutional committees perform review of individual proposals or regularly provide to the DHEW summaries or certifications of such committee actions. The Policy did not specify who should perform the reviews. Under a different action, there was a requirement to keep records of committee activities for at least 3 years after the budget period. Assurance statements had to list the facilities and components of the program and committee. Activities included periodic facility inspections and reports to responsible officials at least once a year. The PHS Policy was revised on January 1, 1979, and now required all animal-using grantee institutions to have a "committee to maintain oversight of its animal care program." The Policy also required an institution to submit an assurance statement to the NIH/Office for Protection from Research Risks (OPRR), now the NIH/Office of Laboratory Animal Welfare (OLAW), and have it found to be acceptable before receiving a PHS grant for studies in which animals or animal facilities were used. In addition, to assure compliance with the edition of the *Guide* then in use, the AWA, and other applicable laws and regulations, it also required that "such assurance must also indicate that the institution has appointed and will maintain a committee to maintain oversight of its animal care program."

AAALAC accreditation was again recommended as the best means of demonstrating conformance with good animal care and use provisions. No explicit name was given to this committee, but the key words and all references to it were *animal care*. Grantee institutions, as in the past, were obliged to keep records of committee activities, including recommendations and determinations, with these records being available for inspection by authorized PHS officials. Review of individual proposals or projects by the institution's committee was encouraged, but not required. Even though the review remained merely a recommendation, the suggestion that the committee act as a reviewing body further strengthened its role. A list of committee members was required on the assurance form and information on each member was to include degrees, position, title, and a short description of the member's relevant background. A sample statement on the form was preceded by the following:

> The samples given below for committee members are not intended to dictate numbers of or qualifications for committee members. However, except in unusual circumstances, the committee should be of at least five members and include at least one veterinarian. Any such unusual circumstances should be explained in a statement accompanying the assurance.

The following was contained in the sample statement:

> We have appointed and will maintain a committee of at least five members to maintain oversight of our animal care program. The members have appropriate education and experience to perform their duties with respect to the types of animals and species used and the kinds of projects to be undertaken. If the conduct of a specific project is to be reviewed, the quorum will not include any member having an active role in the project. Changes in membership will be reported annually to the Office for Protection from Research Risks (OPRR), National Institutes of Health (NIH), and PHS.

Three options were provided for reporting an institution's degree of compliance: AAALAC accreditation, certification by its own institutional committee, or committee recommendations of improvements needed for compliance.

If the committee recommended any immediate or future improvements, the NIH/OPRR would expect to receive an annual report of the progress of these improvements toward compliance. The Policy now required that the committee would review its facilities and procedures at least once a year and that responsible officials would receive and consider all reports from the committee. They also would make an annual review of committee activities for compliance and would keep records of committee meetings and related administrative actions. A new assurance statement was required every 5 years. After allowing 5 years to study the effectiveness of the assurance system, it was concluded in 1984 that the committees frequently seemed "less than fully assertive" in carrying out their responsibility.[7] A recommendation after this study suggested that "the PHS Policy should be further modified to define more precisely the responsibilities of the awardee institutions, particularly the role of the animal care committee. It is imperative that the experience and expertise of the members of such committees be used to conduct full and effective reviews of proposals involving research with animals. The appointment of a nonscientist and an individual unaffiliated with the institution should be given serious consideration."[7] This study resulted in the latest revision of the PHS Policy, which further defines and outlines requirements of an animal care committee. This Policy now includes provisions of the Health Research Extension Act of 1985, which was enacted on November 20, 1985, as Public Law 99-158. The most significant changes required by this action are that the Policy now applies to intramural research that the PHS conducts, and that the IACUC is appointed by the chief executive officer of the institution.[6] The PHS Policy requires that the program description include an explanation of the training or instruction available to scientists, animal technicians, and other personnel involved in animal care, treatment, or use. This training or instruction must include information on the humane practice of animal care and use and the concept, availability, and use of research or testing methods that minimize the number of animals required to obtain valid results and minimize animal distress. The IACUC must now evaluate and prepare reports on all of the institution's programs and facilities (including satellite facilities) for activities involving animals at least twice instead of once each year. The IACUC, through the institutional official (IO), is responsible for reporting requirements; minority views filed by members of the IACUC must be included in reports filed under this Policy.

Department of Defense

While the scientific community and the public sector were gradually evolving guidelines and policies to assure animal welfare and good animal care,[8-17] research conducted by the Department of Defense (DOD) had similar concerns. As early as 1961, the DOD issued the Policy on Experimental Animals,[18-20] which directed that "all aspects of investigative programs involving the use of laboratory animals and sponsored by Department of Defense agencies will be conducted according to the Principles of Laboratory Animal Care as promulgated by the National Society for Medical Research." While this early Policy provided few specifics, a number of revisions followed and included all animals used in DOD laboratories, both in the United States and abroad. A joint regulation issued by the Army, Navy, Air Force, Defense Nuclear Agency, and Uniformed Services University on June 1, 1984, Use of Animals in DOD Programs, required that "all DOD organizations having animals (other than military working, recreational, and ceremonial) will seek accreditation by AAALAC."[21] It further required that "the local commanders of each DOD organization conducting or sponsoring activities involving animals in Research, Development, Teaching and Education (RDTE), clinical investigations, diagnostic procedures, or instructional programs will form a committee to oversee the care and use of animals." Such committees were to be appointed by the local commander, include a doctor of veterinary medicine, be made up of at least three members, review protocols and assure compliance with policies, standards, and regulations. The concept and practice of such committees to review and assure appropriate animal care and use, while not known by the acronym *IACUC*, were in place at many military installations prior to their being regularly formed at most academic institutions.

Evolution of the IACUC

The 1986 PHS Policy first described the IACUC in its present form and membership and only provided general guidance on how such a committee should be formed and operated. With an amendment to the AWA effective October 30, 1989, USDA regulations required a similar IACUC. Most institutions now have a single committee that satisfies both PHS and USDA requirements. The Scientists Center for Animal Welfare (SCAW) was instrumental in providing early guidance to institutions on IACUC functions and organization through regional conferences and workshops, culminating in a special AALAS publication, *Effective Animal Care and Use Committees*.[22] Since 1983, training and regular conferences of this type are provided by annual animal care and use conferences sponsored by the Public Responsibility in Medicine and Research (PRIM&R), regional workshops supported by the NIH/OLAW, and numerous similar activities. As described in the AWAR, this committee is to "prepare evaluation reports of reviews and inspections and submit them to a designated institutional officer."[3] While originally borrowed from the human Institutional

Review Board structure, the establishment of local animal care committees (i.e., IACUCs) to review and assure animal welfare and well-being is now common practice in the animal research community. The goal of each committee is compliance with guidelines and regulations while allowing for the flexibility of individual institutional tailoring to meet unique needs of the institution most effectively. Active participation by research scientists allows for the unique needs of research investigators to be voiced, participation by unaffiliated members protects the public conscience, and veterinarians assure appropriate medical care and provisions. This committee has become the primary oversight mechanism for animal care and use, responsible through an IO and reporting directly to regulatory and granting agencies on animal care and use matters. There is a continuing need for education to assure that this concept works as well as possible.

References

1. Rozmiarek, H., Current and future policies regarding animal welfare, *Invest. Radiol.*, 22, 175, 1986.
2. Committee to Revise the Guide for the Care and Use of Laboratory Animals, Guide for the Care and Use of Laboratory Animals, National Academy Press, Washington, D.C., 1996.
3. Office of the Federal Register, Code of Federal Regulations, Title 9, Animals and Animal Products, subchapter A, parts 1, 2, and 3, Animal Welfare, Washington, D.C., 2002.
4. Fraser, C., The Raid at Silver Spring, *The New Yorker*, 66, April 19, 1993.
5. McCabe, K., Who will live, who will die, *The Washingtonian*, 112, August 1986.
6. Office of Laboratory Animal Welfare, National Institutes of Health, Institutional Animal Care and Use Committee Guidebook, 2nd ed., Bethesda, MD, 2002.
7. Whitney, R.A., Jr., Animal care and use committees: history and current national policies in the United States, *Lab. Anim. Sci.*, special issue, 18, 1987.
8. *Guide for the Care and Use of Agricultural Animals in Agricultural Research and Teaching*, 1st rev. ed., Federation of Animal Science Societies, Savoy, IL, 1999.
9. Committee on Animal Research and Experimentation of the Board of Scientific Affairs, *Guidelines for Ethical Conduct in the Care and Use of Animals*, American Psychological Association, Washington, D.C., 1991.
10. U.S. Department of Health and Human Services, National Institutes of Health, Institutional Administrators' Manual for Laboratory Animal Care and Use, NIH Publication No. 88-2959, Washington, D.C., 1988.
11. ILAR Committee on the Use of Laboratory Animals in Neuroscience and Behavioral Research, Guidelines for the Use of Mammals in Neuroscience and Behavioral Research, National Research Council, National Academy Press, Washington, D.C., 2003.
12. ILAR Committee on the Use of Animals in Precollege Education, Principles and Guidelines for the Use of Animals in Precollege Education, Institute for Laboratory Animal Resources, Washington, D.C., 1989.
13. Joint AAMC-AAU Committee, *Recommendations for Governance and Management of Institutional Animal Resources*, Association of American Medical Colleges and the Association of American Universities, Washington, D.C., 1985.
14. U.S. Department of Health and Human Services, National Institutes of Health, Preparation and Maintenance of Higher Mammals during Neuroscience Experiments, NIH Publication No. 91-3207, Washington, D.C., 1991.
15. Universities Federation for Animal Welfare, *The UFAW Handbook on the Care and Management of Laboratory Animals*, 7th ed., South Mimms, Potters Bar, U.K., 1999.
16. International Air Transportation Association, *Live Animal Regulations*, Montreal, 2004.
17. National Association for Biomedical Research, *State Laws Concerning the Use of Animals in Research*, National Association for Biomedical Research, Washington, D.C., 2004.
18. Department of Defense Directive 5129.1, Department of Defense Instruction, Washington, D.C., 1959.

19. Department of Defense Directive 51225.5, Department of Defense Instruction, Washington, D.C., 1961.
20. Policy on Experimental Animals in Department of Defense Research, Department of Defense Instruction, Washington, D.C., September 1, 1961.
21. Army Regulation 70-18, SECNAVINST 3900.388, AFR 100-2, DARPAINST 18, DNAINST 3216.1B, USUHSINST 3203. The Use of Animals in DOD Programs, United States Army, Washington, D.C., June 1, 1984.
22. Orlans, F.B., Simmonds, R.C., and Dodds, W.J., Effective animal care and use committees, *Lab. Anim. Sci.* special issue, January 1987 (published in collaboration with the Scientists Center for Animal Welfare).

2

Circumstances Requiring an IACUC

Richard C. Van Sluyters and Marcy Brown*

Introduction

All of the human components of an animal care and use program — the IO, the AV, the animal care staff, and the IACUC — have important roles to play. Additionally, federal law, regulation, and policy also place the IACUC in a pivotal position for ensuring animal welfare. This chapter reviews the various circumstances in which institutions are required to appoint an IACUC. The fact that the USDA and the PHS have separate mandates to appoint an IACUC can be a major source of confusion since the circumstances under which these two agencies require IACUCs differ. In an effort to alleviate this confusion, the reader is directed to the specific sections of the AWA, AWAR, and PHS Policy that describe the regulatory requirements for the IACUC. These regulatory requirements are then summarized. The remainder of the chapter provides the answers to a series of questions designed to reveal whether IACUCs are always required at research institutions, and whether they are ever required at grade schools, secondary schools, science fairs, zoos, aquaria, animal shelters, or humane societies.

2:1 What regulatory agencies require the appointment of an IACUC?

Reg. There are two: the USDA and the PHS (see 2:2). The USDA, through APHIS/AC, requires an IACUC at any institution that uses animals in research, provided the species is among those listed under the definition of "animal" in §1.1 of the AWAR (see 12:1). The PHS, through NIH/OLAW, requires an IACUC at any institution that conducts PHS-supported activities involving any live vertebrate animal.

Opin. Within APHIS/AC, IACUC activities are monitored by yearly (or more frequent) unannounced inspections by Veterinary Medical Officers of APHIS/AC. Within the PHS, they are monitored by occasional random site visits and for-cause site visits by an ad hoc team assembled by the NIH/OLAW.

* The authors thank Jane E. Simpson and Ralph B. Dell for their contribution to this chapter in the first edition of *The IACUC Handbook*.

2:2   What specific documents describe the regulatory requirements for the IACUC?

Reg.   The AWA, enacted in 1966, is found in Title 7 of the U.S. Code (7 USC) §2131 et seq. The HREA (Public Law 99-185) directs the Secretary of Health and Human Services to establish guidelines for proper care and treatment of animals used in research, and for organization and operation of the IACUC. The 1985 amendments to the AWA require the appointment of an IACUC, mandate its minimum composition, and require it to perform at least semiannual inspections of animal facilities and study areas. Detailed regulations and standards for implementing the AWA are set forth by the USDA in Title 9 of the Code of Federal Regulations (9 CFR), Chapter 1, Subchapter A, Parts 1, 2, and 3.

Opin.   Although most of the AWAR that specifically govern the IACUC are found in §2.31, references to the IACUC are scattered throughout the AWAR as follows:

Part 2 — Regulations

Subpart C — Research Facilities

- §2.31   Institutional Animal Care and Use Committee
- §2.33,a,3   Attending Veterinarian and Adequate Veterinary Care
- §2.35,a–§2.35,f   Recordkeeping Requirements
- §2.36,b,3   Annual Report
- §2.37   Federal Research Facilities
- §2.38,f,2,ii; §2.38,k,1   Miscellaneous

Part 3 — Standards

Subpart A — Specifications for the Humane Handling, Care, Treatment, and Transportation of Dogs and Cats

- §3;6,d   Primary Enclosures
- §3.8,b,1; §3.8,c; §3.8,d,2   Exercise for Dogs

Subpart B — Specifications for the Humane Handling, Care, Treatment, and Transportation of Guinea Pigs and Hamsters

- §3.28,c,3   Primary Enclosures

Subpart C — Specifications for the Humane Handling, Care, Treatment, and Transportation of Rabbits

- §3.53,c,3   Primary Enclosures

Subpart D — Specifications for the Humane Handling, Care, Treatment, and Transportation of Nonhuman Primates

- §3.80,b,2,iii; §3.80,c   Primary Enclosures
- §3.81,c,3; §3.81,d; §3.81,e,ii   Environment Enhancement to Promote Psychological Well-Being
- §3.83   Watering

The PHS policy sets forth detailed requirements governing IACUCs. References to IACUCs are found primarily in the following sections:

- IV,A,3   Institutional Animal Care and Use Committee
- IV,B   Functions of the IACUC
- IV,C   Review of PHS-Conducted or PHS-Supported Research Projects
- IV,E   Recordkeeping Requirements
- IV,F   Reporting Requirements

The PHS policy requires institutions to use the *Guide* as a basis for developing and implementing an institutional program for activities involving animals. The regulations that govern the appointment, composition, and responsibilities of the IACUC are summarized in the *Guide* (pp. 9–10).

2:3   Are the U.S. Government Principles for the Utilization and Care of Vertebrate Animals Used in Teaching, Research, and Training considered part of the PHS Policy, or are the Principles outside of the regulatory environment?

Reg.   The U.S. Government Principles are incorporated into PHS Policy. The Preface to the 2002 reprint of the PHS Policy states, "Also included in this reprint are the U.S. Government Principles for the Utilization and Care of Vertebrate Animals Used in Testing, Research and Training. The U.S. Principles were promulgated in 1985 by the Interagency Research Animal Committee and adopted by U.S. Government agencies that either develop requirements for or sponsor procedures involving the use of vertebrate animals. The Principles were incorporated into PHS Policy in 1986 and continue to provide a framework for conducting research in accordance with the Policy."

2:4   What is the difference between the Animal Welfare Act and the Animal Welfare Act regulations in relation to the need for an IACUC?

Opin.   Both the AWA and the AWAR require research facilities to appoint an IACUC (see 2:1; 2:2). However, the AWAR provide directives for the implementation of the AWA and sets forth a more extensive list of duties for the IACUC. The AWAR were written so as to harmonize the USDA requirements for IACUCs with those required by PHS Policy.

2:5   What are the regulatory charges of the IACUC?

Reg.   The regulatory charges of the IACUC are described in the AWA (Sect. 13,b), the AWAR (§2.31,c), PHS Policy (IV), and the *Guide* (pp. 9–10).

Opin.   In summary, the above regulations require the IACUC to:

- Review, at least once every 6 months, the institution's program for animal care and use, using the AWAR and the *Guide* as a basis for evaluation.
- Inspect, at least once every 6 months, all of the institution's animal facilities, including animal study areas and satellite facilities, using the AWAR and the *Guide* as a basis for evaluation.

- Submit to the institutional official (IO) reports of the above evaluations which distinguish significant deficiencies from minor deficiencies, contain a reasonable and specific plan and schedule for correcting each deficiency, describe and justify any departures from the *Guide* and PHS Policy, and which are signed by a majority of the committee and include any minority views.
- Review and investigate concerns involving the care and use of animals at the institution resulting from public complaints or from reports of noncompliance received from personnel at the institution.
- Make recommendations to the IO regarding any aspect of the institution's animal program, facilities, or personnel training.
- Review and approve, require modifications in (to secure approval), or withhold approval of those components of proposed activities related to the care and use of animals.
- Review and approve, require modifications in (to secure approval), or withhold approval of proposed significant changes regarding the care and use of animals in ongoing activities.
- Suspend an activity involving animals if it does not comply with the AWAR, the *Guide*, PHS Policy, or the institution's Animal Welfare Assurance (Assurance) approved by NIH/OLAW.

2:6   Is an IACUC always required where animals are used for research, teaching, or product safety evaluation?

Reg.   No. An IACUC is required only if one of the following applies (see 8:4; 8:5):

- The species used for research, teaching, or testing is covered by the definition of animal given in §1.1 of the AWAR. The AWAR definition of animal specifically excludes birds, rats of the genus *Rattus*, and mice of the genus *Mus*, bred for use in research, and farm animals used in agricultural research. (See 12:1.)
- The research, teaching, or product safety evaluation is supported by the PHS, or the activity will be performed at an institution with an Animal Welfare Assurance that commits the institution to comply with the PHS Policy, regardless of the source of funding.
- The institution is receiving support for animal research, teaching, or testing from an agency that requires compliance with PHS Policy (e.g., the National Science Foundation).

Opin.   If an institution is or desires to be accredited by AAALAC, it must have an IACUC.

2:7   A pet dog is used in a hospital as nothing more than a model in an art therapy class. Is it necessary for the hospital to have (or establish) an IACUC to approve the use of this animal?

Opin.   No. A hospital that allows dog owners to bring their pet animals into the hospital for use in an art therapy class is not considered an exhibitor under the definitions in §1.1 of the AWAR because it does not purchase animals and exhibit them to the public for compensation. Accordingly, the regulations do not apply and an IACUC is not required. Assuming this activity is not taking place as part of a PHS-funded research project, an IACUC also is not required under PHS Policy, since it applies only to institutions that receive support for vertebrate animal activities from an agency of the PHS or from an agency that requires compliance with PHS Policy. (See 2:6.)

2:8   Do elementary schools need to have an IACUC to keep pet animals?

Opin.   No. A grade school that keeps pet animals is not considered an exhibitor under the definitions in §1.1 of the AWAR because it does not purchase animals and exhibit them to the public for compensation. Furthermore, both the AWAR (§1.1, Research facility) and the AWA (Sect. 2,e) specifically exclude elementary and secondary schools from being designated as research facilities. Accordingly, the regulations do not apply and an IACUC is not required. An IACUC also is not required under PHS Policy, since it applies only to institutions that receive support for vertebrate animal activities from an agency of the PHS or from an agency that requires compliance with PHS Policy. (See 2:6 and Reference 1.)

2:9   Is it necessary to have IACUC oversight of elementary or secondary school science fair projects or other educational activities?

Opin.   No. Elementary and secondary schools are specifically excluded from the definition of research facility in §1.1 of the AWAR and Sect. 2,e of the AWA. Accordingly, the regulations do not apply and an IACUC is not required. An IACUC also would not be required under PHS Policy, since it applies only to institutions that receive funds for vertebrate animal activities from an agency of the PHS or from an agency that requires compliance with PHS Policy.

2:10   What guidelines exist for use of animals in elementary or secondary school educational activities?

Opin.   Although elementary or secondary schools are not covered by federal regulations or policies governing the use of animals, they can use the guidelines contained in two National Research Council documents: the *Guide* and the *Principles and Guidelines for the Use of Animals in Precollege Education*.[1] The latter document states:

> A plan for conducting an experiment with living animals must be prepared in writing and approved prior to initiating the experiment or to obtaining the animals. Proper experimental design of projects and concern for animal welfare are important learning experiences and contribute to respect for and appropriate care of animals. The plan shall be reviewed by a committee composed of individuals who have the knowledge to understand and evaluate it and who have the authority to approve or disapprove it.

The above guidelines also describe what the plan should include. Another set of guidelines that may be useful are the *Principles for the Ethical Care and Use of Animals* developed by the National Aeronautics and Space Administration.[2]

In addition to these national guidelines, individual state governments or even local school districts may develop standards that are either guidelines or requirements. Also, in at least 30 states there are nonprofit biomedical research societies that schools can consult for information on guidelines. In general, these societies recommend the formation of a review committee that is composed of three people: a teacher (preferably a biologist), a parent, and a veterinarian.

2:11 Do colleges or universities with live mascots need to have the care and use of these animals approved by the IACUC?

Reg.   Technically, institutions using AWAR regulated species as mascots fall under the definition of exhibitor (§1.1 of the AWAR) and, accordingly, should be licensed as such. Although the technical definition requires mascots fall under the definition of exhibitor, common sense says it would not effectuate the purposes of the AWA to license most universities for this purpose. To the extent the mascot is housed in university facilities, it should be cared for and housed in a manner consistent with the AWAR, and the housing and care would be subject to APHIS/AC inspection.[3]

Opin.   The question of whether the care and use of a mascot must be approved by the IACUC depends upon an institution's policies since it is not required under the AWAR. For example, it is the policy at some institutions that all uses of live vertebrate animals must be approved by an IACUC, regardless of whether the animals are an APHIS/AC-regulated species or are used in PHS-supported activities. Other institutions state in their NIH/OLAW Assurance that all uses of animals at the institution will be conducted in accord with the *Guide*. At such institutions, both NIH/OLAW and AAALAC expect the care and use of mascots to be covered by an IACUC-approved protocol.

2:12 Do zoological gardens and aquariums fall under the jurisdiction of the AWAR or PHS Policy?

Reg.   Yes. Zoos and some aquariums (e.g., those that maintain AWAR-regulated species, such as marine mammals) are defined as exhibitors under §1.1 of the AWAR and must comply with all regulations governing exhibitors.

Opin.   Strictly speaking, only zoos and aquariums that receive support for research using animals from an agency of the PHS, such as the NIH, are required by NIH/OLAW to comply with its policy. However, other federal and private funding agencies (e.g., the National Science Foundation) may require adherence to the PHS Policy as a condition of receiving their support. (See 2:6.)

2:13 Under the AWAR, are zoological gardens and aquariums expected to adhere to the same research criteria as biomedical research institutions?

Reg.   Zoological gardens need to be registered as a research facility if they are conducting invasive procedures in a research context, i.e., to gather and disseminate data and information. Typical behavioral and observational studies do not include invasive procedures and, therefore, do not require registration as a research facility.[3] In early 1999 APHIS/AC initiated a process to define research for AWA purposes as it applies to zoos, wildlife studies, and euthanasia, but at the time of this publication the policy has not been finalized.[3]

Opin.   Behavioral training or studies designed to improve the management or care of animals in a zoo or aquarium setting are not considered regulated research. Observational studies conducted at zoos—no matter who is conducting them—are not covered by the AWAR. However, if a nonobservational procedure is performed on an animal for the purpose of obtaining data not directly related to the management or care of the animal or its species in a zoo or aquarium setting, the activity is considered regulated research. In such cases, the zoo, aquarium, or one of their subunits must register as a research facility with the USDA (in addition to being licensed as an exhibitor), and must meet the requirements stipulated by the AWAR for research facilities. (For further information on this topic, see reference 4.)

2:14 Is research conducted at zoological gardens and aquariums required to come under the purview of an IACUC?

Reg.   (See 2:13.)

Opin.   If the activity involves an APHIS/AC-regulated species and is considered regulated research by APHIS/AC (see 2:13), it comes under the purview of an IACUC. Similarly, if the activity is supported by the PHS or other agencies that require compliance with PHS Policy, it must be reviewed and overseen by an IACUC (see 2:6).

2:15 Do private and municipal animal shelters and humane societies fall under the jurisdiction of the AWA and its regulations? Of the PHS Policy?

Opin.   Generally, no. Animal shelters and humane societies do not fall under the jurisdiction of the AWAR unless they sell dogs or cats to a research facility, licensed dealer, exhibitor, or sell animals wholesale as pets (AWAR §1.1, Dealer). If a private shelter provides animals to a research facility, it must be licensed with APHIS/AC as a dealer and must comply with all AWA regulations governing dealers. A municipal animal shelter or humane society that provides animals to a research facility is not defined as "a person" by APHIS and, consequently, does not meet the definition of dealer under §1.1 of the AWAR. Accordingly, municipal shelters and humane societies are exempt from the regulations governing dealers. However, any shelter (private or municipal) that provides animals to a research facility must comply with animal holding and recordkeeping requirements stipulated under the AWA (§28) and the AWAR (§2.101). This section requires the shelter to hold any dog or cat for at least 5 days before selling it to a research facility, and to provide the research facility with a certification that:

- The person who gave the dog or cat to the shelter was notified that the animal may be used for research or education.
- The shelter satisfied the 5-day holding requirement.

The original certification must accompany the dog or cat to the research facility, and the shelter is required to retain a copy of the certification for at least 1 year. APHIS/AC is empowered to review these records at shelters and may assess criminal or civil penalties against any shelter that does not comply with these requirements.

Animal shelters and humane societies fall under the jurisdiction of PHS Policy only if they receive support from the NIH or some other agency that requires compliance with PHS Policy.

2:16 The AWA (§2.31,d,1) exempts studies defined as field studies from the requirement for IACUC review. However, the AWA definition of a field study *excludes* any study that involves an "invasive procedure." In determining whether a field study is exempted from the requirement for IACUC review, what is meant by an invasive procedure?

Reg.   The AWAR (§1.1) state, "*Field study* means a study conducted on free-living wild animals in their natural habitat. However, this term excludes any study that involves an invasive procedure, harms, or materially alters the behavior of an animal under study." The AWAR do not further define the phrase "invasive procedure."

Opin. Although the phrase "invasive procedure" is not further defined, the context in which it appears in the AWAR gives guidance as to its intended meaning. Also excluded from the exemption for field studies are projects that have the potential to harm animals or materially alter their behavior. Clearly, the intent of this regulation is to require IACUC review of any projects that have the potential to disrupt animals or to alter their environments. Included in these would be field studies that require animals to be trapped or confined in any way, handled (see AWAR §1.1 for a definition of handling), anesthetized, tagged or marked in any other way, or subjected to blood or tissue sampling. This leaves strictly observational studies—ones in which neither the animals nor their environments are disturbed or permanently altered—as being exempted from the requirement for IACUC review.

2:17 **The AWAR (§2.31,d,1) exempt field studies from IACUC review. The definition of a field study includes the phrase, "does not harm or materially alter the [animals'] behavior." What constitutes materially altering the behavior of an animal?**

Reg. (See 2:16.)
Opin. (See 2:11.)

2:18 **The AWAR (§2.31,d,1) exempt studies defined as field studies from the requirement for IACUC review. However, the AWA definition of a field study (Sect. 1.1) includes the phrase "free-living wild animals in their natural habitat." In determining whether a field study is exempted from the requirement for IACUC review, can an animal be kept for any length of time outside of its natural habitat before it is no longer considered free living?**

Reg. (See 2:16.) The AWAR definition of a field study makes it clear that any study that has the potential to materially alter an animal's behavior is not exempt from IACUC review.
Opin. Removing an animal from its natural environment for any length of time would appear always to have the potential to materially alter its behavior. Such removal would typically involve capture, trapping or herding, handling, restraint and/or physical confinement. These activities, no matter how carefully they were performed, would carry a high likelihood of altering an animal's behavior in a significant fashion, at least in the short term.

References

1. Principles and Guidelines for the Use of Animals in Precollege Education, Institute for Laboratory Animal Resources, Washington, D.C., 2005. Available on the World Wide Web at: http://dels.nas.edu/ilar_n/ilarhome/reports.shtml.
2. Principles for the Ethical Care and Use of Animals, National Aeronautics and Space Administration, NPD 8910, Washington, D.C., 1996. Available on the World Wide Web at: http://lifesci.arc.nasa.gov/PECA.html.
3. DeHaven, W.R., personal communication, 1999.
4. Kohn, B. and Monfort, S.L., Research at zoos and aquariums: regulations and reality, *J. Zoo Wildlife Med.*, 28, 241, 1997.

3

Creation of an IACUC

Richard C. Van Sluyters and Marcy Brown*

Introduction

This chapter provides information on how institutions can form an IACUC. It describes whose interests the IACUC should serve and whose responsibility it is to appoint the members. Since most IACUCs conduct their meetings in private, the nonaffiliated member of the committee can be the general public's only link to the federally mandated oversight process for animal welfare. Accordingly, information is provided on the critical role of the nonaffiliated member and how institutions can go about finding an effective one. For an IACUC to fulfill its regulatory requirements properly, its members must be informed of their responsibilities under federal laws, regulations, and policies. In addition to these federally mandated obligations, IACUCs may be charged with additional responsibilities by their state, city, or institution. Given the complexity of this environment, institutions have a responsibility to provide their IACUC members with specialized training. This chapter describes some methods by which this can be done. The latter part of the chapter suggests means by which larger institutions can utilize multiple IACUCs to reduce the workload associated with reviewing hundreds of animal use protocols and inspecting dozens of animal facilities. Finally, there is a discussion of the record-keeping requirements for IACUCs and the use of computerized databases to store and retrieve these records.

3:1   In general, what is expected of the IACUC, and whose interests does it represent?

Reg.   The AWA (Sect. 13,b,1) charges the members of the IACUC with representing "society's concerns regarding the welfare of animal subjects." In addition, the nonaffiliated (outside) member is specifically charged with representing general community interests in the proper care and treatment of animals (AWA, Sect. 13,b,1,B,iii; AWAR, §2.31,b,3,ii; *Guide*, p. 9). The PHS Policy (IV,B) references the IACUC as "an agent of the institution" that will "oversee the institution's animal program, facilities, and procedures" (PHS Policy IV,A,3,a).

Opin.   The IACUC is expected to oversee and evaluate the institution's animal care and use programs to ensure that they are consistent with the recommendations of the *Guide*, the AWAR, and PHS Policy. It represents multiple interests, including those

* The authors thank Jane E. Simpson and Ralph B. Dell for their contribution to this chapter in the first edition of *The IACUC Handbook*.

of the institution and the community. It serves as the local oversight arm of federal agencies and accrediting bodies, such as APHIS/AC, NIH/OLAW, and AAALAC.

3:2   Who is responsible for appointing the IACUC at an academic institution?

Reg.   The Chief Executive Officer (CEO) at an academic institution is charged with appointing an IACUC (AWA Sect. 13,b,1; AWAR §2.31,a; PHS Policy IV,A,3,a). The CEO is defined in PHS Policy (IV,A,3,a, footnote 5) as the highest operating official of the organization, such as the President of a university. PHS Policy allows the CEO to delegate the authority to appoint the IACUC but requires that such delegation be specific and written.

    The *Guide* requires institutions to comply with the AWAR and PHS Policy, which both require the CEO to appoint the IACUC. However, the *Guide* (p. 9) refers to the individual who is responsible for appointing the IACUC as the "responsible administrative official."

Opin.   In larger institutions it is not unusual for the CEO to delegate authority for appointing the IACUC to a senior administrator who is more directly responsible for the institution's research program, such as a Vice Chancellor, Dean, or Vice President for research. This individual also frequently serves as the organization's IO (PHS Policy III,G) and signs its Animal Welfare Assurance. It is important to note that the CEO and the IO do not have to be the same person.

Surv.   For practical purposes, who *really* appoints the IACUC at your institution (i.e., the person or group who actually makes the choice, not the person who rubber stamps that choice)?

• The CEO	24/175
• The Institutional Official (if other than the CEO)	42/175
• The IACUC Chairperson	47/175
• The IACUC or an IACUC subcommittee makes the real decision	36/175
• A Dean or other administrative person(s) who is neither the CEO nor the Institutional Official	11/175
• Other	7/175
• No response	8/175

3:3   Who is responsible for appointing the IACUC at an industrial setting?

Opin.   The AWAR and PHS Policy do not discriminate between academic and industrial settings in describing who is responsible for appointing an IACUC. In both settings, it is the CEO or the CEO's designee. (See 3:2.)

3:4   For practical purposes, should the IACUC Chair be given the option of accepting or rejecting a proposed appointment to the committee?

Reg.   Both the AWAR (§2.31) and the PHS Policy stipulate that the IACUC must be appointed by the CEO.

Opin.   Allowing the IACUC Chair the authority to appoint the committee members would be a violation of the AWAR and PHS Policy, unless the CEO chose to delegate (specifically and in writing) authority to appoint the members to the Chair (PHS Policy footnote 5). (Note that the AWAR are silent regarding the ability of the CEO to delegate authority to appoint IACUC members, but such delegation is common at larger institutions and not cited by APHIS/AC veterinary medical officers as a violation of the AWAR during their semiannual inspections of research facilities.) Whereas there might be some situations in which this arrangement could work without compromising the separation of authorities and responsibilities envisioned by the writers of the AWAR and PHS Policy, in most instances it is a bad idea. The potential for conflict of interest and bias in the selection of IACUC members would just be too great. On the other hand, survey results indicate a significant number of institutions (27%) report that the IACUC Chair "really appoints" the committee (see 3:2). The phrase "really appoints" is likely key here, since it suggests that in these institutions the "actual" appointments are still being made by the CEO, but with significant input from the Chair. This form of consultation no doubt takes place even more widely than indicated by the survey results and is usually a positive attribute of an animal care and use program. For example, the Chair can inform the CEO of specific needs of the IACUC in order to ensure it possesses the expertise necessary to fulfill its responsibilities. In addition, many readers will be familiar with the tactic of appointing to the IACUC PIs who have been identified by the Chair as being somewhat refractory with respect to animal welfare regulations, policies, and guidelines as a method for rendering them more tractable.

3:5 What is the definition of the nonaffiliated (outside) member of the IACUC?

Reg.   The nonaffiliated member is defined as an individual who represents general community interests in the proper care and use of animals, is not a laboratory animal user, is not affiliated with the institution, and is not an immediate family member of a person who is affiliated with the institution (AWA Sect. 13,b,1,B; AWAR §2.31,b,3,ii; PHS Policy IV,A,3,b,4; *Guide*, p. 9; APHIS/AC Policy #15).[1] (See 5:31–5:36.)

3:6 How might IACUCs find individuals to serve as nonaffiliated members?

Opin.   Nonaffiliated members can be found in a number of ways. Perhaps the most common source is through personal contacts of the CEO, members of the IACUC, AV, or other personnel at the institution. Other useful sources for recruiting nonaffiliated members may include professional societies (e.g., lawyers, ethicists, clergy, teachers, librarians, health care professionals), local humane societies, and nonprofit service organizations. Some institutions run classified advertisements in local newspapers seeking interested individuals.

3:7 What should be the duration of the IACUC membership for the Chair and individual committee members?

Opin.   Serving as an effective member of an IACUC is a difficult task. To perform their duties competently, the members of an IACUC must become familiar with a large body of regulations, policies, and guidelines. They also must learn about the history, special problems, research strengths, goals, objectives, and idiosyncrasies of their institution and its investigators. Experience has shown that a new IACUC member requires 6 months to a year to gain the experience required to function as a fully effective member of an IACUC. In light of this, most institutions ask members to serve for at least 2 years, and many for 3 years. Some institutions also ask capable and willing members to serve a second 2- or 3-year term or longer.

Serving as an effective IACUC Chair requires all of the knowledge and expertise needed to become an effective committee member, plus the additional administrative skills required to manage the committee's staff and interact with the institution's administration. For these reasons, the person selected to chair the IACUC has usually served at least one term as a committee member. Given the sensitivity of the IACUC's activities, many academic institutions require not just the Chair but also the committee's members to be tenured. Many institutions ask the IACUC Chair to serve multiple terms in order to retain someone with a high level of regulatory and institutional knowledge in this position. (See 5:10.)

Surv. 1 What is the typical length of service for an IACUC Chairperson at your institution?

- Cannot say because we have only had one Chairperson
  serving now for _____ years   23/175
- 1–2 years   18/175
- 3–4 years   42/175
- Over 4 years   76/175
- Other   9/175
- No response   7/175

Note: Range of years reported for the first answer was 1.5–25, with one "no response."

Surv. 2 What is the typical length of service for an IACUC member (Nonchairperson) at your institution?

- Cannot say because we have not had any turnover
  now for _____ years   9/175
- 1–2 years   12/175
- 3–4 years   68/175
- Over 4 years   75/175
- Other   7/175
- No response   4/175

Note: Range of years reported for the first answer was 1–11.

3:8 How can institutions train IACUC members?

Reg. Recognizing the complexity of the responsibilities that new IACUC members are asked to fulfill, the *Guide* (p. 9) states, "It is the institution's responsibility to provide suitable orientation, background materials, access to appropriate resources, and, if necessary, specific training to assist IACUC members in understanding and evaluating issues brought before the committee." Accordingly, AAALAC-accredited institutions are expected both to provide adequate training for IACUC members and to document that such training has been provided.

Opin. There are many ways that institutions can train IACUC members. A common method is to prepare an IACUC member's handbook that contains copies of relevant regulations, policies, and guidelines, as well as copies of the institution's forms and SOPs. New members typically have one or more orientation sessions under the guidance of the IACUC staff, the Chair, or the AV. New committee members also may be assigned a mentor who is a more experienced member of the IACUC, and they may be invited to attend one or two IACUC meetings as observers before their official term starts. In addition, members may attend an ever-growing number of meetings and workshops that address animal care and use issues of importance to IACUCs. Other useful information for training new IACUC members includes a tutorial on the PHS Policy and an online version of the NIH/OLAW–ARENA publication *Institutional Animal Care and Use Committee Guidebook*.[2]

The American Association for Laboratory Animal Science (AALAS) sponsors a World Wide Web page called The IACUC Training and Learning Consortium,[3] which is an excellent source of training materials for IACUC members, and the Scientists Center for Animal Welfare (SCAW) hosts IACUC Talk,[4] a forum for IACUC members to voice their opinions, questions, and concerns.

Several Institute for Laboratory Animal Research reports and training manuals also contain information of value to IACUC members.[5-14] Each of these documents was written by a committee of experts and carefully reviewed. They all are available from the National Academy Press.[15]

3:9   Can an institution have more than one IACUC?

Opin.   Yes. There is no regulatory language that precludes establishment of multiple IACUCs at one institution. In fact, the AWA (Sect. 13,b,1) states, "The Secretary shall require that each research facility establish at least one committee," clearly leaving the door open for appointing more than one committee at an institution. Interestingly, the AWAR (§2.31,a), which implement the AWA, read, "The Chief Executive Officer of the research facility shall appoint an Institutional Animal Care and Use Committee (IACUC)," language that is echoed in PHS Policy (IV,A,3,a) and the *Guide* (p. 9). In practice, a number of institutions have opted to appoint more than one IACUC, and their systems have been approved by both NIH/OLAW and APHIS/AC. Multiple IACUCs are found most commonly at large institutions that have diverse scientific missions, many animal facilities, and large numbers of investigators and protocols.

Surv.   Does your institution have more than one IACUC?

• Yes	11/175
• No	161/175
• Other	0/175
• No response	3/175

3:10   Under what circumstances would multiple IACUCs be advantageous?

Opin.   Multiple IACUCs are advantageous at large institutions where the sheer volume of research activity exceeds the capacity of a single IACUC to review the associated protocols and program in an effective and timely manner. Dividing the work among multiple IACUCs, each appointed to oversee a discrete portion of the research program, can be the only way to handle the workload. An added advantage of dividing the workload is that it allows committees to gain expertise in reviewing protocols in particular research areas, since they are usually appointed within academic units or for coverage of specific research disciplines. Existence of more localized IACUCs also makes it easier for IACUC members to maintain a personal relationship with investigators and a sense of ownership and pride in the portion of the program they oversee.

Another situation in which the existence of more than one IACUC may be advantageous is in institutions that have bipartite research programs. In some cases, this occurs when an organization engages in two very distinctly different types of research. In others, it occurs when the research program is physically divided between two locations. In both situations, the organization may find it effective to appoint individual IACUCs to review protocols from and oversee the separate components of its overall program.

3:11 What might be the disadvantages of multiple IACUCs?

Opin. A major challenge for a multiple IACUC system is to develop mechanisms for maintaining consistency and quality in program and protocol review across the institution. If there is not an overarching coordinating group or committee, the institution might become fragmented and inconsistent, or conflicting practices might arise. Another disadvantage is the need for increased IACUC staffing and funding to sustain the necessarily redundant components of multiple committees. The cost of maintaining an IACUC is considerable and, in the absence of centralized support, can place a large burden on smaller administrative units such as colleges or departments.

Another possible problem with the multiple IACUC system is that each committee must meet all the membership requirements for an individual IACUC (e.g., veterinarian, nonaffiliated member, nonscientist), including the requirement that no more than three members from the same administrative unit may serve on the IACUC (AWAR §2.31,b,4). It can be difficult to appoint a properly constituted committee from within a small administrative unit. Moreover, difficulties can arise with locally constituted IACUCs because no member may participate in the IACUC review or approval of a research project in which the member has a conflicting interest, nor any member who has a conflicting interest contribute to the constitution of a quorum (PHS Policy IV,C,2; AWAR §2.31,d,2). Even when there is no obvious conflict of interest, it is important to recognize that there can appear to be a conflict when IACUC members drawn from a small administrative unit review each other's protocols. Finally, when IACUC members are drawn from a limited pool, individuals may find that they have an excessive time commitment to the IACUC. A number of the difficulties cited, in particular issues related to consistency of actions, can be mitigated by having one or more individuals serve on every IACUC.

3:12 Should multiple IACUCs all report to a central IACUC, advisory committee, or other authority?

Opin. Yes. Experience at institutions with multiple IACUCs has shown that it is essential to have the committees report either to the same IO or to an institution-wide IACUC that then reports to the IO according to PHS Policy (IV,B,3; IV,B,5) and the AWAR (§2.31,c,3). An institution could designate a central policy or advisory body, but all IACUCs must have a direct reporting channel to the IO. This organizational structure is needed to establish uniform policies and procedures and to maintain consistent performance standards across the institution. (See 3:13; 4:1.)

3:13   What are some ways in which multiple IACUCs can effectively interact?

Opin.   In institutions where multiple IACUCs report to a single IO, the IO can appoint an IACUC advisory committee to address general issues that affect the entire animal care and use program. The committee should consist, at a minimum, of the IO or the IO's designee, the Chairs of the individual IACUCs, and the AV. Issues can be raised before this committee by any of its members.

Another way institutions with multiple IACUCs can ensure effective interaction is for the IO to appoint an overarching institutional IACUC to which the individual IACUCs report. That is, the individual IACUCs function more as a prereview and advisory group, providing recommendations to the investigator and the institutional IACUC. In this model, the composition of this institutionwide committee meets the regulatory requirements for and serves as the institution's official IACUC. The products of the individual IACUCs (e.g., prereviewed protocols, facility inspection reports) flow to the institutionwide IACUC for final review and approval as needed. A simple strategy for helping ensure effective interaction among multiple IACUCs is to have some members serve on more than one committee.

In these situations, the goal of the institutionwide IACUC is to make certain that policies, procedures, and performance standards are consistent throughout the institution's animal use program. Examples of the types of activities in which it might engage include development or modification of standardized forms and guidelines (e.g., animal use protocol form, SOPs, animal record forms), compilation or preparation of semiannual reports for submission to the IO and annual reports for submission to APHIS/AC and NIH/OLAW, coordination of AAALAC site visits, development and maintenance of centralized information resources, and evaluation of facility improvement requests.

3:14   What strategies can be used to ensure that protocol review is consistent throughout an institution with multiple IACUCs?

Opin.   The most important step in ensuring consistent protocol review is to develop a single animal use protocol form that can be used throughout the institution. The form should be designed to allow sufficient flexibility in describing proposed procedures, so as to encompass the full range of activities at the institution (e.g., antibody production, survival surgery, product safety testing, field research, breeding) while requiring all investigators to provide the information that is required for any proposed use of animals (e.g., rationale for animal use, justification for animal species and numbers, alternatives to painful procedures, humane end points).

It is also important for the IO to make certain that the membership of each IACUC not only meet the minimal regulatory requirements, but include individuals who have sufficient expertise to evaluate the protocols they receive competently. Each committee needs to have scientists familiar with the kinds of animal use they will be asked to review and veterinarians knowledgeable in the health and husbandry of the proposed animal species. Depending on the nature of the activities to be reviewed, sufficient technical expertise must be available for each IACUC to evaluate potential risks associated with protocols (e.g., biosafety, radiation safety, occupational health) and to ensure that appropriate safeguards are proposed to mitigate these risks. This can be accomplished by including a health and safety specialist on every committee or by providing a centralized health and safety resource for use by all IACUCs.

It is important to note that absolute consistency is unobtainable, even in programs with a single IACUC. Inevitably, criteria change over time and mistakes are made. It is important to maintain a consistent approach to important issues that have the potential to impact the entire program.

3:15   What mechanism might be used to assure consistent IACUC semiannual inspection and review of animal care facilities and programs at institutions that have multiple IACUCs?

Opin.   One way for institutions that have multiple IACUCs to ensure that semiannual inspections and reviews are consistent is to utilize an institutionwide committee either to perform these functions or to review the products of the individual IACUCs that perform them. The development of institutionwide forms to be used in conducting inspections and preparing reviews also can help assure consistency. Another strategy is for personnel from a central unit (e.g., AV, institutional compliance officer) to participate in the facility inspections conducted by each IACUC. Finally, institutions that require all of their components to be accredited by AAALAC should have an independent method for ensuring uniform compliance with applicable regulations, policies, and guidelines.

3:16   What is the least common denominator (e.g., school, college, department, individual investigator) around which an IACUC can form?

Opin.   Institutions with multiple IACUCs may form them around:

- Discrete units within the institution's administrative structure (e.g., medical school, veterinary school, graduate school, college of agriculture)
- The nature of the proposed research (e.g., agricultural, biomedical, field research, product testing)
- The animal species being used (e.g., farm animals, laboratory animals, nonhuman primates, exotic species)
- Geographically separate facilities

When selecting the size of a unit upon which to base an IACUC, consideration needs to be given to:

- The potential workload of the IACUC (e.g., number of protocols per year; number, size, and location of animal facilities)
- The distance of the proposed IACUC from the IO in the reporting lines of the institution
- Whether the unit has the resources to support an IACUC
- Whether the unit has sufficient breadth of animal use to prevent frequent potential conflicts of interest within the IACUC
- Whether the pool of animal users from which IACUC members can be drawn is sufficiently large
- The AWAR requirement (§2.31,b,4) that no more than three members of the IACUC can be from the same administrative unit

3:17   How might records for IACUCs be effectively maintained?

Opin.   Institutions are obliged to maintain various records related to the animal care and use program. The AWAR (§2.35) and PHS Policy (IV,E) require each institution to maintain:

- Minutes of IACUC meetings, including records of attendance, activities of the committee, and committee deliberations (see 6:25)
- Records of applications, proposals, and proposed significant changes in the care and use of animals, and whether IACUC approval was given or withheld
- Records of semiannual IACUC reports and recommendations, including minority views, that have been forwarded to the IO

In addition, PHS Policy (IV,E) requires each institution to maintain:

- A copy of its Assurance
- Records of accrediting body determinations

The AWAR (§2.35,f) and PHS Policy (IV,E) stipulate that all records must be maintained for at least 3 years and that records relating to applications, proposals, and proposed significant changes must be maintained for 3 years after completion of the activity. All records must be accessible for inspection and copying by authorized representatives of APHIS/AC, NIH/OLAW, other PHS representatives, and (for accredited institutions) AAALAC site visitors.

The IACUC should either maintain these records or ensure that they are maintained in a secure and accessible location at the institution. Additional records that should be kept include:

- IACUC correspondence, including records of electronic mail
- Institutional policies, guidelines, and SOPs for animal care and use
- Reports of IACUC investigations of concerns involving animals
- Records of the training provided to animal care personnel, research personnel, and IACUC members

Methods of maintaining records will vary, depending upon the organization of the institution and the size and complexity of its animal use program. However, any method that is used to maintain records must be efficient, reliable, and secure.

In addition to traditional paper records, many institutions use computerized databases to maintain records of approved animal use protocols. Such databases may contain a minimum of information—such as PI name, protocol number and expiration date, approved animal species and numbers—or may contain a complete summary of the protocol, including a list of all procedures, drugs, and personnel. Computerized databases have many advantages, including the ability to sort and retrieve information quickly by various characteristics (e.g., species used, type of procedure, protocol expiration date, location of procedure/housing). The database also can be shared with the unit that orders animals to ensure that only approved animal species and numbers are ordered for a particular investigator. Information stored in the database also can facilitate preparation of the institution's annual report to APHIS/AC and the Animal Use Summary that is required as part of the application for AAALAC accreditation or reaccreditation.

3:18   Can a small institution without an IACUC use the services of a nearby institution with an IACUC to review and approve research supported by a PHS grant?

Reg.   PHS Policy (IV,A,3) stipulates that the IACUC must be appointed by the CEO or the CEO's designee and mandates the number and qualifications of the IACUC members (see Chapter 5). The AWAR (§2.31) essentially echoes PHS Policy, although the minimal number of IACUC members required differs between the two. Neither the PHS Policy nor the AWAR stipulate that any of the nonveterinary IACUC members must also be employees of, or in any other way affiliated with, the institution (see 5:19). However, PHS Policy (IV,A,3,b) and the AWAR (§2.31,b,3,i) state that the IACUC veterinarian must have "direct or delegated program authority and responsibility for activities involving animals at the institution (PHS Policy) [or] research facility (AWAR)." The AWAR (§2.33,a,1) further stipulate, "Each research facility shall employ an attending veterinarian under formal arrangements. In the case of a part-time attending veterinarian or consultant arrangements, the formal arrangements shall include a written program of veterinary care and regularly scheduled visits to the research facility." Finally, the AWAR note (§2.33,a,3), "The attending veterinarian shall be a voting member of the IACUC; *Provided however,* That a research facility with more than one Doctor of Veterinary Medicine (DVM) may appoint to the IACUC another DVM with delegated program responsibility for activities involving animals at the research facility."

Opin.   Although there are some potential pitfalls to this approach (e.g., see 5:19), it might be worth considering in certain situations, such as when a small institution with only a few animal users has as its neighbor a larger institution with a more extensive and experienced animal care and use program. In such a case, the larger institution could agree to include the smaller institution under its NIH Assurance. The CEO of the smaller institution could agree to appoint the larger institution's AV to serve as the smaller institution's AV and the larger institution's IACUC to serve as the smaller institution's IACUC, in each case with the proper program responsibilities. Other variations of this approach are also possible, but in each situation care must be taken by both institutions to ensure that the arrangements they make do not compromise their overall compliance with the requirements of the PHS Policy and the AWAR regarding the responsibilities of the IACUC, AV, CEO, and IO (see also Chapter 5). Institutions considering any such relationship are well advised to consult representatives of NIH/OLAW and APHIS/AC about the appropriateness of their proposed arrangements.

3:19   How might an IACUC be effectively constituted at a very small institution where all or most personnel are likely to be directly or indirectly involved with the research to be reviewed?

Opin.   As pointed out in 3:18 and 5:31, the use of outside members on the IACUC is compatible with AWA and PHS Policy. In this situation, supplementing the affiliated members with a number of nonaffiliated members whose expertise is well suited to the needs of the institution would prevent the problems inherent in trying to populate an IACUC with the institution's entire staff of animal users.

References

1. USDA-APHIS, Animal Care Policy Manual. Available on the World Wide Web at: http://www.aphis.usda.gov/ac/polmanpdf.html.
2. ARENA/OLAW, Institutional Animal Care and Use Committee Guidebook, 2nd ed., 2002. Available on the World Wide Web at: http://grants.nih.gov/grants/olaw/Guidebook.pdf.
3. American Association for Laboratory Animal Science, The IACUC Training and Learning Consortium. Available on the World Wide Web at: http://www.iacuc.org.
4. Scientists Center for Animal Welfare, IACUC Talk, Scientists Center for Animal Welfare, Greenbelt, MD. Available on the World Wide Web at: http://www.scaw.com/iacuctalk.html.
5. Committee on Educational Programs in Laboratory Animal Science, Education and Training in the Care and Use of Laboratory Animals: A Guide for Developing Institutional Programs, National Research Council, National Academy Press, Washington, D.C., 1991.
6. Committee on Dogs, Laboratory Animal Management: Dogs, National Research Council, National Academy Press, Washington, D.C., 1994.
7. Committee on Rodents, Laboratory Animal Management: Rodents, National Research Council, National Academy Press, Washington, D.C., 1996.
8. Committee on Infectious Diseases of Mice and Rats, Infectious Diseases of Mice and Rats, National Research Council, National Academy Press, Washington, D.C., 1991.
9. Committee on Occupational Safety and Health in Research Animal Facilities, Occupational Health and Safety in the Care and Use of Research Animals, National Research Council, National Academy Press, Washington, D.C., 1997.
10. Committee on Regulatory Issues in Animal Care and Use, Definition of Pain and Distress and Reporting Requirements for Laboratory Animals: Proceedings of the Workshop Held June 22, 2000, National Research Council, National Academy Press, Washington, D.C., 2000.
11. Committee on Occupational Health and Safety in the Care and Use of Nonhuman Primates, Occupational Health and Safety in the Care and Use of Nonhuman Primates, National Research Council, National Academy Press, Washington, D.C., 2003.
12. International Perspectives: The Future of Nonhuman Primate Resources, Proceedings of the Workshop Held April 17–19, 2002, National Research Council, National Academy Press, Washington, D.C., 2003.
13. Committee on Guidelines for the Use of Animals in Neuroscience and Behavioral Research, Guidelines for the Care and Use of Mammals in Neuroscience and Behavioral Research, National Research Council, National Academy Press, Washington, D.C., 2003.
14. International Workshop on the Development of Science-Based Guidelines for Laboratory Animal Care Program Committee, The Development of Science-Based Guidelines for Laboratory Animal Care: Proceedings of the November 2003 International Workshop, National Research Council, National Academy Press, Washington, D.C., 2004.
15. Committee to Update Science, Medicine, and Animals, National Research Council, Science, Medicine, and Animals, National Research Council, National Academy Press, Washington, D.C., 2004.
16. Publications of the National Academy Press. Available on the World Wide Web at: http://www.nap.edu.

4

Reporting Lines of the IACUC

Richard C. Van Sluyters and Marcy Brown*

Introduction

Federal law, regulation, and policy stipulate that the IACUC must report to the individual who has been designated to serve as the IO. This chapter clarifies what it means to be the IO by reviewing this individual's many important responsibilities under the PHS Policy and the AWAR. Examples of common job titles for IOs in various types of institutions are used to provide the reader with a sense of the level of power or authority that this person should have within an institution's overall administrative organization. The chapter concludes with brief discussions of useful methods for educating the personnel within an institution about the functions of its IACUC, and the circumstances in which it may be desirable for an institution's internal policies with respect to animal care and use to go beyond merely adhering to the federal requirements.

4:1   To what person or organization does the IACUC report?

Reg.   The IACUC reports to the institution's IO and, in some circumstances, through the IO to APHIS/AC, NIH/OLAW, and other federal funding agencies. According to PHS Policy (IV,B,3; IV,B,5) and the AWAR (§2.31,c,3), the IACUC is required to submit reports of its semiannual program evaluations and facility inspections to the IO and to make recommendations to the IO regarding any aspect of the institution's program, facilities, or personnel training. PHS Policy (IV,F,1; IV,F,2) requires that the IACUC, through the IO, submit an annual report to NIH/OLAW. This report must describe any changes in the program, facilities, or IACUC membership and list the dates when the IACUC conducted its semiannual evaluations of the institution's program and facilities and submitted the evaluations to the IO. In addition, PHS Policy (IV,C,7; IV,F,3) requires the IACUC, through the IO, promptly to provide NIH/OLAW with a full explanation of the circumstances and actions taken with respect to any serious or continuing noncompliance with PHS Policy, any serious deviation from the *Guide*, or any IACUC suspension of an activity.

The AWAR (§2.31,c,3) require the IACUC to report through the IO to APHIS/AC and any federal funding agency, in writing, within 15 days whenever the

---

* The authors thank Jane E. Simpson and Ralph B. Dell for their contribution to this chapter in the first edition of *The IACUC Handbook*.

institution fails to adhere to a reasonable plan and schedule for correcting a significant deficiency in the institution's program or facilities (see 25:13; 26:16). In addition, if the IACUC suspends an activity involving animals, PHS Policy (IV,C,7) and the AWAR (§2.31,d,7) require that the IO, in consultation with the IACUC, review the reasons for suspension, take appropriate corrective action, and report that action with a full explanation to NIH/OLAW and/or APHIS/AC and any federal agency funding the project if the animal activity is regulated under the AWAR.

4:2   What is meant by the *Institutional Official*?

Reg.   The AWAR (§1.1, Institutional Official) state that the IO is the individual at a research facility who is authorized legally to commit on behalf of the research facility that it will meet the requirements of the AWAR. Similarly, PHS Policy (III,G) defines the IO as the individual who signs and has the authority to sign the institution's Assurance, which commits the institution to meet the requirements of PHS Policy. The *Guide* (p. 9) refers to the "responsible administrative official" as the individual who must appoint the IACUC.

Opin. If the Chief Executive Officer (CEO) does not designate an IO, it is the authors' opinion that the CEO becomes the IO by default and must be identified as such on the NIH/OLAW Assurance and any other pertinent documents. NIH/OLAW will not approve an Assurance without the signature of an individual identified as the IO. (See 3:2; 3:3.)

4:3   What are the responsibilities of the IO?

Reg.   The IO is the individual who is responsible for ensuring that an institution complies with all applicable animal welfare laws, regulations, and policies. The IO signs forms, reports, and letters on behalf of the institution and interacts with the IACUC in overseeing the institution's animal care and use program. According to PHS Policy, the IO:

- Signs and has the authority to sign the Assurance (III,G)
- Commits the institution to meet the requirements of PHS Policy (III,G)
- Receives inspection reports and recommendations (IV,B,3; IV,B,5)
- In consultation with the IACUC, determines whether deficiencies are significant or minor (IV,B,3)
- Receives notification of the IACUC's decision to approve or withhold its approval of animal activities (IV,F,3,C)
- Receives and transmits annual reports to NIH/OLAW (IV,F)
- Consults the IACUC regarding suspensions and corrective actions; reports to regulatory and funding agencies (IV,C,7)
- May subject protocols that have been approved by the IACUC to further review and approval but may not approve an activity that has not been approved by the IACUC (IV,C,8)
- Ensures that the institution maintains required records for the specified period (IV,E)

According to the AWAR, the IO:

- Legally commits the institution to meet the requirements of the AWAR (§1.1)
- Signs and submits the registration form and is responsible for notifying APHIS/AC of any changes in the institution's registration (§2.30,a–§2.30,c)
- If not done by the CEO, signs and submits the annual report to APHIS (§2.36,a)
- Receives inspection reports and recommendations from the IACUC (§2.31,c,3; §2.31,c,5)
- In consultation with the IACUC, determines whether deficiencies are significant or minor (§2.31,c,3)
- Forwards IACUC reports of uncorrected significant deficiencies to APHIS/AC and any federal agency funding that research (§2.31,c,3)
- Receives notification of the IACUC's decision to approve or withhold its approval of animal activities (§2.31,d,4)
- Consults the IACUC regarding suspensions and corrective actions; reports to regulatory and funding agencies (§2.31,d,7)
- May subject protocols that have been approved by the IACUC to further review and approval but may not approve an activity that has not been approved by the IACUC (§2.31,d,8)
- Ensures that all personnel involved in animal care, treatment, and use are qualified to perform their duties and that training and instruction in specific areas are provided to those personnel (§2.32,a; §2.32,c)
- Ensures that training and instruction are made available and that the qualifications of personnel are reviewed with sufficient frequency to fulfill the research facility's responsibilities (§2.32,b)
- Ensures that the institution has an AV who provides adequate veterinary care to its animals in compliance with the AWAR (§2.33)
- Ensures that any part-time AVs are employed under a formal written program of veterinary care (§2.33,a,1)
- Ensures that the institution maintains required records for the specified period of time (§2.35)
- If applicable, certifies to APHIS that any outside facilities holding animals for the institution are recognized animal sites under the institution's research facility registration (§2.38,i,3)

4:4 Who can serve as the IO if that role is delegated by the Chief Executive Officer?

Reg. (See 4:2.)
Opin. The AWAR and PHS Policy describe no restrictions nor do they stipulate any minimum qualifications for individuals who serve as IOs. However, to be effective, an IO must have the level of authority described in 4:6. (See 3:2; 3:3; 4:5.)

4:5 Who typically fills the role of IO in academia and in industry?

Opin. Some common titles of individuals who serve as IOs in academia and industry include Chancellor; President; Provost; Vice Provost; Vice Chancellor for Health

Affairs; Vice Chancellor, Administration; Vice Chancellor for Academic Affairs; Vice Chancellor for Research; Executive Director, Research Resources; Director, Research Administration; Director of Sponsored Research; Dean, School of Medicine; Associate Dean, Research and Sponsored Programs; Chief Executive Officer; Executive Vice President for Health Affairs; Senior Vice President and Vice Provost for Health Affairs; Vice President for Health Affairs; Vice President for Research; Vice President for Research and Graduate Studies; Vice President for Clinical Affairs; Vice President for Worldwide Toxicology. (See 3:2; 3:3.)

4:6   What institutional power or authority should the IO have?

Reg.   (See 4:2.)

Opin.   The IO must be authorized to commit legally on behalf of the institution that the requirements of the AWAR and PHS Policy will be met. To be effective, the IO must have sufficient administrative authority to promulgate, implement, and enforce policies across departmental lines. In addition, the IO must have sufficient fiscal authority to approve and fund a level of staffing that is adequate to meet the needs of the program, as well as any needed program improvements, facility repairs, and renovations. (See 4:3.)

4:7   How involved does the IO need to be in the general activities of the IACUC?

Opin.   The IO's level of involvement in the activities of the IACUC varies widely from institution to institution. Some factors that can influence the IO's involvement include the size and complexity of the institution's research program, whether the IO has a background in biomedical research, and the management styles of the organization and the IO. In some institutions, the IO or the IO's direct representative actually serves as a member of the IACUC and is directly involved in the general activities of the committee. NIH/OLAW recommends that the IO not serve on the IACUC, since the IACUC reports to the IO,[1] and this is sound advice. At a minimum, the IO must understand the functions of the IACUC as they are defined by the AWAR and PHS Policy. Furthermore, IOs that are organizationally distant from the IACUC must ensure that there is a mechanism by which they are promptly informed of any potential threats to animal welfare, or any violations of the AWAR, PHS Policy, the *Guide*, or the institution's PHS Assurance.

4:8   What means might be useful for educating personnel with respect to the functions of the IACUC and related policies of the institution?

Opin.   Personnel who should be educated about the functions of the IACUC include research and teaching personnel who use animals, animal care staff, and administrative personnel in departments or units that use animals or administer research grants or projects. Some useful means for educating these personnel include seminars, the development and promulgation of institutional policies and guidelines, training handbooks, Web pages, video tutorials, and posters or brochures. (See Chapter 21.)

4:9   Under what circumstances might institutional policy go beyond federal regulations or policy?

Opin.   Many institutions choose to adopt uniform policies that apply to all live vertebrate animals used in research and teaching, regardless of whether the species are covered by the AWAR or the activities are funded by federal agencies. This approach has a number of advantages. It prevents the development of nonuniform standards for the care and use of animals. It promotes the development of standardized procedures, forms, and records, thereby making it easier to monitor for compliance and simplifying record keeping. Finally, it helps assure the public that the institution adheres to the highest standards of animal care and use, regardless of the species of animal being used or the funding source.

4:10   Is it appropriate for a research facility to establish guidelines on who may contact either the NIH/OLAW or APHIS/AC for guidance on IACUC-related issues?

Opin.   It is wise for institutions to have designated spokespersons for communicating with external entities such as federal regulating agencies and the media. Such organization helps prevent inadvertent errors and miscommunications. Note that in situations in which an institution is making an official report of an animal welfare incident to APHIS/AC, NIH/OLAW, and AAALAC, the AWAR (§2.31,d,7), PHS Policy (IV,E), and AAALAC's Rules of Accreditation (Sect. 2) stipulate the mechanisms for submitting a formal report. Adherence to these requirements is, of course, not optional. Efforts to designate official spokespersons must not, however, infringe upon the rights granted to individuals to report alleged animal welfare incidents under institutional, state, and federal whistleblower and whistleblower protection policies.

4:11   The AWAR (§2.31,d,1,vi) require that the housing, feeding, and nonmedical care of animals are to be directed by the AV or another scientist who has the appropriate training and experience with the species being maintained or studied. In a typical multispecies animal facility, can a person who has research expertise with only one species that is housed direct the overall care of animals and delegate the day-to-day responsibility to persons who have appropriate training and experience?

Reg.   Whereas the AWAR (§2.31,d,1,vii) stipulate that *"medical* [emphasis added] care for animals will be available and provided as necessary by a qualified veterinarian," the AWAR make no such stipulation regarding the provision of nonmedical care. Accordingly, under the AWAR the AV may direct a scientist to provide the daily husbandry for animals, including their housing, feeding, and nonmedical care.

Opin.   Delegation of the day-to-day responsibilities for animal husbandry by the AV is common practice in animal care and use programs. In larger, more centralized programs this delegation is usually made to the staff employed by the laboratory animal care unit, while in smaller, decentralized programs, responsibility for daily care may be delegated to researchers and their staff. The critical factor in each situation is to be clear about the lines of responsibility and authority—delegating the provision of daily care to a researcher does not absolve the AV of the responsibility for ensuring that adequate care is provided; nor does it diminish the AV's overall authority in this area.

Reference

1. Garnett, N., personal communication, 1999.

5

General Composition of the IACUC and Specific Roles of the IACUC Members

Christian E. Newcomer

Introduction

The general composition requirements for IACUCs, established by federal regulations since 1985, give institutions considerable latitude in fashioning their IACUCs to reflect the scientific expertise and meet the needs of their animal care and use programs. Although the specific IACUC composition requirements vary according to the regulatory oversight agency involved, both the AWAR and the PHS Policy require the IACUC to have a diversified membership, including a veterinarian and a member unaffiliated with the institution. The PHS Policy further extends the diversification by requiring the institution to include a nonscientist on the IACUC. Membership diversification on the IACUC is intended to broaden the perspective and add depth to the important IACUC review processes.

Most institutions recognize that developing and sustaining an effective IACUC require planning and an ongoing commitment. The selection of individuals to meet specific membership requirements, new IACUC member orientation or education activities, and successful integration of new IACUC members as contributing members demand considerable effort. Even in a fully functional and effective IACUC, self-assessment and improvement efforts are necessary for continued high-quality service to the institution and for maintainance of an appreciation for and understanding of the emerging trends in animal care and use programs.

To assist in IACUC development and self-review efforts, this chapter is intended to examine the overall composition of the IACUC and provide information on the roles, responsibilities, and issues involving the IACUC and its various member categories. Some of the trend information provided was acquired through the author's review of approximately 120 institutions during the conduct of AAALAC site visits (or similar review activities) over the course of two decades, and from the author's personal experience as an IACUC member at ten institutions during this same period.

5:1 How many members must an IACUC have?

Reg.   The number of IACUC members required depends upon whether the institution is a recipient of funding from the PHS or other cooperating federal agencies, such as the National Science Foundation or the Department of Defense. In these instances, the institution must have at least five IACUC members (PHS Policy IV,A,3,b). Institutions that use animal species covered by the AWAR and funded through internal or private sources are required to have at least three IACUC members (AWAR §2.31,b,2). The *Guide* (p. 9) states, "The size of the institution and the nature and extent of the research, testing and educational programs will determine the number of members of the committee."

Opin.   The organizations visited by this author usually exceeded the minimal membership requirements set by the federal regulations, and this impression is affirmed by the data provided by 172 institutions queried about the characteristics of their IACUC membership. As shown in Table 5.1, the majority of institutions in both the academic and nonacademic categories reported that the committee size was most commonly between six and ten members. In academic institutions, larger committees of 11–15 members were the second most prevalent category, whereas in nonacademic institutions, smaller committees of three to five members ranked second in size order. Considering institutions of both types, only 10% continue to operate with a committee size hovering at the regulatory minimums. Although a minimum of five IACUC members may be adequate for a small organization with a focused research program and a limited spectrum of animal use, generally a minimum of seven or eight members is required in larger, more complex programs. This facilitates adequate representation of the various areas of research expertise in the organization and fulfills the IACUC's protocol and program review responsibilities.

Surv.   How many members are there on your IACUC?

TABLE 5.1

Number of Voting Members on the IACUC, by Type of Institution

	3–5 Members (%)[a]	6–10 Members (%)	11–15 Members (%)	>15 Members (%)
Academic institutions (*n* = 115)	9 (8)	58 (50)	38 (33)	10 (9)
Nonacademic institutions (*n* = 57)	9 (16)	37 (65)	8 (14)	3 (5)
Total (*n* = 172)	18 (10)	95 (55)	46 (27)	13 (7)

[a] Percentages are expressed according to the institutional category total.

5:2   What specific members are required for the IACUC?

Reg.   The AWAR (§2.31,b) state that the committee shall be composed of a Chair and at least two additional members. The AWAR does not describe the qualification of the Chair, but it states that the other two members of the committee will include at least one veterinarian, and at least one shall not be affiliated in any way with the facility and shall not be a member of the immediate family of a person who is affiliated with the facility.

The AWAR (§2.31,b,4) state that if the committee consists of more than three members, not more than three members shall be from the same administrative unit of the facility. The *administrative unit* is defined as the organizational or management unit at the departmental level of a research facility, which, for example, may include the office of research administration and the university (institutional) laboratory animal resources.

The *Guide* (p. 9) states that the IACUC should have a veterinarian "who is certified by the American College of Laboratory and Animal Medicine (ACLAM)" or one who has the experience "in the use of the species in question." In addition, there should be "at least one practicing scientist experienced in research involving animals." Finally, there should be "at least one public member to represent general community interests in the proper care and use of animals. Public members should not be laboratory animal users, be affiliated with the institution, or be members of the immediate family of a person who is affiliated with the institution."

PHS Policy (IV,A,3,b), on the other hand, states that membership must consist of not fewer than five members. The PHS Policy (IV,A,3,b,1–IV,A,3,b,4; IV,A,3,c) states that the committee shall include at least one veterinarian, one practicing scientist experienced in research involving animals, one member whose primary concerns are in a nonscientific area (for example, ethicist, lawyer, member of the clergy), and one individual other than a member of the IACUC who is not affiliated with the institution in any way and is not a member of the immediate family of a person who is affiliated with the institution. One individual may fulfill more than one requirement. However, no IACUC constituted under the PHS Policy may consist of fewer than five members. Also note that PHS Policy II requires institutions to comply with the AWAR, as applicable, and to follow the *Guide*.

Opin.   Organizations involved in agricultural research and teaching should be aware that there are guidelines for the composition of animal care and use committees operating in this setting.[1] These guidelines have no regulatory empowerment and closely parallel the PHS Policy, differing only in that they specify two types of scientific individuals on the committee. One should be a scientist from the institution who has experience in agricultural research or teaching involving agricultural animals, and the other should be an animal, dairy, or poultry scientist who has training and experience in the management of agricultural animals. This document further recommends that a separate committee not be established for this purpose, but that the composition of the IACUC be modified according to the recommendations just noted to provide for the centralized and uniform oversight of the institution's animal care and use program. In an institution that conducts both agricultural research and PHS-funded research, the requirement for two committee members to have the agricultural expertise specified could be met within the five-person membership requirement of the IACUC under PHS Policy. However, in the author's experience, the IACUCs serving these dual arenas have had at least seven to ten members and often more.

5:3   Must an IACUC report to NIH/OLAW and APHIS/AC the names and titles of all members of the IACUC?

Reg.   The general requirements to NIH/OLAW in the Letter of Assurance include the provision for the reporting of IACUC members' names, position titles, and credentials; however, a footnote clarifies that institutions may, at their discretion, represent the names of members other than the IACUC Chair and veterinarian with numbers or symbols (PHS Policy IV,A,3,b, footnote 6). The AWAR do not contain a comparable provision.

5:4   Can one person fill more than one position on the IACUC?

Reg.   The PHS Policy (IV,A,3,c) provides a written assent to dual representation as long as the IACUC totals five members. The AWAR offer no direct discussion of this matter and do not preclude this practice.

Opin.   It is permissible for an individual to fill more than one of the membership categories on an IACUC under both the AWAR and the PHS Policy. While it is permissible,

APHIS/AC strongly discourages the practice of one person's filling more than one role, citing the "potential for conflicts of interest and/or undue influence by one person over the facility's program."[2] Very few institutions have exploited dual category representation as a long-term strategy for their IACUC, and, when it occurs, it frequently involves the unaffiliated member serving in the dual capacity as a nonscientist. It is also conceivable, but exceedingly rare in the author's experience, that an entrepreneurial veterinarian in a small contract laboratory setting could serve as the IO, scientist, and AV within the AWAR. In each of the foregoing circumstances, the objectivity and independence of the IACUC's decisions could be compromised, particularly if the committee is small. In the author's experience, dual capacity appointments as the nonscientist and unaffiliated member are most commonly used only transiently to bridge a gap in the mandated membership after the departure or resignation of a member.

5:5   What problems can occur if the number of IACUC members is too large or too small?

Opin.   There can be significant problems related to the size of the IACUC. IACUCs that have a small membership often have a narrower base of expertise and, depending upon the size of program they service, may encounter an onerous workload. Also, small IACUCs may often have a more difficult time making a quorum, and, even in the presence of a legitimate quorum, member absences can have a marked effect on decision making. The challenges for IACUCs that have large memberships are very different. In large IACUCs, the principal problem is that members may become disengaged from the activities of the IACUC. This might be due to impediments to open communication in the IACUC or to the daunting scope and size of the program and its challenges. Thus, members have no sense that their efforts are having any impact on program quality or progress. In the depths of ennui, these members contribute to the quorum in number only. The challenges for the IACUC Chair are to ensure that all members have the opportunity for real contribution and an opportunity to be heard on issues, to help members feel a sense of accomplishment, and, when necessary, to inform the IO that the IACUC may be too large for the tasks at hand. Training opportunities are important for the maturation and effective operation of all IACUCs and can help keep committee members engaged.

5:6   What institutional constituencies should be represented on the IACUC?

Opin.   The majority of organizations go beyond the federally mandated composition of the IACUC (see 5:2) to diversify the IACUC membership. The first priority in this effort generally is to enlarge the complement of scientific members on the IACUC to match the predominant areas of scientific expertise in the program. In addition to the potential political benefits this has for the IACUC within the scientific community it serves, it confers tangible benefits for the IACUC's deliberations on protocol matters. It also may improve the quality of correspondence exchanged between the IACUC and PIs.

Other institutional constituencies that this author finds to be favored (in approximate order of frequency) are the following:

- Health and safety personnel (including occupational health and safety)
- Senior animal management or supervisory personnel

- Research laboratory technicians (with extensive animal involvement)
- Legal counsel
- Public relations personnel
- Grants and financial personnel
- Student body representatives

The merits of having these people on the IACUC should be evaluated in the context of the program for which they are intended. In the author's opinion, health and safety personnel, legal counsel, and public relations personnel should be given first consideration. Health and safety personnel are comfortable working at the interface of science and society and can make a significant contribution to the IACUC by keeping it apprised of the status of health and safety regulations, and recommendations relating to the animal care and use environment.[3] Legal counsel and public relations personnel also can be of great assistance in the consideration of issues relating to science, policy, perception, and the public.

5:7 Should the selection of IACUC members depend upon the institution's research goals and expertise?

Opin.   The two most important functions of the IACUC, protocol review and programmatic review, should be major driving forces for the IACUC to select members with a share in the institution's research goals and expertise. Protocol review should be insightful, thorough, and objective. Members who have scientific expertise in the areas under review can help focus the discussion of relevant issues, improving the quality of the review. IACUC programmatic review is a mandated responsibility of the IACUC (PHS Policy IV,B,1–IV,B,5; AWAR §2.31,c,1–§2.31,c,5), but it really is best performed as a bidirectional process involving the community served. This implies that the IACUC should foster conduits for the flow of information from all participants in the animal care and use program. The appointment of IACUC members who represent specialty interests or areas of expertise can prove advantageous to this process. Most institutions make a concerted effort to staff their IACUCs with members who reflect the research expertise of the program, at the same time making sure to prevent possible conflicts of interest.

5:8 What is the typical IACUC composition in a large versus a small institution, and are the members' roles different in these settings?

Opin.   There are different trends in the composition of the IACUC and in the ways the IACUC uses its members in large versus small institutions. In small commercial institutions or independent research organizations based around a particular area of scientific inquiry, the IACUCs tend to be smaller (five to seven members) because the number and diversity of issues requiring consideration in the IACUC are limited. The reader should note that although institutional size was not queried, 81% of nonacademic institutions had fewer than ten members, whereas only 58% of academic institutions had this profile. Small academic institutions tend to have slightly larger IACUCs (seven to ten members). Although the research portfolio may be limited compared to that of a large academic institution, the relative diversity of the research activities can be quite high even though each department may only have a few active faculty researchers. The intimacy of the small institutional environment sometimes imposes a higher level of expectation that the IACUC members will be able to speak authoritatively about the research and programmatic needs of their colleagues. IACUC members in small institutions also are more likely to be active participants in developing and implementing programmatic components (e.g., biosafety review and IACUC educational efforts) that may fall under the IACUC's purview.

Large institutions sometimes have 15 to 20 or more IACUC members simply to provide ample representation of the animal user groups and investigators (e.g., 150 to 200) in the program. Table 5.1 indicates that 13 (8%) of institutions surveyed operate with IACUCs of more than 15 members. In this author's experience, these IACUCs tend to be more formal, impersonal, and insular. Frequently, institutionally provided administrative support obviates the need for the personal involvement of the IACUC members in ushering new initiatives forward. As noted in 5:5, a significant challenge for IACUC members in this position is to combat these tendencies and convince the faculty that the IACUC is a receptive, helpful, and responsive body that works for the institution in support of the scientific mission by encouraging improvement of the animal care and use program. Another issue germane to extremely large IACUCs is the number of veterinarians appointed to that committee. Although federal law has stipulated that an IACUC need only have one veterinary member, most institutions that have large IACUCs (greater than 15 persons) have increased the number of veterinarians appointed. This action helps ensure that the veterinary perspective will be represented consistently and affords the veterinarians an opportunity to divide the workload and concentrate on specialty areas of interest and needs.

5:9   Should the IACUC Chair, individual IACUC members, or their departments receive compensation for their efforts?

Opin.   In the academic setting, the compensation of the IACUC Chair has continued to be more common since the first edition of this book, and the open discussion of this matter at national meetings has likely accelerated this trend. Reimbursement for the Chair's time commitment occurs in approximately one half of the academic institutions visited by or reported on anecdotally to the author. Salary support, generally ranging from 10–25% provided to the academic department of the Chair, remains the most common approach. However, funding to support professional development, research program development, research technical support, and a variety of other approaches have emerged from the faculty–administrator negotiations. With less success than the Chair to date, other faculty IACUC members are also beginning to pursue and win some compensation for their efforts. Examples include travel support to national meetings, reduced instructional commitments, or release from ancillary academic obligations (e.g., other committee appointments), and recognition of their service in considering promotion and tenure. The author cannot confirm that any institutions are providing salary support for faculty members of the IACUC; however, this subject has been broached in several institutions. Compensation for the unaffiliated member is discussed in 5:36.

5:10   How can the performance of the IACUC be evaluated?

Opin.   Several mechanisms are available and should be utilized to evaluate the performance of the IACUC on an ongoing basis. The IO should meet periodically with the Chair of the IACUC to discuss the content, style, and timeliness of IACUC reports and correspondence with members of the institution's research community. This is the most common method reported by the institutions known to this author, although few institutions couch such meetings in terms of an evaluation. Also, without the inducement of a particular incident, the IO should consider soliciting the comments of the faculty or staff who use laboratory animals as well as of the members of the IACUC. The assessment of the IACUC as a working group seems particularly important because of the profound potential impact of group dynamics on IACUC morale, attention to detail, and, ultimately, decision making.

Participation in the voluntary peer review process of the AAALAC entails a substantial review of IACUC function through interviews and the review of written materials. Generally, AAALAC site visits are conducted at triennial intervals and should be supplemented by the institution's internal evaluation mechanisms. In some instances, expert consultants also have been used by institutions.

5:11   Who can remove the IACUC Chairperson or other IACUC members?

Reg.   The PHS Policy (IV,A,3,a) requires the Chief Executive Officer (CEO) to appoint the IACUC, and the AWAR (§2.31,a) specifically directs the CEO to appoint the Chair. Thus, although once established the IACUC possesses autonomous review and reporting functions, service on the IACUC is at the discretion of the CEO. Under the PHS Policy (IV,A,3,a, footnote 5, 1996 reprint), the CEO may delegate the authority to appoint the IACUC if the delegation is specific and in writing.

Opin.   It is very fortunate that institutions are rarely, if ever, confronted with this quandary because very few institutions have clearly written bylaws for their IACUCs that stipulate who has jurisdiction or what the process is in this matter, and no clarification is offered in the PHS Policy or the AWAR. For the most part, informal discussions originating from the Chair, in the case of other IACUC members, or from the CEO, in the case of the Chair, apparently have proved adequate for the silent and amicable departure of IACUC members in the vast majority of cases. In the absence of bylaws clearly delineating who has authority in this matter, the CEO retains that authority. In a related matter, without bylaws the extent to which and the situations in which parliamentary procedure should be applied in IACUC meetings and deliberations and its ability to self-regulate through the use of censure or expulsion procedures are also questionable.[4]

5:12   What criteria can be applied for removing the IACUC Chairperson or other IACUC members?

Opin.   Most organizations agree that inappropriate personal conduct, poor attendance at or participation in IACUC activities, or repeated inadequate preparation for assigned IACUC duties might constitute sufficient reason to seek the removal of a member. Ideally, these broad areas would be included in the bylaws developed for the IACUC, and the IACUC members would be informed during their initial IACUC training of the general performance criteria for committee members. The IACUC Chair or IO should inquire whether there are any mitigating circumstances for those members who cannot regularly attend two thirds of the IACUC meetings held annually and consider the replacement of these individuals if better attendance is not forthcoming.

5:13   What is the procedure for selecting and appointing the IACUC Chair?

Opin.   Most IACUCs attempt to prepare for a transition in the position of Chair through the appointment of a Vice Chair or an ad hoc Chair-in-Training position (see 3:7). In this regard, the IACUC usually has the opportunity to serve in an advisory capacity to the CEO, who ultimately has the responsibility of appointing the Chair (AWAR §2.31,a; §2.31,b). (See 3:2; 3:3.) Other groups that can prove useful in identifying and supporting candidates for the position of IACUC Chair include the faculty senate and faculty research advisory committees.

5:14   What traits or qualifications are desirable for the IACUC Chair?

Opin.   In most institutions, a concerted effort is made to recruit a Chair who is a respected scientist and who has had significant experience using laboratory animals in research. Regardless of whether or not the Chair has a scientific background, he or she should be regarded as a good colleague in the context of the institution's scientific mission. The Chair should be patient, tolerant, diplomatic, tactful, and efficient in the handling of the IACUC's business and sensitive issues. He or she should be able to dedicate the time necessary with sufficient flexibility to plan, oversee, and/ or participate in all critical IACUC functions and activities. The notion of adequate participation in critical IACUC functions presupposes that the Chair will be knowledgeable and conversant in current regulatory interpretations, contemporary practices, and trends pertaining to animal use and welfare and committed to leading a successful institutional animal care and use program. In addition to providing a continuing and stabilizing presence, the Chair is likely to encounter occasional urgent matters that require immediate attention.

   Some institutions have been slow to recognize that they have appointed a Chair who has not been performing adequately or may be constitutionally unfit for this important position. A partial list of problems indicative of such an appointment includes disorganization of IACUC meetings and activities; inattention to issues raised by the faculty, IACUC members, or veterinary staff in IACUC meetings; resistance to the adoption of needed programmatic changes as new standards emerge; proclivity for unbalanced communications (written or verbal) with investigators without the collaborative input or endorsement of the IACUC; or inappropriate and unilateral supplication, or alternatively, rigidity in dealing with investigators.

5:15   What is the Chair's role in the oversight of IACUC activities?

Opin.   The Chair should play an active role in the oversight of all IACUC activities regardless of his or her role as an actual participant. The Chair serves five important constituent groups:

- The senior administration (embodied in the CEO and IO)
- The scientific community
- Other members of the IACUC
- The federal government
- The public

   The immense commitment within the institution of time, resources, and interest to the IACUC's activities should compel the Chair's enthusiastic involvement in IACUC oversight. (See 3:1.)

5:16   Can the AV serve as the IACUC Chair?

Opin.   While there is no prohibition in the PHS or AWAR of the AV's serving as the IACUC Chair, this practice is not recommended.[2] Only a few institutions among the approximately 120 known to this author have taken this approach. NIH/OLAW has issued an explanation that strongly suggests that having either the AV or IO serve as the IACUC Chair may be inappropriate because of real or perceived conflicts of interest and the disruption of the necessary checks and balances intended in the institutional reporting structure and IACUC review processes.[5] The programmatic review activities of the IACUC encompass laboratory animal management and veterinary care, both of which are often under the direct purview of the AV. Thus, the appointment of the AV as the Chair affords the AV an opportunity to influence the IACUC openly or subtly. Similar efforts to dissuade institutions from continuing the use of this practice have been advanced routinely by AAALAC during site visits. AAALAC has made the argument that the responsibilities of the AV are already sufficiently demanding and should not be compounded by adding the complex job of the IACUC Chair. Moreover, AAALAC has recognized some instances in which the AV does not carry the title of IACUC Chair but is functioning de facto as the Chair and carrying the workload of the IACUC.[6] Institutions should be wary of this devolutionary development.

5:17   Should the manager of the institution's laboratory animal resource unit serve as the IACUC Chair?

Opin.   The appointment of the animal facility manager as the IACUC Chair is much more common than the appointment of the AV in this capacity. Facility managers have a broad understanding of the various aspects of the animal care and use program but generally have a narrower scope of responsibilities than the AV. This lessens the conflict of interest issue discussed in 5:16; the issue is not entirely eliminated, however, since the review of facilities and the diverse provisions for laboratory animal management are critical elements of the IACUC's programmatic review that generally fall within the domain of the facility manager. The involvement of the facility manager in the financial management of the resource may be another deterrent to the appointment of this individual as the Chair in some institutions.

5:18   In an academic setting, what should be the faculty rank of the IACUC Chair? Is a person who has academic tenure preferred?

Surv.   We queried the faculty appointment status of the Chair in 103 institutions. Resulting data are provided in Table 5.2.
    Institutions demonstrated an overwhelming preference for the appointment of a faculty member to chair the IACUC, and a majority of the faculty appointed had

TABLE 5.2

Faculty and Tenure Status of the Chair in Academic Institutions

	Nontenured	Tenured	Not Specified	Total
Nonfaculty	N/A	N/A	N/A	6
Assistant Professor	6	1	5	12
Associate Professor	7	11	17	35
Full Professor	3	33	14	50

achieved a rank of Full Professor. Of the Full Professors serving in this capacity, 66% were tenured. In an additional 28% the tenure status was not indicated. Thus, the actual percentage of tenured Full Professors serving as the Chair is likely higher. Also, more Associate Professors occupy the position of Chair than Assistant Professors (34% versus 12%), and approximately one third of the Associate Professors are also tenured. Among all faculty positions reported, 62% of the Chairs were either Full Professors or of a lower rank but had attained tenure.

Opin.   The data support the interpretation that most academic institutions believe that high-ranking faculty who have achieved tenure will best command the respect and wield the authority to lead the IACUC through its federally mandated review activities impacting institutional resources and faculty satisfaction. In many large academic institutions, the appointment of a tenured Full Professor (and one who has an active research program or prior history of laboratory animal use) is deemed desirable. Academic rank, per se, is a less critical factor than an individual's scientific stature, collegiality, and reputation for fairness within the institution's scientific community. Institutions should recognize that the faculty often directs its antagonism toward adverse decisions made by the IACUC to the Chair. Hence, the appointment of a nontenured faculty member (in a tenure track position) as the IACUC Chair conceivably might enhance that individual's vulnerability during tenure review decisions. Most institutions (or faculty members) appear to recognize this and avoid placing lower-ranking academic faculty in this precarious position.

5:19   Can an IACUC be composed entirely of persons not employed by the university (or other institutions)?

Reg.   Both the AWAR and the PHS Policy (see 5:2) require that at least one member of the IACUC not be affiliated with the institution in any way other than as a member of the IACUC. Other members are not required to be affiliated except that the veterinarian must have direct or delegated program authority (PHS Policy IV,A,3,b,2) and responsibility (AWAR §2.31,b,3,i) for activities involving animals at the institution. Also, pursuant to PHS Policy (IV,B), the IACUC functions as an "agent of the institution."

Opin.   Neither the PHS Policy nor the AWAR stipulate that some (or any) of those appointed must be institutional employees. Nevertheless, the author has never encountered an IACUC that did not contain at least one member who was an employee. (See 3:18.) The pivotal issue in the matter of IACUC appointment is not whether the CEO holds any financial sway over the committee members through employment, but rather, whether the CEO is capable of disbanding the IACUC if it proves to be incompetent or inattentive to animal welfare issues or fails to discharge its federally mandated responsibilities with due diligence. It is conceivable that one organization participating in a cooperating research partnership with another organization may elect to coappoint a knowledgeable IACUC that serves the research missions of both organizations, irrespective of the employment associations of the membership. This type of arrangement is rare and, in the experience of the author, is likely to be carefully examined by federal oversight and accreditation bodies.

5:20   Can persons who do not have prior experience or expertise with laboratory animals be appointed as members of an IACUC?

Reg.   The regulatory requirements for personnel with experience with laboratory animals are minimal (PHS Policy IV,A,3,a; IVA,3,b; AWAR §2.31,a; §2.31,b). Both require that the veterinarian have experience with laboratory animals, and the PHS

Policy stipulates that, in addition, one practicing scientist has experience in research involving animals. Members who do not have laboratory animal service can be called into IACUC service.

Opin. Most organizations easily exceed the regulated number of those with laboratory animal experience, but IACUCs often have many members who have laboratory animal experience related to one or a few species involving applications in a particular area of research. Much is to be said for the vicarious experience these individuals garner from talking with colleagues, participating in active research programs, reading the literature, and gaining experience on the IACUC. However, introspection by the IACUC is also desirable to allow the self-identification of areas of IACUC weakness and foster the recruitment of internal or external consultants in areas of need.

5:21   Can or should an outside consultant serve as the IACUC Chair?

Opin. No institution known to this author uses an outside consultant as the Chair of the IACUC. This practice is not prohibited by the PHS Policy or the AWAR. The difficulty in appointing an outside consultant as the Chair is that such an individual lacks an appreciation for the scientific landscape and key players of the institution and will less effectively respond to issues in a timely and situationally appropriate manner. These difficulties might be overcome by an outside consultant with a long history of involvement with the institution.

5:22   What are the specific duties of the IACUC Chair?

Opin. The IACUC Chair has the responsibility for overseeing the coordination and implementation of effective, efficient systems for protocol review and program review by the IACUC in compliance with the PHS Policy and the AWAR. These review activities can only be performed at a properly convened meeting of the IACUC (see Chapter 6). Thus, the Chair should:

- Ensure that a quorum of the IACUC is present
- Declare the loss of a quorum, resulting in the end of official business if a sufficient number of members depart
- Prepare or oversee the preparation of meeting minutes and reports and submit these documents to the IO in accordance with PHS Policy (IV, E; IV,F) and the AWAR (§2.31,c,3; §2.31,d,2; §2.31,d,4; §2.35,a)
- Report to the IO any activities that have been suspended by the IACUC for noncompliance with PHS Policy or the AWAR (see Chapter 29)
- Establish a sound system of written communication for the IACUC with investigators concerning the approval status of protocols and the steps necessary to secure approval

Beyond these relegated duties that the Chair performs on behalf of the committee, most institutions expect the Chair to keep abreast of new regulatory trends and interpretations and evaluate and champion policy and practice initiatives (e.g., new training and educational programs) to improve the animal care and use program. This process involves the Chair's regular interaction with other areas of expertise within the organization, ranging from other institutional committees to occupational health and safety, human resources, and the physical plant.

5:23   When a quorum is lost because of the mandated nonparticipation of a member that results from a conflict of interest, may that person's alternate reconstitute the quorum and vote?

Opin.   Situations such as this should provide a strong impetus for institutions to articulate the role, responsibility, and authority of alternate members clearly in a charter document. In the author's view, such a substitution to make a quorum would be appropriate because the alternative member represents a redundancy in position and not of perspective. Alternative member appointments are acceptable only if they correspond one to one with the primary members. The underlying assumption should be that an alternative member will be knowledgeable about the facts under review and evaluate them autonomously, understand all relevant IACUC issues, and vote his or her conscience. *This interpretation is at variance with the opinions that have been offered by others on this matter,* who have cited the following regulatory guidance:

- "An IACUC member and his/her alternate may not contribute to a quorum at the same time or act in an official IACUC member capacity at the same time. An alternate may only contribute to a quorum and function as an IACUC member if the regular member for whom they serve as alternate is unavailable."[7] Further, in an unpublished opinion, NIH/OLAW has written, "OLAW guidance permits an alternate IACUC member to participate in a meeting if the regular member for whom they serve as alternate is unavailable. If a regular IACUC member must excuse her/himself from a meeting due to a conflict-of-interest, that member is not unavailable (rather, they are conflicted) and therefore an alternate may not step in to take the regular member's place at the meeting."[8]
- "A regular member of the IACUC is considered 'unavailable' if they are entirely unable to perform their required duty (e.g., review proposed animal activities, vote at a convened meeting, or inspect the research facility's animal facilities)."[9]

In the author's view, the caution against the alternative member's contribution "at the same time" should be taken quite literally to mean "simultaneously" and not "at the same convened meeting" and the regular member can indeed be regarded as "unavailable" for the quorum and for the review of the proposed activity by virtue of regulatory stipulation.

5:24   Can alternate members of the IACUC perform semiannual inspections that involve AWA-regulated species?

Reg.   APHIS/AC and NIH/OLAW have agreed in a guidance document that alternate members can participate by sharing their expertise and perspectives in IACUC meetings and review activities even when their primary IACUC member is present. However, they may only contribute to a quorum in place of, and not as a supplement to, the primary member they are intended to represent.[7]

Opin.   Appropriately appointed alternate members would be eligible to participate in semiannual inspections involving USDA-covered species. In the author's view, it would be inadvisable, if not impermissible, for a regular member and his or her alternate to constitute the two-person team required under the AWAR (§2.31,c,3) during the facility inspection process, because they represent only a single vote on the IACUC. As suggested in 5:23, creation of a charter document governing these activities would be advisable to clarify the institution's expectations of its IACUC members. (See 23:16.)

5:25   An IO appoints a scientist as an alternate member of the IACUC, to fill in when any scientific member is "unavailable." Under what conditions can this individual participate in IACUC functions? Should institutions place any conditions on what constitutes the "unavailability" of a member?

Opin.   According to an NIH/OLAW-APHIS/AC joint statement on this subject, the alternate member is intended to serve in an official capacity when his or her corresponding regular member is unavailable.[7] Alternates may not be assigned randomly for shifting absences among the IACUC membership. The author does not believe it would be advisable for institutions to attempt to define the planned commitments and uncontrolled conditions in the lives and schedules of its members that result in their unavailability. However, institutions may find it useful to communicate a performance standard indicating that members who are regularly "unavailable" at the last moment or otherwise appear to be avoiding participation are not meeting their IACUC commitments. Alternate IACUC members should resist the capricious or compromised decision making that might be associated with short notice circumstances.

5:26   Must the institutional AV also serve as the veterinary member of the IACUC?

Reg.   The PHS Policy (IV,A,3,b,1) and the AWAR (§2.31,b,3,i) state that the veterinarian on the IACUC shall have training or experience in laboratory animal science and medicine and have direct or delegated program authority and responsibility for activities involving animals at the institution. The AWAR (§2.33,a,3) state, "The attending veterinarian shall be a voting member of the IACUC; *Provided, however,* That a research facility with more than one Doctor of Veterinary Medicine (DVM) may appoint to the IACUC another DVM with delegated program responsibility for activities involving animals at the research facility."

Opin.   The AV is not required to serve on the IACUC, although serving on the IACUC is the strategy exercised in most institutions. Some institutions have deliberately avoided placing the AV or the veterinary program director on the IACUC, to dissociate the overall program of veterinary care and its associated scientific interactions with investigators from the monitoring and policing functions inherent in IACUC activities. Other institutions have used this allowance to rotate different veterinary members onto the IACUC periodically.

The wording of AWAR §2.31,b,3,i, as indicated in the Regulatory paragraph, describes the IACUC veterinarian with the wording found in AWAR §1.1, Attending Veterinarian. Although this suggests that the AV must be an IACUC member, AWAR §2.33,a,3 states that if the research facility has more than one veterinarian, then a veterinarian other than the AV can serve on the IACUC in place of the AV as long as that other veterinarian has delegated program responsibility for animal activities. PHS Policy appears to be in concurrence since it allows the IACUC veterinarian to be one who has "direct or delegated program authority and responsibility." That is, under both the AWAR and PHS Policy, if the AV is not the IACUC veterinarian, then at the least the IACUC veterinarian and the AV collectively must have the appropriate direct or delegated responsibility and authority.

The question arises, What happens in a large animal facility where there is a rotation of veterinarians on and off of the IACUC, often on a monthly basis? In this situation, it is suggested that the IACUC proceed with caution. Both APHIS/AC and NIH/OLAW have opined that under this circumstance problems can arise since it is unlikely that each month a different IACUC veterinarian can speak with the required authority and responsibility for the program of animal care and use.[10,11]

5:27   What training or experience is useful for the veterinary member of the IACUC?

Reg.   The PHS Policy (IV,A,3,b,1) and AWAR (§2.31,b,3,i) stipulate that the veterinary member of the IACUC should have training or experience in laboratory animal medicine and science. (See 27:2.) The AWAR (§1.1) further clarify under the definition of the *attending veterinarian* that this individual should have either graduated from a veterinary school accredited by the American Veterinary Medical Association Council on Education, acquired a certificate issued by the American Veterinary Medical Association's Education Commission for Foreign Veterinary Graduates, or received equivalent formal education as deemed appropriate by the APHIS administrator.

Opin.   The intent of the PHS Policy and AWAR is to help ensure that the IACUC can rely upon the veterinary member not only for competent clinical insights, but also for information on ancillary areas such as unusual animal models, zoonoses and other occupational health and safety concerns, hazard containment, genetics, and unique nutritional and husbandry requirements of laboratory animal species. Proficiency in these diverse subject areas can be demonstrated formally by board certification in the American College of Laboratory Animal Medicine (ACLAM), and it is advisable for one or more of the veterinarians responsible for a large and complex program of research animal use to be board certified. However, the needs of research animal care and use programs that are smaller or of a more limited scope may be met by veterinarians who are not ACLAM board certified. Regardless of their ACLAM certification status, all veterinary personnel involved in the oversight of research animal use should be aware and stay abreast of the central research, clinical, and regulatory concerns pertaining to the species under their care. This can be accomplished by reviewing the literature and attending the national meetings sponsored by various organizations, including the American Association for Laboratory Animal Science, ACLAM, American Society of Laboratory Practitioners, Institute for Laboratory Animal Research, Association of Primate Veterinarians, American Veterinary Medical Association, and AAALAC.

5:28   What is the role of the veterinarian on the IACUC?

Reg.   The role of the veterinarian within the context of the IACUC is defined under the AWAR (§2.31,d,iv,B; §2.31,d,vi; §2.31,d,vii; §2.31d,ix; §2.31,d,x,B) and by the *Guide* (pp. 12–13), which serves as an extension and amplification of the PHS Policy. In summary, the AWAR include provisions for:

• Veterinary consultation on the recognition and palliation of pain
• Direction of animal care and use (a specific role of the AV)
• Medical care
• Aseptic surgery and postoperative care

- Oversight of multiple major survival surgery resulting from a veterinary condition in an animal that also had experimental surgery

    Specific provisions within the *Guide* include:

- Advising the IACUC on new procedures or procedures with the potential to cause pain and distress that cannot be reliably controlled
- Ensuring that veterinary care is available to mitigate the illnesses, lesions, or behavioral abnormalities associated with animal restraint

These sources also expand on the responsibilities of the AV to the institutional animal care and use program that are unrelated to IACUC membership per se. (See Chapter 27.)

Opin.   Veterinarians play a unique role on IACUCs, as a result of their expertise in laboratory animal medicine and science and broader understanding of the physiological, behavioral, nutritional, and husbandry needs of laboratory animal species. Although by no means a lone voice on the IACUC on matters of animal health and welfare, the veterinarian should make the concerted effort to address these areas with precision and authority. Special areas of emphasis for the IACUC veterinarian should include the discussion of the use of proper anesthesia and analgesia in laboratory animals in the relief of pain and distress, the analysis of possible iatrogenic complications to the procedures used or the disease model proposed, and a review of the plans for appropriate and timely medical intervention. (See 27:6.)

5:29   Under what conditions can the veterinarian delegate some of his or her responsibilities to other IACUC members?

Reg.   According to the AWAR (§2.31,d,1,iv,B), the AV may delegate his or her responsibility for playing a role in the planning and consultation on procedures that may cause more than momentary or slight pain or distress.

Opin.   Depending on the background, training, and experience of the other IACUC members, and the size, complexity, and maturity of the animal care and use program, many of the veterinarian's responsibilities can be delegated to other members for defined periods (e.g., a particular IACUC meeting or semiannual review). Very few of the institutions known to this author would proceed with an IACUC meeting in the absence of the veterinarian, but none is prepared to neglect significant issues because of insufficient veterinary input. The strategy most often used when the IACUC veterinarian is absent is to invite another veterinarian, knowledgeable in laboratory animal medicine, to participate as a guest in an advisory capacity. Secondarily, IACUCs can table issues requiring the veterinarian's special analysis and comment. (See 6:7; 27:8.)

5:30   Does the attending veterinarian have any mandated training responsibilities under the AWAR or PHS Policy?

Reg.   Training is not assigned as a specific requirement to the AV in either the AWAR or the PHS Policy. The AWAR (§2.32,a; §2.32,b) indicate that training "shall be the responsibility of the research facility." The PHS Policy (IV,C,1,f) places the onerous responsibility of assuring that personnel are appropriately qualified and trained to conduct the procedures proposed on research animals on the IACUC, through its protocol review process. Many institutions choose to address the training initiative in a collaborative venture with the veterinary staff, but most use a multidisciplinary approach involving institutional scientists, capable technicians with advanced skills, dedicated training personnel within the IACUC, or veterinary resource group as well as the veterinarian(s). In the author's view, verification of training and documentation of proficiency continue to be areas of vulnerability in many research animal programs. AVs, or their designee, however, must provide "*guidance* to principal investigators or other personnel involved in the care and use of animals regarding handling, immobilization, anesthesia, analgesia, tranquilization and euthanasia; and … adequate pre-procedural care and post-procedural care in accordance with current established veterinary medical and nursing procedures," as stipulated in AWAR §2.33,b,4 and §2.33,b,5. IACUCs should work collaboratively in support of the program of veterinary care to ensure that ample resources are available to the veterinarians to meet this critical objective.

5:31   What is meant by the *nonaffiliated (outside) member* of the committee, and what is the intent of including such an individual on the committee?

Reg.   The intent of including a nonaffiliated member on the IACUC is to ensure that someone who is not affiliated with (or beholden to) the institution in any manner is involved in the review of the institution's animal care and use activities. This individual also may not be a member of the immediate family of a person affiliated with the institution (AWAR §2.31,b,3,ii; PHS Policy IV,A,3,b,4; *Guide*, p. 9). The *Guide* also stipulates that public members should not be laboratory animal users. According to the AWAR (§2.31,b,3,ii), this individual is intended to represent the "general community interests in the proper care and treatment of animals." (See 3:4.)

Opin.   Subsequent clarification on this matter by NIH/OLAW has indicated that it is inappropriate for an institution to have an individual who currently uses laboratory animals serving in this capacity.[12] NIH/OLAW does not rule out individuals who have used laboratory animals in the past. Ostensibly, because the nonaffiliated member is not expected to benefit personally from any of the activities proposed and is not dependent on the institution for his or her livelihood, she or he enhances the public's confidence in the unfettered objectivity of the IACUC review processes. The unaffiliated member should be regarded as a full IACUC member and should be afforded opportunities to participate equivalent to those of all other IACUC members. However, in the author's view, the unaffiliated member should not be asked to take the lead in areas in which he or she may not be comfortable (e.g., as the primary reviewer in the Designated Review process for a protocol).

5:32   As an extension and example of 5:31, if two colleges, A and B, reciprocate in course admissions for students and exchange faculty for course instruction, would it be appropriate for the unaffiliated IACUC member at College A to be a spouse of a faculty member at College B?

Opin.   In the author's view, institutions should exercise common sense in the application of appointments when the "affiliation" is extremely tenuous as it is in this case. If the faculty member at College B is not personally involved in some aspect of the animal care and use program at College A or is not dependent upon the success of College A for salary and livelihood, there is no compelling reason to believe that the spouse would not be able to perform this important public service admirably.

5:33   What are the backgrounds of individuals who typically serve as nonaffiliated members of IACUCs?

Opin.   There is no typical background of these individuals. Institutions have chosen a wide variety of individuals to serve in this capacity. Without implying that the following categories are mutually exclusive, the range of individuals includes broadly educated, erudite humanists; businesspersons; public servants; educators; persons involved in other types of animal care and use; physicians, veterinarians, scientists, and technicians who do not use laboratory animals; and concerned citizens who seek to make a public contribution.

5:34   Can the nonaffiliated member be a scientist, veterinarian, ethicist, or biostatistician who is not affiliated with the institution?

Reg.   Yes.
Opin.   Any of these individuals would qualify as long as he or she is not involved in the care and use of laboratory animals, either directly or through consulting and similar tangential service. (See 5:31.)

5:35   Can a retiree from an institution serve as the unaffiliated member of the same institution's IACUC?

Opin.   In the author's experience, very few institutions exercise the opportunity to appoint either retirees or alumni as unaffiliated members perhaps because they are disinclined to create the perception of a conflict of interest. Such appointments are not prohibited by the regulations nor discouraged in guidance documents. In the few cases of this type known to the author, fidelity to the institution appeared to engender robust participation, yielding very honest and critical questions, review, and commentary.

5:36   Can or should the nonaffiliated member be compensated for service on the IACUC?

Opin.   Neither the PHS Policy nor the AWAR prohibit the compensation of the outside member. NIH/OLAW has stated that nominal compensation is permissible without jeopardizing a member's nonaffiliated status if compensation is only in conjunction with service on the IACUC and is not so substantial as to be considered an important source of income or to influence voting on the IACUC.[13] (See 5:9.)

The results of a survey done to characterize the range of compensation practices used by institutions to acknowledge the efforts of the unaffiliated member on behalf of the IACUC are given in Table 5.3.

Most institutions (77%) provided either no compensation or compensation (financial or nonfinancial) regarded as minimal such as travel and parking reimbursement, complementary lunch, institutional library privileges, or inexpensive athletic tickets. However, 20% of institutions provided an honorarium. Honoraria were more often given by nonacademic institutions (26%) than by academic institutions (16%), and nonacademic institutions were more likely to give honoraria greater than $100 per meeting (10% versus 3%). Institutions in the survey were not asked to indicate the actual dollar amount when the honoraria exceeded $100 per meeting nor the system by which their honorarium rate was set. Institutions

TABLE 5.3

Compensation Practices for Service of the Nonaffiliated Member by Type of Institution

Level of Compensation	Academic—All Types	Nonacademic	Total Responding
None	35	38	73
Minimal nonfinancial[a]	27	6	33
Minimal financial[b]	18	8	26
Honorarium <$100[c]	13	12	25
Honorarium >$100[c]	3	7	10
Other[d]	2	2	4

[a] Examples of minimal nonfinancial compensation provided to the respondents were assured parking, lunch, library use, or inexpensive athletic tickets.
[b] Examples of minimal financial compensation provided were travel and parking reimbursement.
[c] The survey specified that the honorarium applied to each meeting attended.
[d] Other compensation described included annual gift certificates, donation to a charity of the nonaffiliated member's choice, payment for travel to and attendance at conferences on IACUC-related topics, or payment for other continuing education.

should be cautioned that compensation of their nonaffiliated members at or exceeding the rate of the member's primary income source is likely to be perceived by the lay public and by regulatory agencies as a conflict of interest and cannot be recommended.

Several other approaches were used by institutions to honor the service of the unaffiliated member. Two organizations gave an annual lump sum payment of $500 or an annual gift certificate. One institution provided full travel support and registration for a training meeting on IACUC matters, with clear benefit to the institution as well as the member. All costs associated with the attendance of the unaffiliated member (a physician) to a continuing educational conference of his or her own choosing were paid. Also, one institution made a charitable donation in the name of the unaffiliated member to an organization selected by that member. These types of innovative approaches illustrate the wide latitude institutions can exercise to acknowledge the valuable contributions and important public service provided by the unaffiliated member.

5:37   Can the nonaffiliated member serve on more than one IACUC?

Opin.   There is no proscription against nonaffiliated members' serving on more than one IACUC if they have an interest in that level of involvement. Skeptics may speculate that the excessive involvement of an individual as a nonaffiliated member on several IACUCs is an indication of a shift in the individual's neutrality and objectivity, which might be a reason for institutions to avoid the "overzealous" contributor. Also, scientific staff may be concerned that the individual could be a potential source for the leakage of scientific ideas to other institutions. This matter could be readily addressed by assuring that the nonaffiliated member has received training in issues concerning confidentiality and nondisclosure of proprietary information. Institutions may also want to consider asking their nonaffiliated members to sign formal agreements to this effect.

5:38   What is the role of the nonaffiliated member on the IACUC?

Opin.   The role of the nonaffiliated member is not really defined other than as described in 5:31. Most institutions are hopeful that this individual will play an active role in all IACUC activities and will be comfortable with and become adept at making persistent, straightforward, and disarming inquiries about matters that are unde-tected by the institutional members on the IACUC.

5:39   What are the background and qualifications of individuals who typically fill the role of the nonscientist?

Reg.   The PHS Policy (IV,A,3,b,3) indicates that the "primary concerns" of individuals serving in this capacity should be "in a nonscientific area," and the specific exam-ples cited (e.g., ethicist, lawyer, member of the clergy) have no obvious connections to any area of science. The AWAR do not specify the need to appoint an IACUC member in this particular category.

Opin.   In many institutions, an individual who has been chosen has some scientific training and perhaps even some responsibilities in a scientific area but clearly does not qual-ify as a "practicing scientist with experience in research involving animals" as noted in PHS Policy (IV,A,3,b,2). The types of individuals seen in this role among the insti-tutions surveyed include lawyers, clergy, health and safety personnel, business and human resources personnel, public relations personnel, quality assurance/control personnel, and technicians not involved with animal care or use.

5:40   Is a biostatistician considered a nonscientist?

Opin.   Surely, statisticians are scientists because statistics is a branch of mathematics, a pure science. However, as noted, most institutions would classify a biostatisti-cian as a "nonscientist" even if his or her work may entail the analysis of animal studies. Although the biostatistician might be involved in the application of a mathematical science to the analysis of animal studies, few people would embrace the proposition that a biostatistician would be tempted to encourage animal studies simply to perpetuate the data needed for the practice of his or her science. (See 5:47.)

5:41   What interests does the nonscientist represent on the IACUC?

Opin.   This individual serves to diversify the IACUC membership further, adding balance for the scientific members who may be regarded as having a vested interest in the promotion of animal studies.

5:42   What is the role of the nonscientist on the IACUC?

Opin.   As with the nonaffiliated member, the nonscientist should participate in and contrib-ute to all of the IACUC's mandated activities. With regard to protocol review, this member can be especially valuable by working to ensure that the approach and justi-fication for the proposed animal studies are understandable to nonscientific persons and that humane care and study end points have been included in the study design.

5:43   Should the scientist on the IACUC have animal research experience?

Reg.   PHS Policy (IV,A,3,b,2) and the *Guide* (p. 9), but not the AWAR, indicate that an IACUC should include at least one practicing scientist who has laboratory animal experience in its membership.

Opin.   This is a universal practice among the institutions visited by this author. In large institutions, typically a number of scientists are appointed to the IACUC to reflect the interests of different user groups in the organization. The appointment of scientists who have laboratory animal experience aids the IACUC's discussion of relevant issues during protocol and program review. It helps the IACUC better understand the selection, use, and limitations of animal models and certain aspects of experimental design.

5:44   Should the scientific member of the IACUC be a senior-level scientist, or is a junior-level scientist acceptable?

Opin.   Scientists at any level are appropriate, but many institutions feel that senior scientists confer more authority and credibility on the IACUC. However, most academic organizations have found the recruitment of senior scientists to the IACUC to be difficult because these individuals usually have already made significant committee contributions to the institution in other areas and have other significant obligations within the institution.

5:45   What is the role of the scientist on the IACUC?

Opin.   The principal roles of the scientist on the IACUC are to ensure that the interests of scientific colleagues are being fairly represented in the review process and to aid in the IACUC's assessment of the relevance, validity, and technical aspects of the studies proposed. Of course, the scientist recognizes the confluence of animal health, welfare, and scientific interests in research animal studies and plays an important role as a proponent for the prudent, ethical, and humane use of animals. In broader issues involving program development and implementation, the scientist can give to the IACUC perspectives on how best to launch new initiatives to engender the support of the scientific community and others involved in the care and use of laboratory animals.

5:46   Should the IACUC include an ethicist as a member?

Opin.   Some institutions have been successful in recruiting one or more individuals knowledgeable in ethics with an education and work focus in the arts and humanities or in business, but few have been able to identify a bona fide ethicist. Clearly, issues in ethics may be important in the IACUC's deliberations, and there may be occasions when an ethicist can be helpful in leading this process.

5:47   Should the IACUC include a biostatistician as a member?

Opin.   Most IACUCs have not perceived the need to include a biostatistician to ensure that investigators are using "the minimum number (of animals) to obtain valid results" (PHS Policy IV,D,1,a–IV,D,1,b; *Guide* [p. 10] and Appendix D; U.S. Government Principle III; AWAR §2.31,e,1; §2.31,e,2) because they expect the scientist and veterinarian on the IACUC to evaluate this area. Indeed, in some cases the scientific, veterinary, or other IACUC members may be capable of performing this function if they can commit the effort and are given sufficient information about the assumptions on the underlying database anticipated in the experimental study. However, confirming that the investigator is proposing to use the minimal number of animals necessary can involve an extensive and potentially redundant effort of the IACUC. If an IACUC routinely encounters egregious explanations for the numbers of animals requested in protocols, the addition of a biostatistician to the IACUC may be helpful.

5:48   Should animal care technicians be considered for IACUC membership?

Opin.   In this author's experience, several institutions have appointed animal care technicians to their IACUCs in an effort to develop a better sense of how the basic animal care program functions. An added benefit is that it may promote a sense of empowerment and inclusion in the IACUC review processes for the entire animal care staff.

5:49   If an institution is performing Good Laboratory Practices (GLP) studies, can a quality assurance (QA) person also serve on the IACUC?

Opin.   In the author's opinion, it would be permissible to have a QA person serve on the IACUC. QA personnel have responsibility for all processes influencing the integrity, quality, validity, and documentation of data. IACUC functions are not included in the QA bailiwick, but some IACUCs might benefit from a heightened scrutiny of their processes as well as the general contributions an individual who has a QA background would be expected to add as his or her experience accumulates.

5:50   Should individuals who have specialty expertise be included as voting, nonvoting, or ad hoc IACUC members?

Reg.   PHS Policy (IV,C,2; IV,C,3) and the AWAR (§2.31,d,2; §2.31,d,3) clearly indicate that voting privileges are reserved for full IACUC members who are appointed by the CEO. Nevertheless, the PHS Policy and AWAR do not preclude the use of nonvoting consultants.

Opin.   Individuals who have specialty expertise might include properly appointed ad hoc members who are willing to participate in the full range of IACUC activities. Individuals who have specialty expertise who are not interested in assuming all of the responsibilities of full membership but who would be helpful periodically to the IACUC can be invited to participate in an advisory capacity as consultants without voting privileges. The types of ad hoc appointees by function who might be considered include librarians for alternative searches; environmental or occupational health and safety personnel for hazardous studies; and ichthyologists, herpetologists, or ornithologists for studies involving species in their respective disciplines. These contingencies should be addressed in the bylaws developed by the institution for the IACUC.

5:51   Should the IACUC seek the advice of consultants for issues that may require special expertise, such as pain and distress concerns?

Opin.   IACUCs have the freedom to exercise their ability to use consultants whenever necessary. However, in most institutions, their use is reserved for truly extraordinary issues as opposed to any areas in which the IACUC feels it might improve its

decision making. In this author's opinion, many IACUCs would potentially benefit from the more liberal use of consultants.

5:52   Should consultants be from inside or outside the institution, and should they be anonymous?

Opin.   Consultants can be selected from either inside or outside the institution, depending upon the availability of relevant expertise and the resolution of concerns about
objectivity. Most organizations have not found it practical, necessary, or desirable
to maintain the anonymity of the consultant. However, if the need for a consultant is
likely to involve a highly contentious and disputed high-stakes issue, anonymity
may be elected by the IACUC or imposed by the consultant as a condition of participation. Regardless of the circumstances, the IACUC has the obligation to foster an
environment conducive for the investigator to respond to the consultant's critique.

5:53   Should consultants to the IACUC be compensated?

Opin.   External consultants frequently are compensated for their efforts, and it seems reasonable that these individuals should be compensated according to the institutional policies established for other types of external academic review.

5:54   Should there be a confidentiality agreement between the consultant and the institution?

Opin.   It is advisable for institutions to have a confidentiality agreement with consultants
to prevent the disclosure of important, and potentially patentable, scientific or
technical information. In addition, confidentiality agreements are important
because they establish or reaffirm a precedent for the manner in which all materials
relevant to the IACUC's deliberations are handled. This issue becomes particularly
important from the legal standpoint if the institution is faced with the prospect of
retrieving documents that may have been obtained illegally by adversarial parties.

5:55   What support staff is useful for the IACUC?

Opin.   Many institutions have developed a support staff to assist in the administrative,
monitoring, and training activities of the IACUC. The composition of this staff
depends upon the size and scope of the animal care and use program and the
extent to which the IACUC members are willing and able to become personally
involved in the fine details of committee function. The types of positions that have
proved to be very useful on the IACUC support staff in different settings include a
director or senior assistant to the IACUC Chair; training, monitoring, and compliance personnel; computer/database specialists; biostatisticians; and clerical staff.
In small institutions with limited programs, a support staff of 0.25 full-time equivalent (FTE) or less may be sufficient, but in large, complex programs it is not
uncommon for the IACUC support staff to include three to five FTEs.

5:56   Who typically provides support staff to the IACUC?

Opin.   The support staff is generally recognized and provided as an administrative cost
that is provided under the authority of the CEO or IO of the organization.

5:57 What are some of the typical responsibilities of the support staff?

Opin. Support staff is often involved in preparing the correspondence related to semi-annual reviews, protocol matters, and policies and procedures of the IACUC and in maintaining the records thereof. The staff also may be responsible for making the arrangements for all IACUC activities. If staff members have appropriate training and technical expertise, they may be involved in conducting institutional training seminars and visiting laboratories and animal use areas to provide instruction and to monitor ongoing activities. This staff may maintain databases of IACUC protocol approval and of the appropriate ancillary approvals in other areas such as use of biohazards, and the enrollment in occupational health and safety activities. In a few institutions, support staff has been available to conduct literature searches for alternatives to animal use and to provide a statistical analysis to confirm that the investigator has properly justified the number of animals requested.

Although many institutions draw their IACUC support staff from technical backgrounds involving experience with laboratory animals, others have had success recruiting from diverse disciplines without this experience. Institutions should promote the development activities of this staff by supporting their attendance at workshops and conferences dedicated to IACUC-related topics, animal care and use program improvements, and/or the training of technical staff. These types of programs are offered by the American Association for Laboratory Animal Science (AALAS), AAALAC, the Laboratory Animal Welfare Training Exchange (LAWTE), Public Responsibility in Medicine and Research (PRIM&R), and NIH/OLAW in partnership with regional academic sponsors and also by commercial enterprises. This investment in staff training will make the staff more effective in general. It can efficiently give staff who do not have prior experience in the laboratory animal arena a quick and balanced understanding of key issues in this new environment.

References

1. Committee to Revise the Guide for the Care and Use of Agricultural Animals in Agricultural Research and Teaching, *Guide for the Care and Use of Agricultural Animals in Agricultural Research and Teaching*, 1st rev. ed., Federation of Animal Science Societies, Savoy, IL, 1999.
2. U.S. Department of Agriculture, Animal and Plant Health Inspection Service, Policy #15, IACUC Membership, March 7, 2006. Available on the World Wide Web at: http://www.aphis.usda.gov/ac/policy15.html.
3. National Research Council, Institute of Laboratory Animal Resources, Occupational Health and Safety in the Care and Use of Research Animals, National Academy Press, Washington, D.C., 1997.
4. Robert, H.M., *The Scott Foresman Robert's Rules of Order Newly Revised*, Robert, S.C., Robert, H.M. III, and Evans, W.J., Eds., Scott Foresman, Glenview, IL, 1990.
5. Division of Animal Welfare, Office for Protection from Research Risks, National Institutes of Health, Frequently asked questions about the Public Health Service Policy on Humane Care and Use of Laboratory Animals, *ILAR News*, 35(3–4), 47, 1993.
6. Newcomer, C.E., Behold! The animal care and use magician! *AAALAC Int. Connect.*, Fall, 1997.
7. Office of Extramural Research, Guidance regarding Administrative IACUC Issues and Efforts to Reduce Regulatory Burden, Notice, NOT-OD-01-017, February 12, 2001. Available on the World Wide Web at: http://grants.nih.gov/grants/guide/notice-files/NOT-OD-01-017.html.
8. Wigglesworth, C., personal communication to Dr. J. Silverman, 2006.

9. Gipson, C.A., A word from USDA, *Lab Anim.* (NY), 33(3), 19, 2004.
10. Willems, R., Personal communication to Dr. J. Silverman, 2006.
11. Garnett, N., Personal communication to Dr. J. Silverman, 1997.
12. Division of Animal Welfare, Office for Protection from Research Risks, National Institutes of Health, Maintenance of Properly Constituted IACUCs, OLAW Reports, No. 97-03, Animal Welfare, June 2, 1997. Available on the World Wide Web at: http://grants2nih.gov/grants/olaw/references/dc97-3.html.
13. Division of Animal Welfare, Office for Protection from Research Risks, National Institutes of Health, Frequently asked questions about the Public Health Service Policy on Humane Care and Use of Laboratory Animals, *ILAR News*, 33(4), 68, 1991.

6

Frequency and Conduct of Regular
IACUC Meetings

Sreekant Murthy

Introduction

A comparison of the general issues for IACUC composition and functions indicates that the PHS endorses the U.S. Government Principles I to XII for the Care of Vertebrate Animals Used in Testing, Research, and Training. Principle I states: "The transportation, care, and use of animals should be in accordance with the Animal Welfare Act (7 U.S.C. 2131 et seq.) and other applicable federal laws, guidelines, and policies." The guidelines include those in the *Guide*. The AWA requires that proposed activities using animals be conducted in accordance with the AWAR. Based on these acts, regulations, policies, and principles, IACUCs are constituted to protect the health and well-being of the animals used in testing, research, education, and training. The AWAR (§2.31,a) state that "nothing in this part shall be deemed to permit the Committee or IACUC to prescribe methods or set standards for the design, performance, or conduct of actual research or experimentation by a research facility." This statement, and U.S. Government Principle II, ensure animal care and use without compromising research for the good of the society. Therefore, a carefully composed IACUC that oversees humane use and well-being of the animals and facilitates animal research is fundamental to a research facility's teaching and research objectives and ultimately to the good of society. This chapter addresses critical issues that are pertinent to the function of the IACUC.

The surveys presented in this chapter originate from questionnaires sent to several IACUC officials. Responses were received from 170 individuals; however, not all of the survey questions were answered by all those who responded.

6:1 How is the frequency of meetings of the IACUC decided?

Reg. The IACUC is required by the AWAR (§2.31,c,1; §2.31,c,2) and PHS Policy (IV,B,1–IV,B,3) to conduct a review of its animal care and use program every 6 months. Thus, the PHS Policy requires at least one IACUC meeting every 6 months to review animal care and use policies. How this is accomplished in various institutions is reflected in the survey responses that follow. OPRR (now OLAW) has stated that "semiannual program review and facility inspection may be conducted in a variety of ways, as described in the PHS Policy and USDA regulations. However, final reports must be reviewed and endorsed by a convened quorum."[1]

The *Guide* (p. 9) states that "the IACUC must meet as often as necessary to fulfill institutional responsibilities, but it should meet at least once every 6 months. Records of committee meetings and of results of deliberations should be maintained. The committee should review the animal-care program and inspect the animal facilities and activity areas at least once every 6 months. After review and inspection, a written report, signed by a majority of the IACUC should be made to the responsible administrative officials of the institution." PHS Policy (IV,A,1) requires compliance with the *Guide*. Therefore, under the PHS Policy it is required that the IACUC meet at least once every 6 months and endorse the semiannual report via signatures.

APHIS/AC requires that a majority of the IACUC review and sign the report of the semiannual facility inspection and program evaluation (AWAR §2.31,c,3), although for this purpose the AWAR do not require a quorum to be present. Thus, for institutions regulated under the AWA but not conducting PHS-funded activities, there is no minimal frequency requirement for full committee meetings.

There is a mandated not-less-than annual review of all IACUC approved protocols (AWAR §2.31,d,5) and a review not less than every 3 years under the PHS Policy (IV,C,5). These reviews do require a quorum of the IACUC to be present if the study requires full committee review. (See Chapter 9 for additional information on designated member review.)

Opin. Each institution must evaluate its responsibilities to the investigator and funding organizations. On the basis of the number of IACUC applications, the institution should conduct as many meetings as necessary to review protocols expeditiously. Many institutions decide on the frequency on the basis of NIH or other sponsor deadlines that may include industry-related and internal applications. It is this author's opinion that animal welfare is best served by having the semiannual review of the program of animal care and use (including facility inspections) at a convened meeting of the IACUC with a quorum present. It is advisable for institutions that do not receive federal funds to establish a policy for the frequency of convened meetings to provide an opportunity for open discussion of various issues pertinent to animal welfare. As shown in the surveys in the following questions, generally all IACUCs have a sufficient number of meetings to meet all regulatory requirements.

6:2 Who determines the frequency of IACUC meetings?

Opin. There is no regulation or policy that authorizes specific individuals to determine the frequency of meetings. Either the Chair alone or the Chair after consulting members of the IACUC, the IO, or the Office of Research Administration can determine the frequency of meetings on the basis of the number of protocols anticipated to be reviewed. In some instances, the institution's internal policy or SOPs (including its bylaws) can authorize the Chair or Office of Research Administration to make such decisions. The scheduling of a meeting is recommended to be performed by the IACUC after consultation with the Chair.

Surv. At your institution, who determines the frequency of IACUC meetings?

- No policy   12/162
- IACUC Chair   53/162
- IACUC Chair after consulting IACUC members   46/162
- IO   10/162

- IACUC administrator   19/162
- Part of standard operating procedures   18/162
- Other   4/162

Note: "Other" included designated member review (1), veterinarian (1), Chair consulting veterinarian (1), and IACUC administrator after consulting Chair (1).

6:3   How often should an IACUC meet as a full committee to review protocols?

Reg.   (See 6:1.)

Opin.   There is no regulation or policy that specifies how often the IACUC should meet as a full committee. However, the *Guide* (p. 9) states that "the IACUC must meet as often as necessary to fulfill institutional responsibilities, but it should meet at least once every 6 months." This semiannual meeting need not necessarily be for reviewing protocols. Because all protocol reviews can technically be performed by designated member review (see 9:21–9:27), there is no minimal requirement for the IACUC to meet to review protocols. The suspension of an animal use activity can be accomplished only by the vote of a majority of a convened quorum of the IACUC. (See Chapter 29.)

Surv.   How often does your IACUC meet as a full committee to review protocols in your institution?

- Never, it is all done electronically (computer, video
  conferences, conference calls, etc.)   7/168
- Annually   1/168
- Every 3 years   0/168
- Usually once a month   93/168
- Usually every other month   0/168
- Usually quarterly   11/168
- Usually twice a year   37/168
- Other   19/168*

Note: "Other" included as required (4), two times a week (1), twice a month (3), weekly (3), ten times a year (2), five times a year (1), three times a year (2), or designated member reviews only (3).

6:4   What parliamentary rules are used in the conduct of IACUC meetings?

Reg.   The PHS Policy (e.g., IV,C,2; IV,C,6) and AWAR (e.g., §2.31,d,2; §2.31,d,6) address some aspects of parliamentary procedure, such as conducting full reviews of protocols at a convened meeting of a quorum of the IACUC and voting in the event of a suspension. NIH/OLAW has noted in a Dear Colleague letter[2] that the validity of IACUC activities is always predicated on the existence of a properly constituted IACUC, although there is no requirement that all of the members be present at all meetings. Institutions may not allow Robert's Rules of Order or other parliamentary rules to supplant those procedures that are specified in PHS Policy or the AWAR.

Opin. Neither the AWAR, PHS Policy, nor the *Guide* prescribes specific parliamentary procedures for the IACUC. Nevertheless, many institutions use either Robert's Rules of Order or their institution's own parliamentary rules as prescribed in the bylaws of that institution.

Surv. At your institution, what parliamentary rules are used to conduct IACUC meetings?

- Robert's Rules of Order   59/152
- Institutional internal rules   39/152
- Robert's Rules of Order and internal institutional rules   50/152
- No rules   4/152
- Other rules (not specified)   0/152

6:5 What information should be provided to IACUC members prior to a meeting?

Reg. The AWAR (§2.31,d,2) and PHS Policy (IV,C,2) state, "Prior to IACUC review, each member of the committee shall be provided with a list of proposed activities to be reviewed. Written descriptions of all proposed activities that involve the care and use of animals shall be available to all IACUC members, and any member of the IACUC may obtain, upon request, full committee review of those activities."

Opin. It is customary for many IACUCs to provide members with all correspondence relevant to the review of a protocol, including protocols, amendments, correspondence concerning allegations of protocol violations or animal mistreatment, and results of subcommittee investigations (particularly those that may lead to suspensions that require review and approval by the full committee). Also included are reports of any task forces that the IACUC has requested to review or modify a policy or procedure, updates of institutional policies and procedures that affect the conduct of an IACUC meeting, membership changes, updates on APHIS/AC and NIH/OLAW regulations, copies of all pertinent publications, and meeting notices that may enhance the review process.

6:6 What additional verbal or written information might be given to IACUC members at the time of a regular meeting, but before research protocols are discussed?

Opin. Verbal information that can be presented at a convened meeting may include reports from the CEO, IO, director of office of research, Chair of the IACUC, AV, and supervisor of the animal care facilities and reports from other committees. Other written materials may include updates on regulatory changes, opinions by external consultants and any other reports that the Chair or the institution deems necessary for the proper conduct of IACUC functions. Written information may include minutes of a previous meeting for approval, the agenda for the convened meeting, summaries of protocol reviews, external consultants' reports, updates on policies and procedures, educational materials, regulatory updates, and any other handouts that are pertinent for the conduct of IACUC functions. Very few IACUCs conduct separate quarterly, semiannual, or annual business meetings to discuss various issues of general concern to the IACUC (see 6:7). When held, such meetings might address

- Evaluation of surveys conducted by the IACUC
- Recruitment of new members

- Review of attendance records
- Replacing a member
- Training new members
- Creating and updating standard operating procedures
- Creating/reviewing institution's emergency/disaster plans
- Policy on whistle-blowers
- Updating IACUC forms
- Educational matters
- Methods to train and certify animal users
- Creation and dissemination of newsletters for animal users
- Security issues
- Ethical issues for the appropriate use of animals in research and education
- Review of USDA/APHIS inspection reports or reports from AAALAC site visits

An additional purpose of such a meeting is to reserve the routine meetings for the IACUC to focus its undivided attention on the review of the institution's animal care and use program and protocols for animal use.

6:7 What specific business should be conducted at regular IACUC meetings? Is a formal vote required for any business conducted?

Reg. The AWAR (§2.31,c,1–§2.31,c,8; §2.31,d,1), PHS Policy (IV,B,1–IV,B,8; IV,C,1–IV,C,8), and the *Guide* (pp. 9–10) require that the following specific business be conducted at the meetings:

- Review and approve reports of an institution's program for the humane care and use of animals and inspections of animal facilities, including animal study areas (laboratories), before submitting them to the IO
- Discuss and distinguish significant deficiencies from minor deficiencies included in the above report
- Discuss and provide a reasonable and specific plan and schedule (with dates) for correcting each deficiency
- Review, and if review warrants, investigate concerns involving animal care and use complaints from all sources
- Make recommendations to the IO on all aspects of the institutional animal care and use program and training of personnel who handle animals
- Review and approve or require modifications or clarifications or withhold approval of proposed and ongoing activities with significant changes related to animal use
- Suspend activities involving animals that compromise the health and well-being of an animal used in research (PHS Policy IV,C,6; AWAR §2.31,d,6)

Opin. Many of the preceding regulatory requirements are pertinent to the semiannual review of the institution's animal care and use program. In the author's opinion, teaching, testing, or research activities involving animals that are reviewed at this meeting should be formally approved by a majority of the quorum present, and all minority opinions also should be recorded. Suspending a previously approved animal activity requires a formal vote of a majority of the quorum present (see 29:39). Though federal regulations do not specifically require internal policy discussions about activities that have affected the health and well-being of the animals, it is advisable to have them so that a consensus opinion can be reached to prevent such problems in the future. A formal vote may or may not be taken to adopt such changes; however, institutions can establish their own policies on when a formal vote is required for changing their internal polices and procedures. Routine meetings are usually reserved for reviewing research protocols. However, there is no reason for not conducting both aspects of IACUC business in a single convened meeting.

Many IACUCs have agendas ranging from discussion of the institutional program of animal care and use to review of protocols and other general business of the IACUC. In the author's opinion all business matters that pertain to institutional program evaluations, protocol reviews, suspensions, punitive actions (if the institution has authorized the IACUC to do so), and any other matters that are part of the AWAR and PHS Policy (or institutional policy) should be approved by the majority of the quorum present. All other business matters should be discussed, and, if the committee is developing a policy, such policies must be approved by a majority of members.

Surv.   What specific activities should be conducted at the IACUC meetings?

- All IACUC business matters   169/170
- Review and approval of protocols only or a separate business meeting that includes protocol review pending from a previous meeting   0/170
- We have a separate business meeting that excludes protocol review   1/170

6:8   Does your IACUC require a formal vote for items other than protocols that are reviewed?

Reg.   The formal votes required for the approval of items other than the protocol itself are described in 6:7. Assurance of compliance with the PHS Policy (IV,B,1–IV,B,8) requires written recommendations to the IO (PHS Policy IV,B,3) regarding any aspect of the institution's animal program, facilities, or personnel training. Neither the PHS Policy nor the AWAR state that the report need be formally voted on and approved.

Opin.   Each institution, acting through its IACUC, is responsible for all animal-related activities at the institution. These activities are not always protocol specific, and some may require discussion, voluntary policies, and formal voting. Examples of these include allegations of mistreatment or noncompliance, suspension of animal use activities, substantiation or rejection of a whistle-blower's allegations, changes in policies and standard operating procedures, all of which may have to be formally approved by the IACUC.

Surv.   Does your IACUC require a formal vote for items other than protocols that are reviewed?

- It depends on the item   77/167
- Usually, but not always   51/167
- Always   30/167
- Rarely   9/167

6:9　In order for an IACUC meeting to occur, what is a reasonable policy regarding the attendance of specific IACUC members?

Reg.　The AWAR (§2.31,b,1–§2.31,b,4) and the *Guide* (p. 9) require that the IACUC consist of at least three members, including a veterinarian and a person not affiliated with the institution. PHS Policy (IV,A,3,b; IV,A,3,c) requires a minimum of five members, including a veterinarian, a scientist, a nonscientist, and a nonaffiliated member. For a quorum of the allowable minimal number of committee members, the minimal attendance is three for the PHS Policy; thus, any of the two members required could be absent, yet the meeting could still be legally conducted. (See 6:14 for a definition of quorum.)

　　　The AWAR (§2.31,d,1,iv,A; §2.31,d,1,iv,B) require that for procedures involving more than momentary or slight pain or distress, either the AV or his or her designee must be consulted in the planning of the research project. If the AV was not involved in the initial planning, then the AV or his or her designee must be present at the meeting to provide advice to the IACUC before a vote is taken.

Opin.　Ideally, the nonaffiliated and nonscientist (under PHS Policy) members should be present at meetings to address ethical issues and appropriate use of animals from the noninstitutional and nonscientific perspectives. The presence of a statistician may add strength in evaluating one of the "three Rs." However, it is not legally required for any specific member to be present at all meetings, as long as the committee membership is properly constituted and a quorum is present.[2]

　　　When research projects involve pain or distress, the veterinary consultation on analgesia or anesthesia is normally provided prior to IACUC review by the AV. Another approach to this requirement is to have the AV identify one or more designees who can be involved in planning and consultation. Another veterinarian with knowledge in this area is an appropriate choice for such designation. Nonveterinarians also may be involved in the consultation at the discretion of the AV. (See 5:29; 27:8.)

Surv.　What is your IACUC's policy in regard to attendance of specific IACUC members at a meeting?

- There is no policy　　　　　　　　　　　　　　　　　　　　　　3/169
- All members as required by APHIS/AC or NIH/OLAW
  must be present　　　　　　　　　　　　　　　　　　　　　　14/169
- Attending veterinarian must be present　　　　　　　　　　　7/169
- At least some of the members as required by regulation or
  policy must be present　　　　　　　　　　　　　　　　　　7/169
- Any members, as long as a quorum is present.　　　　　　　94/169
- Any member and AV　　　　　　　　　　　　　　　　　　　24/169
- All members as required by APHIS/AC or NIH/OLAW
  and AV　　　　　　　　　　　　　　　　　　　　　　　　　6/169
- At least some members and AV　　　　　　　　　　　　　　6/169
- Other　　　　　　　　　　　　　　　　　　　　　　　　　8/169

　　　Note: "Other" generally included a combination of either the veterinarian to be present or at least some of the members required by regulation or any member to be present.

6:10　What is your IACUC's expectation with regard to the frequency of attendance of an IACUC member at meetings?

Opin.　Attendance of a committee member at the scheduled meetings is an important consideration in selecting a member. As evidenced in the survey that follows, a majority of institutions do not have a formal policy. Attendance of certain members is critical for the proper review of a protocol or a policy. The absence of specific members may result in unnecessary delays in the approval of a study because of lack of expertise to review a protocol. Determination of the legitimacy of the absence should be at the discretion of the institution's officials, and particularly the IACUC Chair. The institution's officials and the IACUC Chair may set guidelines for members missing consecutive or too many meetings. Likewise, the institution in their bylaws and SOPs may set up policies specifying when and how far in advance of a committee meeting a member should notify the committee of his or her absence and the policy for unacceptable absences and reporting of absences to the institution's officials and administrators who monitor committee attendance.

Surv. What is your IACUC's expectation with regard to the frequency of attendance of an IACUC member at meetings?

- There is no formal policy   103/168
- Attendance at at least one third of all meetings expected   1/168
- Attendance at at least one half of all meetings expected   9/168
- Attendance at at least two thirds of all meetings expected   20/168
- Attendance at at least more than two thirds of all meetings
  expected   30/168
- Other   5/168

6:11 What actions might be taken if an IACUC member does not attend an adequate number of meetings?

Opin. A majority of institutions do not have a formal policy on actions to be taken for lack of attendance by a member. It is important that the institution or the IACUC establish a policy on action to be taken on a member who fails to attend two consecutive meetings or groups of meetings without an explanation acceptable to the committee Chair. Institutions should establish policies on what constitutes the grounds for removal upon request by the Chair or the IACUC to the IO or other person responsible for appointment of IACUC members. It is not unusual that some members may typically attend only those meetings in which their own or a departmental colleague's protocol is being reviewed. In such instances, careful consideration should be given to assure that there exists no conflict of interest for any member voting on such protocols. If there is a conflict of interest, it is appropriate for the member to abstain from voting. An abstention does not negate an approval as long as the quorum is maintained.

Surv. What actions are taken if an IACUC member does not attend an adequate number of meetings?

- There is no formal policy   101/169
- There is an informal notification before any formal
  action is taken, and improved attendance is required   41/169
- The member is asked to resign from the committee   13/169
- The member is notified that he or she is no longer on
  the committee   9/169
- Other   5/169

Note: "Other" included a combination of formal notification of the member about continued absence, request that the member resign, notification that member is no longer on the committee, or no clear explanation.

6:12 If a member were asked to leave the committee, who would formally take this action?

Opin. Removing a committee member is also a relatively uncharted and controversial territory unless the evidence is clear that the member has violated certain IACUC policies including attendance requirements. Generally, the Chair or the IACUC as a whole may make a recommendation to the appointing official to remove or replace a member

Surv. If a member were asked to leave the committee, who would formally take this action?

• There is no formal policy	49/169
• The IO	86/169
• The IACUC chairperson	25/169
• The IACUC administrative official	6/169
• Other	3/169

6:13 When should the IACUC veterinarian first have the opportunity to evaluate pain and distress on a protocol and provide any input back to the PI?

Reg. The AWAR (§2.31,d,1,iv,A; §2.31,d,1,iv,B) require that for procedures involving more than momentary or slight pain or distress, either the AV or his or her designee must be consulted in the planning of the research project. The AWAR (§2.33,b,4; §2.33,b,5) state that the AV should provide "guidance to principal investigators and other personnel involved in the care and use of animals regarding handling, immobilization, anesthesia, analgesia, tranquilization, and euthanasia" and "adequate pre-procedural and post-procedural care in accordance with current established veterinary medical and nursing procedures."

Opin. When research projects involve pain or distress, the veterinary consultation on analgesia or anesthesia is normally provided by the AV prior to IACUC review. It is generally preferable that the institution establish a policy when this consultation should occur before the meeting and before a protocol is submitted for review to prevent unnecessary delays in approval. If, for some reason beyond control, a veterinarian is unavailable, another method to fulfill this requirement is to request that the AV identify one or more designees who can be available for planning and consultation. Another veterinarian with knowledge in this area is an appropriate choice for such designation. Nonveterinarians also may be involved in the consultation at the discretion of the AV. (See 3:18; 5:29; 27:10.)

The author endorses the American College of Laboratory Animal Medicine (ACLAM) position, which states that "the institution bears responsibility and must assure, through authority explicitly delegated to the veterinarian or to the IACUC that only facilities with programs appropriate for the intended surgical procedures are utilized and that personnel are adequately trained and competent to perform the procedures. The veterinarian's inherent responsibility includes monitoring and providing recommendations concerning preoperative procedures, surgical techniques, the qualifications of institutional staff to perform surgery and the provision of postoperative care."[3] ACLAM also recommends that "the veterinarian must be involved in the review and approval of all animal care and use in the institutional program. This includes advising on the design and performance of experiments using animals as related to model selection, collection and analysis of samples and data from animals, and methods and techniques proposed or in use. This responsibility is usually shared with investigators, the IACUC, and external peer reviewers."[3]

Surv.   When does your IACUC veterinarian first have the opportunity to evaluate pain and distress on a protocol and provide any input back to the PI?

- Not applicable: our protocols never include pain or distress   2/168
- At the same time that all other members get the protocols   56/168
- A reasonable time before other members get the protocols, but even if the veterinarian provides input to the PI, the PI does not have time to update the protocol   11/168
- In sufficient time for the veterinarian to contact the PI, and the PI has time to update the protocol before it reaches the committee   92/168
- Other   7/168

6:14 What constitutes the quorum needed to conduct an IACUC meeting?

Reg.   A quorum is an assembly of a majority (more than 50%) of the voting members of the IACUC (AWAR §1.1, Quorum; PHS Policy III, I). For the AWAR (§2.31,b) the minimal number is two and for PHS Policy (IV,A,3,b) the minimal number is three (provided the IACUC has only three and five members, respectively). In any event, it requires more than 50% of members to be present at a convened meeting to constitute a quorum.

No member may participate in the IACUC review of an activity in which that member has a conflicting interest (e.g., is personally involved in the activity) except to provide information requested by the IACUC; nor may a member who has a conflicting interest contribute to the constitution of a quorum at which a proposed activity relevant to the conflict is being considered for approval (AWAR §2.31,d,2; PHS Policy IV,C,2). The IACUC may invite consultants to assist in the review, but they may not approve or withhold approval of an IACUC activity (AWAR §2.31,d,3; PHS Policy IV,C,3); therefore, they cannot vote or be included as part of the meeting's quorum. NIH/OLAW and APHIS/AC accept the use of designated alternates for IACUC members, if the alternates are formally appointed by or through the CEO and identified in the Animal Welfare Assurance (the latter for NIH/OLAW). In these instances, the individual(s) serving as alternate(s) does not add to the total number of committee members. (See 5:23.)

Opin.   Achieving a quorum is a frequent problem for some IACUCs. In this regard, IACUCs apply many methods to attain or maintain quorum at convened meetings. These methods can include reminding members of the committee shortly in advance of the meeting or arranging the meeting on a day that is suitable for all or a majority of the members. Other approaches include designating alternative members and providing food at the meetings so that the members can devote their breakfast, lunch, or dinner time for the meeting. All of these methods are workable, and there are no AWAR or PHS Policy limitations for applying these methods.

Generally it is advisable to have the IACUC membership at more than the minimal number mandated by either the AWAR or PHS Policy. For example, it is possible that one or more members may leave the meeting at an inopportune time because of prior commitments or unpreventable circumstances. This might result in the loss of a quorum if only the minimal number of persons is at the meeting initially. Conflicting interest and abstentions are difficult to resolve if an IACUC membership is small. (See 5:1.)

The simple polling of IACUC members outside a convened meeting does not satisfy the definition of a meeting of a convened quorum and should not be used for conducting IACUC business that requires the vote of a convened quorum. For example, polling the votes of members outside a convened meeting of the committee should not be considered a valid method of voting under the "full committee review" method of protocol review and is not an acceptable substitute for having a vote of a convened quorum to vote on the suspension of a previously approved activity involving animals. E-mail, polling, conference calls, and so on, are typically not acceptable methods for full committee review; however, recent NIH guidance states that some forms of telecommunication are acceptable for full committee review (see 6:17; 6:18). It is acceptable to poll members to see whether any wants a full committee review of an individual protocol.

Often, IACUCs encounter problems when members abstain from voting as a result of conflict of interest or other valid reasons. In such cases, adaptation of Robert's Rules would be of some benefit since Robert's Rules of Order, Article VI, Section 38, defines a *majority vote* as "a majority of the votes cast, ignoring blanks, is sufficient for the adoption of any motion that is in order." This means that only votes for and against are counted to pass a motion as long as the quorum is maintained. If institutions do not follow Robert's Rules, their SOPs should define how votes are counted; however, institutional SOPs cannot override federal regulations (see 9:48; 9:49).

Surv.   What constitutes the quorum needed to conduct an IACUC meeting?

- Above 50% (a quorum) of the membership   165/168
- Just those people who are there for the meeting   0/168
- Other (unspecified)   3/168

6:15   If a quorum is present at the beginning of an IACUC meeting, but the quorum is lost during the meeting, is there any way that the meeting can continue?

Opin.   There are no regulations that stipulate whether a meeting should be suspended if the quorum is lost in the middle of a meeting. However, the AWAR and PHS Policy (see 6:14) state that a quorum of members must be present to approve, withhold approval, or perform other IACUC actions; therefore, all IACUC activities that relate to voting by a majority of the quorum will become suspended. In such exceptional circumstances, the Chair may decide to review a protocol via the designated member method (see 9:21–9:27). Alternatively, to prevent issues related to the maintenance of a quorum, the institution may amend its assurance to conduct a review by electronic means; however, even such electronic reviews require stringent policies, which are described in 6:17 and 6:18 (see 5:23).

6:16 Should any or selected investigators be invited to the IACUC meeting?

Opin. Some IACUCs invite investigators to respond to questions in a face-to-face meeting. Because the committee has the dual responsibilities of thoroughly reviewing protocols and executing decisions, the PI or his or her designee can answer specific questions raised by the IACUC. This often saves time for the IACUC and the PI. The PI should leave the meeting before a vote is taken. If the protocol involves a committee member who has a conflicting interest, the member is asked to leave the meeting while the committee is discussing that protocol (although the person may give factual information to the committee if requested to do so). The member who has a conflicting interest should leave the meeting when the committee votes.

6:17 Can electronic methods be used for conducting IACUC meetings?

Reg. PHS Policy IV,C,2 and IV,C,6 require, respectively, that full committee approval of a proposed research project or suspension of an activity by the IACUC occur only after review of the matter at a convened meeting of a quorum of the IACUC, and with the approval or suspension vote of a majority of the quorum present. AWAR §2.31,d,6 and §2.31,d,6 have the same wording.

Guidance on interpreting the requirement for a convened meeting is provided in articles published by NIH/OLAW concerning IACUC functions and the use of technological advances in communication technology.[1,4] In the most recent NIH/OLAW guidance,[4] APHIS/AC is noted as being in concurrence with NIH/OLAW. That guidance is meant to clarify the statements in reference 1 (and see the following).

Opin. Recent advances in electronic communication methods have raised questions about the acceptability of teleconferences, audio-visual conferences, fax transmissions, electronic mail, and postal mail as alternate methods to face-to-face IACUC meetings. The NIH/OLAW does not fully endorse the use of these alternative methods to a traditional meeting because IACUCs conduct diverse activities, and some of these activities require direct interactions, voting, and good record keeping of minutes. Be that as it may, NIH/OLAW considers that there may be circumstances in which one or more of the aforementioned options could be used. In a recent notice,[4] NIH/OLAW, with the concurrence of APHIS/AC, states that certain methods of telecommunications (e.g., telephone or video conferencing) are acceptable for the conduct of official IACUC business requiring a quorum, provided the following criteria are met (see 6:14; 6:18; 8:20; 8:21):

- All members are given notice of the meeting.
- Documents normally provided to members during a physically convened meeting are provided to all members in advance of the meeting.
- All members have access to the documents and the technology necessary to participate fully.
- A quorum of voting members is convened when required by PHS Policy.
- The forum allows for real time verbal interaction, equivalent to that occurring in a physically convened meeting (i.e., members can actively and equally participate and there is simultaneous communication).
- If a vote is called for, the vote occurs during the meeting, and it is taken in a manner that ensures an accurate count of the vote. A mail ballot or individual telephone polling cannot substitute for a convened meeting.

- Opinions of absent members that are transmitted by mail, telephone, fax, or e-mail may be considered by the convened IACUC members but may not be counted as votes or considered as part of the quorum.
- Written minutes of the meeting are maintained in accord with PHS Policy IV,E,1,b.

The original 1995 guidance provided by NIH/OLAW is as follows:[1]

- The Animal Welfare Assurance file with the NIH/OLAW is to have a description of the procedures that the IACUC will follow to fulfill the requirements of the PHS Policy, which may include description of an alternate method to conduct a meeting under exceptional circumstances.
- NIH/OLAW expects that all approved Assurances are complete and reflect the use of the nontraditional procedures used to conduct IACUC business, if such procedures are used.
- Conference calls, audio-video conferences, and possibly some forms of highly interactive online computer discussion groups may qualify in exceptional circumstances. [The exact definition of *exceptional circumstances* is not provided.]
- The conveyance, by fax or electronic mail, of information such as the institutional Assurance, animal study proposals, agendas and minutes of meetings, institutional policies and standard operating procedures, reports, announcements or correspondence from oversight or regulatory agencies, and other matters related to the institutional animal care and use program for consideration and review by IACUC members, would be regarded as appropriate.

Surv. What are the methods your IACUC uses to formally review protocols requiring a vote? Indicate the approximate percentage of each method if more than one is used.

- We review by one or more of the methods but only vote at a face-to-face meeting of the full committee. 27/168
- Face-to-face meetings (100%) 70/168
- Face to face 95% and 5% standard electronic mail 8/168
- Face to face 90% and 10% standard electronic mail 14/168
- Face to face 80% and 20% standard electronic mail 3/168
- Face to face 75% and 25% standard electronic mail 1/168
- Face to face 50% and 50% standard electronic mail 1/168
- Face to face 40% and 60% standard electronic mail 1/168
- Face to face 30% and 70% standard electronic mail 1/168
- Face to face 20% and 80% standard electronic mail 1/168
- Face to face 10% and 90% standard electronic mail 2/168
- Face to face 5% and 95% standard electronic mail 4/168
- Standard electronic mail (100%) 3/168
- Conference call (100%) 1/168
- Face to face 90% and 10% conference call 2/168

- Face to face 80% and 20% conference call   2/168
- Face to face 75% and 25% conference call   1/168
- Face to face 50% and 50% conference call   1/168
- Video conferencing   1/168
- Customized electronic review/voting system   0/168
- Face to face 80% and 20% customized electronic review   1/168
- Face to face 50% and 50% customized electronic review   1/168
- Face to face 98% and 2% designated member review   2/168
- Face to face 95% and 5% designated member review   1/168
- Face to face 90% and 10% designated member review   1/168
- Face to face 75% and 25% designated member review   1/168
- Face to face 70% and 30% designated member review   1/168
- Face to face 60% and 40% designated member review   1/168
- Other in combination with face to face meeting   7/168
- Standard mail   3/168
- Other   5/168

6:18   What are the requirements for conducting other than face-to-face IACUC meetings ("exceptional circumstances" meetings)?

Reg.   The guidance from NIH/OLAW[1,4] is as follows: although the traditional meeting is still seen as the optimal environment for fulfilling the intent of the PHS Policy, OPRR (now NIH/OLAW) recognizes the new communications tools available and the need for flexibility in the ways that institutions may comply with the PHS Policy in the many diverse settings encountered. For these reasons, several criteria have been provided for establishing alternate methods that may be considered functionally equivalent to meetings of a convened quorum under exceptional circumstances. Two of those criteria are that the alternate approach must include a high degree of interactivity and allow for careful deliberation of sensitive issues. Another is that a quorum of IACUC members must be in direct communication with each other and be given full opportunity to participate for the duration of the meeting. Institutions are reminded that details of their IACUC procedures, especially those that may vary from those outlined in the PHS Policy, should be thoroughly described in the institutional Assurance and submitted for NIH/OLAW review. Details are provided in 6:17.

Opin.   In certain circumstances, telephone and audio-visual conferencing may be appropriate alternatives to face-to-face meetings for full committee review of protocols and other IACUC matters (see 6:17). The requirements for such meetings are described in 6:17.

6:19   How can a designated member review (see 9:21–9:27; 11:9; 11:10) of protocols be accomplished using the electronic media meeting format?

Reg.   (See 9:21.) Both the PHS Policy (IV,C,2) and the AWAR (§2.31,d,2) recognize a method of designated-member review that is widely implemented by research institutions. Some institutions refer to this review process as *expedited review,* but *designated member review* is the appropriate term. Designated member review must conform to the following process: Written descriptions of research projects that involve the care and use of animals must be made available to all IACUC members, and any member of the IACUC must have the opportunity to request full-committee review of those research projects. "If full committee review is not requested, at least one member of the IACUC, designated by the chairperson and qualified to conduct the review, shall review those activities, and shall have the authority to approve, require modifications in (to secure approval), or request full committee review of any of those activities" (PHS Policy IV,C,2; AWAR §2.31,d,2).

Opin.   A designated member review requires an IACUC Chair–appointed reviewer(s) to review an IACUC protocol. Designated member review also can be conducted using electronic media to facilitate the review process as long as the requirements in the PHS Policy (IV,C,2) and AWAR (§2.31,c,2) are met. (See 6:20; 8:20; 9:21.)

6:20   Can IACUC reports be endorsed by electronic methods and can signatures of members be obtained by using electronic methods?

Opin.   Since most reports, including semiannual program review and inspection reports, require IACUC approval (see 6:1), alternate methods that exactly follow those used at a legally convened face-to-face meeting of the IACUC are appropriate. (See 6:17; 6:18.) IACUC activities generally require documentation of votes or verification of committee approval. They also may require that signatures be legally binding. Electronic methods of voting are conducted quite rapidly, but signatures are difficult to obtain by this method. Therefore, in an "electronic meeting" the Chair calls for a vote, the IACUC coordinator records the vote, and the coordinator then prepares and distributes the final document by mail to solicit original signatures for the permanent record.

6:21   Must the minutes of IACUC meetings be recorded? If so, what precautions should be taken to maintain confidentiality?

Reg.   PHS Policy (IV,E,1,b) and the AWAR (§2.35,a,1) require records of IACUC meetings to be kept, whether the meeting is face-to-face or held alternatively using electronic communications (see 6:11; 6:12). They must be compiled and maintained on file for 3 years (AWAR 2.35,f; PHS Policy IV,E,2). Minutes must include records of attendance, activities of the committee, and committee deliberations (PHS Policy IV,E,1,b; AWAR §2.35,a,1).

Opin.   While recording meeting minutes satisfies regulatory needs, special care must be taken to ensure proper usage of words because meeting minutes can, under certain circumstances, be obtained through a state's Freedom of Information Act (FOIA) or any state open record laws (the federal FOIA is at present only applicable to documents in the possession of a federal agency). (See Chapter 22.) The minutes should record all controversial issues discussed and any consensus opinion of the committee on controversial issues. Minutes should also record any minority opinion expressed on issues pertinent to a protocol or IACUC issue. An example of an uncontroversial item that may be included in the minutes is approval of a breeding colony, especially determining colony management practices such as which animals are to be included in the estimated number of animals used and the estimated total number of animals that are kept for breeding purposes but not subject to experimental manipulations.

It is imperative that each IACUC takes into account potential security problems of any privileged and trade-secret confidential information that may be protected under the AWA. Because minutes are often generated by using computers, special care should be taken to prevent unauthorized access to computer records. The mandatory enforcement of passwords, controlled access, and encryption should be practiced to prevent unauthorized access. IACUC members should be advised to return copies of the minutes to the IACUC secretary once the minutes are ratified by the committee. Unwanted and used copies of the minutes should be destroyed by appropriate methods.

Many IACUCs record the meeting on an audiotape. This practice raises a question as to whether these tapes should also be saved as part of record keeping. This is internal policy unique to each IACUC—whether to save or erase the tape. In general, these tapes also contain irrelevant discussions; therefore, it is customary for the IACUC to transcribe portions of the discussion that are relevant and erase the tape for future use. There are no regulations that explicitly deal with audiotaping and archiving of such tapes. (See 8:18; 8:19; 22:4; 22:5; 22:8; 22:9.)

Surv.   In a survey conducted for the first edition of this book, 61 of 63 respondents stated that only discussions that are pertinent to IACUC business are recorded in meeting minutes and 2 of 63 stated that all items discussed are recorded in the minutes.

6:22   What voting information should be recorded by the IACUC?

Reg.   The AWAR (§2.35,a,1) and PHS Policy (IV,E,1,b) state that the minutes of IACUC meetings should include records of attendance, activities of the committee, and committee deliberations. The AWAR and the PHS Policy do not state whether the names of individuals who voted, approved, opposed, or abstained need to be recorded.

Opin.   It is customary to record the names of individuals attending a meeting and to record the actual number of individuals approving, withholding approval, or abstaining in a vote.

Surv.   What voting information is recorded by your IACUC?

• All votes are recorded with names	25/166
• All votes are recorded without names	130/166
• Nothing recorded	2/166
• Other	9/166

6:23   Should each IACUC protocol be voted on individually?

Opin.   In general, each protocol is unique and it is best to approve them individually as they are discussed. This method inherently prevents problems that arise from maintaining a quorum. Exceptions for collective approval arise if two similar (or dissimilar) protocols from the investigator are discussed, and the investigator is called to answer specific questions pertaining to one or more of the protocols.

6:24   Who records IACUC meeting minutes?

Opin.   Persons assigned to this task should be knowledgeable about federal and other requirements and the institution's animal care and use program. The institution should assume responsibility for selecting a person to record meeting minutes, on the basis of input from the IO, the office of research, and the Chair of the IACUC.

Surv.   Who records the minutes of your IACUC meetings?

- We do not record minutes   0/168*
- Research administration staff   79/168
- A recording secretary other than the research
  administration staff   41/168
- IACUC Chairperson   27/168
- One of the voting members on the committee, but not the
  Chairperson   18/168
- Other   3/168

Note: Varies from response to 6:22.

6:25   What should be included in the meeting minutes?

Reg.   (See 6:21.)

Opin.   Since the AWAR (§2.35,a) and PHS Policy (IV,E,1,b) state that the minutes require
the records of attendance, activities of the committee, and committee deliberations,
it is reasonable that the following items be included in IACUC meeting minutes:

- Inspection reports and reports of the program review distinguishing signif-
  icant and minor deficiencies with plans and schedules for correcting each
  deficiency
- Investigation of complaints and issues of noncompliance received from the
  public or research facility personnel
- Approvals, conditions such as modifications and clarifications, withheld
  approvals, and suspensions of protocols
- Approval or withheld approval of proposed significant changes to protocols
- Training and certification guidelines approved by the committee
- Consultants' reports on protocols

The minutes should include sufficient detail of the discussions and use language
that is generally understood by the nonscientific and lay members of the committee.

6:26   How might the IACUC use the minutes of its meetings?

Reg.   The IACUC minutes must be maintained for the duration of the project plus at
least 3 years and be accessible for inspection and copying by authorized NIH/
OLAW or other PHS representatives at reasonable times and in a reasonable manner (PHS Policy IV,E,2). The AWAR (§2.35,f) have similar wording for APHIS/AC
or federal funding agency representatives.

Opin.   There are no guidelines for the way the recorded minutes of the meetings are to be
used. In general, minutes are maintained to:

- Meet regulatory requirements
- Resolve any questions that may arise from a program or protocol review
- Resolve an ethical issue that was previously discussed
- Refer to a previous discussion

Surv.   How might the IACUC use the minutes of the meeting?

- As an official record of the meeting   22/167
- For APHIS/AC inspections   0/167
- For AAALAC site visits   1/167
- To record animal welfare issues that are discussed in the
  meeting   0/167
- To record issues pertinent to pain and distress that required
  consensus of the IACUC   0/167
- As an official record, for APHIS/AC inspections, for AAALAC
  site visits, as a record of animal issues discussed at the meeting,
  and as a record of issues pertinent to pain and distress   58/167
- As an official record, for AAALAC site visits, as a record of
  animal issues discussed at the meeting, and as a record of issues
  pertinent to pain and distress   23/167
- As an official record, for APHIS/AC inspections, as a record
  of animal issues discussed at the meeting, and as a record of
  issues pertinent to pain and distress   11/167
- As an official record and for AAALAC site visits   8/167
- As an official record, for APHIS/AC inspections, and for
  AAALAC site visits   5/167
- As an official record, as a record of animal issues discussed at the
  meeting, and as a record of issues pertinent to pain and distress   5/167
- As an official record, for APHIS/AC inspections, for AAALAC
  site visits, as a record of animal issues discussed at the
  meeting, as a record of issues pertinent to pain and distress, and
  for other issues   4/167
- Miscellaneous: minutes always used as an official record in
  combination with either use for APHIS/AC inspections, use
  for AAALAC site visits, as a record of animal issues discussed at
  the meeting, and as a record of issues pertinent to pain and
  distress alone or multiple combinations   30/167

6:27   Can the Freedom of Information Act (FOIA) or a similar state law impact what is placed in the IACUC minutes?

Opin.   *The Institutional Animal Care and Use Committee Guidebook*[5] cautions that the person
assigned to prepare reports should be aware of the FOIA and any state open record
laws.

The FOIA does not require a private organization or business to release any
information directly to the public, whether it has been submitted to the federal
government or not. However, information submitted to the federal government by
such organizations or companies can be available through a FOIA request if it is
not protected by a FOIA exemption, such as the one covering trade secrets and confidential business information (Freedom of Information Act 5 U.S.C. §552). Many
of the reports written may be accessible under such laws and particular care must
be taken to prevent use of language that may be misconstrued by the lay public.

Further, institutions that use computers to prepare and store meeting minutes and reports must be aware of potential security problems that could lead to unauthorized retrieval, tampering of reports, or obtaining of information on trade secrets protected by law (AWA Sect. 27). Thus, the general advice is that meeting minutes should be written carefully with appropriate usage of words so that the lay public will not be alarmed by reports obtained through the FOIA or pertinent state laws. (See Chapter 22.)

Surv.   Does the FOIA have an impact on your IACUC minutes?

- It has no effect   90/166
- It has a significant effect   5/166
- It may or may not have an effect, depending on the nature of discussions   40/166
- We do not include proprietary research information in our minutes   10/166
- It may or may not have an effect, depending on the nature of discussions, and we do not include proprietary research information in our minutes   13/166
- Other   8/166

In the first edition of this book some respondents said that the FOIA or some states' open record laws have an impact on the meeting minutes. One respondent stated that meeting minutes may become the subject of the FOIA only if the APHIS/AC inspector copies them and it becomes a federal government document. The remaining respondents said that conscious or carefully chosen words and the elimination of minute details of discussions (ethical or confidential) may help prevent unnecessary problems linked to the FOIA.

In the present survey, responders who are categorized as "others" generally indicated that it is worthwhile to include proprietary information whether it had no effect or significant effect.

6:28   IACUCs often review many protocols at a meeting. The concern is that those protocols considered at the beginning of the meeting may be reviewed more thoroughly than those at the end of the meeting. How can IACUCs avoid this problem?

Opin.   The AWAR (§2.31,c,4–§2.31,c,7) and PHS Policy (IV,C) provide appropriate guidelines for reviewing research projects, but the guidelines do not provide a schedule for discussing reports and research projects in a meeting. The *Guide* (p. 9) suggests that each IACUC should meet as often as necessary to fulfill its responsibility and provides topics (p. 10) that should be considered in the review of animal care and use protocols.

Surv.   How does your IACUC assure that all protocols undergo a thorough review?

- We limit the number of protocols that can be reviewed in a meeting   2/168
- We limit the time spent discussing any one protocol   1/168
- We lengthen the meeting to accommodate all protocols without losing a quorum   58/168

- If we are short on time, we postpone discussions to the next meeting if doing so does not affect the research   7/168
- We lengthen the meeting to accommodate all protocols without losing quorum, *or* if we are short on time, we postpone discussions to the next meeting if doing so does not affect the research   30/168
- This is not an issue for us   58/168
- Other   12/168

6:29   **Should the IACUC have a written policy on what constitutes a conflict of interest that may lead to a member's being excused from voting?**

Reg.   The AWAR (§2.31,d,2) and the PHS Policy (IV,C,2) state that "no member may participate in the IACUC review or approval of a research project in which the member has a conflicting interest except to provide information as requested by the IACUC."

Opin.   Conflict of interest occurs in many forms. A common potential or apparent conflict of interest occurs when a PI is a member of the same department as the IACUC member. It is not considered acceptable to serve as a designated member reviewer in those circumstances as the member has sole say on approving the protocol. In full committee consideration a member must decide individually whether to vote or abstain. However, if there is any apparent, perceived, or potential conflict of interest, the member should declare so before joining any review, discussion, or vote. If assigned as one of a team of reviewers, the member can either provide that service or ask to be relieved from that duty. In making that statement it is recognized that a member from the same discipline can often provide the best review of a protocol, but may then have to abstain or be absent from any vote. The overall guiding principle is that members should act professionally and impartially in considering IACUC matters.

     Likewise, resolution of conflict of interest may also occur in many ways. An SOP to resolve the conflict is critical for the growth of a research enterprise. The SOP should describe the procedures for resolving conflicts arising from the investigator as well as a committee member. Investigator-initiated conflicts generally arise at the time of submitting a protocol. Here the PI may request that a member be excluded from the review of the protocol, provided that there is evidence to substantiate the claim that a conflict of interest exists. If that member's expertise is essential for the review, the IACUC can exercise the option of using a consultant for the review to ensure animal welfare. Likewise, if a member thinks that he or she has a conflict with a protocol, the member should notify the IACUC Chair. That member may not participate in the IACUC review or approval, except to provide information when requested by the committee. Potential conflict is implicit if the member is conducting competing research or if the protocol provides access to funding information that may give an unfair advantage to the member. Occasionally, personal bias against an investigator or an investigator's research can occur.

Surv.   Should the IACUC have a written policy on what constitutes a conflict of interest that may lead to a member's being excused from voting?

- Yes   100/168
- No   61/168
- Other   7/168

Acknowledgment

The author thanks Mr. Amol Kumar Korva for his assistance in preparing this chapter.

References

1. Garnett, N., and Potkay, S., Use of electronic communications for IACUC functions, *ILAR J.*, 37(4), 190, 1995.
2. Ellis, G., and Garnett, N.L., Maintenance of Properly Constituted IACUCs, OPRR Reports, No. 97-03, June 2, 1997. Available on the World Wide Web at: http://grants.nih.gov/grants/olaw/references/dc97-3.htm.
3. American College of Laboratory Animal Medicine, Public Statements, Adequate Veterinary Care. Available on the World Wide Web at: http://www.aclam.org/pub_adquate_care.html.
4. National Institutes of Health Notice NOT-OD-06-052, Guidance on Use of Telecommunications for IACUC Meetings under the PHS Policy on Humane Care and Use of Laboratory Animals. March 24, 2006. Available on the World Wide Web at: http://grants.nih.gov/grants/guide/notice-files/NOT-OD-06-052.html.
5. National Institutes of Health, Office of Laboratory Animal Welfare, Institutional Animal Care and Use Committee Guidebook, 2nd ed., Bethesda, MD, 2002, p. 33.

7

General Format of IACUC Protocol Forms

Christian E. Newcomer

Introduction

Protocol review was mandated by federal law for most institutions using animals in research, teaching, and testing by the passage in 1985 of an amendment to the AWA known as the Improved Standards for Laboratory Animals Act (Public Law 99-198) and by a component of the legislative reauthorization for the NIH, known as the Health Research Extension Act of 1985 (Public Law 99-158). In advance of this requirement, and as early as 1980, several academic institutions already had begun to implement the basic concepts of protocol review. The veterinarians in these institutions had made a compelling case for needing some information about the nature of the ongoing animal use to be able to sign their institutional APHIS/AC annual report in good conscience. In the context of the contemporary standards for protocol review in most institutions, the forms used were rudimentary. The information retrieved in these pioneering efforts was extremely scant and generally would now be insufficient for the issuance of an IACUC approval. Nevertheless, these tentative and incomplete efforts were important because they introduced the setting, players, and ethos of an evolving process that eventually would be adapted and embellished at diverse institutions across the country as the federal mandates for protocol review took effect.

Many of the factors that influenced the early approaches to protocol review continue to be important today. These have shaped the character and content of the forms used in the protocol review process according to institutional preference and stimulated organizations to meet regulatory and ethical considerations with clarity, consistency, and completeness of all documents. For example, most institutions and investigators want the protocol review process to be objective, consistent, and efficient. This orientation fosters the support of research investigators and IACUC members who must commit precious time to fulfill this institutional requirement. However, the actual protocol review forms used to facilitate this process are markedly different from institution to institution. Institutional philosophy, the types of research conducted, pragmatic considerations related to administrative and regulatory detail, and many other ancillary factors can influence the format of the IACUC protocol review form.

The purposes of this chapter are to examine the various strategies used by institutions in the information collection phase of the protocol review process and to highlight the strategies that appear to be the most successful and widely adaptable. The chapter also discusses alternative approaches that have proved to be well suited to particular institutional environments. Information for the preparation of this chapter was acquired through the review of approximately 120 institutions during the conduct of AAALAC site visits or similar activities over the course of two decades, and from the author's personal experience as an IACUC member at ten institutions during this same period.

A well-designed protocol form can assist investigators in their efforts to provide the clear, concise, and comprehensive information necessary for the review process. It is very important to the smooth functioning of the IACUC. Most IACUCs have revised their protocol forms several times since the inception of the protocol review process in response to new or augmented regulatory requirements, the need to improve the quality and detail of the information provided, or the necessity to enhance the retrieval and sharing of information contained in these documents. While the protocol review form should be considered a living document and an IACUC should not be reluctant to change the protocol review format for due cause, format changes are likely to be disruptive and burdensome during a transition period. For this reason, format changes should be carefully planned and coupled with investigator outreach and education efforts.

7:1   Should there be a standard protocol form for submission to the IACUC?

Opin.   The large majority of institutions require investigators to submit their animal care and use protocols to the IACUC for review on an institutionally developed, standardized protocol review form. This approach is usually regarded as an essential first step in the fulfillment of the IACUC's mission to establish a review process that is perceived as unbiased, consistent, efficient, and, under optimal conditions, user-friendly. Most institutions have PIs who file multiple studies with the IACUC, and the use of a standardized form allows these individuals to develop a familiarity with the type of information required by the IACUC for successful completion of that committee's review process. Most institutions have found that the quality of information provided by PIs using standardized protocol forms improves over time. Nevertheless, when changes are incorporated into existing forms to meet newly defined needs of the IACUC, a period of coaching and reacclimatization for investigators is necessary. Additionally, standardization is generally regarded as helpful and critical to the IACUC members in meeting their review obligations. The compartmentalization of information into predictable areas enables the IACUC members to focus their review and discussion on sensitive areas with greater alacrity and facilitates the retrieval of information pertinent to any subsequent IACUC monitoring and compliance activities.

It was quite unusual for institutions to conduct IACUC protocol review without the use of a standardized form, but this practice was used by some institutions at the time of the first edition of this book. The author has not encountered additional institutions using this method in the intervening years and cannot determine to what extent this approach continues to be utilized. It was used mostly by academic institutions with smaller animal care and use programs and a limited research profile. It usually involved ten or fewer PIs. In these instances, the IACUC reviewed the PI's grant application and a supplemental narrative concerning specific areas deemed necessary for IACUC review. The narratives were responsive to a list of queries provided by the IACUC to ensure that all of the federally mandated aspects of protocol review were covered. While it is conceivable that this approach could occasionally produce a superbly integrated, relevant, and coherent document for the IACUC review process, in practice this rarely happened. In addition to requiring IACUC members to demonstrate a unique flexibility and commitment, this approach to protocol review challenged IACUC members to wade through much extraneous information and maintain a running inventory of the quality and location of the investigator's responses (or nonresponses) to important issues. In order for the information gleaned from the grant application and protocol at the time of the review to remain conveniently accessible to the IACUC for later use, a synopsis of the review becomes necessary. For these reasons, a free-form protocol IACUC review process will likely remain limited, if indeed it continues, to animal care and use programs in small institutions imbued with a sense of collegial courtesy, reciprocity, and the good faith principle.

7:2 What general information is necessary for the IACUC to review a protocol adequately?

Opin. Several authoritative documents reiterate the basic components of an animal care and use activity that should be assessed by the IACUC during the protocol review process. These documents include the *Guide*, the PHS Policy, and the AWAR. The AWAR impose specific review requirements on covered animal species, and many institutions have chosen as a matter of policy to extend these requirements to all vertebrate species. Other institutions have maintained a dichotomy in the review criteria based upon exclusions provided in the definition of the term *animal* in the AWAR (§1.1). (See 12:1.) The essential broad areas of review that can be derived from these documents include the following:

- Procedures conducted on animals
- Justification for using animals
- Rationale for using the animal species selected and number of animals proposed
- Consideration and adoption of applicable alternatives to animal use where appropriate
- Assurance that the studies conducted are not unnecessarily duplicative
- Efforts to minimize animal pain and distress through the use of anesthetics, sedatives and analgesics, or timely intervention
- Provisions for aseptic survival surgery
- Provisions for postsurgical and postprocedural care
- Appropriate justification for multiple major survival surgery
- Provision of appropriate living conditions for animals including the unique requirements of unusual laboratory animals
- Method of animal euthanasia and disposal
- Provisions for a safe working environment for personnel
- Training of personnel in the aforementioned topics commensurate with their assigned responsibilities and areas of involvement in the conduct of the protocol
- Assurance or certification statement affirming that the investigator accepts full responsibility for abiding by regulations and implementing the protocol as approved

The reader should refer to other chapters of this book where many of these topics are explored in detail.

7:3   Is there an optimal format for an IACUC protocol form?

Opin.   A central question in the protocol review process is how to optimize information
retrieval from the areas identified in 7:2, in order for the IACUC protocol review
process to be sufficiently rigorous to ensure high-quality animal care, safety, and
other considerations and still improve a PI's prospect for receiving protocol
approval on a first attempt. There is no apparent consensus on this matter as evidenced by the many different approaches taken to protocol review by IACUCs,
which in turn are mirrored by the variety of forms used by different IACUCs. The
forms used by institutions generally fall into the following categories:

- Those that state a requirement or use a question (regarding the areas defined in
  7:2) to elicit a free-form narrative response and consolidate all relevant information onto a single form

- Those that use a free-form narrative approach but are divided into modules allowing the form to be tailored to the particulars of the submission
  (e.g., those not involving surgery would delete the sections aligned with this
  activity)

- Those that emphasize a checklist and fill-in-the-blank approach whenever
  possible, either in the modular or consolidated format: forms generally
  intended to expedite completion and review of straightforward protocols

Institutions using a free-form narrative for protocol review simply ask questions
related to each of the items in 7:2, eliciting the investigator's response. The protocol
form includes questions covering all areas, and answers are typically provided only
for the segments of the protocol that apply. Thus, in studies involving minimal animal procedures, the majority of the form may not be used. However, in protocols
involving complicated studies with extensive animal use and entailing specialized
expertise (e.g., surgery, postprocedural care, or the use of biohazardous agents), all
information deemed necessary for the IACUC review is provided. Most institutions
realize that investigators cannot be relied upon to provide the kind of qualitative and
quantitative information necessary for protocol approval without assistance, even
though this subject may have been extensively covered in institutional training and
education activities. Therefore, many institutions using this format generally make
an effort to enhance the PI's appreciation for the "hot button" issues or nuances particular to each area of review. This is accomplished through the distribution of guidelines for the completion of a protocol review form, example (mock) protocols,
IACUC position statements, regulatory announcements on issues, or articles emphasizing the IACUC's interest or perspective on a topic. In a variant of this approach,
some IACUCs incorporate this type of information into sidebar or parenthetical
comments in the protocol form to channel the investigator in the right direction.

Institutions using the modular version of the free-form narrative protocol work
toward paperwork reduction by capitalizing on the fact that many studies only
involve a limited, and somewhat predictable, spectrum of activities around which
the "core" protocol form is designed. This core form is linked to appendices that
are designed for specific topics such as surgery and postoperative care, prolonged
restraint, multiple major survival surgery, other procedures potentially involving
significant pain and distress for laboratory animals, and the use of biohazards. The
psychological benefits attendant to the reduction in paperwork for investigators
and IACUC members are additional reasons given by institutions that use this
approach.

Checklists and "fill-in-the-blanks" formats are used by some IACUCs as the predominant modes of information collection in a form in an effort to expedite completion of the forms through a series of discrete, informative inquiries into aspects of the protocol. Some areas of inquiry and some research settings are better suited to this approach. For example, organizations having a narrow research mission in which many of the studies follow a predictable format and involve techniques that are well described and repetitive often use this format to allow investigators to respond quickly. In particular, pharmaceutical companies and commercial laboratories involved in testing compounds in conformance with federal regulations often have extensive standard operating procedures for animal care and use. Many aspects of their IACUC protocol form are addressed simply by citing these procedures or other documents used in project development if these primary sources are relevant and have been reviewed and approved by the IACUC. Some larger animal care and use programs, with diversified research endeavors, have incorporated this approach into their protocol form to service research activities that can benefit. A significant pitfall in this approach, when it is applied to complicated studies, is that it tends to disperse the information about the different types of procedures being performed on the animals and can limit the reviewer's appreciation for the temporal sequencing and impact of procedures on the animals. Flowcharts illustrating the various uses of animals in an application often can remedy this problem. A second problem is that these forms tend to be lengthy because they are structured to capture numerous small bits of information. If the forms are poorly worded, the information gathered may be voluminous but disconnected and confusing.

Many institutions have blended the narrative, checklist, and fill-in-the-blank strategies in their protocol review efforts to optimize these efforts. This is the approach to protocol review that the author deems optimal for most animal care and use programs. The experience of the IACUC often leads to identifying portions of the protocol form that investigators repeatedly do not answer correctly or adequately. These areas of the form can be transformed into a checklist format or narrative blocks with supplemental instruction on the appropriate content of the blocks. For example, in the segment on the search for alternatives to painful procedures, the form might include a list of commonly used databases, along with prompts for the keywords used and the date of the search. As another example, in the discussion on the provisions for postoperative care, the questions might stimulate responses for each important aspect of care (e.g., physiological monitoring, analgesic therapy, fluid and antibiotic therapy, record keeping). A final example are procedures such as polyclonal antibody production; here, the questions could channel the investigator into methods that conform to IACUC guidelines.

An approved IACUC protocol form should be regarded as the investigator's contract with the institution to conduct animal care and use activities in conformance with applicable laws, regulations, and sound clinical and scientific practices. In summary, any approach, or mix of approaches, has merit if it satisfies the investigators and IACUC members and produces a well-defined contract through a thorough and appropriate IACUC review and approval process.

7:4   How specific should the information on the protocol form be to secure an approval? Is it acceptable for an IACUC to approve a generic protocol?

Reg.   For PHS Policy purposes, IACUC approvals must be project specific in order to address the required review criteria at PHS Policy IV,C,1. The use of a more generic protocol, with project-specific amendments to cover significant changes, is one method that meets this requirement.

Opin. There are many investigators who mistakenly view the IACUC protocol review and approval function only as an approval of the animal procedures. Thus, if they use the same animal procedures over an extended period, they prefer to work under a generic approval to reduce their "unnecessary" interactions with the IACUC "bureaucracy." Nevertheless, most IACUCs have not acquiesced to this preference because protocol approval is not simply an approval of procedures. In general, IACUCs require PIs to submit very specific information on their protocol forms, to allow the IACUC to link the scientific objectives with the justification for a particular animal model, the number of animals requested to complete the studies, and the particular procedures that are involved. Accordingly, in the few IACUCs that do use this approach for certain types of protocols, the PIs are required to submit supplementary information for each study conducted under the procedure-based protocol to define its objectives, assure the validity of the animal model, assure the lack of alternatives, and explain the need for the number of animals requested. Information pertaining to any new animal welfare or personnel safety concerns that may have been introduced by new variables in the studies proposed also must be provided and reviewed.

There are times when the approval of a generic protocol, abiding by the contingencies stated, may be appropriate and efficient. An example is a protocol with a standard set of procedures that is used repetitively to test compounds and materials in a particular animal model with recognized validity of the study of a class of compounds. The only variable in these protocols is the material or compound under investigation.

7:5 Should the protocol form be used as a vehicle for the continued instruction of investigators in aspects of the PHS Policy, AWAR, and ethical considerations in animal research?

Opin. Most IACUCs use protocol forms that refer to the PHS Policy, AWAR, and ethical issues in one or more areas. Some include a statement signed by the investigators assuring that all of the activities described will be conducted in conformance with the aforementioned regulations. These references to PHS Policy and the AWAR serve to remind PIs of the regulatory underpinnings of the protocol review process. They also emphasize the contractual nature of an approved protocol. Similarly, many IACUC protocol review forms attempt to evoke the discussion of ethical considerations in response to a variety of experimental practices such as prolonged restraint, food and water restriction, use of aversive stimuli, surgery, multiple major survival surgery, disease induction, postprocedural care, and the selection of study end points. While use of the protocol form in this fashion to give these issues continued visibility to the investigators is well intended, in this author's opinion using the protocol form as a vehicle for this instruction, per se, has little merit. Most IACUCs would likely agree that investigators tend to focus their attention on filling out protocol forms for the purpose of securing of approval, not for the purpose of personal edification or enlightenment.

7:6 Could inclusion of a questionnaire on the form on animal welfare considerations be useful to the IACUC?

Opin. Fewer than half of the IACUCs this author is familiar with have attempted to evaluate animal welfare considerations in a separate section or questionnaire. When this is done, it usually is limited to querying the investigator about the allocation of animals to the various APHIS/AC pain categories to aid the PI's understanding of the AWAR requirement for the search for alternatives and scientific justification for conducting USDA Category E animal studies. Even at this basic level, most institutions encounter occasions in which the IACUC does not agree with the PI's assessment or the IACUC is not able to reach a full consensus on an animal welfare matter. On most IACUC protocol forms, this type of information is integrated into the sections involving the discussion of procedures performed and the overall management of the animal model. If the information given in these areas is insufficient for the IACUC to identify and assure the PI's attention to animal welfare issues, changes in the protocol form or the training of investigators is warranted.

7:7   Should IACUC protocol forms require the PI to include a statement written in language understandable to the lay public about the purpose and relevance of the proposed studies?

Reg.   This inquiry has become standard on most IACUC protocol forms. The basis for making this request is U.S. Government Principle II, which states, "Procedures involving animals should be designed and performed with due consideration for relevance to human or animal health, the advancement of knowledge, or the good of society."

Opin.   In the age of highly specialized research, most IACUCs regard this type of statement as essential for the lay member of the IACUC, institutional public relations or development personnel, and scientists on the IACUC who work in areas unrelated to the protocol under consideration. In the author's experience, weak responses to this inquiry on the IACUC protocol form often reflect the general inexperience of scientists in addressing a scientifically unsophisticated public. Some institutions do not provide this type of inquiry on the form because all animal protocols are subsumed under an organizational or corporate research mission that assures this matter.

7:8   What information is appropriate to be included in the nontechnical description of the project(s) and the statement on the significance of the animal use to the scientific objectives of the project(s)?

Opin.   The composition of an effective lay summary can be a challenging task, and the effort required is well invested in the author's view. The lay summary provided by the investigator should include a brief and general overview of the intended use of animals. The function of the lay summary is to provide the nonscientific members of the IACUC, and particularly the nonaffiliated member, with an understanding of the scientific question being asked. The lay summary should indicate how the animal model contributes to the hypothesis being tested, the exploration of a new and potentially important area of science, or the resolution of an important scientific dispute. Impenetrable scientific jargon and the lack of effort to connect the studies to conditions or phenomena that are widely known or easily appreciated by the lay audience are still the most common weaknesses in these statements that have been observed by the author. To alleviate problems related to the use of scientific terms that are unfamiliar to lay members of the committee, some institutions have elected to provide a glossary of lay terms as a reference (e.g., http://www.med.nyu.edu/irb/glossary/gloss_3_10_03.pdf). Investigators and their institutions should keep in mind that ultimately science competes for the attention, recognition, and continuous stream of financial support from a public that has other avid interests and recognizes the competing needs of other worthy endeavors.

7:9 Should IACUC protocols submitted for educational activities have the same basic structure as a research protocol?

Opin.   Of the institutions known to this author, very few actually have experience with the approval of protocols involving the use of animals in educational activities, at either the undergraduate, graduate, or professional school level. Most institutions use the same protocol form for all activities involving animal care and use. Several institutions have devised separate protocol form attachments to address areas regarded as unique to the use of animals in this capacity. The areas of special emphasis include a review of:

- Course syllabus to understand how and why animal studies are integrated into the course
- All relevant handouts provided to students pertaining to animal use
- Student profile information (e.g., preparatory coursework and degree major)
- Summary of animal care and use educational discussion or materials provided to the students
- Backup provisions for students not fulfilling their obligations to laboratory animals
- Occupational health and safety provisions
- Summary of students' written critiques of the educational activity from the previous year
- Opportunities or plans for the course director to develop alternatives to the use of animals for subsequent sessions of the course

7:10 Is a PI required to provide information on the protocol review form that may be proprietary in nature?

Opin.   PHS Policy and the AWAR do not require IACUCs to review proprietary information. Most IACUCs have designed their protocol forms and implemented procedures to prevent the unwitting and unnecessary disclosure of proprietary information by investigators. On the other hand, IACUCs must have sufficient information about all aspects of the study to make an informed decision about the assurances for animal welfare during all phases of animal care and use. Where PHS Policy is concerned, the IACUC must be able to identify compounds and other materials, equipment, practices, and procedures that may have an impact on occupational health and safety considerations for personnel involved in the project (*Guide*, pp. 14–18). Hence, most IACUCs do not require the disclosure of the name of a proprietary compound but do require information about the chemical class of the compound, toxicity data in the target animal species or other animals, special handling requirements, and the basis for dosage selection. The development of new techniques and the use of new instrumentation in animals for application in humans pose more of a quandary. For example, if a new surgical technique is an integral component of a proprietary package involving a new device, an IACUC would be remiss to waive its responsibility to assess the impact of the procedure on experimental animals. Also, most IACUCs require sufficient information about the purported advantages of the new equipment to support the claim of potential benefit. (See 22:3–22:7.) The training of IACUC members should include the importance of nondisclosure of proprietary information available to them through the review process.

7:11 Should information such as the PI's name, title of the application, and research sponsor be included on the IACUC protocol form, or could that information bias the IACUC's decisions?

Reg. PHS Policy (IV,C,1,f) and the AWAR (§2.31,d,1,viii) require the IACUC to verify that personnel conducting procedures are appropriately qualified and trained. This would be difficult, if not impossible, without the identification of the PI.

Opin. While there is some variation in the content of the protocol signalment among the institutions known to this author, all identified the PI by name and included a title for the proposed activities. Most protocol forms also asked the PI to provide information about the funding source, particularly when the IACUC was responsible for sending the letters of protocol approval to funding agencies. Nevertheless, in some institutions a response to this inquiry is at the discretion of the PI. In institutions that fund their research programs internally, this inquiry is not relevant. Virtually none of the institutions believe that this type of information is likely to compromise their objectivity, notwithstanding the admission by several institutions that internally funded projects have to provide evidence of peer scientific review via an internal mechanism or consultants, whereas externally funded projects do not. Other information about the PI and associated laboratory personnel often collected in the protocol form includes position or academic title; office, laboratory, and home (emergency) phone numbers; and office and laboratory location. In the author's view, emergency contact information is of paramount importance for critical correspondence on animal welfare matters and should always be provided. In many institutions the ability to serve as a PI is reserved for individuals with particular positions or academic titles.

7:12 What are some methods used by IACUCs to identify protocols for the purpose of correspondence, timely periodic reviews, and aiding of the oversight of ongoing animal-based research activities?

Opin. Most institutions use a numerical or alphanumerical identification system for the protocols that are submitted to and approved by the IACUC. This type of system offers several advantages and there are numerous variations in this approach. It provides a brief, content-neutral reference that links both the animals and the investigator's grant application or award to a set of approved animal care and use procedures. Many institutions indicate the year of protocol submission in their log or accession system. This method provides rapid identification of those protocols that are due for annual or triennial renewal. Other numbers are added to reflect the sequence of protocol submission within the year, and sometimes letter or decimal extensions are used to indicate the renewal year within the triennial period. As an example, 06-38.09 might indicate the 38th protocol submitted to the IACUC in 2006, with a full review of the protocol scheduled in the year 2009.

It is not uncommon for the approval of problematic protocols submitted at or near the end of the calendar year to be delayed until the following year. Computer databases are used routinely to track the actual dates of submission and approval to ensure that the requirements for periodic review or renewal are met in a timely fashion. Also, organizations vary in their requirements for the number of animal species that are permitted in a particular protocol application; some allow multiple species, whereas others allow only one. Most institutions are now following the species-dedicated protocol approach to allow ease of activity tracking and postapproval monitoring efforts. Institutions that allow only one species per protocol application sometimes retain the same number but follow it with a letter designator for species (e.g., R = rat, Rb = rabbit, M = mouse) to associate the procedures described with a particular grant application or award. The same tracking number may be retained, along with a modifying extension, to indicate that animal care and use procedures are identical to those described in the parent protocol, but an alternate project title or funding source is associated with the protocol. (See 7:13.)

7:13   When protocols are due to expire, should the IACUC retain the original number assigned or assign a new number to the protocol?

Reg.   The AWAR (§2.31,d,5) require review of IACUC protocols at least annually, while the PHS Policy (IV,C,5) requires the same at least every 3 years.

Opin.   Many institutions apply a modifying extension to the same protocol number when an annual review of animal care and use activities is mandated for a species, such as occurs under the AWAR. However, most institutions that receive PHS support for their research programs prefer to retire protocol numbers when the protocols have reached the limit of their 3-year PHS approval period. It should be noted, though, that at least a general review of each protocol involving AWAR-covered species must be performed on an annual basis. This approach is convenient in that it ensures individuals involved in animal care and use (and research oversight) that the review process is current. In a few cases in the author's experience, an institution has elected to retain the original protocol number to indicate the continuity and longevity of a research program, both noting the consistency of the animal model and procedures used as well as emphasizing with distinction an investigator's funding history. This approach is also satisfactory as long as the IACUC has performed a rigorous review at the appropriate review cycle. (See 7:12.)

8

Submission and Maintenance of IACUC Protocols

Kathy Laber and Alison Smith*

Introduction

The authors would like to preface this chapter with the observation that there are *no* laws that require that a *protocol* be written and approved by the institution's IACUC, but rather that specific components related to the care and use of animals in biomedical research, biological testing, and research training meet specific requirements detailed in applicable regulations and laws. Nevertheless, the most common approach to obtaining, documenting, and reviewing the required information is, as recommended in the *Guide for Care and Use of Laboratory Animals*, through the completion of a standardized protocol form that the IACUC then reviews. Questions often arise about when and how best to utilize a protocol review system. This chapter helps to address some of the more commonly asked questions on the submission and maintenance of animal care and use protocols.

8:1   When is an IACUC-approved protocol needed?

Reg.   The laws from which the animal care and use protocol arose are the AWA and the HREA. As defined by these laws and their associated explanatory regulations and policies, IACUC-approved protocols are required for the following:

All live or dead warm-blooded animals used in research, teaching, or testing in registered research facilities as defined by the AWAR (§1.1 and §2.31). Specific animals excluded from the AWAR are birds, rats of the genus *Rattus*, and mice of the genus *Mus*, bred for use in research, and farm animals used as a source of food or fiber or for research designed to improve the quality of food or fiber. Specific research studies excluded from the requirement for a protocol are field studies in which wild animals are not engaged in invasive procedures and the study does not alter animal behavior.

Vertebrate animals used in research, research training, and biological and testing activities conducted or supported by any PHS agency (PHS Policy, II). Note that the PHS Policy expands the definition of *animals* to include all vertebrates, not just warm-blooded vertebrates.

Opin.   An IACUC-approved protocol is also needed when projects supported by other public or private agencies have a requirement for IACUC approval (see 8:4–8:6).

* The authors thank Farol N. Tomson for his contribution to this chapter in the first edition of *The IACUC Handbook*.

Alternately, the individual institution may require IACUC approval even when the animal is not a USDA-regulated species, and the study is not supported by a PHS agency. Most institutions with Assurances accepted by NIH/OLAW do not differentiate between vertebrate animal activities supported by PHS funds and those that have other funding sources regarding the need for an IACUC-approved protocol. It is common practice for institutions with Assurances to require IACUC-approved protocols for all non-PHS-supported activities involving vertebrate animals. For example, a breeding colony of mice may not be funded through the PHS, and mice are not currently a regulated species covered by the AWAR. Therefore, the colony is not required by any of the regulations to have an IACUC approval. Nevertheless, it is recommended that institutions extend IACUC review to projects that fall in this category to ensure that an appropriate standard of animal care and use is uniformly applied.

8:2   What activities are specifically exempt from the requirement for an IACUC-approved protocol?

Opin.   IACUC-approved protocols are *not* required if the animals to be used are:

- Not under the auspices of the AWAR or PHS Policy (see 8:1), that is, not a regulated species (e.g., a horseshoe crab, not supported by PHS funding, or not used for research, biological testing, or research training)
- Not supported by any additional private or public agency requiring IACUC approval
- Not required by the home institution to be IACUC reviewed

Institutions may own and use animals for a variety of reasons that do not include research, biological testing, or research training. Animals can play a role as institutional mascots, as key participants in rodeos, as featured stars in a theatrical play, or the focus of a student's art project. People often keep companion animals on site as pets or in cages or aquariums in their offices or dormitory rooms. Institutions may house animals for specific purposes such as guard dogs, pet therapy animals, mounted police horses, or classroom pets. Institutions also deal with pest animals, such as wild rodents and certain birds. If these animals are not involved in biomedical research or intended to be used in a research project, they are exempt from IACUC approval. (See Chapter 2.)

8:3   What information is required on an IACUC protocol?

Reg.   The AWA and the HREA do not detail protocol content requirements; however, the AWAR and the PHS Policy are very specific.
    1. PHS Policy (IV,C,1,a–IV,C,1,g):

> In order to approve proposed research projects or proposed significant changes in ongoing research projects, the IACUC shall conduct a review of those components related to the care and use of animals and determine that the proposed research projects are in accordance with this policy. In making this determination, the IACUC shall confirm the research project will be conducted in accordance with the Animal Welfare Act insofar as it applies to the research project, and that the research project is consistent with the *Guide* unless acceptable

justification for a departure is presented. Further, the IACUC shall determine that the research project conforms to the institution's Assurance and meets the following requirements:

- Procedures with animals will avoid or minimize discomfort, distress, and pain to the animals, consistent with sound research design.
- Procedures that may cause more than momentary or slight pain or distress to the animals will be performed with appropriate sedation, analgesia, or anesthesia, unless the procedure is justified for scientific reasons in writing by the investigator.
- Animals that would otherwise experience severe or chronic pain or distress that cannot be relieved will be painlessly killed at the end of the procedure or, if appropriate, during the procedure.
- The living conditions of animals will be appropriate for their species and contribute to their health and comfort. The housing, feeding, and nonmedical care of the animals will be directed by a veterinarian or other scientist trained and experienced in the proper care, handling, and use of the species being maintained or studied.
- Medical care for animals will be available and provided as necessary by a qualified veterinarian.
- Personnel conducting procedures on the species being maintained or studied will be appropriately qualified and trained in those procedures.
- Methods of euthanasia used will be consistent with the recommendations of the American Veterinary Medical Association (AVMA) Panel on Euthanasia,[1] unless a deviation is justified for scientific reasons in writing by the investigator.

2. PHS Policy (IV,A,1) requires institutions to use the *Guide*[2] as a basis for developing and implementing an institutional program for activities involving animals. The *Guide* is supported by the National Research Council and provides guidance on caring and using animals in keeping with the highest scientific, humane, and ethical principles. This document is the reference resource used by AAALAC, the primary agency that accredits programs that utilize animals in biomedical research and biological testing. To be in compliance with PHS Policy when AWA-regulated species are used, both the AWA and the recommendations in the *Guide* must be followed (PHS Policy II). Additional requirements in the *Guide* (p. 10) that are not specifically detailed in PHS Policy include the following:

- Rationale and purpose of the proposed use of animals
- Justification of the species and number of animals requested
- Statistical justification of animal use (recommended)
- Consideration of the use of alternatives, which may include less invasive procedures, other species, *in vitro* methods, or computer simulation
- Unusual housing and husbandry requirements
- Attention to prevention of unnecessary duplication of experiments
- Conduct of multiple major operative procedures
- Criteria and process for timely intervention, removal of animals from a study, or euthanasia if painful or stressful outcomes are anticipated
- Postprocedure care

- Method of euthanasia or disposition of animal
- Safety of working environment for personnel

3. In addition, the AWAR state that the IACUC assures that the following requirements are met:

- The PI must document consideration of alternatives to painful procedures and provide a written narrative description of the methods and sources used to determine that alternatives were not available (§2.31,d,1,ii).
- Deviations from any AWA standards or regulations must be identified and reviewed and approved by the IACUC. The IACUC must report these exceptions to the APHIS/AC in their annual reports (§2.36,b,3).
- Protocols include a description of the procedures and drugs used to provide relief from pain or distress (§2.31,e,4).
- Protocols include a complete description of the proposed animal use (§2.31,e,3).
- Veterinary consultation is required during planning of procedures that may cause more than momentary pain or distress (§2.31,d,1,iv,B).

Opin.   *The Institutional Animal Care and Use Committee Guidebook*[3] summarizes the information and items that are legally required to be included on IACUC forms. However, the mechanisms that institutions use to solicit the appropriate information in a concise manner from the investigative community are as varied as the institutional cultures. Each operating IACUC should determine for itself the most effective way of obtaining and then evaluating the legally required information. In addition, institutions may opt to require information about animal-related activities that extends beyond the legal requirements mentioned, such as Drug Enforcement Administration registration numbers, location of controlled substances, and individual responsible for them. Many IACUCs have found that the use of a protocol form helps research investigators to delineate the information that the IACUC requires to review a proposal and helps the IACUC to achieve greater consistency in its review. Many of these forms are available from institutional Web sites.

A sample animal study protocol is provided on the NIH/OLAW Web site at www.grants.nih.gov/grants/olaw/sampledoc/animal_study_prop.htm. This form is based on a form used by intramural NIH investigators and was modified as the result of review of many different extramural institutional forms in order to anticipate a variety of research scenarios.

8:4   Do government agencies conducting or funding animal research require IACUC-approved protocols?

Reg.   (See 2:6; 8:1.)
Opin.   In 1985, an Interagency Research Animal Committee, with representatives from various federal agencies, developed a set of principles (U.S. Government Principles for Utilization and Care of Vertebrate Animals Used in Testing, Research, and Training) that applies to institutions and persons using animals in any type of federally funded program.[4] These principles do not specifically mention or require IACUC reviews. The applicability of PHS Policy requirements to IACUC review and approval is therefore determined by the various non-PHS federal agencies that fund research. Many federal agencies, such as the National Science Foundation

and the Veterans Administration, embrace the PHS Policy requirements. Regardless of funding source, when AWAR species are used, the requirement for IACUC-approved protocols as detailed in the AWAR applies.[5]

8:5   Do private agencies funding biomedical research require IACUC-approved protocols?

Reg.   (See 8:1 for animals and animal activities requiring IACUC approval.)

Opin.   If the animal is used for research, biological testing, or research training and is an AWAR-regulated species, then an IACUC approval is required by law. However, there are no federal or state laws that require private funding agencies to obtain IACUC approval of animal use projects if the animals are not AWAR-regulated covered species. Some, but not all, private agencies require IACUC approval before releasing funds for research; for example, the American Heart Association, American Cancer Society, and American Lung Association require IACUC approval for all vertebrate animals.[6–9] (See 2:6.)

8:6   How can institutions require IACUC approval of projects funded from agencies that do not require it or that involve animals that are not covered by either the AWAR or the PHS Policy?

Opin.   Savvy institutions recognize—and others have been reminded by the public, which holds them accountable—that the ethical responsibility for the use of animals extends well past the legally defined types of research and animal species boundaries assigned by regulatory agencies. It is the option of each institution to determine how best to shoulder that responsibility. If an institution chooses to have its IACUC review and approve nonbiomedical (i.e., agricultural) research using animals, or research using animals not regulated under the limitations of the AWAR or PHS Policy (e.g., invertebrates), they can simply make an IACUC review of those species an institutional policy. In addition, they may choose to announce this policy to NIH/OLAW, the oversight office for the PHS Policy, as part of their Assurance statement. NIH/OLAW advises institutions that the maintenance of uniform and consistent standards is an essential component of the development and implementation of a high-quality animal care and use program.[10]

8:7   Must the IACUC review and approve protocols using animal carcasses and parts of carcasses?

Reg.   As stated in 8:1, AWAR (§1.1, Animal) define *regulated animals* as "live or dead." A *research facility* (AWAR, §1.1, Research Facility) is defined as one using or intending to use *only live* animals. There is no reference to using dead animals in research facilities. It is only in the definition of a dealer (§1.1, Dealer) that the AWAR refer to live or dead animals. (See 12:14; 14:28.)

Only live animals are mentioned in PHS Policy III,A. A proposal involving animals to be killed for the purpose of using their tissues or one that involves project-specific antemortem manipulation is not exempt from protocol review.

PHS grant applicants using shared animal tissues or slaughterhouse materials are advised to specify the origins of the tissues when describing their proposed use in an application, especially if the "no" box is checked on the vertebrate animal block of the face page. Any reference to the use of animal tissues in the application is likely to trigger questions about IACUC approval. Therefore, an explanation in the application that the tissues will come from dead animals as a by-product of other IACUC-approved studies or from a slaughterhouse will help avoid delays in the peer-review process.[11]

Opin.   Ordinarily, neither the AWAR nor the PHS Policy protocol review requirements apply to dead animals in the research facility setting; however, IACUCs often choose to review the use of carcasses and carcass parts. This practice may best serve the interests of the institutions for a variety of nonregulatory reasons such as improved public relations, occupational safety, and decreased institutional liability. The review of carcasses and carcass parts can also help to protect the investigators from a public that is critical about the PIs' obtaining certain samples, for example, researchers who receive biopsy material (or carcasses) from endangered species. The IACUC review is an effective way of informing the institution of these projects. IACUC members can discuss and advise the institution on projects that may become controversial and warrant additional consideration.

8:8   Must the IACUC review and approve protocols using animal tissues, cells, or biological fluids derived from living animals?

Reg.   (See 8:1.)

Opin.   The IACUC must address issues related to animal care, animal handling and distress that can be associated with obtaining the specimens, and occupational health and safety issues for the personnel collecting the specimens. If the specimen collection technique is not invasive and does not require the animal to be restrained (e.g., feces on the floor, remnants of placentas, preserved tissues), occupational health and safety issues may still require consideration. A greatly simplified protocol format addressing only the potential issues related to specimen collection could be easily processed via a designated reviewer on the IACUC. Often, these types of collections are "add-ons" from an investigative group that does not own the animal or is not involved with the primary reason the animals are in the facility. As an alternative, the "specimen collector" could simply be added on as an amendment to the PI's protocol.

8:9   Must the IACUC review and approve protocols using tissues, cells, or biological fluids acquired from sources other than the IACUC's institution?

Reg.   (See 8:1.)

Opin.   This issue is not directly addressed in the AWAR, but a similar issue is addressed by the NIH/OLAW as it relates to purchasing antibodies from a commercial source (see 8:10). The outside source (vendor) providing the tissues, cells, or fluids must comply with the conditions of 8:1 and must have a NIH/OLAW Assurance if live animals are being used to produce these products specifically for the institution, versus a standard "off-the-shelf" product the company sells commercially.

8:10   Does the purchase of antibodies from commercial sources require IACUC approval?

Reg.   NIH/OPRR (now NIH/OLAW) has made the following interpretation of PHS Policy:[10]

In the case that standard reagent antibodies (e.g., mouse-antihuman) are produced by a commercial supplier using their own resources and offering them for general sale, for example, through a catalog, the institution may consider the antibodies to be "off-the-shelf" reagents and the supplier is not required to file an Assurance with OPRR. If, on the other hand, a supplier or contractor produces custom antibodies using antigen(s) provided by or at the request of a principal investigator, the antibodies are considered "customized" and the vendor or subcontractor must file an Assurance with OPRR.

Usually it is known in advance that someone intends to perform this kind of work under a PHS grant. In such cases, the applicant must mark the PHS Grant Application (PHS Form 398) "yes" for vertebrate animal involvement and include the appropriate Animal Welfare Assurance number(s), verification of project-specific date of IACUC protocol review, and the identification of all project performance sites. All animal-related activities supported by the PHS must be conducted at Assured institutions and must be reviewed and approved by an IACUC. When both the PHS Grantee and its Contractor hold OPRR-approved Assurances, some latitude is allowed in determining which IACUC (if not both) will review the proposal. However, the institution which subcontracts or subgrants any animal activity retains partial accountability for providing effective oversight mechanisms to ensure compliance with the PHS Policy. Part of that responsibility includes ensuring that subgranted/subcontracted animal-related activities are conducted only at an Assured institution.

Opin. There are no AWAR requirements for a research facility to review protocols of researchers using commercial companies producing their antibodies. However, as noted in APHIS/AC Policy 10,[12] if that commercial company uses rabbits (or any other regulated species) to produce antibodies, it must be registered as a research facility and have its own IACUC review relevant procedures. It is advisable, although not necessary, for the research facility to have a record of the commercial company's NIH/OLAW Assurance and/or USDA registration number. (See 19:15.)

8:11 Does the use of fertilized eggs (birds, reptiles, and amphibians) or fetuses (mammalian) require an approved protocol?

Reg. If the eggs and fetuses remain as part of the female (dam), it is the dam that is the object of IACUC interest. If the dam is under the auspices of the AWAR or PHS Policy (see 8:1), then IACUC approval is required. If they are chicken eggs, NIH/OPRR (now NIH/OLAW) has made the following interpretation of the PHS Policy for governmental biomedical research:[13] "The PHS Policy is applicable to proposed activities that involve live vertebrate animals. While embryonic stages of avian species develop vertebrae at a stage in their development prior to hatching, OPRR has interpreted 'live vertebrate animal' to apply to avians (e.g., chick embryos) only after hatching."

Opin. If the eggs are allowed to hatch and if the animals are under the considerations described in 8:1 (i.e., birds, reptiles or amphibians in addition to mammals), then IACUC approval is required. The AWAR does not regulate most birds, or cold-blooded (poikilothermic) animals. Therefore, at this time there are no AWAR requirements for the IACUC to review the use of eggs (unhatched or hatched) from these species.

Use of mammalian fetuses or fertilized eggs cannot be done without use of the fetus or egg-producing female. An IACUC approval is needed if the female is included in the conditions described in 8:1, 8:4, or 8:5. (See 13:11; 14:19; 14:23; 16:36.)

8:12   Is an IACUC protocol required for breeding colonies, independently of any experimental procedures performed on the animals?

Reg.   If the breeding animals are included in the considerations described in 8:1, then IACUC approval is required.[14] All PHS-supported activities involving animals must be reviewed and approved by an IACUC (PHS Policy II). (See 8:14; 14:22.)

Opin.   In some instances, such as the production of genetically unique rodents, some IACUCs might consider the breeding itself as an animal "research" project and require IACUC approval. In other instances, breeding colonies (rodents, fish, birds, domestic livestock, etc.) that are *not* involved with any activities described in the regulatory section of 8:1 (i.e., are species not covered by the AWAR and not involved in PHS-supported activities) are not required to have IACUC approval unless the institution requires it (see 8:6; 8:14; 14:22).

Surv.   Does your institution have a separate breeding colony protocol form, or is breeding included on your standard IACUC protocol form?

- Not applicable   37/170
- Separate breeding protocol/required use   35/170
- Separate protocol/optional use   2/170
- No separate breeding protocol   92/170
- Miscellaneous approaches   4/170

8:13   What information should be included for a protocol describing the maintenance of a breeding colony?

Reg.   If IACUC approval is required of a breeding colony (see 8:12), then colony managers or responsible investigators must submit certain information to the IACUC (see 8:3). The *Guide* (pp. 47–48) notes that it is important to maintain the genetic integrity of the colony whether the animals are inbred or outbred and that the use of standard nomenclature is important.

Opin.   In addition, IACUC members may want to know more about the colony husbandry practices. This information may be requested in special forms or formats, or can be submitted as a separate document for review. Conventional IACUC protocol forms can also be used, but because research is usually not being performed, the forms include many unrelated questions.

Breeding animal programs require special attention to certain animal welfare concerns, such as male-to female-ratios, artificial insemination, chemicals synchronizing estrus, length of time the male is left with the female, and caging type. Additional variables influence breeding results and the IACUC may want very detailed information on file (e.g., light cycles, nutrition, sanitation, bedding material, cage size and type). Other information often requested involves the age of the breeders, when to cull old breeders, number of litters permitted per dam, and ways birthing difficulties are managed. The information requirement will vary from IACUC to IACUC and from species to species.

Some institutions have developed forms to be used for animal breeding programs. Additional information related to breeding colonies may include some or all of the following (in addition to items mentioned previously):

- The name of the veterinarian to be contacted if and when needed
- The source of the breeding animals
- The number of offspring per year produced or number of breeding pairs per year maintained
- The disposition of animals when no longer useful
- The method of euthanasia when needed
- The justification for maintaining animals
- A description of painful or stressful husbandry procedures (if any)
- A list of personnel who will handle animals and their training and experience

The Institutional Animal Care and Use Committee Guidebook[3] is a useful resource for IACUC review of breeding colony operations.

8:14   Do sentinel animals or blood donors need IACUC approval? Since these animals are not used in specific research projects, is it necessary to develop protocols for the review and approval of these projects by the IACUC?

Reg.   If the animals in question fall under the considerations described in 8:1, then IACUC approval is required. NIH/OPRR (now NIH/OLAW) has made the following interpretation of the PHS Policy:[14]

> PHS Policy applicability is not limited to research. It also includes all activities involving animals including testing and teaching. OPRR has determined that although animals used as sentinels, breeding stock, chronic donors of blood and blood products, or for other similar objectives may not be part of specific research protocols, their use for these purposes contributes significantly to the institutional research program and constitutes activities involving animals. Consequently, the IACUC must receive and approve of protocols and appropriate systems to monitor the use of animals prior to the commencement of such activities, and should then perform reviews at the appropriate intervals (IV.C. of the PHS Policy).

Opin.   The AWAR (§1.1) broadly define activities involving animals as "those elements of research, testing, or teaching procedures that involve the care and use of animals." The regulations apply to those animals involved in biomedical and/or other types of research projects. By inference, the AWAR cover animals that contribute to an institution's research program. There are, however, many animal uses that are not considered to be research. For example, blood donor animals used for veterinary clinical procedures (dogs, cats, birds, ferrets, etc.) in veterinary clinics (within registered research facilities) may not be part of any research or teaching project and are not required by regulations to be reviewed by the IACUC. Many management-related decisions are based on sample trials of products on animals to determine whether they serve the intended purpose; for example, different bedding materials, caging, foods, or enrichment devices are often tested on resident animals. Data collection consists of observations of the animals with which they are used, and the results aid management to make purchase decisions. Small animal projects that impact on housing and management of the animals are typically not required by regulation to be reviewed by the IACUC. If these projects, however, have a true experimental design that has potential physiological impact on the animals and the results are submitted for publication, such projects are more equivalent to typical research projects that require IACUC review. (See 8:31.)

8:15   If an investigator wishes to conduct research using animals at another institution, must the IACUC at the investigator's home institution approve the protocol? Must the other institution's IACUC approve it?

Reg.   There are many circumstances that involve partnerships between collaborating institutions or relationships between institutional animal care programs. OLAW and APHIS agree that review of a research project or evaluation of a program or facility by more than one recognized IACUC is not a federal requirement.

It is imperative that institutions define their respective responsibilities. PHS Policy requires that all awardees and performance sites hold an approved Animal Welfare Assurance. OLAW negotiates Inter-institutional Agreement Assurances of Compliance when an awardee institution without an animal care and use program or IACUC will rely on the program of an Assured institution. Assured institutions also have the option to amend their Assurance to cover non-assured performance sites, which effectively subjugates the performance site to the Assured institution and makes the Assured institution responsible for the performance site.

If both institutions have full PHS Assurances, they may exercise discretion in determining which IACUC reviews research protocols and under which institutional program the research will be performed. It is recommended that if an IACUC defers protocol review to another IACUC, then documentation of the review should be maintained by both committees. Similarly, an IACUC would want to know about any significant questions or issues raised during a semiannual program inspection by another IACUC of a facility housing a research activity for which that IACUC bears some responsibility or exposure.[15]

8:16   If an investigator wishes to conduct research using animals at a *foreign* institution, must the IACUC at the investigator's home institution approve the protocol? Must the other institution's IACUC approve it? What about samples that are obtained by citizens of the foreign country and sent to our institution?

Reg: All animal activities supported by the PHS must be reviewed by the IACUC of the domestic-assured awardee institution that receives such support. Foreign institutions that serve as performance sites must also have Assurances on file with OPRR [now OLAW]. The OPRR considers institutions whose scientists are engaged in such collaborative work accountable for the animal-related activities from which they receive animals or animal parts. When a foreign institution holds a PHS Assurance, it also is expected that the institution will conduct the study in accordance with the applicable host-nation's policies and regulations. In the specific case of sample collection, the review should take into account the species involved, nature of the specimen, and the degree of invasiveness of the procedure, giving appropriate consideration to the use of anesthetics and analgesics. In cases in which samples are obtained directly by citizens of a foreign country for subsequent shipment, recipient PHS-supported investigators should determine the proposed methods of collection and present that information to their IACUC for review.

Prior to sample collection and regardless of whether specimens are obtained by an awardee institution's investigator directly or by persons in a foreign country, the OPRR strongly recommends that each awardee institution consult with other agencies of the U.S. government concerning importation requirements. Depending on the species involved and the nature of the specimen, the following may be of assistance: the U.S. Fish and Wildlife Service, Department of the Interior (for compliance with the International Convention on Trade in Endangered Species of Fauna and Flora [CITES]), APHIS, U.S. Department of Agriculture (regarding potential animal pathogens), and the Centers for Disease Control and Prevention (concerning importation of nonhuman primates and potential pathogens of human beings).[16]

Opin.   In collaboration with a domestic PHS-assured institution, duplicative IACUC review is not required; however, as indicated, a domestic IACUC must review protocols conducted in foreign locations.

8:17 What minimal qualifications does a person need in order to submit a protocol to the IACUC? For example, can students, postdoctoral fellows, medical residents, or visiting scientists submit a protocol?

Reg.   Personnel conducting procedures on the species being maintained or studied will be appropriately qualified and trained in those procedures (AWAR §2.31,d,1,viii; PHS Policy IV,C,1,f). Furthermore, it is the responsibility of the institution to ensure that all personnel involved in animal care and use are qualified to perform their duties (AWAR §2.32,a).

Opin.   There is a difference between the person submitting the protocol and the person actually performing the animal work. There are no federal or state minimal requirements or qualifications for anyone to "submit" a protocol to the IACUC. Such requirements are left to the discretion of each institution. Some institutions allow only their own scientists or faculty to submit protocols. Others may allow outside scientists to submit protocols. Some find a middle ground and have local faculty "sponsor" nonfaculty members.

In some instances, the PI who submits a protocol may never see or handle the animals. The research staff performs the actual work. Nevertheless, the PI is held responsible for the conduct of the study. It is important for the IACUC to know who is working with the animals and who is ultimately responsible for the conduct of the study.

Review of research personnel qualifications is under the purview of the IACUC. The IACUC is best equipped to evaluate individuals' qualifications when very specific information is submitted regarding their duties and training pertinent to their roles in the protocol. The IACUC has the authority to accept or reject an individual's credentials or to require further training if warranted.

8:18 Who typically maintains protocols and how long should they be kept?

Reg.   For AWA-regulated species and PHS-funded animal projects, protocol records must be held for 3 years after completion of a study, as are the IACUC minutes and records of deliberations. These records must be made available to APHIS/AC or NIH/OLAW when requested (AWAR §2.35,a; §2.35,f; PHS Policy IV,E,1; IV,E,2).

Opin.   A separate IACUC office or a research administration office usually maintains the protocol files. Investigators should also keep a copy. These records are also considered institutional documents and subject to whatever institutional requirements are established (above and beyond the federal regulations).

8:19   Who should have access to IACUC protocols?

Reg.   For AWA-regulated species and PHS-funded animal projects, protocol records must be made available to APHIS/AC or NIH/OLAW when requested (AWAR §2.35,f; PHS Policy IV,E,2).

Opin.   Activities conducted at public institutions inclusive of IACUC operations may be covered by "sunshine" laws (open meetings or open records laws). (See Chapter 22.)

IACUC documents can be subject to the state's public records law, and, except when exempted under the law, IACUC documents are available to the public. For example, Florida's Statute Section 240.241(2) exempts the following information from being released:[17] "[M]aterials that relate to methods of manufacture or production, potential trade secrets, potentially patentable material, actual trade secrets, business transactions, or proprietary information received, generated, ascertained, or discovered during the course of research conducted within the state universities shall be confidential and exempt from the provisions of s.119.07(1), except that a division of sponsored research shall make available upon request the title and description of a research project, the name of the researcher, and the amount and source of funding provided for such project."

In addition, IACUC meetings may also be subject to the state's open meetings law; therefore, students, faculty, staff, media, and the public at large can have access to them. Individual states' freedom of information acts govern who has access to IACUC records and meetings. Questions that arise regarding the legal access to this information should be addressed by legal counsel from the institution. (See Chapter 22.)

In privately funded institutions not subject to sunshine laws, access to meetings and access to records become policy matters for the institution. Researchers, certain administrators, as well as the IACUC, veterinary, and husbandry staffs are most involved and affected by this information and, therefore, should be afforded access. Conducting open meetings and distributing nonexempt information to the public can be in the best interest of animal research. This is an effective means of public education and helps to negate the veil of secrecy that often hinders the public image of animal research.

Surv.   At your institution, who has access to IACUC protocol forms other than the IO, IACUC office, and IACUC Chairperson? Check all that apply.

• Nobody else	20/170
• Some additional high-level IACUC members	23/170
• Animal facility administrators	66/170
• Other PIs	22/170
• The AV	95/170
• Anybody within the institution	41/170

Note: "Anybody within the institution" included veterinary technicians, the animal facility staff, and all members of the IACUC.

8:20   May protocols be submitted electronically to the IACUC and can protocols be electronically distributed to IACUC members for review?

Reg.   The PHS Policy concept of institutional self-regulation with local IACUC oversight evolved from similar mechanisms applied to the protection of human subjects in research. As with the Institutional Review Board (IRB) for human subjects, assumptions regarding the general environment in which oversight of animal related research takes place were included in the design of prescribed oversight mechanisms. One of those assumptions was that committees normally function through periodic meetings where members would consider, deliberate, and vote on matters within their purview. This model did not anticipate the technological advances that would make it possible to approximate the traditional meeting environment from greater distances and with greater ease and speed than was the norm at that time.

Although the traditional meeting is still seen as the optimum environment for fulfilling the intent of the PHS Policy, OPRR [now OLAW] recognizes the new communications tools available and the need for flexibility in the ways that institutions may comply with the PHS Policy in the many diverse settings encountered. For these reasons, several criteria have been provided for establishing alternate methods that may be considered functionally equivalent to meetings of a convened quorum under exceptional circumstances. Two of those criteria are that the alternate approach must include a high degree of interactivity and allow for careful deliberation of sensitive issues. Another is that a quorum of IACUC members must be in direct communication with each other and be given full opportunity to participate for the duration of the meeting. Institutions are reminded that details of their IACUC procedures, especially those that may vary from those outlined in the PHS Policy, should be thoroughly described in the institutional Assurance and submitted for OPRR review.[18]

A newly released NIH statement, detailed in 6:17, further clarifies the NIH/OLAW and APHIS/AC position.

Opin.   Neither the AWAR nor PHS Policy prohibits protocols from being electronically submitted to the IACUC and electronically posted for members to review. Technology is available for protocols to be completed and submitted to secure Web sites. Members can then gain access to this site to read or download the information and return e-mail messages with their results. It is emphasized that the final approval of protocols submitted in this fashion must not vary from the regulations and requirements set forth by the AWAR and PHS Policy regarding full committee review and designated member review. The use of electronic communication simply speeds up the review process by replacing the slower mail delivery systems. (See 6:17–6:20.)

Several problems can occur with electronic reviewing. For example, reading from a computer monitor can be difficult when scrolling through pages of text. Signature requirements of the investigator and administrators (if they are required) need to be resolved. The following comments from NIH/OLAW lend perspective to the issue of signature requirements: "Many IACUC activities involve the need for documenting votes or verifying committee approval. Letters and reports often require signatures in order to be legally binding. Techniques for providing such legally acceptable 'electronic signatures' are being developed by the computer industry to address this problem. Currently, electronic communication is frequently used to facilitate the rapid conduct of business with signed original documents to follow for the permanent record."[18]

Surv.   Does your institution allow for the electronic submission and distribution of IACUC protocols?

- Yes, for both submission and distribution   94/170
- Yes, but only for submission, not distribution   23/170
- No for both submission and distribution   44/170
- Other   9/170

Note: Other responses included that hard copy with signature must follow electronic copy, option available, no for submission but yes for distribution, for preview by attending veterinarian, and that e-copies can be distributed but hard copy must be obtained by IACUC office.

8:21   What are some recommendations for maintaining confidentiality if electronic media are used for submission and review?

Reg.   The unauthorized release of confidential IACUC information by members is already prohibited by law (AWA Sect. 27a; Sect. 27b):

> It shall be unlawful for any member of an Institutional Animal Committee to release any confidential information of the research facility including information that concerns or relates to (1) the trade secrets, processes, operations, style of work, or apparatus; or (2) the identity, confidential statistical data, amount or source of any income, profits, losses, or expenditures, of the research facility. It shall be unlawful for any member of such Committee (1) to use or attempt to use to his advantages, or (2) to reveal to any other person any information which is entitled to protection as confidential information under subsection (a).

Opin.   Security is always a concern when dealing with animal research records, especially those protocols under discussion and not officially approved. NIH/OLAW provides the following perspective: "Institutions utilizing innovative modes of communication must be aware of the potential security problems inherent in the method chosen. Some material considered by IACUC's should be treated as privileged or confidential, especially prior to final committee action. In the case of trade secrets, such information may be protected by law (7 USC 2157, Sect. 27). Because of widespread reports of unauthorized access to computer records, the use of available computer security measures such as passwords, controlled access, and encryption should be considered."[18] Use of the electronic mail at public institutions is not private. Files of public employees may be considered public documents and may be subject to inspection under a state's public records law.

8:22   What is meant by "just-in-time" review of an IACUC protocol?

Reg.   Effective September 1, 2002, the NIH changed the PHS Policy to permit institutions with PHS Animal Welfare Assurances to submit verification of IACUC approval for an application *subsequent to peer review but prior to award.* Prior to this date, approvals had to be received by NIH within 60 days after the grant was submitted. The former process, which has been in use since May 2000 for review of human subjects in research, is referred to as *just-in-time.* The purpose is to reduce the burden on investigators and IACUCs, allowing resources to be focused only on those projects likely to be funded. Key to this change are the following expectations:

- PHS still requires that the institution hold an approved Assurance and the animal use component must receive IACUC approval prior to the release of grant funds.

- Institutions may still choose to review the animal protocol at any time during the grant submission process.

- Under *no* circumstances may an IACUC be pressured to approve a protocol or be overruled on its decision to withhold approval.
- NIH peer review of the animal component of the grant will continue and will supplement, but not replace, the IACUC review and approval.
- It is incumbent upon investigators to convey to their IACUCs all modifications in animal usage that may result from NIH peer review feedback.
- It is incumbent upon institutions to communicate to the NIH any IACUC-imposed changes that impact on the scope of the funded study.
- It is incumbent upon the NIH to ensure that institutions are given adequate notice to allow for IACUC review prior to award.[19]

Opin. Many funding agencies (private, nonprofit organizations, etc.) still require animal use approval (or pending approval) prior to accepting a grant proposal. Some, however, apply the "just-in-time" process, akin to the NIH funding procedure. As the approach required by the individual funding agency may change, the authors advise investigators to check with the funding agency routinely prior to submitting a request for funds.

8:23   Is it necessary to submit an IACUC application at the time a grant is submitted to PHS?

Reg.   (See 8:22.)
Opin. With the change by the NIH to just-in-time review, technically the answer is no. Nevertheless, the majority of grant applications that propose to use animals contain preliminary data based on animal use. Therefore, an approved IACUC protocol reflecting the preliminary data collection must be on file at the institution.

8:24   With the just-in-time process in place, after an NIH grant is approved for funding, how much time does an investigator have to submit a protocol to the IACUC, and how much time does the IACUC have to review the protocol?

Reg.   (See 8:22.) The regulations do not specify a time frame; they only state that an IACUC approval must be obtained prior to release of funds.
Opin. NIH will send a request to both the institution and the PI for just-in-time information after the grant review has occurred and a percentile of less than 20% has been awarded. If the grant is funded, the release of funds usually occurs within 2 months of NIH's request for the information. Additionally, the investigator receives a Summary Statement sheet after the peer review process that signals whether or not the application is in the fundable range. Given that both investigators and their supporting institutions are eager to receive the grant monies, investigators are advised to submit their animal use protocol for IACUC approval at this time. On rare occasions, a grant with a score that initially appears not to be within the fundable range is funded. When this happens the institution and PI receive notification of the need for just-in-time review of the animal use protocol congruent with the release of funds notification. In this situation, no grant-related work involving the use of animals can be initiated until the IACUC approval has been verified by NIH.[20] (See 9:51; 9:52.)

8:25   Investigator Jones has a PHS grant and permanently moves from Institution A to Institution B. The PHS allows him to transfer his grant to Institution B. He has an IACUC-approved protocol at institution A but will complete the animal-related research at Institution B. Must Institution B's IACUC review and approve the proposed research before it can begin?

Reg.   Transfer of PHS-supported research must follow specific procedures. A transfer requires the prior approval of the PHS funding agency, which is dependent on the receiving institution's (Institution B's) both having an approved animal Assurance as well as having approved the animal component of the work proposed. In addition, the institution relinquishing the grant must do so prior to the transfer of the award. If the request to transfer the grant is approved by PHS, the new grantee, with an approved Assurance, must assure that an IACUC is established and certify that the protocol has been reviewed and approved.[21]

Opin.   Because many of the policies and regulations that govern the use of animals in research are performance based (e.g., scientific justification for not providing pain medication), individual institutions vary in their approaches to approving animal care and use. Because accountability transfers to the receiving institution, the outcome of protocol review at Institution B may require the investigator to modify his protocol to meet the institution's own animal care policies.

8:26   Which people should be listed on an IACUC protocol?

Reg.   (See 8:3.)

Opin.   The IACUC must fulfill its charge that all personnel conducting procedures on animals are adequately trained and qualified, and that they are enrolled in applicable occupational health and safety programs. Therefore, the funded PI as well as all individuals who will conduct animal related procedures on behalf of the investigator should be identified on the information provided to the IACUC.

Surv.   Which people are typically listed on your IACUC protocol form?

• Only the PI	8/170
• The PI and other persons handling or performing procedures on animals (other than animal care and veterinary services)	121/170
• All persons, including veterinary technicians and/or animal care personnel	24/170
• The PI and only those persons performing procedures on animals (not just handling them)	14/170
• Only those persons handling or performing procedures on animals (may exclude PI)	0/170
• Other	3/170

8:27   What are some ways for the IACUC to minimize the potentially overwhelming amount of information within the protocols that investigators and IACUC members need to complete or review?

Opin.   The process of protocol review can be streamlined for IACUCs by utilizing a preview and/or designated reviewer system. An administrative review by IACUC support staff can help to pick up glaring errors or information omissions. A secondary review, by a qualified veterinarian and investigator who are familiar with the protocol format, further helps to reduce errors and makes the reviewing process less onerous for the rest of the committee members. Finally, utilization of a designated reviewer for a specified classification of protocols (e.g., those that do not require invasive procedures, those that do not require the use of endangered species) helps to minimize the intensity of review for which each committee member is responsible. (See 9:21–9:27.)

The process of protocol completion can be streamlined for investigators by utilizing a protocol format that is clear and concise, thereby supporting accurate completion of the necessary information. It is also helpful to have the IACUC and veterinary unit provide online standard operating procedures, policies, and veterinary information for the more common, universally applied animal use techniques (see 8:29). Investigators can then, if applicable, "cut and paste" information such as anesthetic and analgesic regimes, blood withdrawal frequencies and volumes, and DNA sampling procedures.

8:28 What is a reasonable sequence of time and events that begins when a PI submits a protocol and ends when the protocol is ready for a vote?

Opin. The time frame for the process of approving protocols hinges largely on the number of times per year the institution's IACUC meets. Although an average seems to be monthly, IACUCs at larger institutions may choose to meet more frequently, and conversely, IACUCs at smaller institutions may meet less frequently. Typically the completed protocol is required to be submitted by the investigator to the IACUC administrative staff at some set date prior to the convened meeting. The interval set between protocol submission and committee review is in part dictated by the number of protocols the committee has to process, the infrastructure support provided to the committee, and the availability of individuals serving on the committee (e.g., institutions can opt to provide dedicated time and support for faculty members to complete committee assignments).

The protocol is distributed through the channels that the institution has established (e.g., some variation of prereview; see 8:27) or to the IACUC Chair for assignment of designated reviewers (see 9:21–9:27) or to the full committee. With a prereview process, comments are returned to the investigator for correction or discussion prior to deliberation by the IACUC. Use of the designated reviewer system can shorten the approval process if modifications to obtain approval are agreed to by the PI. If not, the protocol must be reviewed at a convened meeting of the full committee. The length of time required to resolve protocol issues is somewhat dependent on the nature of the committee's concerns, but in the authors' opinion, it is more dependent on the communication skills of the PI and the working relationship between the committee members and the investigative community.

Surv. 1 How long does it typically take from the time a PI first submits an IACUC application until it receives an initial review (or prereview) by an IACUC member or other nonadministrative person?

- Less than 1 week   68/170
- One to 2 weeks   62/170
- More than 2 but less than 4 weeks   37/170
- More than 4 weeks   2/170
- Other   1/170

Surv. 2 How long does it typically take from the time a protocol receives an initial review (or prereview) by an IACUC members(s) until it is ready for an initial vote on its acceptance (either by full committee or by the designated member review process)?

- Less than 1 week   31/170
- One to 2 weeks   71/170
- More than 2 but less than 4 weeks   54/170
- More than 4 weeks   2/170
- Other   12/170

Surv. 3 Typically, how long does it take from the time an IACUC protocol is first submitted to the IACUC office until it is finally approved?

- Less than 1 week   7/170
- One to 2 weeks   28/170
- More than 2 but less than 4 weeks   82/170
- More than 4 but less than 8 weeks   41/170
- Over 8 weeks   6/170
- Other   6/170

8:29 If the exact same procedure is repeatedly used (such as a surgical procedure), can the IACUC or PI have an approved "procedure bank" so a PI can reference the approved procedure without having to write it out in detail every time? (See 8:30.)

Opin. Yes. SOPs describing, for instance, a surgical procedure can be used within a protocol. It is recommended that these SOPs be physically attached to or inserted within the protocol so the IACUC can review the protocol in total. It is required that these SOPs also then be reviewed and approved for changes or concerns by the IACUC on a annual basis (AWAR §2.31,d,1,xi,5), with a 3-year *de novo* review for institutions following the PHS Policy (IV,C,5).

Surv. Does your IACUC have a collection of standard research techniques for your institution that a PI can reference in his or her IACUC application?

- No   62/170
- Yes, it can be referenced as being the approved standard, but it still must be written out on the protocol form   53/170
- Yes, and it is not necessary for the PI to rewrite it on the protocol form   52/170
- Other   3/170

8:30 If a procedure is repeatedly performed, but with a single variable (e.g., a new drug to be tested), can the IACUC have an approved "procedure bank" so a PI can reference the approved procedure without having to write it out in detail every time? If yes, how should the IACUC address the issue of the single variable?

Opin. As stated in 8:29, SOPs can be used within a protocol to streamline the process, such as the approach to drug delivery. Nevertheless, a new protocol must be reviewed and approved with each new drug that is proposed for use in the animal. An alternative approach to attaching or inserting SOPs within a newly formatted protocol would be to continuously amend a standard protocol with the necessary information about the new drug. The committee can then provide appropriate review for the amendments. Some have argued that if the compound administered is in the same class (e.g., an antigen), the PI and staff are the same, and the type of animals is identical, there should be the ability to have one "global approval" for work done. This approach does not address animal usage or the potential for variability in drug reactions even within a similar class of agents.

8:31   A PI wishes to use a very small amount of blood (for research) that will be taken for clinical purposes by a veterinarian from a pet dog. Is IACUC approval needed for the veterinarian to give the blood to the researcher?

Opin.   The purpose for which the blood is drawn is a critical component in the response to this question. Because a portion of the blood is drawn to obtain data for research purposes, an IACUC-approved protocol is necessary. It is recommended for the protection of the institution that an informed consent be obtained from the client as the representative of the animal (the subject). (See 14:42.)

8:32   A veterinarian draws blood from a cow that is to be slaughtered the next day. The blood is specifically destined for use in an *in vitro* biomedical research project. Is IACUC approval needed to draw and use the blood?

Opin.   Yes. This is considered antemortem specimen collection and the time of death after the blood collection is irrelevant. If, however, the blood was drawn after the animal was slaughtered, then the discussion found in 8:7 is relevant.

8:33   Live mice are fed to snakes as a source of food, and the snakes are the research subject supported by PHS funds. The mice arrive in the morning and are fed by the end of the day. Does the use of the mice as food require IACUC approval?

Opin.   Some oversight by the IACUC is very necessary. The source of the incoming mice could have a significant impact on the health of other rodents housed within the facilities. The source of food for the snakes should be reviewed and described in the IACUC protocol that covers the snakes (i.e., an unusual husbandry issue) or may be covered in a husbandry SOP that the IACUC reviews as part of the overall animal care and use programmatic review.[22] (See 12:24.)

References

1. American Veterinary Medical Association, 2000 Report of the AVMA Panel on Euthanasia, *J. Am. Vet. Med. Assoc.*, 218(5), 669, 2001.
2. Institute of Laboratory Animal Resources, National Research Council, Guide for the Care and Use of Laboratory Animals, National Academy Press, Washington, D.C., 1996.
3. Office of Laboratory Animal Welfare, National Institutes of Health, Institutional Animal Care and Use Committee Guidebook, 2002. Available on the World Wide Web at: http://grants1.nih.gov/grants/olaw/GuideBook.pdf.

4. Office of Science and Technology Policy, Interagency Research Animal Committee, U.S. Government Principles for Utilization and Care of Vertebrate Animals Used in Testing, Research, and Training, Federal Register, Washington, D.C., May 20, 1985. Available on the World Wide Web at: http://www.grants.nih.gov/grants/olaw/references/phspol.htm#USGovPrinciples (agencies represented include Veterans Administration; Department of Energy; National Aeronautics and Space Administration; Environmental Protection Agency; Department of the Interior; Department of State; Department of Defense; National Science Foundation; U.S. Department of Agriculture, and Consumer Product Safety Commission; Department of Health and Human Services, including the National Institutes of Health, Fogarty International Center, Centers for Disease Control and Prevention, Office of International Health, Health Research Services Administration, and Food and Drug Administration).

5. DeHaven, W.R., personal communication, 1998.

6. American Heart Association Policy Statement regarding the Use of Animals. Available on the World Wide Web at: http://www.americanheart.org/presenter.jhtml?identifier=4437.

7. American Cancer Society, Inc., Atlanta, GA. Available on the World Wide Web at: http://www.cancer.org/docroot/RES/RES_0.asp (provides access to grant applications and instructions related to IACUCs).

8. American Lung Association. Available on the World Wide Web at: http://www.lungusa.org/site/pp.asp?c=dvLUK9O0E&b=33309 (provides access to grant applications and instructions related to IACUCs).

9. American Heart Association. Available on the World Wide Web at: www.americanheart.org/presenter.jhtml?identifier=1200033 (provides access to grant applications and instructions related to IACUCs).

10. Potkay, S., et al., Frequently asked questions about the Public Health Service Policy on Humane Care and Use of Laboratory Animals, *Lab Anim.* (NY), 24(9), 24, 1995. Available on the World Wide Web at: http://www./grants.nih.gov/grants/olaw/references/laba95.htm.

11. Garnett, N.L., and DeHaven, W.R., OPRR and USDA Animal Care Response on Applicability of the Animal Welfare Regulations and the PHS Policy to Dead Animals and Shared Tissues, *Lab Anim.* (NY), 26(3), 21, 1997.

12. U.S. Department of Agriculture, Animal and Plant Health Inspection Service, Policy 10, Licensing and Registration of Antibodies, Sera and/or Other Animal Parts and Pregnant Mare Urine (PMU), April 14, 1997. Available on the World Wide Web at: http://www.aphis/usda.gov/ac/policy10.html.

13. Division of Animal Welfare, Office for Protection from Research Risks, National Institutes of Health, The Public Health Service responds to commonly asked questions, *ILAR News*, 33(4), 68, 1991. Available on the World Wide Web at: http://grants.nih.gov/grants/olaw/references/ilar91.htm.

14. Potkay, S., et al., Frequently asked questions about the Public Health Service Policy on Humane Care and Use of Laboratory Animals, *Lab Anim.* (NY), 24(9), 24, 1995. Available on the World Wide Web at: http://grants.nih.gov/grants/olaw/references/laba95.htm.

15. Office of Extramural Research, Guidance regarding Administrative IACUC Issues and Efforts to Reduce Regulatory Burden, Notice NOT-OD-01-017, February 12, 2001. Available on the World Wide Web at: http://grants.nih.gov/grants/guide/notice-files/NOT-OD-01-017.html

16. Potkay, S., et al., Frequently asked questions about the Public Health Service Policy on Humane Care and Use of Laboratory Animals, *Contemp. Topics Lab. Anim. Sci.*, 36(2), 47, 1997.

17. Florida Statutes, Sect. 240.241(2).

18. Garnett, N., and Potkay, S., Issues for IACUCs: Use of electronic communications for IACUC functions, *ILAR J.*, 37 (4) 190, 1995. Available on the World Wide Web at: http://www./grants.nih.gov/grants/olaw/references/ilar95.htm.

19. *Fed. Regist.*, Vol. 67, No. 152, Wednesday, August 7, 2002.

20. NIH/OER/OPERA, Request for just-in-time information. Available on the World Wide Web at: http://www./grants2.nih.gov/grants/peer/jit.pdf.

21. Silverman, J., A pain in the rat, *Lab Anim.* (NY), 30(3), 21, 2001. Available on the World Wide Web at: http://www./grants.nih.gov/grants/olaw/references/laba01v30n3.htm.

22. Garnett, N.L., A word from OLAW, *Lab Anim.* (NY), 32(10):19, 2003.

9

General Concepts of Protocol Review

Ernest D. Prentice, Gwenn S.F. Oki, and Michael D. Mann*

Introduction

Reviewing protocols involving the use of animals is one of the most important responsibilities of the IACUC. Although the format employed for this review varies across institutions, the criteria used in reviewing and approving protocols should be as consistent as possible. Consistency of review is a particularly important consideration with regard to animal welfare issues, such as the use of nonanimal model alternatives and the application of refinement techniques to reduce animal pain, discomfort, and distress.

The purpose of this chapter is to present general concepts of protocol review that will help IACUC administrators, IACUC members, AVs, and other interested individuals gain an appreciation for how IACUCs conduct protocol review. Particular attention is given to issues such as scientific merit review, preview, use of consultants, designated member review versus full committee review, and IACUC actions. Because the answers to the questions posed in this chapter are designed to be as succinct as possible, the reader is encouraged to consult the PHS Policy, the AWAR, and the *Guide*.

9:1 What guidelines do the AWAR, the PHS Policy, and the *Guide* provide with respect to IACUC review of protocols?

Reg.   PHS Policy (IV,C,1,a–IV,C,1,g; IV,D,1,a–IV,D,1,e) and the AWAR (§2.31,d; §2.31,e) specify criteria that must be met before an IACUC can approve a proposed research protocol. These requirements address issues such as the following:

- Prevention or minimization of discomfort, distress, and pain to the animals
- Appropriate living conditions for species used in the project
- Availability of medical care to be provided by a qualified veterinarian
- Euthanasia consistent with the recommendations of the American Veterinary Medical Association Panel on Euthanasia[1]
- Qualifications of personnel conducting procedures on the species being studied

* The authors thank Molly Greene, Eifaang Li, Michael Mann, Joy A. Mench, and Philip Tillman for their contribution to this chapter in the first edition of *The IACUC Handbook*.

PHS Policy (IV,A,1) requires that Assured institutions also base their IACUC review of protocols on the *Guide* and that they comply with the AWAR that apply to APHIS/AC-regulated species.

The AWAR (§2.31,d,1,i–§2.31,d,1,xi; §2.31,e,1–§2.31,e,5) list a total of 16 requirements that must be met before the IACUC can approve a protocol. Some of the requirements mirror the PHS Policy, whereas others do not. For example, the AWAR require an investigator to consider alternatives to painful procedures and to describe the methods and sources (e.g., electronic literature search) used to determine that alternatives are not available. The AWAR also require the investigator to provide written assurance that the animal-related activities do not unnecessarily duplicate previous experiments. The PHS Policy requires that the PI consider alternatives but does not require a description of the methods and sources used.

The PHS Policy also includes the nine U.S. Government Principles for the Utilization and Care of Vertebrate Animals, which provide additional guidance concerning the concepts of replacement, reduction, and refinement (3 Rs). Of particular note is Principle II, which states that "procedures involving animals should be designed and performed with due consideration of their relevance to human or animal health, the advancement of knowledge, or the good of society." Principle II represents the scientific and societal justification for using animals in research.

Opin. It should be noted that neither the PHS Policy nor the AWAR provide specific guidelines in all areas of protocol review. It is, therefore, left to the institution to develop specific guidelines for implementation of federal requirements. Many institutions have, therefore, developed investigator handbooks that include information concerning local policies for anesthetics, analgesics, blood sampling, tail sampling, antibody production, pre- and postoperative care for nonrodent animals, and other aspects of animal research.

Finally, it should be mentioned that NIH periodically publishes guidance on selected topics in "Dear Colleague" letters and in notices in the NIH *Guide for Grants and Contracts* that provide information concerning NIH/OLAW's interpretation of the requirements of the PHS Policy.[2] In a somewhat similar vein, APHIS/AC issues its *Animal Care Policy Manual*, which provides further guidance for IACUCs.

9:2 What are some useful pathways to disseminate protocols for review once they have been submitted to the IACUC?

Reg. PHS Policy (IV,C,2) requires that each IACUC member be provided with a list of proposed projects to be reviewed, and that written descriptions of projects be available to all IACUC members. The AWAR (§2.31,d,2) have the same requirement.

Opin. There are many methods used to disseminate protocols for review. Most IACUCs meet in a common location to review protocols, but some committees communicate by fax, electronic mail, phone, or video conferencing (see 6:17). It is common practice for IACUCs to assign each protocol to a primary and secondary reviewer (chosen on the basis of their expertise in the subject matter of the protocol), who are given principal responsibility for in-depth review. With this method, all members of the IACUC receive either a complete copy of the protocol or a protocol summary for review. Some IACUCs, particularly at institutions that conduct extensive research using animals, utilize a prereview system in which protocols are clarified and modified as necessary prior to full committee review (see 9:5; 9:6). Many IACUCs also utilize a designated member review system (also referred to as *expedited review,* although the use of this term is strongly discouraged by the APHIS/AC, NIH/OLAW, and AAALAC), which in turn helps to decrease the workload during full committee meetings (see 9:22). In general, the method used to disseminate protocols for review is dictated by the size and specific needs of the institution.

9:3   What are the IACUC and institutional obligations for reviewing a protocol for an internally funded project?

Reg.   If an institution has elected, in its NIH/OLAW Assurance, to apply the requirements of the PHS Policy to all activities (regardless of the source of funding), then the IACUC obligations for reviewing internally funded projects do not differ from those obligations under the PHS Policy.

Opin.   Internally funded projects may pose special problems for the review process, particularly with respect to scientific merit review. Even IACUC members who maintain steadfastly that the committee has no responsibility or right to review scientific merit agree that the IACUC has some obligation to do so when the proposal will not be subjected to external peer review. At least a minimal level of merit that justifies animal use must be assured before such protocols are approved. Needless to say, animal welfare considerations should be the same, and the same rigorous review process with regard to regulatory compliance must be uniformly applied to both internally and externally funded projects.

9:4   At what point in the review of a protocol are consultants best included?

Reg.   PHS Policy (IV,C,3) and the AWAR (§2.31,d,3) allow the use of consultants for the review of a protocol. Consultants may not vote with the IACUC, and the IACUC remains responsible for its actions and decisions.

Opin.   The use of consultants is highly variable among IACUCs. In some institutions, it is common practice for IACUC reviewers to use colleagues within the institution as consultants for clarification of issues raised during prereview of a protocol (see 9:5; 9:6). This prereview clarification can greatly speed the final review process. Outside consultants are seldom employed unless there is disagreement among the IACUC members as to the merit of a given proposal or concerns related to animal welfare are unresolved. Consultants also are useful if an investigator asks the committee to reconsider its decision (see 9:56; 9:57; 29:30; 29:31).

9:5   What is meant by *prereview* of a protocol?

Opin.   *Prereview* refers to review of a protocol by one or more individuals prior to formal review by the IACUC. Neither the PHS Policy nor the AWAR make any specific mention of the term or concept of prereview, although many institutions find it useful. The AWAR (§2.31,d,1,iv,B) require veterinary consultation during the planning of a procedure that might cause more than momentary pain or distress. (See 9:6; 16:4.)

9:6   What are the reasons for a prereview of protocols?

Opin.   Protocols are sometimes incomplete when they are submitted for IACUC consideration. Information that the IACUC needs in order to review the protocol effectively might be absent or unclear. For example, the investigator may be proposing to use a method of anesthesia involving injectable agents without specifying the dose. Clearly, this information must be provided before the IACUC can approve the protocol. Prereview saves the IACUC time, because IACUC members do not have to read protocols that are obviously not ready to be reviewed. Prereview also should be helpful to the investigator in terms of speeding up the review process, because the protocol is less likely to be tabled and carried over to the next IACUC meeting pending further information and clarification.

9:7 Is prereview of protocols required by any regulation or policy?

Opin. Prereview of protocols is not a PHS Policy requirement, but the AWAR (§2.31,d,1,iv,B) require PIs to consult a veterinarian when potentially painful procedures are planned. Prereview of the protocol by a veterinarian is one way that such a consultation can be provided. (See 9:5; 16:4.)

9:8 Can the prereview team be an IACUC subcommittee composed of IACUC members, non-IACUC members, or a combination of these?

Opin. Because prereview is not a mandated function of an IACUC, the prereview process can be structured in a way that is most helpful to the IACUC and investigators. Prereview can be carried out by designated IACUC members, qualified IACUC office staff, campus veterinarians, or any other experienced individuals who are able to review the protocol adequately.

9:9 What is a workable mechanism for the prereview of protocols?

Opin. The best mechanism depends on the institutional structure and the composition of the prereview team. For example, an institution with multiple vivaria may choose to have incoming protocols immediately sent to a prereviewer. Each prereviewer handles all of the submitted protocols from the particular vivarium on the campus, so the prereviewer becomes very familiar with the investigators, facilities, and types of research conducted in that unit. The prereviewer reviews the proposal, notes any points of inadequacy in the document, and communicates with the PI. The prereviewer may be able to suggest minor corrections or addition of clarifications to the protocol (for example, changing the route, dose, or type of anesthetic agent) or addition of a sentence to the protocol (for example, indicating the volume of blood to be withdrawn). In the case of more substantial corrections, the prereviewer discusses the problems with the PI and may recommend that the investigator rewrite the protocol and submit a corrected version. There is a potential problem that the PI may interpret prereview as actual committee review, and changes made may be interpreted as IACUC approval. The PI should clearly understand that prereview is not full committee review and that additional changes may be required by the IACUC.

9:10 What should the PI do after prereview?

Opin. Assuming that the prereviewer suggests changes that are in accordance with IACUC policies and concerns, it is in the PI's best interest to make the necessary corrections before the IACUC meeting so that the protocol is suitable for review.

9:11   What happens to a protocol after it is prereviewed?

Opin.   After a protocol is prereviewed, it must be processed for formal review by the IACUC. This may be accomplished by either a designated reviewer (see 9:21–9:26) or review at a fully convened meeting of the IACUC. (See 9:28–9:32.)

9:12   Are the results of the prereview presented to the full IACUC? If so, by whom?

Opin.   In institutions that still use paper submissions, the IACUC may be presented with a completely revised application form including all of the changes made by the PI as a result of the prereview process. Alternatively, the changes may be presented as a separate document along with the original application form. The former method is preferred because it yields a "clean," complete protocol application, which is a benefit to both the IACUC and the investigator and his or her staff. In institutions that use electronic submissions, the PI can simply alter the online form, especially if the submission software is able to track all the changes. Unfortunately, not all of the commercial software packages allow such tracking. There are, however, separate software packages that provide this capability. From the point of view of minimizing the workload of the IACUC it is preferable to have a single document for review.

9:13   Are prereview comments binding on investigators?

Opin.   No. Prereview is simply a method for giving investigators advice on their protocols. A possible exception may arise when the AV determines during prereview that a particular anesthetic or surgical procedure is not appropriate (AWAR §2.33,b,4). Investigators may occasionally disagree with the prereviewer's suggestions and choose to provide an explanation and send an unmodified or a partially modified protocol to the IACUC for formal review. However, an investigator who insists on presenting an inadequate document to the IACUC is likely to find that the IACUC will not approve the protocol, and the protocol will be returned for modifications or clarification for the same reasons identified by the prereviewer. This process will result in unnecessary delays in obtaining IACUC approval of the project.

9:14   Should the IACUC review protocols for scientific merit? Do the AWAR or the PHS Policy specifically require (or prohibit) review of scientific merit?

Opin.   In our experience, many IACUCs review protocols for "scientific merit" or "scientific relevance." It is not clear, however, that these terms have the same or similar meaning to all IACUCs. Whether IACUCs should perform such review is open to debate. Prentice and associates[3] and, more recently, Mann and Prentice[4] reviewed the PHS Policy and found that the term *scientific merit* is not used. Rather, it refers to "relevance" in U.S. Government Principle II, in a manner that strongly suggests the terms are synonymous. The PHS Policy (IV,D,1,d) also uses the terms *sound research design* (IV,C,1,a) and *scientifically valuable research*. When all of these terms are considered together, they support the position of NIH/OLAW that an IACUC should at least consider the scientific relevance of a proposal.[5] Indeed, NIH/OLAW[5] states:

The primary focus of the IRG is scientific merit, whereas the primary focus of the IACUC is animal welfare. It is evident, however, that there is some overlap of function between the two bodies.

Although not intended to conduct peer review of research proposals, the IACUC is expected to include consideration of the "U.S. Government Principles for the Utilization and Care of Vertebrate Animals in Testing, Research and Training [PHS Policy, p. 27] ..." in its proposal review process. Principle II calls for an evaluation of the relevance of a procedure to human or animal health, the advancement of knowledge, or the good of society. Other references (sections IV.C.1 and IV.D.1) include language such as "consistent with sound research design," "rationale for involving animals," and "in the conduct of scientifically valuable research,"which presumes that the IACUC will consider in its review the general scientific relevance of the proposal. The presumption is that a study that could not meet these basic tests would be inherently invalid or wasteful and, therefore, not justifiable.

The AWAR appear inconsistent in their reference to "scientific merit review." In the Public Comment Section of the regulations, APHIS/AC states, "we added the term 'animal care and use procedure'... to avoid any misunderstanding or implication that APHIS intends to become involved in the evaluation of the design, outlines, guidelines, and scientific merit of proposed research."[6] On the other hand, the AWAR (§2.31,e,4), as does the NIH/OLAW, make reference to "scientifically valuable research." Also, the AWAR (§2.31,a) and the AWA itself (Sect. 13,a,6,A,i–Sect. 13,a,6,A,ii) state that "except as specifically authorized by law or these regulations, nothing in this part shall be deemed to permit the Committee or IACUC to prescribe methods or set standards for the design, performance, or conduct of actual research or experimentation by a research facility."

It appears that NIH/OLAW and APHIS/AC do not want to commit themselves explicitly to a requirement of IACUC review of scientific merit. Nevertheless, according to NIH/OLAW,[7] an institution cannot defer scientific relevance review to the funding agency. IACUC approval, using criteria stated in the PHS Policy, must (in the original policy statement,[5] not under the just-in-time criteria noted later) precede NIH peer review. Therefore, approval by an IACUC of a proposed activity that is conditional upon successful peer review by the funding agency would not be in keeping with the PHS Policy requirements. Such conditional action does not constitute IACUC approval required by the PHS Policy prior to review by the NIH initial review group (IRG) or study section. The NIH review of merit should, therefore, be viewed as additional assurance rather than the only assurance that the research has value.[3,7] On this basis, Prentice and colleagues[3] concluded that the IACUC does have a responsibility to review the scientific relevance of animal projects that are subject to the requirements of the PHS Policy. There is not, however, general agreement with this conclusion. Black[8] argues that merit and relevance are not the same: that IACUC review does not constitute peer review, that the consequences of IACUC review are different from those of external review, and that IACUC review may constitute a violation of researchers' academic freedom. Prentice and coworkers[9] and Mann and Prentice[4] offer counterarguments. Clearly there is not general agreement. (See 12:17.)

The advent of just-in-time certification of IACUC approval of research proposals (see 8:22) seems to add another level to the arguments. It would seem that the NIH now allows the IACUC to refrain from reviewing scientific merit or relevance. On the other hand, referring to the just-in-time process, the NIH wrote, "The fundamental PHS Policy requirement that no award may be made without an approved Assurance and without verification of IACUC approval remains in effect. This change only affects the timing of the submission of the verification of that review." Given this, it seems that the PHS still expects the IACUC to do reviews for scientific merit or relevance.

9:15 Can the IACUC approve judicious use of animals without consideration of scientific merit?

Opin.   See 9:14 for a general discussion. Prentice and associates[3] note that there are two levels of review for scientific merit. They refer to a "fundamental level" of review in which scientists form "basic judgments about the adequacy and appropriateness of experimental design in terms of the ability to test the hypothesis, use controls, sample size, statistical analysis, and the training and experience of investigators." This first-level merit review is necessary for the IACUC to approve the judicious use of animals. The other level of review, the "knowledge-based level," requires "an assessment be made of the scientific importance of the study." An assessment of merit at the fundamental level could be made by any appropriately constituted IACUC, even in the absence of special expertise in the topic of the protocol. It would seem that a judgment of the merit at the knowledge-based level would be necessary to determine the appropriateness of the proposed animal use. Of course, the latter judgment requires that the IACUC have suitable expertise in a range of subject matters.

9:16 The AWAR (§2.31,a) state that the IACUC shall not prescribe methods or set standards for the design, performance, or conduct of actual research or experimentation by a research facility. What does this actually mean since the IACUC has the authority to approve or withhold approval of animal research activities?

Reg.   Actually, that section reads, "Except as specifically authorized by law or these regulations, nothing in this part shall be deemed to permit the Committee or IACUC to prescribe methods or set standards for the design, performance, or conduct of actual research or experimentation by a research facility."

Opin.   This apparent contradiction in the AWAR was discussed in detail by Prentice and colleagues (1992)[3] and Mann and Prentice (2004).[4] Consider an animal use protocol that did not contain suitable controls. One could argue that if such a protocol contained no animal welfare issues, it should be approved by the IACUC because the quoted text seems to say that scientific merit should not be considered. On the other hand, §2.31,e,3 says that a protocol should contain "a rationale for involving animals and for the appropriateness of species and numbers to be used." In this case, the numbers would be inappropriately low because of the missing controls, and the IACUC should not approve it.

      It seems that the phrase "Except as specifically authorized by law or these regulations" is critical. The IACUC should not "prescribe methods or set standards for the design, performance, or conduct of actual research or experimentation" except as specifically stated in the AWA itself or other relevant laws. (See 9:4; 9.15.)

9:17 Is it appropriate for the IACUC to use consultants to review scientific merit?

Reg.   (See 9:4.)

Opin.   If the IACUC assesses the scientific merit of a study (see 9:14; 9:15), then it is appropriate for the IACUC to use consultants, particularly if the IACUC does not have the requisite expertise to assess scientific merit. Also, outside reviewers would likely be most useful when there is disagreement among the IACUC members as to the scientific merit of a protocol or when an investigator can appeal the decision of the IACUC (see 9:56; 9:57; 29:31). Indeed, the IACUC could seek advice from consultants with regard to any aspect of the project that is problematic.

9:18 What types of consultants might be useful for review of scientific merit? Who picks them?

Opin. Consultants can be the following:

1. Experts on the topic of the protocol under review or
2. Experts in the use of animals, with respect to a particular procedure proposed in the protocol, or
3. Both

An example of the former is an expert who is asked to review a protocol because the IACUC is uncertain whether the use of the ascites method for making monoclonal antibodies is required in a particular project as the PI claims. Or, the consultant can be asked to judge whether any useful information will be forthcoming once the antibody is made. An example of the second use occurs when an expert in primate behavior and care is asked to review a protocol because the IACUC is uncertain whether a rhesus monkey may reasonably be restrained continuously for more than 24 hours. Or the expert can be asked whether any useful information could be derived from the experiments that require the monkey to be restrained for that period.

Consultants should be selected by the IACUC, but it may also be reasonable to allow the PI to suggest possible experts. This arrangement allows the committee to maintain objectivity in the review process while allowing the PI to feel that his or her point of view is being presented. Some institutions require the consultant to sign a confidentiality disclosure agreement (CDA) to protect the proprietary nature of the proposed study. The CDA will be useful in terms of protecting anonymity of the consultant and the confidential nature of the study.

9:19 Should the identity of consultants remain unknown to the investigator?

Opin. The same arguments apply to anonymity of consultants and the anonymity of grant proposal reviewers. It has been argued that reviewers refrain from making negative comments about a protocol if their identity is known to the investigator and may fear reprisals. It also is possible to argue that reviewers may make unfair statements about a protocol because their identity is withheld. Whether this will happen probably depends upon who the reviewer and investigators are. In any case, if the reviewers of a protocol remain anonymous, then certainly consultants should as well. On the other hand, anonymity may be difficult if the investigator is allowed to suggest possible experts.

9:20 An IACUC retains the services of a consultant to help evaluate a complex protocol. Partly on the basis of the consultant's recommendation, the protocol is approved. The IACUC subsequently learns that the consultant was a former student of the PI who submitted the protocol. Should the IACUC revisit the protocol's approval?

Opin. Clearly the PI should have informed the IACUC of this consultant's former relationship. But even if consultants are used in the review process, it is still the IACUC that must make the decision for approval. In this case, the IACUC must have determined that the protocol should be approved. The occurrence of the conflict of interest would not change the fact that the IACUC had found the protocol acceptable. Therefore, it would be unnecessary to revisit the protocol's approval. On the other hand, if the protocol had not been approved on the basis of the consultant's input, then the IACUC might well want to revisit the withholding of approval.

9:21 What is "designated member review" and what is its intent?

Reg. Both the PHS Policy (IV,C,2) and the AWAR (§2.31,d,2) recognize a method of "designated member review" that is widely implemented by research institutions.[6,10,11] Some institutions refer to this review process as *expedited review*; however, *designated member review* is the appropriate term. Designated member review must conform to the following process: Written descriptions of research projects that involve the care and use of animals must be made available to all IACUC members, and any member of the IACUC must have the opportunity to request full committee review of those research projects. "If full committee review is not requested, at least one member of the IACUC, designated by the chairperson and qualified to conduct the review, shall review those activities, and shall have the authority to approve, require modifications in (to secure approval), or request full committee review of any of those activities" (PHS Policy IV,C,2; AWAR §2.31,d,2).

Opin. Although the exact procedures frequently vary among institutions, the designated member review process can enable IACUCs to review and approve protocols faster than those presented for full committee review. It is important to mention, however, that designated member review in no way implies that the quality of review is less stringent than a protocol review by the full committee. A successful designated member review process allows institutions, particularly larger institutions that process a large number of protocols, the opportunity to reduce the workload of the IACUC at convened meetings, thereby allowing members to focus on protocols that may warrant more time and attention. (See 11:9.) It should also be noted that the designated member review process does not reduce the IACUC office staff workload, which may include follow-up reminders to the designated reviewer, correspondence with the PI, and other tasks.

There is no regulation that limits the number of protocols that may be reviewed by the designated review process. It is expected that an IACUC will meet in a convened meeting at least twice per year to approve the semiannual program review and inspections. However, there is no required number of meetings. Clearly, progressive IACUCs meet on a regular basis to conduct whatever business is appropriate to the program. This is most often spelled out in the PHS assurance.

In the designated review process, the members of the IACUC do not vote. A member who is not assigned as a designated reviewer may only refer the protocol to the full committee for review. If a full committee review is not called for, the designated reviewer has the authority to approve a protocol, require modifications to secure approval, or refer it to the full committee. If there is more than one designated reviewer, they must agree on a course of action or refer the protocol to the full committee.

9:22 Under what circumstances might a protocol be assigned to a designated member review process?

Reg. A protocol may be assigned to one or more designated reviewers only after all IACUC members have been provided the opportunity to call for full-IACUC review (PHS Policy IV,C,2; AWAR §2.31,d,2). (See 9:21.)

Opin. Because the AWAR and the PHS Policy do not define the criteria for assigning protocols to the designated member review process, each institution must develop an
internal policy, tailored to the individual institution's needs and idiosyncrasies. Even
if an institution develops criteria to determine the kinds of activities or protocols that
it will permit to use this review method, all IACUC members must still be provided
with the opportunity to call for full-IACUC review of each individual project.

An institution should first decide whether it needs to establish a designated
member review process. Then, the IACUC should create a list of well-defined criteria to determine the types of protocols qualifying for designated member versus
full committee review. For example, some institutions restrict designated member
review to protocols involving noninvasive or acute procedures that do not cause
more than momentary or slight pain or distress to the animals (e.g., the procedures
only involve euthanizing the animals in order to harvest tissues) or to protocol
amendments that involve minor changes to an approved study.

If an institution elects to adopt a designated member review process, the IACUC
should document the assignment criteria and the procedures for processing the
protocols. The IACUC should develop a set of SOPs in order to ensure that the designated member review process is not misused. The SOPs should be included in
the institution's Animal Welfare Assurance, approved by NIH/OLAW, and
included in the institution's IACUC Policy and Procedures Manual.

9:23 Under what circumstances is the designated member review not appropriate?

Opin. Although each institution must determine the criteria appropriate for its own program, designated member review may be unacceptable under many circumstances, including the following:

- IACUC members are not provided with sufficient information concerning
  the proposed research activities.
- IACUC members are not given an opportunity to call for full committee
  review prior to the approval of the proposed protocol via a designated
  member review process (PHS Policy IV,C,2; AWAR §2.31,d,2; NIH/OLAW
  "Dear Colleague" letter, May 21, 1990).[12]
- Any IACUC member requests full committee review (PHS Policy IV,C,2;
  AWAR §2.31,d,2).
- One or more members of the subcommittee authorized to conduct designated member review and to approve research activities have a conflict of
  interest (PHS Policy IV,C,2; AWAR §2.31,d,2).
- The proposed protocol does not meet the requirements for designated mem-
  ber review delineated in the institution's NIH/OLAW-approved Animal
  Welfare Assurance or the institution's IACUC Policy and Procedures Manual.
  For example, the invasive nature of the research may not permit the protocol
  to be reviewed by the designated member review method according to the
  institution's policies and its Animal Welfare Assurance.

9:24 Should the entire IACUC be involved in the designated member review process, or can a subcommittee conduct designated member review?

Reg. (See 9:21; 9:22.)

Opin.   The entire committee must have the opportunity to review the information pro-
           vided on each designated member protocol review list and to request full commit-
           tee review. A subcommittee of two or more members who are qualified to review
           the protocol (e.g., the Chair and the AV) can be assigned to review the protocol in
           its entirety. Questions and concerns raised by any member of the subcommittee
           should always be addressed prior to approval.

9:25   What is required for approval of a protocol via the designated member review?

Reg.   The designated reviewer(s) must use the same criteria that are applicable to proto-
           cols undergoing full-IACUC review.
Opin.   Although the intent of the designated member review process is to conduct rapid
           review of protocols, by no means does this process require less information than
           the full committee review process. The designated reviewer(s) must ensure that the
           review method and criteria for approval are in full compliance with all of the appli-
           cable requirements of the PHS Policy (IV,C,1,a–IV,C,1,g; IV,D,1,a–IV,D,1,e) and the
           AWAR (§2.31,d,1,i–§2.31,d,1,xi; §2.31,e,1–§2.31,e,5). Before approval may be
           granted via a designated member review process each member of the IACUC must
           have the opportunity to review the protocol and the opportunity to request full
           committee review. If a subcommittee is being utilized for this review process, all
           designated reviewers *must* agree on the decision; it is not acceptable to use a major-
           ity vote.[11] If one of the reviewers disagrees, the protocol should be referred to the
           full committee for review.

9:26   What is an effective means for administering a designated member review?

Opin.   One significant difference between designated member review and full committee
           review is that the designated member review process does not require a fully con-
           vened meeting of the IACUC. Despite this fact, after a protocol is assigned to des-
           ignated review, at least one or two IACUC members should review the entire
           protocol with the same degree of thoroughness that is given to a protocol reviewed
           by the full committee.
                  Another difference between the two processes involves the way in which proto-
           col information can be disseminated to the IACUC members. Unlike the full com-
           mittee protocols, the information pertinent to protocols presented for designated
           review can be more easily disseminated through the use of electronic means (e.g.,
           fax or e-mail) in order to facilitate the most expeditious type of review. (See 6:17.)
                  The following is an example of a designated member review process. This par-
           ticular example goes beyond the requirements of the AWAR and PHS Policy in that
           it requires that a summary be sent to all members and each designated reviewer's
           decision with regard to designated member review be documented:

           • The IACUC administrator reviews the protocol submitted to the committee
             and decides whether it qualifies for designated member review on the basis
             of the IACUC's predetermined criteria.

           • A summary of the protocol qualifying for designated member review (includ-
             ing title of project; species, number of animals requested; type of experimental
             procedures) is forwarded to members of the IACUC via postal mail, fax, or
             e-mail (see 6:17). In some instances, additional information may be included
             (e.g., supporting grant materials or parts of the original protocol).

- Each IACUC member reviews the summary and, if necessary, requests a copy of the complete protocol to determine whether clarification or changes are needed and has the opportunity to call for full committee review.

- If no member calls for full committee review in a reasonable specified period, a subcommittee, composed of the IACUC Chair and the AV, is authorized to review and approve the protocol. In a case in which the Chair or the AV has a conflict of interest (e.g., is personally involved in the project), he or she must be replaced by another member qualified to conduct the review (PHS Policy VI,C,2; AWAR §2.31,d,2). In the authors' experience, a week is sufficient time for the designated reviewer process to occur. The protocol can be reviewed and either approved or referred to the full committee.

- Documentation, including review sheets, is maintained as evidence of each designated reviewer's decision with regard to these protocols. The designated reviewer sheets are kept in the protocol file in the event the designated member review of a protocol is ever questioned. *No matter what process an institution adopts, the designated member review procedure must be documented from assignment to approval each time it is used.*

9:27   An IACUC Chair appoints designated member reviewers. Can an IACUC member who was not chosen as a designated reviewer demand to be made a designated reviewer?

Reg.   Both AWAR §2.31,d,2 and PHS IV,C,2 specify that one member of the IACUC is designated *by the Chair* and qualified to conduct the review.

Opin.   There is no requirement for the Chair to honor such a request. Although such a request may be unusual, an IACUC Chair would be wise to find out why an IACUC member has made this demand and perhaps refer the protocol to full committee review.

9:28   What is "full committee" review?

Reg.   Full committee review of an IACUC protocol is one that is conducted by a quorum of the IACUC at a regularly scheduled or specially convened meeting (PHS Policy IV,C,2; AWAR §2.31,d,2).

Opin.   Because of the large volume of protocols reviewed at many institutions, a good portion of the actual work involved in protocol review may be done before the meeting via a mechanism such as prereview (see 9:5–9:13). Thus, at the meeting, the committee members are able to concentrate on the proposed protocol, its "scientific merit," and the care and use of laboratory animals without the necessity of obtaining clarifications related to incomplete information and/or confusing points.

9:29   What is the purpose of a full committee review?

Opin.   The purpose of full committee review is to have all IACUC members involved in reviewing and making decisions regarding the disposition of protocols during an interactive meeting. This in turn allows the IACUC to utilize the expertise of its members in a discussion-based format, thus facilitating the resolution of protocol-related issues.

9:30   When is full committee review of a protocol appropriate?

Reg.   Full committee review is appropriate at any time and required when requested by any member of the IACUC (PHS Policy IV,C,2; AWAR §2.31,d,2). Each member must have the opportunity to review any protocol before approval may be granted.[6,10,11] There are no other federal requirements specifying when a protocol should receive full committee review or the type of protocol that should receive full committee review.

Opin.   Many IACUCs have developed criteria to determine the types of protocols requiring full committee or designated member review. Items to be considered in developing such criteria include the following:

- Invasiveness of procedures
- Level of pain or distress
- Species and number of animals requested
- Experimental design
- The nature of the animal use, e.g., research project versus educational use
- "Controversial" procedures (e.g., death as an end point)
- Procedures that request exceptions to regulations (e.g., multiple survival surgeries), that require appropriate justification, and for which the IACUC may want to review only at the time of a fully convened meeting
- Whether the protocol will receive peer review by the funding agency or other group prior to funding

9:31   What is an effective full committee review process?

Opin.   Each IACUC should establish SOPs regarding conduct of full committee review of protocols. One effective method of full committee review includes the following process:

- Preview (see 9:5–9:13)
- Review by IACUC staff or of the committee for completeness of the application and compliance with information requirements, as indicated in PHS Policy (IV,C,1,a–IV,C,1,g; IV,D,1,a–IV,D,1,e; AWAR §2.31,d,1,i–§2.31,d,1,xi; §2.31,e,1–§2.31,e,5)
- Review by the AV in addition to the consultation required during design of the study (AWAR §2.31,d,1,iv,B)
- Review by outside consultants if the IACUC members do not possess relevant expertise
- Review by committee members assigned as primary and secondary reviewers
- Primary and secondary reviewers communicate with the investigator prior to the meeting, during which the reviewers may recommend modifications in the protocol (PHS Policy IV,C,2; AWAR §2.31,d,2)
- Presentation of the protocol to either a protocol review subcommittee or the full committee by the primary and secondary reviewers followed by a general discussion of the protocol
- If reviewed by a subcommittee, it is recommended that a synopsis of that group's deliberations be presented to the full committee to allow for any additional comments

- Every member of the committee should be provided with a copy of the protocol, or, at a minimum, a summary of the protocol to be reviewed at the meeting
- After discussion at a convened quorum of the committee, a vote is taken to determine final disposition of the protocol

9:32   What are the PHS Policy and AWAR requirements for approval of a protocol by full committee review?

Reg.   The PHS Policy and AWAR require the following:

- Each member of the IACUC must be provided with, at the minimum, a list of protocols to be reviewed (PHS Policy IV,C,2; AWAR §2.31,d,2).
- Both the PHS Policy (IV,C,1,a–IV,C,1,g; IV,D,1,a–IV,D,1,e) and the AWAR (§2.31,d,1,i–§2.31,d,1,xi; §2.31,e,1–§2.31,e,5) have specific requirements about information that must be included in the protocol as well as the items the IACUC must consider. These are described later.
- No member of the IACUC may participate in the review or approval of a research project in which the member has a conflict of interest (e.g., is personally involved in the project) except to provide information requested by the IACUC (PHS Policy IV,C,2; AWAR §2.31,d,2). Some IACUCs require this member to leave the room during final discussion and voting on the protocol in question.
- Approval of a protocol considered by the full committee "may be granted only after review at a convened meeting of a quorum of the IACUC and with the approval vote of a majority of the quorum present" (PHS Policy IV,C,2). The AWAR (§2.31,d,2) have the same requirement.
- A member who has a conflict of interest in a protocol under review may not contribute to the constitution of a quorum (PHS Policy IV,C,2; AWAR §2.31,d,2). This person may not vote on the protocol in question.
- Institutions must maintain written documentation of committee deliberations (PHS Policy IV,E,1; IV,E,2; AWAR §2.35,a,1; §2.35,a,2). Whereas most IACUCs include this information in the minutes, such documentation can be maintained instead on file with the protocol. Filing records of deliberations along with the protocol is of importance to institutions subject to state open records laws. It is the responsibility of the IO, IACUC Chair, or IACUC staff to explain to either APHIS/AC, NIH/OLAW, or AAALAC the rationale for the institution's record keeping methods. This should be documented in the PHS Assurance and in the institutional Policy and Procedures Manual.
- Both the investigator and the institution must be notified in writing of the committee's decision (PHS Policy IV,C,4; AWAR §2.31,d,4).

9:33   What constitutes "administrative review," and when may it be used?

Reg.   The PHS Policy (IV,C,1) states, "In order to approve proposed research projects or posed significant changes in ongoing research projects, the IACUC shall conduct a review of those components related to the care and use of animals and determine that the proposed research projects are in accordance with this Policy."

It goes on to say that such review must use either the designated reviewer or full committee review method. But, the policy does not say what should be done if there are nonsignificant changes. The AWAR (§2.31,c,6–§2.31,c,7) contain essentially the same requirements. Neither the PHS Policy nor the AWAR use the term *administrative review* or discuss anything to which the term could reasonably be applied. In the June 6, 2003, NIH Dear Colleague Letter,[13] it is stated that "IACUCs may, by institutional policy, classify certain proposed additions or changes in personnel, other than Principal Investigator, as 'minor' provided an appropriate administrative review mechanism is in place to ensure that all such personnel are appropriately identified, adequately trained and qualified, enrolled in applicable occupational health and safety programs, and meet other criteria as required by the IACUC."

Opin.   According to the PHS, administrative review is possible in some cases provided the process and its application are documented. It is essential that the IACUC establish an SOP for the application of administrative review, and that this be documented in both the institutional Assurance and the Policy and Procedures Manual.

When can it be applied? Clearly, it is inappropriate for initial review of all animal projects. It is also inappropriate for all continuing review.[14] Such administrative review may be applied to requests for change that are not significant. This would include changes in personnel other than the PI, as indicated previously. NIH/OLAW has no written endorsement of these, but it seems that replacement of a small number of animals due to a technical, vivarium, or vendor-related problem; a slight increase in the amount of blood to be drawn; correction of the dosage of pain-relieving drug; or a change in a procedure that occurs after terminal anesthesia but before euthanasia could all be approved by administrative review. (See 10:3.) NIH/OLAW has indicated that changes in the objectives of a study; proposals to switch from nonsurvival to survival surgery; changes in the degree of invasiveness of a procedure or discomfort to an animal; changes in species or in approximate number of animals used; changes in anesthetic agents, use or withholding of analgesics or methods of euthanasia; and changes in the duration, frequency, or number of procedures performed on an animal would not qualify for administrative review.[15] Wolff and associates,[15] speaking for NIH/OLAW, have clearly stated that administrative review may not be used to grant continuation of a project the approval for which has expired because the investigator failed to seek continuing review in a timely fashion.

Administrative review, when it is appropriate, may be performed by the IACUC Chair or administrator, by the AV, or by an IACUC member designated by the IACUC Chair.

9:34 An IACUC protocol was properly approved. Can an investigator begin research with animals on the basis of an oral approval from the IACUC office, or must there be written documentation of the approval?

Opin.   Once approval of the protocol has been documented in the IACUC records, which can mean written in the IACUC minutes by the recording secretary, recorded on a summary check sheet, or documented through some other process approved by the IACUC, oral notification can be given to the investigator and the research may begin although an official approval letter has not been issued to the investigator. However, the official approval letter, with the same date as the oral approval, is required for the establishment of a paper trail.

9:35 Is it important for the IACUC to know whether a protocol is new or a resubmission with a change in title?

Opin.   If a protocol is submitted and disapproved or tabled (approval withheld), it should be treated as a new protocol upon resubmission regardless of whether the title or funding source is changed or not. It should be handled as a new protocol, reviewed by either designated member review or full committee review as specified in the PHS Assurance or institutional Policy and Procedures Manual. If a protocol has been previously reviewed and approved, then the title or funding source may be changed without rereview if it is the policy of the institution to handle changes in title administratively.

9:36 What procedures can the IACUC use to determine whether a protocol is new or a resubmission with a change in title?

Opin.   If a protocol is determined to be a resubmission with a change in the title only, the IACUC may decide to verify this by comparing the resubmission with the previously approved protocol on file. Investigators sometimes inadvertently modify protocols during the course of resubmission without recognizing that the IACUC must approve all proposed significant changes (PHS Policy IV,B,7; AWAR §2.31,c,7). It is perhaps a better procedure to make changing the title of a protocol a separate process with its own form or treat it as just one of several types of amendments to a protocol. That way the changes in the protocol would be obvious.

      For IACUCs that use electronic submissions of amendments, it is possible to detect changes in the protocol electronically through use of software with this dedicated purpose.

9:37 Should IACUC decisions be influenced by the source or size of a research grant?

Reg.   The IACUC must review grants under the requirements of PHS Policy IV,C,1, regardless of the size of the grant. Non-PHS-supported research may or may not be covered by the PHS Policy, depending on the statement of applicability in the institution's Animal Welfare Assurance.

Opin.   Most IACUC members would probably reflexively answer "No" to this question. In general, all protocols should be reviewed with the same rigor regardless of how much money is involved or the source of funding. Whereas political pressures for IACUC approval of large or prestigious grants are real, studies should be approved on the basis of appropriateness of the proposed research. Certainly, animal welfare should never be compromised in the interest of increasing grant funds. Nevertheless, in one study,[16] results of a survey indicated that IACUC deliberations potentially can be influenced by the size of a grant as well as by pressure and perceptions that the committee may not even recognize. IACUCs must be constantly vigilant for inconsistencies in their deliberations and decisions, especially those affected by such pressures or perceptions.

9:38 Should the decisions of the IACUC be influenced by the potential scientific importance of a project proposed in a protocol?

Reg.   The IACUC has the authority to approve a proposed project. NIH/OLAW and APHIS/AC have stated that under no circumstances is an IACUC required to approve a project against its will.[17] (See 16:9.)

Opin.   IACUC decisions should not be influenced by the investigator, species, funding source, or "hotness" of the research topic. Review should be based on animal welfare issues in consideration of the regulatory requirements. Nevertheless, both the dollar value of the grant and a species-specific view of an animal's societal worth potentially could but should not affect IACUC deliberations.[15]

9:39   If a protocol is completely novel and contains untested surgical or experimental procedures, how could such a protocol be reviewed and approved when the procedures cannot be referenced?

Opin.   Few protocols actually contain untested surgical procedures. Those that do can be handled in one of two ways. In some cases, the IACUC could recommend that a pilot study involving only the part of the protocol using the untested procedures be conducted. With these pilot data derived from successful use of the previously untested procedures, the investigator could then obtain IACUC approval for the complete protocol. Pilot studies must be reviewed and approved by the IACUC.

      Alternatively, the IACUC could approve the protocol as submitted but with a reduced number of animals, the remainder to be approved when the investigator has tested the new procedure. Both of these actions allow the researcher to proceed with the study while the animal subjects are being protected. (See 13:8.)

9:40   What are some possible actions an IACUC can take with respect to protocols that have been reviewed?

Reg.   Both the PHS Policy (IV,B,6) and the AWAR (§2.31,c,6) allow the IACUC to review and approve, require modifications in (to secure approval), or withhold approval of proposed activities related to the care and use of animals.

Opin.   In determining the disposition of a protocol submitted for review by the full committee review method, an IACUC has several options, including the following:

- Approval
- Requirement of modifications to secure approval (see 9:42)
- Tabling of the protocol with a request for major revisions
- Withholding of approval
- When the designated member review method is used, the only options open are the following:
  - Approval
  - Referral to full committee for review
  - Requirement of modifications to secure approval

9:41   What constitutes "approval" of a protocol? Does this action necessarily mean that no further change or information is needed?

Reg.   For PHS Policy and the AWAR, IACUCs either approve, require modifications in (to secure approval), or withhold approval of protocols (AWAR §2.31,c,6; PHS Policy IV,B,7). Designated reviewers may approve, require modifications to secure approval, or request full committee review. Anything short of final approval is not adequate for initiation of animal activities or submission of an IACUC approval date to the NIH as part of a grant application.

Opin.   Generally, an approved protocol is one that contains all the required information and has been judged by the IACUC to be acceptable.

9:42   What constitutes "approval with conditions"?

Reg.   (See 9:41.)

Opin.   The IACUC may require modifications of a protocol in order for a PI to secure approval when it is determined that no major revisions or clarifications are required. Often, the Chair or the AV is assigned to review the PI's response or revised protocol and is empowered by the IACUC to approve the protocol without further review by the full committee. However, depending upon the conditions imposed, the entire committee may wish to see the response to conditions or a subcommittee may be formed and empowered to review the response and approve the protocol. Conditional approval does not, however, mean that the study can be initiated. The PI must first comply fully with all conditions arising from the IACUC's review, and then final approval can be granted. The reader is cautioned that terms such as *conditional approval*, *provisional approval*, or *approved pending clarification* frequently cause confusion. NIH/OLAW and APHIS/AC prefer that IACUCs either avoid these terms or describe them in sufficient detail to be fully understood. (See 9:47.)

    IACUCs might determine that a protocol is approvable, contingent on receipt of a specific modification (e.g., receipt of assurance that the PI will conduct the procedure in a fume hood). The IACUC can handle this modification (or clarification) as an administrative detail that is documented in the protocol record. On the other hand, protocols that are missing substantive information necessary for the IACUC to make a judgment (e.g., justification for withholding analgesics in a painful procedure) are incomplete. If the protocol is incomplete, it is not possible to satisfy the protocol review criteria. NIH/OLAW and APHIS/AC recommend that IACUCs devise effective ways of differentiating between substantive omissions and administrative issues.[18]

9:43   What constitutes a *tabled* or *postponed* protocol?

Opin.   The terms *tabling* and *postponing review* of a protocol are often used interchangeably by IACUCs but are not used as outlined in the more formal structure of Robert's Rules of Order. (See 6:4.) The IACUC may "table" or "postpone" further review of a protocol pending receipt of additional substantive information or a significant revision of the protocol. The action of tabling a protocol is generally used during the process of full committee review when the IACUC decides it is necessary for the whole committee to rereview the protocol before further action can be taken. Tabling a protocol is usually reserved for proposals that do not contain sufficient information or those in which the IACUC has identified a serious animal welfare concern.

9:44   What might cause a protocol to be disapproved (approval withheld)?

Reg.   Approval may be withheld if any of the PHS Policy (IV,C,1) or AWAR (§2.31,d,1; §2.31,d,2,e) criteria is not met.

Opin.   Withholding of approval of protocols is uncommon. Generally, approval is withheld when the PI and the IACUC cannot agree on fundamental aspects of the proposed study such as the protocol design or animal welfare issues, or the PI will not agree to comply with the IACUC's requirements.

Both the PI and the IO must be notified in writing of the committee's decision. This written notification should include a statement of the reasons for the decision and provide the PI with an opportunity to respond in person or in writing (PHS Policy IV,C,4; AWAR §2.31,d,4). The IACUC may want to seek outside review in the case of an appeal by the PI of withholding of approval for a protocol but perhaps only if it feels that it lacks expertise to make a decision.

9:45 How should an investigator respond to questions or conditions from the IACUC?

Opin.   Responses from the PI to questions or concerns raised by the IACUC should be in such a format as to result in a protocol that contains a complete, easy to discern description of the proposed activities. The response from the PI should clearly answer each issue raised. This can be achieved via a point-by-point letter that serves as a means for the PI to respond to the committee's questions and concerns accompanied by a revised protocol that incorporates all required changes and clarifications. Requiring the PI to provide a revised protocol has the advantage of ensuring that a complete and up-to-date master protocol is embodied in one document. Copies of the final, approved IACUC protocol should be maintained in the IACUC administrative office, in the animal facility, and in the investigator's laboratory. This master protocol also has the advantage of facilitating the management of the protocol by the animal care staff and for compliance review purposes.

9:46 Should there be a time limit to receive responses to queries raised by the IACUC?

Opin.   Establishment of time limits and other constraints should be considered when the IACUC develops an SOP. Specific time constraints and deadlines can be of benefit to both the investigator and the committee when the expectations are known. Ironclad deadlines and inflexible staff, however, can present a major point of contention for faculty already juggling multiple projects and attempting to comply with much paperwork and many deadlines.

9:47 Once a response from an investigator is received, can the IACUC Chair or his or her designee approve protocols if conditions are met?

Opin.   Generally, this should be determined when the decision is made to grant "conditional" approval. That is, approval is granted on condition that the PI meet certain requirements, and the committee agrees that the assigned reviewers or the Chair can determine whether the conditions have been satisfactorily met and subsequently grant approval (PHS Policy IV,C,2; AWAR §2.31,d,2). It should be noted that "conditional approval" is not mentioned in either the AWAR or PHS Policy. Technically, the IACUC should initially "withhold approval" and then grant approval via the designated member review process. A distinction should be made between conditions based on administrative details and information that has a bearing on the IACUC's decision. (See 9:41; 9:42.)

9:48 Is a majority vote required for any IACUC actions related to protocol approval?

Reg.   Approval of a protocol considered under guidelines for full committee review "may be granted only after review at a convened meeting of a quorum of the

IACUC and with the approval vote of a majority of the quorum present" (PHS Policy IV,C,2). The AWAR (§2.31,d,2) has the same requirement. (See 9:21 for designated member review considerations.)

Opin. These regulations represent the minimal requirements, and an institution may implement more stringent requirements. These should be clearly stated in the PHS Assurance and in the institution's Policy and Procedures Manual.

9:49   At a full committee meeting of a 20-member IACUC, 15 members are present. Six vote to approve a protocol, six abstain, and three vote against approval. Is the protocol considered approved according to the AWAR and PHS Policy?

Opin. No. In this example, the 15 members present constitute the quorum. In order to approve a protocol, a simple majority of those members present, that is, eight, would have to vote for approval of the protocol. Because only six members voted to approve, the vote count fell short by two votes, and the protocol cannot be approved according to the AWAR and PHS Policy.[19] (See 9:48.)

9:50   How should the IACUC handle minority opinions to IACUC actions on the review of protocols?

Reg. PHS Policy (IV,E,1,d) requires institutions to maintain copies of minority views of semiannual reports (not protocol reviews) but does not specify how the IACUC should handle them. There is also a PHS Policy (IV,F,4) requirement that minority views filed by IACUC members be forwarded (via the IO) to NIH/OLAW along with the annual report to NIH/OLAW (but again, this does not refer to protocol reviews). If the minority views relate to an IACUC action that is required to be reported promptly (PHS Policy IV,G,3), they should be provided to NIH/OLAW at that time. The AWAR (§2.31,c,3; §2.35,a,3) also require maintaining records of semi-annual reports, including minority views.

Opin. Because neither the PHS Policy nor the AWAR address this issue in relation to specific IACUC protocols, it should be considered when the IACUC develops an SOP. A common practice is to record dissenting opinions in the minutes. However, another option is to allow the dissenter to write a letter expressing her or his opinion. This is then filed along with the IACUC protocol.

9:51   A protocol has been approved with changes requested by the IACUC. The protocol is associated with a grant application to the NIH. Does the approval letter sent by the research administration office to the NIH have to detail the changes made in the animal use portions of the grant?

Reg. The original PHS Policy (IV,D,2) requires "verification of approval (including the date of the most recent approval) by the IACUC of those components related to the care and use of animals. . . . If verification of IACUC approval is submitted subsequent to the submission of the application or proposal, the verification shall state the modifications, if any, required by the IACUC." For competing applications or proposals, verification of IACUC approval may be filed at a time not to exceed 60 days after the proposal application deadline (PHS Policy IV,D,2).

Opin. In 2002, this policy was amended to allow verification of IACUC approval to be submitted upon request from the PHS (just-in-time verification). (See 8:22; 8:24.)

The PHS issues the request when it determines that a grant proposal is likely to be funded. Institutions are not required to use just-in-time and may still follow the original requirements as stated earlier.

Whereas the original 60-day grace period and the new just-in-time verification may reduce workloads of both the PI and the IACUC, it may create a situation in which the IACUC requires changes to a protocol that would result in a change in the grant proposal after the proposal has been approved.[20,21] As pointed out by Mann and Prentice,[4,22] it is not clear what the granting agency will do then.

9:52   Should the IACUC review only the IACUC protocol, or should the IACUC also review the animal care and use sections of an associated grant proposal?

Reg.   Verification of IACUC approval submitted to the NIH means that the IACUC has reviewed and approved those components of the grant application related to the care and use of animals: "Applications or proposals (competing and non-completing) covered by this Policy from institutions which have an approved Assurance on file with OLAW shall include verification of approval . . . by the IACUC of those components related to the care and use of animals" (PHS Policy IV,D,2). The signature of the institutional representative on the PHS 398 form is a legally binding statement that the IACUC has approved all animal activities covered in the grant application. The submission of just-in-time verification also constitutes a binding statement.

Opin.   If the IACUC protocol accurately reflects the information contained in the grant application, then the verification is obviously valid. The IACUC, however, cannot be assured of such validity unless it also reviews the animal care and use sections of the associated grant application.[22] With the advent of the just-in-time review and approval in 2002 (see 8:22), review of the grant proposal need not take place until it appears that the proposal will be funded. This delay may have as yet unknown consequences for funding.[4,22]

Many institutions do not review the grant proposals at all. They simply rely upon a statement by the PI that the protocol accurately reflects what is in the grant proposal. However, if the PI's certification is inaccurate, the resulting institutional verification to the PHS will be invalid.[22]

9:53   Should the IACUC accept an approval statement from an IACUC at another institution?

Reg.   An IACUC may accept the approval of an IACUC at another institution if that institution has an approved NIH/OLAW Animal Welfare Assurance. This practice is normally limited to collaborations or subgrant/subcontracts involving performance sites outside the awardee institution. In most instances, the IACUC of the performance site assumes responsibility for animal activities in its facilities. Both institutions should have a clear understanding of what their respective responsibilities are in this situation, particularly if one IACUC agrees to abide by the determinations of another IACUC.[23] (See 8:10.)

Opin.   This is an area for which the IACUC should establish an SOP. The IACUC should have some "comfort level" about the other institution, its committee, and the quality of its research. To prevent problems, a collaborative protocol approved by an IACUC at another institution should probably receive at least the equivalent of designated member review by the local IACUC. In some instances, full committee review is warranted. Issues to consider in accepting an approval statement from an IACUC at another institution include the following:

- Does the institution have an approved NIH/OLAW Animal Welfare Assurance?
- Is the institution a USDA-registered research facility?
- Is the institution accredited by AAALAC?

For example, this situation may arise when an institution's faculty conducts research at a nearby Veterans Administration (VA) facility. Because the VA facility may not apply for or receive funding from other federal agencies, the grants must be made to the investigator's home institution. The problem is created because VA IACUC approval must be granted before research is initiated at the VA facility. The home institution has jurisdiction by virtue of receiving the grant. Thus, either redundant application or this kind of "external" approval is necessary.

Acceptance of an external IACUC review and approval also includes the responsibility for continuing review. The external institution should agree to supply all continuing review documents relative to the protocol as they are reviewed and approved. Failure to do this can result in withholding of a notice of approval to the granting agency and possible loss of grant funds. In some cases the home institution's sponsored programs office may require an assurance of continuing IACUC approval for the grant funds to be released. This can be another source of problems for the PI's research.

9:54 Can an investigator receive "conditional approval" for a protocol approved at another institution until the IACUC can review and approve the protocol? What should be the conditions and limitations for this conditional approval?

Reg.   (See 9:41.)

Opin.   An IACUC may choose to grant conditional approval for a protocol approved by a different institution. However, any conditions regarding acceptance of an approval statement from an IACUC at another institution should be described in the IACUC's written policies. For example, in order to facilitate research, an IACUC may choose to allow animals to be purchased or transferred to its institution, but procedures that involve animals cannot be performed until the local IACUC has unconditionally approved the protocol. (See 9:42.) Because local institutional requirements vary, automatic approval is not advisable.

9:55 An investigator subcontracts part of a research project to another institution where animals will be used. What oversight and paperwork responsibilities does the primary institution have relative to the PHS Policy and the AWAR?

Reg.   NIH/OLAW requires that any subcontracted work involving animals that is supported by PHS funds be conducted only at other NIH/OLAW-Assured institutions (PHS Policy V,B). If the performance site (collaborating institution or subcontractor) is not NIH/OLAW-Assured, then NIH/OLAW requires that an assurance for the performance site be negotiated in order for the work to go forward (PHS Policy V,B). Alternatively, an institution can choose to accept responsibility for the work under its own Assurance with a written agreement between the institutions. It should, however, be recognized that the latter arrangement means that the NIH/OLAW-Assured institution assumes full responsibility for ensuring that all the animal work conducted at the collaborating institution is in full compliance with the PHS Policy and the AWAR. This arrangement necessitates that the NIH/OLAW-Assured institution's IACUC conduct semiannual program reviews and facility inspections of the collaborating facility. (See 8:10.)

Opin.   When an institution subcontracts part of a research project to another institution, there should be a clearly defined written agreement concerning the responsibilities of each institution to comply with the PHS Policy and AWAR regardless of whether both institutions have approved Assurances on file with NIH/OLAW. If a serious compliance problem arises at a subcontract site, it will likely impact the primary institution. However, if the subcontract site is registered with APHIS/AC and is the facility whose IACUC approved the protocol, that site would be held primarily responsible. Written agreements can help minimize and resolve potential problems.

9:56   Can an investigator appeal the decision of the IACUC about a protocol review decision?

Reg.   The PHS Policy (IV,C,4) and the AWAR (§2.31,d,4) require that written notification of withheld approval include a statement of the reasons for the decision, "to give the investigator an opportunity to respond in person or in writing." Nevertheless, IACUC decisions to withhold approval may not be overturned by a higher authority (PHS Policy IV,C,8; AWAR §2.31,d,8).

Opin.   The preceding regulatory information suggests that a reconsideration by the IACUC should be an option. Furthermore, the "Supplementary Information" that accompanies the August 31, 1989 *Federal Register*, containing 9 CFR Parts 1, 2, and 3 Final Rules (p. 36132), states that "on the basis of the response, the committee may reconsider its decision." This is another area for which the IACUC is advised to establish a written SOP. (See 29:30; 29:31.)

9:57   What is an appropriate and effective mechanism for appeal of IACUC decisions? (See 9:56; 29:30; 29:31.)

Opin.   Suggestions to consider in developing an SOP for appeals include the following:

- Assignment of IACUC members to work with the investigator to facilitate the process
- Assignment of an IACUC member to be a "hearing" officer
- Involvement of the dean for research or institutional equivalent as a "hearing" officer
- Participation of the IO as a "hearing" officer

In any case, the SOP must include a reconsideration of the protocol by the IACUC in light of the findings of the "hearing" officer. According to both the AWAR and PHS Policy an institutional official may not overturn a decision of the IACUC. (See 9:59.) Therefore, the IACUC must alter its own decision, as it can (according to the AWAR) by a vote of a majority of a quorum at a convened meeting or via the designated member review process, although the latter method may not do justice to an appeal that really should be considered by the full committee.

9:58 Can a protocol for which approval has been withheld be resubmitted with modifications?

Reg. Yes, neither PHS Policy nor the AWAR preclude resubmission.

Opin. Resubmission of a modified protocol should be encouraged. A resubmission provides the investigator an opportunity to respond to the IACUC's review as allowed by PHS Policy IV,C,4 and the AWAR §2.31,d,4. The IACUC also may consider assigning a member to work with the investigator to develop a protocol that can be approved by the committee and allow the investigator's research program to progress. (See 9:56.)

9:59 Is there any institutional authority that can reverse the decision of the IACUC?

Reg. No. The AWAR (§2.31,d,2) state that "officials of the research facility . . . may not approve an activity involving the care and use of animals if it has not been approved by the IACUC." The PHS Policy (IV,C,8) is nearly identical. (See 9:56; 29:30–29:31.)

9:60 At times, PIs who do drug testing may receive a test drug, but the manufacturer does not want to reveal what the drug is. Can the IACUC approve the use of the drug in animals?

Opin. It is not uncommon that a pharmaceutical company contracts testing of a new drug to an investigator in a research institution. Because such drug development is highly competitive, the company may not want to divulge the exact identity of the drug. This condition poses a problem for the IACUC charged with evaluating the effect of the drug on the welfare of animals. In order to approve the use of the drug, the IACUC must know at least the general class of the drug as well as its dose and route of administration and any known risks to the animals. With this information it is possible for the IACUC to approve use of the drug. However, a binding confidentiality agreement signed by all IACUC members may obviate this problem. (See 9:61.)

9:61 Sometimes PIs do not know which of a class of drugs they will use, either because there are many or because they do not know which will be available. Can the IACUC approve the use of a class of drugs in animals?

Opin. Sometimes an investigator wants to use a particular class of drug in a study, but he or she is unsure whether or when a given drug will be used. The IACUC can approve the use of a general class of drugs if a list of drugs that might be used along with dose and route of administration for each one is provided. It would also be good to know the conditions under which each would be used. The IACUC may also require that the PI notify the committee of which drugs are actually used either at time of continuation or in a special interim report. (See 9:60.)

9:62 The IACUC has a responsibility for initial review and continuing oversight of animal research projects. Sometimes projects involve adverse events or unexpected problems. What actions should the IACUC take, if any, when notified of these adverse events?

Opin. If the unexpected problem results in the deaths of animals, the IACUC may require the PI to submit interim reports, that is, to report more often than annually as required by AWR and PHS Policy. The nature and duration of such reporting would be determined by the severity of the problem and its frequency. The IACUC may be required to increase its frequency of oversight when a federal agency determines that such increased oversight is necessary to ensure animal health and welfare.

9:63   An IACUC member requests that the IACUC reopen the discussion of a previously approved protocol. He was away when the protocol was approved, and he has concerns that were not considered during the review process. Can an IACUC reexamine the approval of a protocol under these circumstances?

Opin.   The IACUC is charged with continuous oversight of research and teaching projects using animal subjects. Therefore, the IACUC *can* reexamine a protocol at any time it considers necessary. If an absent IACUC member raises valid concerns that were not addressed at the time of initial or continuing review, these should be presented to the full IACUC. It should be noted, however, that the project may have already been started and therefore the IACUC should include this fact in its reconsideration of a project.

9:64   Can an IACUC develop and/or accept SOPs in lieu of the investigator's describing procedures in detail within a protocol? What special concerns might be generated by doing so?

Opin.   The animal welfare regulations do not specifically preclude the use of SOPs. Nevertheless, there are some issues that should be considered. First, if the SOP is to be part of an animal use protocol, then it should be specifically reviewed together with the protocol that references its use. This is an important facet of the review process and clarifies what will happen to an animal from the beginning to the end of the study. If not, then all aspects of the protocol have not been reviewed by the IACUC. Second, when the SOP is modified, this would constitute a change in every protocol that references its use, and all associated protocols should be amended each time the SOP is modified. In addition, every investigator who uses the SOP should be aware of the change and adopt it. Changing an SOP without adoption by the investigator might well have the effect of making him or her noncompliant.

It might be possible for the institution to maintain different versions of the same SOP, but it seems to us that keeping track of which version accompanies which protocols would be a difficult process.

9:65   For commonly used, well-established clinical procedures, such as taking a blood sample from a dog's cephalic vein, should the IACUC request details of how the procedure itself will be performed?

Opin.   The IACUC should ask for information about preparation of the skin for the sample, the volume of blood to be drawn, and the frequency of draws. It would not be necessary to ask for the angle of needle insertion and such items unless the IACUC was not convinced that the person doing the blood draw knew how to do it. Training should be required if personnel lack sufficient expertise.

9:66   **Rather than providing details of an experimental procedure on an IACUC form, a PI provides literature references that clearly describe the details of what he will do. Should the IACUC consider approving this study with just the literature reference, or should the actual procedure be part of the IACUC protocol?**

Opin.   The actual procedure should be part of the IACUC protocol so that what will be done as part of the experimental procedures is clear. As necessary, the IACUC reviewer may choose to read the literature references.

References

1. 2000 Report of the AVMA Panel on Euthanasia, *J. Am. Vet. Med. Assoc.*, 218(5), 669, 2001. Available on the World Wide Web at: http://www.avma.org/resources/euthanasia.pdf.
2. NIH/OPRR Reports, Dear Colleague Letter. Available on the World Wide Web at: http://grants.nih.gov/grants/olaw/references/dearcolleague.htm. (Documents also can be ordered by fax: (301) 594-0464.) Since March 30, 2000, OPRR has been OLAW, so these should be properly termed *NIH/OLAW Reports* and *Dear Colleague Letters*.
3. Prentice, E.D., Crouse, D.A., and Mann, M.D., Scientific merit review: the role of the IACUC, *ILAR News*, 34, 15, 1992.
4. Mann, M.D., and Prentice, E.D., Should IACUCs review scientific merit of animal research projects? *Lab Anim.* (NY), 33(1), 26, 2004.
5. Office for Protection from Research Risks, National Institutes of Health, The Public Health Service responds to commonly asked questions, *ILAR News*, 33, 68, 1991.
6. *Fed. Regist.*, 54, 36114, 1989.
7. Miller, J.G., personal communication, 1991.
8. Black, J., Letter to the editor, *ILAR News*, 35, 1, 1993.
9. Prentice, E.D., Crouse, D.A., and Mann, M.D., Letter to the editor, *ILAR News*, 35, 2, 1993.
10. Garnett, N. and Potkay, S., Use of electronic communications for IACUC functions, *ILAR J.*, 37(4), 190, 1995.
11. Wolff, A., Correct conduct of full committee and designated member protocol review, *Lab Anim.* (NY), 31(9), 28, 2002.
12. OPRR Report, Dear Colleague Letter, May 21, 1990. Available on the World Wide Web at: http://www.nih.gov:80/grants/oprr/faxcall.htm.
13. Dear Colleague Letter, NIH Office of Extramural Research Guidance Regarding IACUC Approval of Changes in Personnel Involved in Animal Activities, June 6, 2003. Available on the World Wide Web at: http://grants.nih.gov/grants/olaw/references/dearcolleague.htm.
14. Oki, G.S.F., et al., Model for performing Institutional Animal Care and Use Committee: Continuing review of animal research. *Contemp. Topics Lab. Anim. Sci.* 35(5), 53, 1996.
15. Wolff, A., et al., FAQs about the PHS Policy on Humane Care and Use of Laboratory Animals, *Lab Anim.* (NY), 32(9), 33, 2003.
16. Silverman, J., Do pressure and prejudice influence the IACUC? *Lab. Anim.*, 26(5), 23, 1997.
17. Garnett, N.L., and DeHaven, W.R., Protocol review: a word from the government, *Lab Anim.* (NY), 27(3), 19, 1998.
18. Garnett, N.L., and DeHaven, W.R., Protocol review: OPRR and USDA commentary, *Lab Anim.* (NY), 27(8), 1998.
19. Silverman, J., Majority rules? *Lab Anim.* (NY), 25(4), 22, 1996.
20. Silverman, J. Grants proposals and animal use protocols: Is the IACUC playing by the rules? *Lab Anim.* (NY), 35(2), 16, 2006.
21. Mann, M.D., Don't ignore the proposal, *Lab Anim.* (NY), 35(2), 17, 2006.
22. Mann, M.D., and Prentice, E.D., Verification of IACUC approval and the just-in-time process, *ILAR J.*, 47(2), E1–E18, 2006.
23. Garnett, N.L., personal correspondence, 1999.

10

Amending IACUC Protocols

Diane J. Gaertner and Kathleen D. Moody

Introduction

Biomedical research plans evolve as data are collected and analyzed. Increased experiences with the animal models themselves and new findings in the scientific literature often stimulate PIs to revise their approved protocols at irregular intervals. Using laboratory animals as models for human disease, basic research, or drug discovery may reveal unanticipated clinical complications or an exacerbation of expected adverse sequelae. In order to facilitate the IACUC's being informed of the current status of a protocol, many institutions have devised methods to streamline the submission and consideration of protocol revisions without necessarily requiring a *de novo* protocol submission. This chapter describes the mechanisms being used to amend protocols and suggests times when amending an approved protocol may be useful and appropriate.

10:1 What is the purpose of an amendment to an existing protocol?

Reg. PHS Policy (IV,B,7) requires PIs to seek IACUC approval for significant protocol changes. The AWAR (§2.31,c,7) have similar language.

Opin. The purpose of a protocol amendment is to allow a PI to revise an approved protocol at times other than when continuing protocol reviews are routinely scheduled. A PI's research plan using laboratory animals may change and evolve while a study is in progress, especially since protocols are often submitted prior to receipt of research funding. Additionally, the protocol amendment process provides an opportunity for an investigator to refine his or her experiment further by reducing or eliminating the animal pain and distress component. From the researcher's perspective, a relatively simple, straightforward, and timely method to make significant changes to a previously approved protocol facilitates research while fulfilling the obligation to keep the IACUC apprised of changes from the original animal use plan.

10:2 What are the regulatory requirements for protocol amendments?

Reg. Both the AWAR and PHS Policy require that the IACUC review and approve proposed significant modifications to ongoing activities using animals prior to initiation.

The AWAR (§2.31,d,1) state that "the IACUC shall determine that the proposed activities or significant changes in ongoing activities meet the following requirements," which are then detailed at §2.31,d,1,i–§2.31,d,1,xi. These requirements include addressing pain and distress, alternatives to painful procedures, animal housing and veterinary care, personnel training and qualifications, surgical standards, and appropriate euthanasia techniques. PHS Policy (IV, C,1,a–IV,C,1,g) states that the IACUC should review the animal-related components and determine that the proposed research projects are in accordance with PHS Policy, the AWA, the *Guide*, and the Institution's Assurance with the NIH/OLAW using criteria similar to those of the AWAR "unless acceptable justification for the departure is presented." (See 10:1.)

Opin. Although the PHS Policy and AWAR require prior IACUC review and approval of significant changes, institutions may have difficulty in enforcing these policies. Both NIH/OLAW and APHIS/AC have found that non-IACUC-approved animal activities occur frequently.[1] Changes in animal activities without IACUC review and approval have been identified during AAALAC site visits,[2] emphasizing the need for investigators to be made maximally aware that procedures on animals must always follow the approved protocol. Initiating significant changes to IACUC-approved protocols without prior IACUC review is considered to be non-compliance with PHS Policy, which should be promptly reported to NIH/OLAW.[3,4]

10:3 What activities can be considered appropriate for an amendment?

Opin. The IACUC must consider its regulatory responsibilities when considering the best methods for reviewing proposed amendments to approved protocols. Whether the activity proposed can be a simple amendment or requires a rewritten protocol with regular IACUC review depends upon whether the proposed change is considered to be minor or significant. Federal regulations and policies have not defined all changes they consider significant, although NIH/OLAW[5] and APHIS/AC[6] have provided guidance with selected examples of significant changes such as the following:

- Changing the objectives of a study
- Changing from nonsurvival to survival surgery
- Changing the degree of invasiveness of a procedure or discomfort to an animal
- Performing a new procedure or changing a procedure being used (USDA)
- Changing the species or the approximate number of animals used
- Changing personnel involved in animal procedures
- Changing anesthetic agent(s), the use or withholding of analgesics
- Changing methods of euthanasia
- Changing the duration, frequency, or number of procedures performed on an animal

Table 10.1 lists major and minor protocol changes that might necessitate a revision process and the suggested level of review needed.

TABLE 10.1

Examples of Major and Minor Changes to Protocols and the Suggested Mechanism for These Types of Changes

Type of Change	Major (Significant)	Minor
Examples	Change in purpose or specific aim of study	Substitution of a qualified student or technician
	Change in principal investigator	Addition of a faculty collaborator
	Change of species	Addition of another strain of the same animal species
	Addition of USDA-regulated species	Change in sex of animal to be used
	Large increase in animal numbers	Small increase in animal numbers
	Addition of survival surgery	Need to repeat an experiment
	Addition of painful procedure	Addition of minor surgery
	Unanticipated marked increase in clinical signs or proportion of animal deaths	Addition of sample collection times[a]
		Additional noninvasive sampling[a]
Suggested mechanism	Rewriting of protocol for IACUC review	Submission as an amendment and following of institutional procedure for review of amendments

[a] These are minor changes provided that total volume collected is in compl ᴣ nce with institutional standards for the specified species.

As previously indicated, the IACUC must distinguish between minor protocol amendments (changes) and proposed significant changes to the protocol that would necessitate a complete revision and resubmission of the protocol. Thus, in the authors' opinion all substantive protocol changes must be reviewed and approved by the IACUC, although less extensive changes may utilize the amendment mechanism rather than necessitating a complete *de novo* submission. An institution can permit alterations in administrative information to be changed without IACUC review (e.g., electronic mail address of the PI or a change in the funding agency provided there are no substantive or minor changes to the protocol or number of animals to be used). In an academic setting, a change in funding agency or funding from a second agency will usually necessitate submission as a new protocol. Industrial organizations, utilizing internal funding, may use standardized protocols for administration of test substances with only minor customization to indicate the change in the tested substance.

Since each institution's research and animal use program is unique, the NIH/OLAW has suggested that each institution develop its own guidelines regarding significant protocol modifications and ensure their availability to investigators to clarify the amendment process.[5,7] Another method to determine the significance of an amendment is to assess the potential or actual reduction in animal welfare and overall ethical cost–benefit ratio of the research. In this view, the proposed amendment must be considered in conjunction with the original protocol to determine adequately whether a significant change has been proposed.[8,9]

Complex protocol changes or submissions of multiple sequential amendments to an existing protocol are not well suited to the amendment mechanism. Ideally, revisions submitted as amendments should be relatively simple and should not affect the existing documentation. Otherwise, the amended protocol can confuse rather than clarify. Complex changes to an approved protocol should be documented by the submission of a completely rewritten protocol, which must be subject to the same approval mechanisms as the original protocol before animal experiments can proceed.

Either the rewritten protocol may be considered a new protocol (and given a new identification number) or its relationship to an earlier protocol may be maintained by utilizing a numbering system that indicates the original protocol number.

Surv. What activities can be considered appropriate for a formal major amendment? More than one response is possible.

- Request for an additional 5% or fewer animals   57/170
- Request for more than 5% additional animals   109/170
- Request for more than 10% animals of the same species   9/170
- Addition of any new species   148/170
- Addition of a new strain or stock of the same species   62/170
- Change in the PI   133/170
- Changes in personnel, but not the PI   38/170
- Any change in protocol for an APHIS/AC-regulated species 75/170
- More severe clinical signs seen than expected   107/170
- Use of a more painful procedure   152/170
- Addition of a small change (SQ rather than IP injection) 50/170

10:4 What format should be used to submit an amendment?

Opin. Amendments may be submitted through formal or informal mechanisms provided that all the necessary information is supplied. Institutions that have large numbers of approved protocols often use an amendment form to document proposed changes in protocols. Depending on the specific changes requested, questions to be answered for each protocol amendment should include the following:

- What is the purpose or rationale for the protocol amendment?
- Will different people use the animals? If personnel have changed, their qualifications (e.g., education, training, and experience) must be documented.
- Will the species, sex, or strain of animal change?
- Will there be specialized housing requirements (e.g., housing that is not standard for the species or the facility, or housing that does not meet the animals' physiological or behavioral requirements)?
- Will more animals be needed? If so, justification for the increased number must be provided.
- Will additional minor surgical procedures or sampling of body fluids or tissues occur?
- Are animals expected to experience more clinical illness, pain, or distress as a result of the procedures proposed in this amendment? Are there alternatives to the use of animals in painful procedures? How will any pain or distress from this new procedure be minimized?
- Will there be new procedures involving the animals? Such changes must be described with the same level of detail as is required in a new protocol.
- Will there be a change in the methods or drugs used to induce anesthesia, analgesia, or euthanasia?
- Will prolonged restraint of conscious animals be required?

- Will new hazardous agents be used?
- Will the surgical plans change (minor to major, multiple survival surgery, additional procedures)?

Surv. Although standardized protocol amendment forms can be used, there are other methods to initiate the process. Indicate which formats are acceptable to your IACUC for amendments (indicate all that are used).

• Use of a letter or memo	76/170
• Amendment form or other means; amendment form not mandatory	43/170
• Amendment form mandatory	70/170
• Acceptance of amendments by electronic mail	49/170
• Acceptance of new protocol submission form as an application for a protocol amendment	2/170

10:5 Do the designated member review and full committee review processes also apply to amendments?

Reg. PHS Policy (IV,C,2) and the AWAR (§2.31,d,2) require the same review procedure for proposed significant changes in ongoing activities (protocols) as they do for new activities.

Opin. The procedures used for new protocol review also can be applied to amendments. All IACUC members must have the opportunity to evaluate proposed significant changes and request full committee evaluation, as in full protocol review.[10] Whether a full review mechanism or a delegated reviewer mechanism is used (see Chapter 9), amendments must be reviewed in the context of the complete protocol, a process that may necessitate discussing the entire approved protocol. This means that the designated reviewer or the full committee may have to review the original protocol, and the full committee may have to discuss the proposed amendment in the context of the complete protocol. Either IACUC approval of protocol amendments must be documented in the IACUC minutes or designated reviewer approval must be documented in the protocol file via a dated signature.

10:6 Can the IACUC Chair or an IACUC subcommittee review and approve significant changes or must the full committee review and approve such submissions?

Reg. The AWAR (§2.31,d,2) and the PHS Policy (IV,C,2) identify full committee review and designated member review as the only approved methods to evaluate animal protocols and proposed significant changes to ongoing protocols. In designated member review (see 9:21–9:27), all IACUC members must be provided with a list of the proposed significant changes to a protocol so that each member has the opportunity to request full committee review. If no member calls for full committee review, the IACUC Chair may designate one or more IACUC members to review the proposed changes and have the authority to approve, require modifications in, or request full committee review. In instances when full committee review is appropriate, the PHS has determined that serial one-on-one meetings, telephone, fax, or electronic mail to poll or obtain members' votes cannot preempt the voting mechanism of a quorum of a convened IACUC;[11] however, a more recent interpretation allows for certain types of telecommunication in lieu of face-to-face full committee meetings (see 6:17–6:18). For those institutions with time constraints, the designated reviewer option using at least one qualified IACUC member to review the amendment, after all voting members have had the option to request a full committee review, would be compliant with the PHS Policy.[12]

Opin. Institutional policy will determine which types of protocols and protocol amendments are eligible for designated member review and which require full review at a convened meeting of the IACUC. Any member of the IACUC, including the Chair, may serve as the designated reviewer. IACUC members, not institutional policy, determine whether new protocols or significant changes to ongoing protocols will be reviewed by the full committee. This is because each member must always have the opportunity to call for full committee review. The exception to this occurs if an institution adopts a policy that all protocols and significant changes to ongoing protocols must be reviewed by the full IACUC; such a policy would obviate the need for an opportunity to call for full committee review.

Surv. Which of the following processes does your IACUC use to amend protocols? More than one response is possible.

- All amendments (major and minor) must go through full committee review process — 52/170
- All USDA Category E amendments must go through the full committee review process — 51/170
- All amendments go through the designated review process — 45/170
- Only major amendments must go through the full committee review process — 76/170
- Only major amendments must go through the designated review process — 13/170
- Only minor amendments go through the designated member review process — 39/170
- An IACUC subcommittee reviews and approves amendments — 22/170
- Some amendments are allowed to be exceptions to these policies — 16/170
- IACUC Chairperson reviews and approves amendments — 6/170

Various exceptions were granted by the preceding IACUCs, including allowing personnel changes or administrative changes to be approved by the Chair without the protocol's having a formal delegation mechanism. Institutions utilize various mechanisms to distribute information about protocols eligible for delegated review, and these mechanisms can be used for protocol revisions and initial submissions.

10:7 How many animals may be added to an ongoing study using an amendment mechanism?

Reg. PHS Policy (IV,D,1,a) and the AWAR (§2.31,e,1–§2.31,e,2) require that proposals to the IACUC specify and include a rationale for the approximate number of animals proposed for use. This is an implicit requirement that institutions establish mechanisms to monitor and document the number of animals acquired and used in approved activities.

Opin.   The cutoff for new animal orders may be the exact number of animals approved by the IACUC or a small percentage (e.g., 5%) in excess of the approved number of rodents.[13] Similarly, some institutions allow an amendment mechanism for ordering up to 10% more animals, with any additional animal requests requiring submission of a new protocol form.[14] In the authors' opinion, requests for additional nonrodent mammalian species, irrespective of the number, require review and approval by the full committee mechanism. Whatever mechanism is used to add animals, it must satisfy the PHS Policy that the number of animals approved must be limited to the number needed to obtain valid results.[13]

Surv.   Can a request for 5% more animals of the same species be approved by the IACUC as an amendment without necessitating submission of a complete protocol revision?

- Not applicable   8/170
- Amendment is acceptable   146/170
- Complete protocol revision is required   9/170
- Not answered   7/170

10:8   The PI on an IACUC protocol does not perform any hands-on animal work. It is done by a research associate. If the PI is changed to another PI who also does no hands-on animal work, is this considered a significant amendment requiring IACUC approval, or is this a minor amendment?

Reg.   Both the PHS Policy (IV,B,7) and AWAR (§2.31,c,7) require the IACUC to review proposed significant changes in ongoing activities using animals. NIH/OLAW identified examples of significant changes, including "changes in personnel involved in animal procedures."[5] In 2003, NIH/OLAW and APHIS/AC jointly published a clarification indicating that IACUCs may establish certain changes in personnel as minor, except for the PI.[15]

Opin.   NIH/OLAW and APHIS/AC have established that institutional IACUCs are responsible for confirming that any personnel changes on an animal use protocol must be reviewed to ensure that new protocol participants have adequate training and qualifications to perform the specified procedures on the selected species. Regardless of the degree of hands-on involvement, a change in PI on an NIH grant is considered to be a significant "change in scope" such that the grantee must obtain prior NIH approval.[16] Also, in the opinion of the authors, PIs of animal protocols must complete the institution's established educational programs regardless of whether or not they handle animals

10:9   A dog is anesthetized and prepped for surgery. The PI then recognizes that the dosage of an experimental drug that she believed to be safe might actually cause a problem. The potential problem is easily correctable by lowering the dosage. Should she make the correction and then notify the IACUC, or stop the research and obtain IACUC approval before proceeding?

Reg.   Both the AWAR (§2.31,c,7) and the PHS Policy (IV,B,7) require that any significant change to an animal use protocol be reviewed and approved by the IACUC prior to initiation of those changes. The APHIS/AC Resource Guide specifically charges the Veterinary Medical Officer to ensure that any significant changes to protocols were approved by using the same procedures as for protocol review, that is, full committee review or designated member review.[17]

Opin.   The current animal use regulations and guidelines for significant changes are directed to emphasizing IACUC deliberation of more painful or distressful procedures. IACUCs can establish their own guidelines regarding whether changes in drug doses are considered to be minor or significant changes. Changes in anesthesia or analgesia are considered to be significant. Minor changes in routine drug doses can be addressed to the IACUC veterinarian as the designated reviewer. It is not clear in this scenario whether the experimental drug is an anesthetic or has another purpose. If it is an experimental drug, changes in its dose would appear to constitute a significant change in procedure in a covered species such that it must be reviewed and approved by the IACUC prior to initiation. Ideally, each IACUC member must have the opportunity to call for full committee review of any significant change. However, if the PI detects a potential serious unforeseen risk during the performance of an experiment, an acceptable alternative would be immediate consultation with the veterinarian and documentation of the consultation and the veterinarian's decision, followed up by amending the protocol utilizing approved methods. This alternative is consistent with the practice of institutional review boards whereby a physician is allowed to make such a change if there are unforeseen, life-threatening situations.

10:10   During the course of research it becomes evident that no pain or distress is occurring in a study that was initially assumed to cause mild pain. The PI wishes to amend his IACUC protocol to indicate no pain or distress. Is this considered a major or minor amendment?

Reg.   The AWAR (§2.36) and APHIS/AC Policies 11 and 17 stipulate the required information to be submitted in an institution's annual report, including the number of animals in each pain category.[18,19] APHIS/AC considers a change in pain classification of a procedure to be a significant change.[6]

Opin.   Each USDA registered research facility using animals must submit an annual report indicating the number of animals in the defined pain and distress categories B, C, D, or E.[19] To facilitate the acquisition of such information, many IACUCs ask the PI to assign a pain and distress level to the animal research proposal (or significant change) at submission, subject to IACUC recategorization at review. Each IACUC may use its own pain and distress categorization or use the current APHIS/AC designations.

The preceding scenario does not identify the laboratory animal as an AWA-covered species and annual reports to APHIS/AC do not require the inclusion of data on species that the AWA does not define as animals.[19] Despite the difference in regulation coverage, most IACUCs follow the same procedures for all species, thereby precluding the perception of a double standard in the animal care program. In this case, then, the IACUC would need to evaluate the change in pain and distress as a significant change.

10:11   At 2 years into a 3-year swine study, a PI submits a protocol amendment. The amended portion of the protocol does not involve any animal pain or distress, although other parts of the original study included more than momentary pain. Does the PI have to review the most current literature to see whether any alternatives to the painful procedures have been found during the previous 2 years?

Reg.   Agricultural animals used for traditional production agricultural purposes are exempt from AWA coverage. Farm animals used as models of human disease (biomedical research) or in development of biologicals for nonproduction or nonagricultural animals (nonagricultural research) are regulated under the AWA.[20] APHIS/AC Policy 12 states that proposed significant changes that may increase the animals' discomfort must be accompanied by "a written narrative description of the methods and sources used to consider alternatives to procedures that may cause more than momentary or slight pain or distress to the animals." Policy 12 goes on to state, "Although additional attempts to identify alternatives or alternative methods are not required by Animal Care at the time of each annual review of the animal protocol, Animal Care would normally expect the principal investigator to reconsider alternatives at least once every 3 years, consistent with the triennial review requirements of the Public Health Service Policy (IV,C,5)."[21] (See 11:6.)

Opin. When submitting an amendment to a protocol that will induce additional pain, the PI must submit an alternatives search. If submitting an amendment that will not produce additional pain and distress, then an additional alternatives search is not required.

In the scenario presented the PI would need to submit a new animal use protocol to the IACUC within the next year if he or she plans to continue the experiments at the institution. Given that timetable, another option for the PI would be to submit a new protocol incorporating the planned amendment, along with a documented search for alternatives for all potentially painful procedures in the entire protocol. Depending upon the complexity and number of amendments to this swine protocol, a new protocol submission may be the most advantageous option.

References

1. Potkay, S., and DeHaven, W.R., OLAW and APHIS: common areas of noncompliance, *Lab Anim.* (NY), 29, 32, 2000. Available on the World Wide Web at: http://grants.nih.gov/grants/olaw/LabAnimal.pdf.
2. Applied Research Ethics National Association/Office of Laboratory Animal Welfare, Institutional Animal Care and Use Committee Guidebook, 2nd ed., 2002, p. 193. Available on the World Wide Web at: http://grants.nih.gov/grants/olaw/GuideBook.pdf.
3. Office of Extramural Research, NIH Guide NOT-OD-05-034, Office of Extramural Research Guidance on Prompt Reporting to OLAW under the PHS Policy on Humane Care and Use of Laboratory Animals, February 24, 2005. Available on the World Wide Web at: http://grants.nih.gov/grants/guide/notice-files/NOT-OD-05-034.html.
4. Garnett, N.L., and Gipson, C.A., A mouse isn't a rat, but what's the big deal? A word from OLAW and USDA, *Lab Anim.* (NY), 32, 19, 2003. Available on the World Wide Web at: http://grants.nih.gov/grants/olaw/references/lab_animal2003v32n9_Silverman.htm.
5. Potkay, S., et al., Frequently asked questions about the Public Health Service Policy on Humane Care and Use of Laboratory Animals, *Lab Anim.* (NY), 24, 24, 1995. Available on the World Wide Web at: http://grants.nih.gov/grants/olaw/references/laba95.htm.
6. USDA/APHIS/AC, Research Manual Part 18. Available on the World Wide Web at: http://www.aphis.usda.gov/ac/researchmanual/18-4PROC.PDF.
7. Division of Animal Welfare, National Institutes of Health, Office for Protection from Research Risks, Frequently asked questions about the Public Health Service Policy on Humane Care and Use of Laboratory Animals, *ILAR News*, 35, 47, 1993. Available on the World Wide Web at: http://grants.nih.gov/grants/olaw/references/ilar93.htm.
8. Oki, G.S.F., et al., Model for Performing Institutional Animal Care and Use Committee: continuing review of animal research, *Contemp. Topics Lab. Anim. Sci.*, 35, 53, 1996. Available on the World Wide Web at: http://grants.nih.gov/grants/olaw/references/contop96.htm.

9. McLaughlin, R., What constitutes significant changes? Protocol Review Panel I, Presented at IACUCs: Improving the Efficiency, Annual meeting of the Applied Research Ethics National Association (ARENA), Boston, March 26, 1998.

10. Wolff, A., Correct conduct of full-committee and designated-member protocol reviews, *Lab Anim.* (NY), 31, 28, 2002. Available on the World Wide Web at: http://grants.nih.gov/grants/olaw/references/laba02v31n9.htm.

11. Garnett, N., and Potkay, S., Use of electronic communications for IACUC functions, *ILAR J.*, 37, 190, 1995. Available on the World Wide Web at: http://grants.nih.gov/grants/olaw/references/ilar95.htm.

12. Wolff, A., et al., Frequently asked questions about the Public Health Service Policy on Humane Care and Use of Laboratory Animals, *Lab Anim.* (NY), 32, 33, 2003. Available on the World Wide Web at: http://grants.nih.gov/grants/olaw/references/lab_animal2003v32n9_wolff. htm.

13. Potkay, S., et al., Frequently asked questions about the Public Health Service Policy on Humane Care and Use of Laboratory Animals, *Contemp. Topics Lab. Anim. Sci.*, 36, 47, 1997. Available on the World Wide Web at: http://grants.nih.gov/grants/olaw/references/faq_labanimals 1997.htm.

14. Doyle, D.J., Oki, G.S.F., and Prentice, E.D., Conducting continuing review and IACUC review of amended protocols, presented at Innovative Biomedical Technologies: The IACUC Response, Annual Meeting of Public Responsibility in Medicine and Research (PRIM&R), Boston, March 28, 1998.

15. Office of Extramural Research, NIH Guide NOT-OD-03-046, Revised Guidelines regarding IACUC Approval of Changes in Personnel Involved in Animal Activities, June 6, 2003. Available on the World Wide Web at: http://grants.nih.gov/grants/guide/notice-files/NOT-OD-03-046.html.

16. NIH Grant Policy Statement. Part II. Terms and Conditions of NIH Grant Awards, 12/2003. Available on the World Wide Web at: http://grants.nih.gov/grants/policy/nihgps_2003/NIHGPS_Part7.htm?Display=Text.

17. USDA/APHIS/AC, Research Manual Part 6. Available on the World Wide Web at: http://www.aphis.usda.gov/ac/researchmanual/6-3PROC.PDF.

18. U.S. Department of Agriculture, Animal and Plant Health Inspection Service, Animal Care, Policy 11, Painful Procedures, April 14, 1997. Available on the World Wide Web at: http://www.aphis.usda.gov/ac/policy/policy11.pdf.

19. U.S. Department of Agriculture, Animal and Plant Health Inspection Service, Animal Care, Policy 17, Annual Report for Research Facilities, March 17, 1999. Available on the World Wide Web at: http://www.aphis.usda.gov/ac/policy/policy17.pdf.

20. U.S. Department of Agriculture, Animal and Plant Health Inspection Service, Animal Care, Policy 26, Regulation of Agricultural Animals, November 17, 1998. Available on the World Wide Web at: http://www.aphis.usda.gov/ac/policy/policy26.pdf.

21. U.S. Department of Agriculture, Animal and Plant Health Inspection Service, Animal Care, Policy #12, Consideration of Alternatives to Painful/Distressful Procedures, June 21, 2000. Available on the World Wide Web at: http://www.aphis.usda.gov/ac/policy/policy12.pdf.

11

Continuing Review of Protocols

Gwenn S.F. Oki and Ernest D. Prentice

Introduction

Initial review and approval of a project by an IACUC represent an informed judgment by the committee that when the protocol is initiated as written it will be conducted in full compliance with all applicable federal requirements contained in the AWA, AWAR, HREA, and PHS Policy. The IACUC review responsibilities do not, however, end with initial approval of the protocol. The IACUC is obligated by both the AWAR and the PHS Policy to conduct ongoing review.

The purpose of this chapter is to present general concepts of continuing review that ideally will be useful to IACUCs. It should be noted that continuing review is just as important as initial review, since it not only serves to ratify the decisions of the IACUC on the current status of the protocol, but also helps the investigator maintain compliance.

11:1 What is meant by *continuing review*?

Opin. APHIS/AC interprets continuing review, performed no less often than annually, as a monitoring process in an effort to determine that the study remains in compliance, that the activities have been "conducted in accordance with the approved protocol," that significant and, at some institutions, minor modifications to the protocol have received prior IACUC approval,[1] and "to ensure that any new requirements of PHS, USDA, or the institution are communicated to the investigator."[2] The NIH/OLAW requires triennial continuing review to be a *de novo* process, meeting all the new proposal review criteria as set forth in PHS Policy IV,C,1–IV,C,4.

11:2 What is the intent of continuing review?

Opin. In general, the intent of a regulation or policy can be found in the preamble. For example, the preamble to the AWAR makes a single statement indicating that the intent of continuing review is "to provide current information to the research facility regarding all ongoing activities so that it can remain in compliance."[3] Although the intent of continuing review was not articulated in the form of a preamble to the PHS Policy, NIH/OLAW has interpreted that continuing review, subject to the AWAR, be performed no less than annually as a monitoring function.[1,4]

11:3   Is a protocol initially considered approved on the date when the research funding begins or on the date when the IACUC has taken final action on the protocol?

Reg.   PHS Policy (IV,D,2) considers a protocol approved on the date of the IACUC approval.

Opin.   While it appears to make sense to synchronize the IACUC initial protocol approval date with that of the funding start date, this practice is not permissible. If the IACUC approves a protocol and submits a letter to the funding agency certifying that the protocol was approved, the actual date of the IACUC's approval must be indicated as required by the PHS Policy IV,D,2. This is not a future date with final approval contingent on actual funding. In addition, the just-in-time process for verification of IACUC approval for the animal use portions of the grant application and any IACUC imposed changes must be noted on the IACUC's verification of approval prior to the grant award. (See 8:22.) The institution should have a mechanism and procedures established to verify concordance of the grant application with the approved IACUC protocol.[2] Despite the availability of the just-in-time process, it should be noted that institutions have the prerogative to require investigators to obtain IACUC approval of the grant application prior to NIH peer review, thereby not utilizing the just-in-time process.[5]

11:4   What is the maximal life of an approved protocol?

Opin.   The AWAR and PHS Policy do not limit the life of a protocol; however, both require that the IACUC conduct continuing reviews of approved studies at specified intervals. (See 11:6; 11:14.) Nevertheless, many institutions require a new submission at least triennially because it is very likely that the science has evolved, that the current protocol may include multiple amendments that have been approved over the 3-year period, and that the IACUC may have new policies or a revised protocol application form that asks new questions that need to be addressed. Since the triennial review constitutes a *de novo* review it may be more efficient to require submission of a new protocol.

11:5   What do the AWAR and PHS Policy indicate with regard to continuing review of protocols?

Reg.   The frequency of IACUC consideration of approved, ongoing activities is one of the few areas in which the NIH/OLAW and APHIS/AC have differing requirements. The AWAR (§2.31,d,5) state that "the IACUC shall conduct continuing reviews of activities covered by this subchapter at appropriate intervals as determined by the IACUC, but not less than annually." PHS Policy (IV,C,5) states that "the IACUC shall conduct continuing review of each previously approved, ongoing activity covered by this Policy at appropriate intervals as determined by the IACUC, including complete review in accordance with IV,C,1–IV,C,4 at least once every three years."

Opin.   Other than the maximal time interval between continuing reviews, the AWAR make no further statements regarding the form or substance of continuing review. The PHS Policy (IV,C,5), however, indicates that review criteria set forth at PHS Policy IV,C,1–IV,C,4 must be satisfied at least triennially. This is the same criterion that IACUCs must use to conduct an initial review of a proposed protocol. Hence, triennial review of a protocol constitutes a *de novo* assessment of a currently approved study. (See 11:6.)

11:6   How often should continuing review of protocols be performed?

Reg.   As mentioned in 11:5, the AWAR require that continuing review be performed at intervals determined appropriate for the study being conducted but no less often than once annually. The PHS Policy requirement is the same, except that the interval is no less than triennially. (See 11:5.)

Opin.   Each IACUC should establish the frequency of continuing review at the time of initial review of the protocol. Many institutions have divided continuing reviews, based on review intervals, to meet the AWAR on an annual basis and to require a new protocol submission at the time of triennial review. Since the PHS Policy necessitates *de novo* review at the end of 3 years utilizing the same assessment criteria established for new protocol submissions and because there will more than likely be changes in the experimental methods and procedures, new institutional policies, and IACUC membership, it may be more efficient to require the investigator to submit a completely new protocol at the time of triennial review. It should, however, be noted that there is no AWAR or PHS Policy requirement for resubmission of a new IACUC application for an ongoing study.

11:7   What is an effective and appropriate way to conduct continuing review of a protocol?

Opin.   Most IACUCs require investigators to use an institution-specific continuing review form. This standardizes the information required for all continuing studies and streamlines the review process. The form should be designed in consideration of the intent of continuing review described previously. One model developed in collaboration with NIH/OLAW and APHIS/AC has been described.[6] In addition, NIH/OLAW has published the following advice:[4]

> A relatively simple mechanism to meet both federal requirements (AWAR and PHS Policy) is to circulate annually to all investigators with IACUC-approved activities a standard form giving current basic information, such as IACUC approval number, IACUC approval date, title of project, and species used. The investigator then notes that either no changes have taken place or he/she describes any changes that have occurred. Responses are reviewed by an IACUC designee for assessment of the changes reported. Any changes to the approved activity that are deemed of sufficient magnitude to merit further consideration may then be presented to the IACUC. All of these dispositions should be documented as official IACUC actions.

It should be noted, however, that significant changes to a protocol must be reviewed and approved by the IACUC prior to implementation. Many institutions also require that minor changes be approved by the IACUC before implementation. This can be accomplished by administrative review performed by the IACUC Chair, IACUC administrator, or AV. In particular, this precludes the possibility that an already implemented protocol change might be unacceptable to the IACUC.

11:8   Is full committee review by the IACUC necessary in order to reapprove a protocol?

Reg.   Review procedures under PHS Policy (IV,C,2) apply to the initial and triennial reviews and IACUC review of proposed significant changes to the use of animals in ongoing activities. Review procedures under the AWAR (§2.31,d,2) apply to the initial and continuing reviews and IACUC review of proposed significant changes to the use of animals in ongoing activities.

Opin. Full committee review is not necessary. Some IACUCs use the designated reviewer process (see 9:21) as indicated in the AWAR (§2.31,d,2) and the PHS Policy (IV,C,2).

- Each IACUC member is provided with a list of continuing protocols to be reviewed. Any IACUC member shall have available the written descriptions of the research protocols.

- Any IACUC member can request full committee review. If full review is not requested, then one IACUC member (the designated reviewer), selected by the IACUC Chair, is provided with the specific continuing review form(s) and the corresponding approved protocol files and is responsible for conducting the continuing review.

- The designated reviewer is authorized to approve required modifications or to request full IACUC review.

In most situations, continuing studies involving no changes or minor proposed changes are most efficiently handled in this manner. Alternatively, some IACUCs conduct full review of all continuing protocols by utilizing a primary reviewer. In this scenario, all IACUC members receive a copy of the continuing review form. The primary reviewer, IACUC Chair, AV, and IACUC administrator also are provided a copy of the currently approved protocol. At the time of the convened meeting, the primary reviewer provides a summary of his or her findings with a recommendation for action. After discussion, a vote is taken.

Depending on the size of the institution's animal program, either the designated member review process or full committee review may be more appropriate. For institutions with smaller animal programs, the latter may be the chosen method; for larger institutions, the former may be more efficient in consideration of time constraints and the volume of continuing reviews.

11:9 Can a designated reviewer be used for continuing review? If so, under what circumstances can the designated reviewer process be used for continuing review?

Opin. As indicated in 11:8, the designated member review process (AWAR §2.31,d,2; PHS Policy IV,C,2) can be utilized for continuing and triennial reviews. IACUCs with a lower volume of protocols may elect full committee review for all aspects of animal research conducted at their institution. Some IACUCs may choose the designated member review process for continuations that involve no changes or minor proposed changes, thus requiring full committee review for all other continuing protocols. Other IACUCs may decide that full committee review is necessary for continuing studies that involve sensitive species or high-profile controversial studies, thus utilizing the designated member review process for all other continuing protocols. Depending upon the institutional policies, protocol volume, and their research portfolio, IACUCs must determine the most appropriate continuing review process. (See 6:19; 11:8; 9:21–9:26.)

11.10 If a designated reviewer is used (see 6:19; 11:8; 9:21–9:26), must the full IACUC still give final approval to the decision made by that person?

Opin. The designated member review process utilized by various IACUCs is meant to save time and to promote efficiency, especially for high-volume IACUCs. As indicated in the AWAR (§2.31,d,2) and PHS Policy (IV,C,2), the designated reviewer is authorized to approve, to require modifications, or to request full IACUC review. If the latter is not requested by the designated reviewer or by another IACUC member, then the designated reviewer has the authority to approve the continuation report or protocol. Requiring that the full IACUC give final approval of the designated reviewer's approval is an unnecessary duplication of effort that defeats the purpose of the designated reviewer process.[7] If the IACUC insists on final ratification of the designated reviewer's approval, then continuation reviews should be conducted at the time of a convened meeting. (See 9:21–9:26.)

11:11 How do investigators change procedures or personnel on their IACUC-approved protocol?

Opin. IACUCs at many institutions require that investigators submit a protocol amendment form when a protocol alteration is needed. This includes modifications in experimental procedures, an increase in the number of animals required, and a change in personnel involved in the study (i.e., direct handling and use of animals, etc.). As discussed in 11:7, significant changes require IACUC approval prior to implementation and can be submitted either before or at the time of continuation review. (See Chapter 10.)

Utilizing the designated member review process (see 11:8), a copy of the original protocol is given to the reviewer who will review the protocol and the continuation review form. Ideally, any questions that arise during this review are addressed through review of the original protocol. Depending on the number of amendments, it may be advisable for the PI to submit a completely rewritten protocol for future reference. When a protocol has undergone a number of amendments, even those that are minor, the reviewer may have difficulty discerning what constitutes the most recent version of the current protocol. Indeed, some institutions have instituted the requirement that any amendment be incorporated into a master protocol. This master protocol would constitute an up-to-date protocol that could be used by the IACUC in their review and approval processes. This method can greatly facilitate the IACUC's understanding of the proposed changes and the way the changes relate to the last approved version of the protocol, facilitate animal care and compliance staffs' management and oversight of the protocol, assist the research staff in the appropriate conduct of the study, and facilitate APHIS/AC inspectors' review of the study.

11:12 What action should the IACUC take if an investigator does not respond to a request for information concerning the continuing review of his protocol?

Reg. If an investigator permits the IACUC approval to "expire" (i.e., the investigator does not respond to a request for updated protocol information in order for the IACUC to fulfill its responsibility for continuing review) and animal work is ongoing, then the research is no longer in compliance with PHS Policy or the AWAR. IACUCs are required to report serious or continuing noncompliance to NIH/OLAW (PHS Policy IV,F,3,a) or APHIS/AC (AWAR §2.31,d,6).

Opin. The AWAR (§2.31,d,5) and PHS Policy (IV,C,5) require that reviews of animal protocols be conducted at least annually and triennially, respectively. If the IACUC's request for information is necessary in order for the continuation report to be approved, the IACUC has no choice but to halt further use of new animals in the study temporarily. If the continuation reapproval date has passed with no response from the investigator, appropriate individuals in the institution should be notified (e.g., the animal facility management, the institution's AV, the PI's department head) with a copy to the PI. Operationally, IACUCs should have established policies and procedures for handling these types of situations. It should be noted that both the PHS Policy and the AWAR are silent on the dispostion of animals already in the process of experimentation when the protocol lapses and a continuation report has not been submitted. There are a wide variety of ways that institutions are handling this issue. The authors of this chapter think that the animals already in the queue should remain in it until protocol procedures for their use have been completed.

11:13   If a protocol is near completion and the continuation report due date is imminent, can the IACUC vote to extend the protocol for a specified short period without submitting a formal report for continuing review?

Reg.   See 11:12.

Opin.   Both APHIS/AC and NIH/OLAW are clear in that the IACUC may not extend the protocol approval dates without the appropriate annual or triennial reviews within the specified period.[8]

11:14   When should an entirely new protocol be submitted? Why?

Opin.   Many institutions require investigators to submit a new protocol every 3 years. Although this procedure fits PHS Policy (IV,C,5) for complete *de novo* review at least triennially well, one must remember that the AWAR and PHS Policy do not limit the life of a protocol. Therefore, it is possible to allow a protocol to continue indefinitely without a new protocol submission as long as both the AWAR and PHS Policy requirements for continuing review on an annual and *de novo* review on a triennial basis are met. For PHS purposes, the triennial review must include a review of all current information relevant to the protocol (e.g., master protocol), although it is not required that an entirely new protocol be submitted. (See 11:4; 11:5.)

11:15   How much information concerning results of a study at the time of continuing review should the IACUC require?

Opin.   Reporting the results of an ongoing study is not an AWAR or PHS Policy requirement. However, at the time of triennial review, the IACUC is required to perform a *de novo* review of the continuing study (see 11:5). This necessitates that the protocol meet all review criteria and study conduct requirements as stipulated in PHS Policy IV,C,5. It, therefore, makes sense for the IACUC to require that the investigator provide the results of the study to date in order to facilitate IACUC review. Indeed, the only way the IACUC can justify use of the animals already included to date in the research project is to obtain information on the progress of the study. Admittedly, investigators encounter "blind alleys" and other problems that may affect study results. In some cases, animal experiments must be repeated. The IACUC should be advised of such issues that arise during the course of continuing review in order to fulfill its responsibilities.

11:16   Is a literature search required at the time of continuing review?

Reg.   The AWAR (§2.31,d,1,ii; §2.31,e) include a requirement that a literature search be conducted to demonstrate that "alternatives to procedures that may cause more than momentary or slight pain and distress to the animals are not available" as a criterion for IACUC approval of a proposed project or significant change(s) in an ongoing project. Although the need for a similar literature search is not specifically indicated, PHS Policy (IV,A,1) requires that applicable AWAR requirements also be met as part of the institution's Assurance. In addition, APHIS/AC Policy 12 provides guidance indicating that "Animal Care would normally expect the principal investigator to consider alternatives at least once every 3 years, consistent with the triennial review requirements of the PHS Policy (IV,C,5)."[9] (See 10:11.)

Opin.   A literature search should, therefore, be conducted at the time of initial review, when there are proposed significant changes to an ongoing protocol, and at the time of triennial review.

References

1. Division of Animal Welfare, Office for Protection from Research Risks, National Institutes of Health, Issues for Institutional Animal Care and Use Committees (IACUCs): Frequently asked questions about the Public Health Service Policy on Care and Use of Laboratory Animals, *ILAR News*, 35(3–4), 47, 1993.
2. ARENA/OLAW Institutional Animal Care and Use Committee Guidebook, NIH Publication No. 92-3415, 2nd ed., 2002, chap. A-3.
3. Office of the Federal Register, 9 CFR Parts 1, 2, and 3, *Fed. Regist.*, 54(168), 36133.
4. Division of Animal Welfare, Office for Protection from Research Risks, National Institutes of Health, Issues for Institutional Animal Care and Use Committees (IACUCs), The Public Health Service responds to commonly asked questions, *ILAR News*, 33(4), 68, 1991.
5. Federal Register, 67(152), 51289-51290DHHS NIH Laboratory Animal Welfare: Change in PHS Policy on Humane Care and Use of Laboratory Animals, Amended Policy Statement, August 7, 2002.
6. Oki, G.S.F., et al., Model for Performing Institutional Animal Care and Use Committee continuing review for animal research, *Contemp. Topics Lab. Anim. Sci.*, 35(4), 53, 1996.
7. Garnett, N.L., and DeHaven, W.R., OPRR and USDA commentary, *Lab Anim.* (NY), 27(8), 18, 1998.
8. Garnett, N.L., A word from OPRR, *Lab Anim.* (NY), 29(3), 19, 2000.
9. APHIS/AC Policy 12, Alternatives to Painful Procedures, Animal Care Resource Guide, June 21, 2000, pp. 12.1–12.3.

12

Justification for the Use of Animals

Larry Carbone*

Introduction

PHS Policy and the AWA and its regulations call on the IACUC to review the investigator's proposal to use laboratory animals. Justification of the type and number of animals must be reviewed, as must exceptions to certain specific guidelines (including multiple major survival surgery, use or nonuse of painkilling drugs, and choice of any nonstandard euthanasia method).

The word *justification* may imply moral or ethical analysis to some, or at least some sort of cost/benefit analysis, but it is used more broadly in the regulations. For the most part, scientific justification (or the scientific rationale for choosing a particular experimental design) is what the regulations require; there is little regulatory guidance in the type of justification expected. The case of multiple major survival surgery stands out in that other types of justification are considered by the *Guide*: it is acceptable for the IACUC and investigator to consider "conservation of scarce animal resources" but not "cost-savings alone" as justification for multiple surgeries.

Survey data in this chapter are presented qualitatively, to illustrate a range of institutional practices. They are not intended as a quantitative representation of the number of institutions that follow the various practices.

12:1 What is the regulatory definition of an *animal*?

Reg.   PHS Policy (III,A) defines an *animal* as "any live, vertebrate animal used or intended for use in research, research training, experimentation, or biological testing or for related purposes."

The AWAR (§1.1, Animal), based on the AWA (Section 2,g), define an *animal* as any live or dead dog, cat, nonhuman primate, guinea pig, hamster, rabbit, or any other warm-blooded animal, which is being used or is intended for use for research, teaching, testing, experimentation, or exhibition purposes, or as a pet. This term excludes birds, rats of the genus *Rattus*, and mice of the genus *Mus*, bred for use in research, and horses not used for research purposes and other farm animals, such as, but not limited to, livestock or poultry used or intended for use as food or fiber, or livestock or poultry used or intended for use for improving animal

---

* The author thanks Joseph S. Spinelli for his contribution to this chapter in the first edition of *The IACUC Handbook*.

nutrition, breeding, management, or production efficiency, or for improving the quality of food or fiber. With respect to a dog, the term means all dogs, including those used for hunting, security, or breeding purposes. (See 8:3; 12:5.) Thus, under the AWA, the overwhelming majority of animals used in laboratories (common laboratory rats and mice) are not considered animals.

12:2 In what ways is it required that the use of animals be justified on an IACUC protocol?

Reg. Institutions using animals regulated by the AWAR (see 12:1) must assure the following:

- That all proposals to the IACUC contain a rationale for involving animals, and for the appropriateness of the species and numbers of animals to be used (AWAR §2.31,e,2)
- That the PI has considered alternatives to procedures that may cause more than momentary or slight pain or distress to the animals (AWAR §2.31,d,1,ii)
- That the PI has provided a written narrative description of the methods and sources (e.g., the Animal Welfare Information Center) used to determine that alternatives to animal use are not available (AWAR §2.31,d,1,ii)
- That the PI has provided written assurance that the research activities do not unnecessarily duplicate previous experiments (AWAR §2.31,d,1,iii) (more detailed discussion of alternatives in 12:6–12:13)

The HREA (Section 495,c,2) requires applicants for grants, contracts, or cooperative agreements involving research on animals to submit "a statement of the reasons for the use of animals in the research to be conducted with funds provided under such grant or contract." Applications and proposals (competing and noncompeting) for awards submitted to PHS that involve the use of animals are required to specify the species and approximate number of animals to be used, the rationale for involving animals, and the appropriateness of the species and numbers to be used (PHS Policy IV,D,1,a; IV,D,1,b).

The *Guide* (p. 10) states, "The following topics should be considered in the preparation and review of animal care and use protocols:

- Rationale and purpose of the proposed use of animals.
- Justification of the species and number of animals requested. Whenever possible, the number of animals requested should be justified statistically.
- Availability or appropriateness of the use of less-invasive procedures, other species, isolated organ preparation, cell or tissue culture, or computer simulation."

12:3 Should the IACUC perform an ethical review of protocols?

Reg. The AWA, AWAR, PHS Policy, and the *Guide* do not use the words *ethical* and *moral*, with the sole exception of the *Guide's* statement (p. 12): "Ethical, humane, and scientific considerations sometimes require the use of sedatives, analgesics, or anesthetics in animals."

Opin. A societal consensus ethic underlies both the use of animals and constraints placed on that use: "It is wrong to inflict harm on individuals without strong justification."[1]

The governmental/societal belief that animal use *can* be justified leads not just to permitting of, but to active funding of animal studies. The belief that such use *must be* justified leads to the establishment of IACUCs and other safeguards to animal welfare.

Animals have limited rights under this system in that their interests must be weighed in any cost–benefit evaluation that occurs. Potential benefits (primarily in the form of human health, welfare, and basic knowledge) are weighed not just against financial costs (the limited funds disbursed by funding agencies or the research dollars in an industrial setting) but also against costs to animal welfare. Funding agencies and corporate administrators may weigh financial costs against potential benefit in distributing funds, but there is only limited weighing of benefit against harm to animals in any setting.

For the most part, the IACUC does not and cannot conduct this explicit ethical review. The IACUC is charged with reviewing the rationale (preferably statistical) for the animal numbers chosen, for instance, but not whether a particular line of research warrants that number. Similarly, the IACUC evaluates a technical claim that nonhuman primates alone are likely to provide the sort of data sought, not whether a particular project ethically merits the use of primates. Because the IACUC does not have the tools (or the regulatory mandate) to conduct a through assessment of the scientific merit (i.e., the potential benefits) of a proposed project, it cannot make a thorough cost–benefit ethical analysis. (See 9:14–9:18.)

The IACUC and researchers operate within a societal ethical context in which several principles are evident. These principles include the ideal that animals should only be used, and especially should only be harmed, for the highest-quality and most important research. Thus, U.S. Government Principle II states, "Procedures involving animals should be designed and performed with due consideration of their relevance to human or animal health, the advancement of knowledge, or the good of society."

Beyond this, harm to animals must be minimized through consideration and implementation of replacement, reduction, and refinement alternatives. An ethical hierarchy of species is built into the regulations and guidelines: human good, even the pursuit of basic knowledge with no evident application, is considered sufficient to justify at least some use of some animals, while a burden is placed on investigators to attempt to replace more sentient species with less sentient species.

Though IACUCs generally avoid putting themselves in the position of being ethical arbiters, most probably venture into that territory when a proposal appears to push ethical boundaries. In studies conducted by Franz Stafleu, Rebecca Dresser, and others, IACUCs are most likely to reject or return protocols when at least two of these three conditions apply: there appears to be a great deal of pain or distress anticipated; the scientific project's worth seems particularly suspect; and especially, when the animal species is one (such as primates) that is of very high concern.[2-4]

12:4   What justifications for animal use are considered appropriate?

Reg.   The AWAR (§2.31,e) require that a proposal to conduct an activity involving animals, or to make a significant change in an ongoing activity involving animals, must contain the following:

- Identification of the species and the approximate number of animals to be used
- A rationale for use of animals and for the appropriateness of the species and numbers of animals to be used

In addition, the IACUC is specifically tasked with reviewing the investigator's consideration of alternatives and assuring that the work is not unnecessarily duplicative, that animal pain is being prevented or treated, that living conditions and veterinary care are appropriate, that standards for sterile surgery are met, and that humane euthanasia techniques are employed.

PHS Policy (IV,D,1,a; IV,D,1,b) requires most of the same information. U.S. Government Principle III states, "The animals selected for a procedure should be of an appropriate species and quality and the minimum number required to obtain valid results. Methods such as mathematical models, computer simulation, and *in vitro* biological systems should be considered."

The *Guide* (p. 10) expands on this by recommending that "the following topics should be considered in the preparation and review of animal care and use protocols:

- Rationale and purpose of the proposed use of animals.
- Justification of the species and number of animals requested. Whenever possible, the number of animals requested should be justified statistically.
- Availability or appropriateness of the use of less-invasive procedures, other species, isolated organ preparation, cell or tissue culture, or computer simulation."

Opin. The IACUC must gain a good sense that the use of animals is justified by the research question being posed and that the methods chosen to answer it are likely to yield success. Peer review of scientific merit addresses most of this set of concerns (i.e., "Is this project worth this use of animals?"). This is generally not performed by the IACUC itself, but the IACUC needs to know whether this expert review has occurred.

It remains that the brunt of the IACUC's work is evaluating the scientific or technical reasons for using animals, especially in ways that may cause pain or distress, and in particular, the investigator's efforts to minimize pain and distress, through his or her search for alternatives to animals and to painful procedures.

12:5 How can the IACUC evaluate the justification for use of a particular species of animal?

Opin. Choice of species is one of the aspects of a project that expert peer reviewers should evaluate, and for the most part, the IACUC places some trust in this review. Legitimate factors include the following:

- Is the species an established model in the literature?
- Is the species an appropriate size for the samples required or the procedures performed?
- Are there particular introduced or natural genes in the species of interest?
- Can the species be maintained easily, safely, and humanely in an animal facility?
- Is the species choice mandated by a regulatory or funding agency?
- Is it an endangered species?

Regardless of the peer review at the funding agency, the PI should explain the choice of species to the IACUC, and the IACUC may have concerns beyond those of the peer review committee. Is the facility appropriate for maintaining a particular species of animal? Do the investigator and support staff have sufficient expertise in working with this species? Though the peer review process may be better suited for evaluating the search for replacement with less sentient or nonsentient subjects, the IACUC is directly charged with reviewing this consideration as well.

12:6   What outside data must an IACUC consider when evaluating an investigator's proposed justification to use animals?

Reg.   Under the AWAR (§2.31,d,1,ii) and the AWA (Section 13,a,3,B), the IACUC is required to assure that "the principal investigator has considered alternatives to procedures that may cause more than momentary or slight pain or distress" and that "the principal investigator has provided written assurance that the activities do not unnecessarily duplicate previous experiments" (AWAR, §2.31,d,1,iii.). Both imply at least some evaluation beyond the investigator's laboratory for comparative information.

The *Guide* and the HREA (Section 495,c,1,B) have similar requirements that an investigator consider alternatives and prevent unnecessary duplication of experiments. On the topic of alternatives, NIH/OPRR (now OLAW) has written, "The federal mandate to avoid or minimize discomfort, pain, and distress in experimental animals, consistent with sound scientific practices, is, for all practical purposes, synonymous with a requirement to consider alternative methods that reduce, refine, or replace the use of animals."[5] The attachment to that statement cites all of the PHS statutory and policy bases for consideration of alternatives. Of particular importance, the *Guide* (p. 10) recommends that animal care and use protocols should consider the "availability or appropriateness of the use of less-invasive procedures, other species, isolated organ preparation, cell or tissue culture, or computer simulation." PHS Policy (IV,C,1,a) states: "In order to approve proposed research … the IACUC shall determine that … procedures with animals will avoid or minimize discomfort, distress, and pain to the animals, consistent with sound research design."

Opin.   The investigator must describe literature searches, consultations, and other sources to assure the IACUC that she or he has done sufficient research to justify the following assertions:

- That the work proposed does not unnecessarily duplicate previous work, whether published or not

- That replacements to the use of live sentient animals, particularly in studies that may cause animal pain or distress, have been sought and considered

- That alternative methods that reduce the numbers of animals used, particularly in studies that may cause animal pain or distress, have been sought and considered

- That refinements to experimental procedures that may cause animal pain or distress have been sought and considered

12:7   What is meant by the search for alternatives?

Reg.   The *Guide* (p. 10) calls on investigators and IACUCs to think about "less-invasive procedures, other species, isolated organ preparation, cell or tissue culture, or computer simulation." The AWA (Sect. 13,a,3,B; Sect. 13,d,3; Sect. 13,e,3) and the AWAR (§2.31,d,1,ii) go one step further by mandating that the investigator not just consider, but actively search for, such alternatives, and that the IACUC review the investigator's search.

Opin. Russell and Burch provided a useful framework for considering alternatives to the use of live sentient animals and to the use of painful or distressful experiments.[6] They elaborated the sometimes-overlapping, sometimes-contradictory "3Rs" framework of replacement, reduction, and refinement alternatives. APHIS/AC Policy 12 clarifies that APHIS/AC adopts this framework and considers "alternatives" to include "some aspect of replacement, reduction, or refinement in pursuit of the minimization of animal pain and distress consistent with the goals of the research."[7] U.S. Government Principles III–V and the *Guide* (p. 10) call for similar consideration without invoking Russell and Burch's language. An investigator searches for alternatives *to* animal research by looking for nonanimal models that would replace animals and alternatives *in* animal research by searching for ways to reduce animal numbers and to reduce animal pain and distress.

12:8   What is meant by replacement alternatives?

Opin. Replacement alternatives are conceptually the most straightforward: find ways to generate research data without using sentient animals at all. Candidates for consideration include studying cells in tissue culture (*in vitro* techniques), developing computer simulations, making better use of human epidemiological data and human volunteers, or using inanimate models in teaching.[1] A paradigmatic current example is the production of monoclonal antibodies in cell culture, as opposed to growing them in mouse ascites fluid. Though animals may still be the source of the cells in culture, the use of whole live sentient animals in a potentially painful way is replaced with the *in vitro* technology.

More problematic is the attempt to replace more sentient animals with less sentient animals (this author strongly resists the terms *higher animal* and *lower animal*). Are frogs, fish, and octopi all reliably known to be less sentient than mammals? Among the mammals, are there any data that rodents are significantly less sentient than dogs, swine, or even primates? For the purposes of this chapter, *sentience* is defined as the capacity to experience pain and suffering. It is more than simply the ability to sense stimuli (even noxious stimuli) and less than the complex self-awareness and self-consciousness possessed by humans. There is no biological reason to assume that it is an all-or-nothing trait, possessed equally, in full, by all animals that possess it. Rather, it exists on a continuum across species, with monkeys, apes, and dolphins, as far as we know, more sentient than fish or squid.[8] (See 13:3.)

Surv. Responding institutions varied in the ways they "rank" animals in terms of sentience. Some say that they have no consistent criteria. Some consider all vertebrate animals equally sentient. Some consider all mammals equally sentient; some differentiate among mammalian species.

12:9   What is meant by *reduction alternatives*?

Opin. Reduction includes all the efforts to lower the numbers of animals used. This often means rethinking statistical tests, to use just the number necessary for statistically valid results. Reduction attempts may rely on refining the study, as when use of healthier, more genetically homogeneous animals lowers in-group variability.[1] Reduction efforts may rely on use of historical controls, when appropriate.

12:10   What is meant by *refinement alternatives*?

Opin.   Refinement alternatives are the most varied, as they comprise all the myriad ways to rethink animal care and use to reduce the potential for pain or distress.[1] Examples are listed in Table 12.1. The author recommends this exercise for investigators seeking refinements in their studies and uses it when reviewing IACUC protocols: First, visualize every step in the animal's life from the initial point of contact (birth at the facility or arrival into the facility) until the animal's euthanasia or departure from the institution. List every reasonable potential source of significant pain or distress, whether related to housing or to the experimental procedures. Each of those potential sources of pain or distress should then be addressed, whether in a targeted literature search, consultation with peers who have published on the methodology, or consultation with veterinarians and animal care specialists, and with as much creative thinking as the investigator can muster.

TABLE 12.1

Some Refinement Alternatives to Reduce Pain or Distress in Animal Research

Choice of experimental endpoints that precede onset of disease or mortality
Improved use of anesthetics and painkillers
Housing social animals in compatible groups
Using flexible tethers to replace rigid restraint devices
Replacing open surgery with endoscopic techniques
Providing supportive veterinary care
Maintaining infection-free animal colonies
Designing cages that allow animals to dig, run, climb, hide
Training animals to cooperate with research procedures
Frequent monitoring of body weight or other indicators of well-being
Using positive reinforcement in behavioral studies
Killing animals using the least painful methods

Source: From Carbone, L., *What Animals Want: Expertise and Advocacy in Laboratory Animal Welfare Policy*, Oxford University Press, New York, 2004. With permission.

12:11   How do the "three Rs" of alternatives complement or conflict with one another?

Opin.   Replacement, reduction, and refinement are not mutually exclusive concepts. Replacing some of the animals in a study reduces the overall numbers. Refining sample collection techniques may reduce the potential for pain as well as the number of animals needed. For example, if an assay requiring 1000 µl of mouse blood can be replaced with one that requires 10 µl, it may become possible to study a small cohort of animals sequentially, rather than euthanizing mice for each data time point needed (1000 µl cannot generally be collected from a mouse on a noneuthanasia basis). Studying animals sequentially may result in less intersubject variability, and thus the statistical need for animals at each time point of data collection. Moreover, 10 µl can be collected via the tail or leg veins in a less invasive procedure than collection of 1000 µl would require.

On the other hand, attention must be paid to the risk of increasing welfare costs when pursuing alternatives. If an investigator reduces animal numbers by more intensively studying a smaller cohort, those animals may be at risk of undergoing more procedures per animal, in a way that may jeopardize their welfare. Efforts to switch species can also lead to challenges. Replacing dogs with mice, for instance, can greatly increase the number of animals needed (for example, studies that require large volumes of blood collection at several time points may require a separate mouse for each time point sample, whereas a single dog could provide samples over several time points), and procedures that can be easily performed with hand restraint and minimal pain in dogs (again, blood collection is a good example) may require anesthesia and invasive techniques in smaller subjects. There are many reasons to replace dogs, or even monkeys, with mice or rats (cost, public relations concerns, regulatory issues, training requirements, health and safety issues, as well as issues of experimental design). This author remains skeptical that mice are sufficiently less sentient than larger mammals for this concern alone to drive the replacement effort, especially when animal numbers will dramatically increase.

12:12 How can the IACUC evaluate the investigator's search for and consideration of alternatives?

Reg. Under the AWAR (§2.31,d,1,ii) and the AWA (Section 13,a,3,B), the IACUC is required to assure that "the principal investigator has considered alternatives to procedures that may cause more than momentary or slight pain or distress" and "has provided a written narrative description of the methods and sources, e.g., the Animal Welfare Information Center used to determine that alternatives were not available."

APHIS/AC Policy 12 expands on this requirement, stating that the USDA believes that "performance of a database search remains the most effective and efficient method for demonstrating compliance with the requirement to consider alternatives to painful/distressful procedures," and that "when a database search is the primary means of meeting this requirement, the narrative must, at a minimum, include:

1. the names of the databases searched;

2. the date the search was performed;

3. the period covered by the search; and

4. the key words and/or the search strategy used."

Policy 12 also includes provision for conditions in which database searching is not the primary route for researching alternatives:

> When other sources are the primary means of considering alternatives, the Institutional Animal Care and Use Committee (IACUC) and the inspecting [USDA] Veterinary Medical Officer should closely scrutinize the results. Sufficient documentation, such as the consultant's name and qualifications and the date and content of the consult, should be provided to the IACUC to demonstrate the expert's knowledge of the availability of alternatives in the specific field of study. For example, an immunologist cited as a subject expert may or may not possess expertise concerning alternatives to *in vivo* antibody production.

Importantly, Policy 12 states: "Regardless of the alternatives sources(s) used, the written narrative should include adequate information for the IACUC to assess that a reasonable and good faith effort was made to determine the availability of alternatives or alternative methods. If a database search or other source identifies a *bona fide* alternative method (one that could be used to accomplish the goals of the animal use proposal), the written narrative should justify why this alternative was not used."

Note that this requirement applies not just to the initial review of a study, but to review of any proposed major modification to an ongoing study. It further states, "Although additional attempts to identify alternatives or alternative methods are not required by Animal Care at the time of each annual review of the animal protocol, Animal Care would normally expect the principal investigator to reconsider alternatives at least once every 3 years, consistent with the triennial review requirements of the Public Health Service Policy."

Opin.   Institutions should comply with APHIS/AC Policy 12, especially for AWA-regulated species, but should also recognize its limitations. The statement that "performance of a database search remains the most effective and efficient method" for searching alternatives assumes that most of the relevant information is found in published, indexed articles. This author does not share that assumption. Very useful information can be found through consultation with others performing similar work, and with veterinarians and animal care specialists who focus on laboratory animals. That said, the author believes it would be the rare project in which consultation with experts were the sole or even primary route to alternatives and would advise most investigators to include the electronic resources described later. (See 12:13.)

Yerk The policy, as written, implies that an investigator decides that "no alternatives were available" although often, in reality, an investigator will retain a painful or distressful procedure in the protocol while still pursuing alternatives that reduce the degree of pain or distress or the number of animals undergoing that procedure.

12:13   What sources, resources, and methods are useful when searching for alternatives to animal use?

Opin.   An information search that fully meets the intent of the AWAR requires several approaches, some of them almost opposites. Assuring, for example, that a proposed study of a particular cytokine's activity in intestinal lymphoid tissue does not unnecessarily duplicate other work requires a targeted search in the relevant immunological literature, with some strategy (such as conference attendance) for learning of abstracts, posters, and other evidence of work not yet published. That same immunological literature might provide some information relevant to the search for replacements to whole live mice for the study. However, if the proposed work requires a surgical approach to the intestines, then a far more general search on mouse surgery, anesthesia, and analgesia is warranted, one that should hardly be confined to the immunological literature.

There are dozens of Internet-based resources for conducting a search for alternatives. Some of those of the most general interest are described in the following; specific disciplines also have specialized resources.

The 1989 AWA amendment that mandated IACUCs and the search for alternatives also established the Animal Welfare Information Center (AWIC) within the National Agriculture Library. AWIC maintains a Web site with many useful links (http://www.nal.usda.gov/awic/). Its staff periodically updates and publishes "quick bibliographies"—literature searches on select topics of fairly general interest, such as "rodents." The site also includes information, backed up by staff information specialists available for consultation, on procedures to conduct literature searches.

For biomedical research, PubMed is certainly the most utilized database search engine. It is maintained by the National Library of Medicine and is accessed on the Web at http://www.ncbi.nlm.nih.gov/entrez/query.fcgi. Its search capabilities include "old Medline" as well as in-process citations not yet available in Medline.

Agricola is the National Agriculture Library's article citation database (http://agricola.nal.usda.gov/). Its agriculture and animal health focuses complement PubMed, and several journals directly related to animal welfare, not indexed in PubMed, are indexed in Agricola.

Altweb (http://altweb.jhsph.edu/) is a project of the Center for Alternatives to Animal Testing at the Johns Hopkins Bloomberg School of Public Health. It has a limited, but growing, number of searches that can be conducted through its online interface (such as the Altweb Anesthesia/Analgesia Database), as well as links to several other databases. The University of California Center for Animal Alternatives (http://www.vetmed.ucdavis.edu/Animal_Alternatives/main.htm) shares many features of Altweb and AWIC. It has some searches on specific topics available, in which the center librarian has set up a search template that the investigator can then run as a live search via the Internet. Most of these search templates are on fairly general topics (such as analgesia, blood collection, or euthanasia) and can be run through both PubMed and Agricola. The site also provides useful guidance on conducting an alternatives search.

The author finds great value in direct communication, especially for refining experimental procedures. Laboratory animal veterinarians and other professionals have access to the American Association for Laboratory Animal Science's Compmed list-serve and archives (http://www.aalas.org/). Online discussions allow shared anecdotal information on research animal care and use, and even shared animals and animal tissues, none of which would be retrieved through a literature search engine such as PubMed.

Investigators developing new experimental techniques should be encouraged not simply to read the Materials and Methods section of published papers, but to contact and visit laboratories conducting the procedure directly, to learn the subtle hands-on knowledge required to perform a procedure or maintain a particular line of animals competently or other applied information.

Finally, commercial search engines (such as Google and Yahoo) also are very useful at times, especially for identifying individuals and institutions that have experience relevant to particular animal models.

12:14 Since the AWAR definition of an animal includes dead animals (see 12:1), should the IACUC request a justification for the use of dead animals?

Reg. The PHS Policy does not require IACUC review and approval of the use of dead animals unless they are killed for the purpose of being used in PHS-supported activities; in that case, the IACUC is really reviewing the use of live animals being euthanized for research, teaching, or testing and the focus is on animal care and handling before and during euthanasia, not after. (See 8:7; 14:28.) The AWAR is less straightforward. While the statutory definition of *animal* includes dead animals used or intended for use in research, the definition of a *research facility* is one using or intending to use live animals (AWAR §1.1, Research Facility). Therefore, reference to committee review of use of animals would seem to apply solely to use of live animals (§2.31). APHIS/AC Policy 28 reinforces this interpretation, by emphasizing that the reference to dead animals is "aimed at preventing stolen pets from being sold for covered purposes."[9]

Opin.   From a legal point of view, justification does not appear to be required. It is an excellent application of alternatives to make maximal use of euthanized animals by distributing needed tissues to several investigators, rather than requiring each to purchase animals separately. IACUC review could slow the process and increase the burden of in-house tissue sharing with no improvement in animal welfare and should not be required. A middle position between total oversight and no oversight would be to allow in-house tissue sharing without IACUC oversight but to require oversight if dead animals are being received through purchase or donation from outside the institution.

12:15   Should the IACUC request a justification for euthanasia of animals?

Reg.   Both the AWAR (§1.1 Euthanasia; §2.31,d,1,i; §2.31,d,1,v) and the *Guide* (pp. 65–66) require the IACUC to review the proposed method of euthanasia and stress the need to minimize pain and distress. Methods likely to cause animal pain or distress must be explicitly approved by the IACUC. Beyond this, there is no special justification required for killing animals (though the numbers of animals killed must be justified, as must the numbers of animals handled and used in any manner). While euthanasia need not be justified to the IACUC, refusal to euthanize animals that experience untreatable pain or distress does require very special justification. The AWA (Sect. 13,a,3,C,v) requires that "that the withholding of tranquilizers, anesthesia, analgesia, or euthanasia when scientifically necessary shall continue for only the necessary period of time." U.S. Government Principle VI states, "Animals that would otherwise suffer severe or chronic pain or distress that cannot be relieved should be painlessly killed at the end of the procedure or, if appropriate, during the procedure." (See Chapter 17.)

12:16   Should the IACUC request justification for the use of animals in field studies?

Reg.   The AWAR (§1.1, Field Study) state that a field study is "a study conducted on free-living wild animals in their natural habitat. However, this term excludes any study that involves an invasive procedure, harms, or materially alters the behavior of an animal under study." Under the AWAR (§2.31,d,1) animals involved in field studies are exempt from IACUC review.

PHS Policy does not distinguish between field studies and laboratory studies and, therefore, requires IACUC review and approval of field studies if covered under an Animal Welfare Assurance.

Opin.   If invasive procedures occur in field studies, the AWAR exemption does not apply. Materially altering animal behavior is left undefined, including duration of any alteration. Studies that require capture and brief restraint for marking or tagging animals may not be invasive, but they may affect animal behavior (especially if anesthesia is used), and the marking or tagging method may itself affect the animals in various ways. Also, assuming the project is performed on vertebrate species, animals in field studies are not excluded from IACUC consideration in institutions covered by the PHS Policy, regardless of the invasiveness or behavioral effects of the study. For consistency, it is a good policy to grant the same review of justifications for field studies as for nonfield studies. Institutional officials should also remember that many tagging systems identify the institution placing the tag and will want a record that the study was duly reviewed if animals with problems are later found and identified as institutional research subjects. This author would recommend that the IACUC, not just the investigator, determine whether a particular field study might pose significant risk to animals to warrant IACUC review.

12:17   Is scientific merit reviewed as a component of the justification of use of animals?

Reg.   Scientific merit includes both the importance of the scientific question being asked and the likelihood that the proposed project will answer it. U.S. Government Principle II states, "Procedures involving animals should be designed and performed with due consideration of their relevance to human or animal health, the advancement of knowledge, or the good of society." It does not state that an IACUC is the body to make this assessment, but rather places it on the IO. The proscription against unnecessarily duplicating existing work relates directly to assessing the importance of the question being asked, and this is a consideration the IACUC is charged with reviewing. PHS policy and AWAR charge the IACUC with reviewing some ancillary components of the project's likelihood of answering the question, such as the qualifications of the personnel involved. Neither requires that the IACUC review whether the basic methodology proposed is likely to answer the research question. (See 9:14–9:18.)

The NIH has provided some guidance on this topic:

> Peer review of the scientific merit of a proposal is considered to be the purview of the PHS funding component, acting through an initial review group (IRG). The PHS Policy requires the funding component to verify that the IACUC has reviewed and approved animal activities before the PHS awarding unit makes an award. Additionally, the IRG has the authority to raise specific animal concerns. The primary focus of the IRG is scientific merit, whereas the primary focus of the IACUC is animal welfare. It is evident, however, that there is some overlap of function between the two bodies.
>
> Although not intended to conduct peer review of research proposals, the IACUC is expected to include consideration of the U.S. Government Principles for the Utilization and Care of Vertebrate Animals in Testing, Research, and Training (PHS Policy) in its proposal review process. Principle II calls for an evaluation of the relevance of a procedure to human or animal health, the advancement of knowledge, or the good of society. Other references (sections IV.C. I and IV.D. 1.) include language such as "consistent with sound research design," "rationale for involving animals," and "in the conduct of scientifically valuable research," which presumes that the IACUC will consider in its review the general scientific relevance of the proposal. The presumption is that a study that could not meet these basic tests would be inherently invalid or wasteful and, therefore, not justifiable.[10]

Opin.   Any review of the ethical justification of animal use includes some cost–benefit reckoning: is the potential pain, death, or distress of animals worth the benefits that will accrue to the potential beneficiaries of the knowledge gained? Thus, scientific merit certainly is an inherent component of the justification of the use of animals, as nonmeritorious research can hardly justify harming animals. However, the IACUC is not necessarily the body to conduct this review. Depending on the field of research under consideration and the makeup of the IACUC, there may not be sufficient expertise to adjudicate at other than a superficial level the value of the research question or the methodology proposed. Even when one or two IACUC members have relevant expertise, this expertise may be inferior to that of a panel of experts reviewing a project. For these reasons, IACUCs rightly defer in most cases to peer review panels at granting agencies.

While the IACUC may not be the body that assesses the scientific merit of a project, it should know that peer review has been performed. The IACUC needs to know that this review was favorable, and done by a body the IACUC finds credible. Negative peer review comments may also be useful.

The IACUC may also review and approve studies that have not had a peer review process at a funding agency. For example, internal funds may be available to help investigators perform preliminary work necessary for writing a successful grant application. In such circumstances, some in-house departmental review committee or ad hoc expert consultation may be sought. The IACUC may decide to require and accept such reviews rather than try to assess the scientific merit on its own.

A valuable source of peer review feedback could be the reviewer comments on failed or unfunded applications, which could be especially useful to a departmental committee reviewing one of their colleagues' proposals. Unlike limited grant dollars that are intended only for the very best proposals, there is no theoretical limit to the number of studies an IACUC or a departmental review committee can approve. With detailed funding agency comments in hand, the departmental committee and the IACUC would know whether they are being asked to approve a project that has not been deemed good enough for external funding and could know whether the protocol at hand has been revised to address the funding agency's concerns.

12:18   Is a requirement by a funding agency to use animals, or to use animals in particular ways, considered adequate justification?

Reg.   APHIS/AC Policy 12 states:

> The written narrative for federally-mandated animal testing (for example, testing product safety/efficacy/potency) needs only to include a citation of the appropriate government agency's regulation and guidance documents. Mandating agency guidelines should be consulted since they may provide alternatives (for example, refinements such as humane endpoints or replacements such as the Murine Local Lymph Node Assay) that are not included in the Code of Federal Regulations. If a mandating agency-accepted alternative is not used, the principal investigator should explain the reason in the written narrative.[7]

This author is not aware of a similar exclusion for agency-mandated animal use of the PHS or NIH/OLAW.

Opin.   Adjudication as to what is proper use of animals is not dictated *solely* by granting agencies. The institution, through its IACUC, must determine whether the justification to use animals is appropriate, and the measure that IACUCs use is *scientific* justification. If a granting agency requires the use of animals, or a particular species or strain, their reasons for doing so could be submitted to the IACUC for its consideration. However, the IACUC must base that decision on the parameters previously noted, not just on the policy of the granting agency. Other details of a protocol may also be mandated either by a granting agency or by a regulatory agency that will review the data. For example, approval of veterinary vaccines may require submission of vaccine trial data that use death as an end point. Though efficacy may be well demonstrated long before mortality, or even before significant morbidity, animal lives would be wasted on such a study if the efficacy would not be accepted by the Department of Agriculture, which licenses such vaccines.

Approving such a study therefore means approving potentially painful procedures when alternatives have been rejected not for scientific but for regulatory reasons.

There may be occasions when the IACUC finds itself in a bind, asked to approve agency-mandated procedures that the IACUC considers unjustified. An IACUC might strike a balance in which it approves some animal use, but also initiates its own correspondence to persuade the agency in question to change its mandate or, at least, better explain and justify it.

12:19 How much detail should the IACUC require relative to animal identification in the IACUC protocol and in records? For example, should the IACUC request that rats be identified as "rats" or *Rattus norvegicus* or "F344/Crl rat"?

Reg.   The *Guide* (p. 46) states that the strain or stock of an animal should be recorded with other information on that animal's identification card, and that standardized strain and substrain nomenclature be used (p. 48). It does not state that the IACUC must review and approve the use of a particular strain.

Opin.   While nothing specific in the AWAR, the PHS Policy, or the *Guide* requires identification of animals by strain on IACUC protocols, this author believes that description of phenotype is essential. Often the reason a given species is used is the special characteristics of a particular strain or construct. Likewise, many of the challenges of maintaining animal welfare relate to genetically based developmental problems or susceptibility to illness. When such strains are to be used, if one is to justify animal use, those special characteristics must be described and plans to manage a phenotype that may cause pain or distress must be outlined.

IACUCs have broad authority and if they want to require that investigators use official nomenclature of strains on IACUC applications, they are free to do so. The strain or construct should be named by using appropriate nomenclature (see *Guide*, p. 48), though this is far less important to the goal of animal welfare than careful characterization of the phenotype.

12:20 Are there any federal requirements dictating that animals or certain species of animals must be used in certain forms of research or product safety testing?

Reg.   The AWA and the HREA do not include mandates for animal testing of any sort. The Food and Drug Administration (FDA) and the Environmental Protection Agency (EPA) are the two federal agencies most responsible for the review of product safety testing. Testing requirements are complicated by the wide range of medicines, medical devices, environmental chemicals, and other products they regulate. For the most part, these agencies provide general guidance on the types of testing data required for licensing a product. In some instances Requests for Proposals or contract stipulations from the FDA suggest the use of a particular species. For example, the FDA *Guidelines for Preclinical and Clinical Evaluation of Agents Used in the Prevention or Treatment of Postmenopausal Osteoporosis* state, "Because no animal species duplicates all of the characteristics of human osteoporosis, it is felt that an examination of bone quality in two species is necessary to adequately investigate the effectiveness and safety of drugs for this indication. One study should be conducted in the ovariectomized rat model and the second in a nonrodent model (i.e., larger, remodeling species) which will be left to the discretion of the sponsor."[11] The FDA similarly recognizes consensus standards on product safety testing, in part to harmonize with other nations' regulatory practices.[12] Some of these standards may be quite explicit; for example, the Association for the Advancement of Medical Instrumentation standard test for skin irritation calls for use of three healthy young adult albino rabbits for initial testing.[13] Companies developing medical devices document their adherence to these standards and justify the applicability of these standards in submitting safety data to the FDA.

Opin.   Individuals not already thoroughly familiar with testing data required for a particular type of product or device will find value in consultation with the relevant regulatory agency before initiating animal studies.

12:21   What is the position of the FDA relative to using animals for cosmetics testing?

Reg.   The FDA has stated:

> The FD&C Act does not specifically require the use of animals in testing cosmetics for safety, nor does the Act subject cosmetics to FDA premarket approval. However, the agency has consistently advised cosmetic manufacturers to employ whatever testing is appropriate and effective for substantiating the safety of their products. It remains the responsibility of the manufacturer to substantiate the safety of both ingredients and finished cosmetic products prior to marketing. ... Animal testing by manufacturers seeking to market new products is often necessary to establish product safety. ... We also believe that prior to use of animals, consideration should be given to the use of scientifically valid alternative methods to whole-animal testing.[14]

12:22   Should an IACUC approve development of an animal model if another well-established model exists, simply because the investigator has no experience with the established model?

Opin.   The IACUC should evaluate each protocol on the basis of how well the investigator has justified the use of whichever species, strain, or construct is proposed. One issue is the appropriateness of the animal model. Crucial questions would include whether development of a new model required large numbers of animals or invasive work solely to validate the model, the costs to animal welfare in the investigator's learning to use the established model, and the likelihood—a question of scientific merit—of the new model's relating to the established literature sufficiently that results can be accepted through a peer review process.

12:23   Since many people believe that cephalopods are sentient and because they have a well-developed nervous system, does an IACUC have the authority to include such animals under its purview? What about other invertebrates?

Opin.   Neither the AWAR nor PHS Policy covers the use of invertebrate animals. Relative to invertebrates, the *Guide* (p. 2) states that "many of the general principles in this *Guide* apply to these species and situations." There are no restrictions that either of these agencies has placed on IACUCs related to oversight of animal care and use activities that exceed the scope of their regulations or policies. Therefore, an institution can confer upon its IACUC the authority to oversee the care and use of invertebrates.

12:24   Must the IACUC review and approve the use of animals destined to be used as a food source for other animals being used in IACUC-approved studies?

Opin.   Studies of animal predation that involve feeding live or euthanized animals to others are, of course, research projects that must receive IACUC review and approval. Beyond that, the AWAR and PHS Policy do not explicitly call for IACUC approval for animals maintained as dietary items for other animals. NIH/OLAW has published guidance that in such a case, if the animals are not covered through the IACUC's protocol review process, the IACUC should oversee this use as "a covered component of the institutional program of animal care and use."[15] (See 8:33.)

12:25   Must the IACUC assure that the animal use activity does not duplicate previously conducted work? If so, how is this done?

Reg.   The AWAR (§2.31,d,1,iii) require that "the principal investigator has provided written assurance that the activities do not unnecessarily duplicate previous experiments." Elaborating in its Policy 12 on alternatives searches, APHIS/AC also includes the investigator's assurance of no unnecessary duplication.

Opin.   The IACUC must not solely accept the investigator's claim that the proposal does not entail unnecessary duplication but apparently must review how the investigator knows this to be true. Though the search strategy for nonduplication differs from the search for alternatives, a simple approach to documenting compliance is to include a statement on the investigator's signed protocol application that the work does not entail unnecessary duplication; in addition, the documentation for an alternatives search should include information on how that would uncover unnecessary duplication.

   The key word here is *unnecessary.* Duplication is permitted if it serves a valid scientific purpose, such as confirming another's work or validating laboratory assay systems. To determine whether the duplication is or is not necessary, the IACUC needs to evaluate the justification of the investigator who proposes the duplication.

Surv.   Responding institutions range from those that claim that "we currently have no way of assuring unnecessarily duplicative work" through those that require simply a signed statement by the PI to those that require a detailed description of search keywords and even search findings.

12:26   Is the use of animals in teaching protocols considered necessary duplication?

Opin.   There are no clear federal laws, regulations, or policies pertinent to this question. A genuine consideration of alternatives, not a simple statement that the protocol is for teaching, is warranted. The important question is not whether duplication (i.e., teaching a new cohort of students using procedures employed in previous semesters) is necessary, but whether animal use, especially invasive animal use, is necessary. A careful alternatives search on this topic would include consideration of published studies on the educational effectiveness of live animal versus other teaching methods and consultation with faculty at other institutions who are employing alternative methods. Though animal laboratories may be retained in the curriculum, refinements may be appropriate. These could include videotaping some animal laboratories to show future generations of students, requiring preliminary skills development with nonliving models prior to handling live animals, or delegating the most challenging part (e.g., pithing or anesthetizing an animal) to an experienced professor rather than an inexperienced roomful of students.

Surv.   The vast majority of responding institutions recognize some circumstances in which duplication might be justified, such as confirmation of previous and others' findings, certain types of testing protocols, and certain teaching protocols.

Acknowledgment

The assistance of Drs. Felix Vega and David Takacs in the preparation of this chapter is greatly appreciated.

References

1. Carbone, L., *What Animals Want: Expertise and Advocacy in Laboratory Animal Welfare Policy*, Oxford University Press, New York, 2004.
2. Dresser, R., Developing standards in animal research review, *J. Am. Vet. Med. Assoc.*, 194, 1184, 1989.
3. Stafleu, F.R., The Ethical Acceptability of Animal Experiments as Judged by Researchers, doctoral thesis, Universiteit Utrecht, Utrecht, the Netherlands, 1994.
4. Stafleu, F.R., et al., The Influence of Animal Discomfort and Human Interest on the Ethical Acceptability of Animal Experiments, in *Welfare and Science: Proceedings of the Fifth Symposium of the Federation of European Laboratory Animal Science Associations*, Royal Society of Medicine Press, Brighton, U.K., 1994.
5. National Institutes of Health, Office of Extramural Research, Office of Laboratory Animal Welfare, OPRR Reports, No. 98-01, Production of Monoclonal Antibodies Using Mouse Ascites Method, 1997. Available on the World Wide Web at: http://grants.nih.gov/grants/olaw/references/dc98-01.htm.
6. Russell, W.M.S., and Burch, R.L., *The Principles of Humane Experimental Technique*, Methuen, London, 1959.
7. United States Department of Agriculture, Animal and Plant Health Inspection Service, Animal Care, Animal Care Resource Guide, Policy 12 (updated June 21, 2000): Written Narrative for Alternatives to Painful Procedures. 2001. Available on the World Wide Web at: http://www.aphis.usda.gov/ac/policy12.html.
8. DeGrazia, D., *Taking Animals Seriously: Mental Life and Moral Status*, Cambridge University Press, Cambridge, U.K., 1996.
9. United States Department of Agriculture, Animal and Plant Health Inspection Service, Animal Care, Animal Care Resource Guide, Policy 28 (updated September 30, 1999): Licensing Sales of Dead Animals, 1999. Available on the World Wide Web at: http://www.aphis.usda.gov/ac/policy/policy28.pdf.
10. Division of Animal Welfare, Office for Protection from Research Risks, National Institutes of Health, The Public Health Service responds to commonly asked questions, *ILAR News*, 33(4), 68, 1991. Available on the World Wide Web at: http://grants.nih.gov/grants/olaw/references/ilar91.htm#7.
11. United States Food and Drug Administration, Guidelines for Preclinical and Clinical Evaluation of Agents Used in the Prevention or Treatment of Postmenopausal Osteoporosis, Division of Metabolic and Endocrine Drug Products, 1997. Available on the World Wide Web at: http://www.fda.gov/cder/guidance/osteo.pdf.
12. Food and Drug Administration, Center for Devices and Radiological Health, Recognition and Use of Consensus Standards: Final Guidance for Industry and FDA Staff, 2001. Available on the World Wide Web at: http://www.fda.gov/cdrh/ost/guidance/321.html.

13. Association for the Advancement of Medical Instrumentation, *AAMI Standards and Recommended Practices. Vol. 4. Biological Evaluation of Medical Devices*, Association for the Advancement of Medical Instrumentation, Arlington, VA, 1997.

14. United States Food and Drug Administration, Center for Food Safety and Applied Nutrition, Animal Testing, 2005. Available on the World Wide Web at: http://www.cfsan.fda.gov/~dms/cos-205.html.

15. Garnett, N.L., A word from OLAW, *Lab Anim.* (NY), 32(10), 19, 2003.

13

Justification of the Number of Animals to Be Used

Ed J. Gracely

Introduction

Choosing an appropriate number of subjects is an important part of any research project. A proper sample size is essential to obtaining valid results and to minimizing the number of individuals exposed to the potential risks and harms of research. For this reason, the IACUC is interested in ensuring that a sufficient but not excessive number of animals are used. A study with too few animals is ethically problematic because the animals are being subjected to potentially painful procedures or loss of life with relatively little likely benefit to the advancement of knowledge. On the other hand, a study using more animals than are truly needed is also problematic, because it unnecessarily exposes some of them to the same harms.

This chapter wrestles with key questions that may arise in determining animal numbers for IACUC purposes. Although the existing regulations and policy are not very explicit and IACUCs differ on many of these issues, experience, common sense, and the results of an informal survey all help provide direction. The survey results emanate from a questionnaire sent primarily to academic institutions.

13:1 What do the AWAR and PHS Policy state with respect to justifying the number of animals requested in the protocol?

Reg.   The AWAR (§2.31,e,1; §2.31,e,2) state that the proposal to use animals must include "identification of the species and the approximate number of animals to be used" and "a rationale for involving animals and for the appropriateness of the species and numbers of animals to be used." U.S. Government Principle III states, "The animals selected for a procedure should be of an appropriate species and quality and the minimum number required to obtain valid results." The PHS Policy (IV,D,1,a; IV,D,1,b) also states that all applications and proposals that are submitted to the PHS and that involve animals must include "identification of the species and approximate number of animals to be used" and "rationale for ... appropriateness of the species and numbers to be used."

     The *Guide* (p. 10) states that "whenever possible, the number of animals requested should be justified statistically." This is the most specific statement of how animal numbers should be justified in any of these documents.

Opin.   The phrase *approximate number* appears in both the AWAR and the PHS Policy, without further explanation. Needless to say, all animal number estimations are approximate in some sense. Animals die, procedures fail, unexpected events occur, and new research directions emerge. Furthermore, it is always possible to amend a protocol to deal with such eventualities or with data more variable than had been anticipated, thus requiring additional animals per group. If the wording in the law and the policy is interpreted as primarily intended to encompass these sorts of uncertainties, then that wording is consistent with requiring an initial estimate that is specific, but understood implicitly to be approximate.

It is possible that the regulations may have been intended to allow a broader type of flexibility (such as "from 200 to 250 animals will be required"), but this interpretation does not follow from the wording.

13:2   What constitutes sufficient justification for a requested number of animals?

Opin.   The AWAR and PHS Policy do not address this issue. Thus, it is left to the institution to develop appropriate policies. Clearly, the IACUC must be able to determine how all of the requested animals will be utilized, which research questions will potentially be answered by the number to be used, and why those questions could not be answered with fewer animals.

Describing animal use is generally straightforward, except in studies that are open ended and have evolving goals. In the latter instance, some attempt at providing a sequence of events and a desired end point should be made. This author is not comfortable with a justification of numbers based solely on an ongoing need for a certain number of animals per year, without an explanation of what is likely to be accomplished in each specific period.

An interesting set of three comments by researchers and veterinarians on the justification of numbers for an open-ended study has been published.[1-4] Four invited commentators (in three articles) responded to a scenario in which an established scientist is planning a 3-year series of studies requiring a total of 5500 mice. The project had already been funded (pending IACUC approval) by the NIH. The discussion centered around the fact that the justification of numbers is brief for such a large number of animals, being basically a statement of animal needs per week multiplied by 3 years × 52 weeks = total number. One commentator was fairly satisfied with the justification, especially given that the NIH review panel was satisfied. This respondent wanted only a few examples of the kinds of studies to be done. The second comment (written by two respondents) listed a variety of questions, mainly about ways the researcher might reduce the required animals or better justify the weekly details (for example, could more cells be obtained from each animal by using improved techniques?). It was not clear whether this pair of commentators would have accepted the unspecified research to be done if these more practical questions had been satisfactorily answered. The third respondent took objection to the whole process, arguing that if the researcher could not explicitly lay out the studies to be done over 3 years, then approval should be requested only for that series that *could* be planned in advance, with amendments to request more animals as needed. Clearly this is a topic that generates strong and diverse feelings!

Justifying the specific animal numbers for individual research questions is often difficult as well, and there is no perfect way to do it. Previous research, while sometimes helpful, is only a crude guide to the number of animals needed in the present study. Even statistical power analysis requires assumptions (such as the magnitude of effects worth looking for) that are often subjective or that must themselves be estimated from other data.

This author believes that sample size justification should be seen as a means to ensure that researchers and IACUCs are aware of the issues and are attempting to prevent serious miscalculations. A moderate amount of detail is sufficient to accomplish this. Consider three specific approaches, each of which may be adequate when it is applicable:

- Report of a power analysis, with enough information provided to show that the researcher knows how to analyze the data and use power analysis techniques. Question 13:4 lists key considerations of a fairly thorough power analysis. A briefer approach may be acceptable, but this author's experience suggests that attempts at an abbreviated power analysis often produce descriptions that are garbled or lack key information.

- Citation of previous research, with sufficient information provided to indicate that the previous research is similar enough in concept and methodology to the present proposal as to make it a reasonable model for sample sizes in the latter.

- Derivation of animal numbers from material needs, when the study has no statistics (e.g., histological characterization of tissues), with a clear indication why the specific amount of material is needed, and why the number of animals requested is appropriate to provide that amount of material.

Survey questions 1 and 2 focus on instances in which inferential statistics are not relevant because the researcher is determining sample size from the amount of material needed for histological or biochemical analysis. The majority (53–55%) of the institutions for whom the question applied indicated an intermediate level of requirements in these situations.

Surv. 1 Some protocols may not require any statistics (e.g., one that focuses on the detailed histological analyses of a certain tissue). When your IACUC is presented with such a protocol, which of the following is the minimum it will accept for sample size justification?

- Not applicable to our institution   15% (25/164)

Of the remaining responses:

- Few details: A statement that N animals are needed to provide enough tissue for all analyses to be performed   22% (30/139)

- Simplified calculations: The PI must at least indicate the total amount of each type of tissue needed, with perhaps a brief explanation; then the amount available from each animal is indicated, and a calculation of total animal numbers is made from these values   53% (73/139)

- Detailed explanation: The PI must justify the number of histological observations required; then the amount of each type of tissue available from each animal is indicated, and a calculation of total animal numbers is made from these values   24% (34/139)

- Other   1% (2/139)

Surv. 2 Another example of a situation in which the determination of sample size may not involve statistical considerations is one focusing on the quantity of materials needed, such as one or more proteins for biochemical analyses. When your IACUC is presented with such a protocol, which of the following is the minimum it will accept for sample size justification?

- Not applicable to our institution   16% (27/165)

Of the remaining responses:

- Few details: A statement that N animals are needed
  to provide enough material for all assays to be
  performed   23% (32/138)
- Simplified material calculations: The PI must indicate
  the total amount of product needed for all analyses and
  the amount that can be acquired from each animal; a
  calculation of total animal numbers is made from these
  values   55% (76/138)
- Full details of material and animal calculations: The PI
  must justify why a particular amount of product is
  required for each assay (or group of assays) and the
  amount of product available from one animal; then a
  calculation of the total animal numbers required
  is made from these values   20% (28/138)
- Other   1% (2/138)

Survey questions 3–6 focus on studies that include some statistical analyses as an integral component. Only about a quarter of committees *require* power analyses when relevant (survey 3), although another 45% prefer them. Of those protocols with a power analysis provided (whether required or not), committees run the gamut of details expected, with the modal requirement being a simplified description of the power analysis (survey 4). About half of all committees will accept a justification based on the PI's experience (survey 5), whereas almost 80% will accept a justification based on similar published studies (survey 6). Since so little specific guidance is given in the regulations, and even the *Guide* only *recommends* the use of statistical methods (interpreting *should be* as weaker than *must be*) the onus is on the individual IACUCs to determine when sufficient justification has been provided.

Unfortunately, there are serious problems with some of these data. For example, 23 of 34 respondents who said that they require a power analysis answered "yes" to survey question 6, indicating that they would accept, in a statistical-type study, a justification based on similar published research. These answers were meant to be mutually exclusive, in that committees that "require" a power analysis in studies involving statistical analyses should not accept anything else. Perhaps the nuances of what is and is not an inferential statistics study or a statistical-type study were unclear to respondents. The data are presented as given, but readers should be alert to possible confusion.

Surv. 3 Does your IACUC require a statistical power analysis for those protocols involving inferential statistics?

- Yes   23% (34/149)
- No, it is preferred but not required   45% (67/149)
- No, other justification methods are fully acceptable with
  no preference for a power analysis   28% (41/149)

- Other: 5% (7/149)
  - Not applicable   (4/7)
  - No policy   (1/7)
  - Case-by-case decisions   (1/7)
  - Other   (1/7)

Surv. 4 Some protocols require inferential statistical analysis (e.g., a study testing whether rats of three different ages have different mean levels of certain plasma proteins). If a statistical power analysis is the primary basis a PI provides for justifying the sample size in a particular protocol, what is the minimum statistical information your IACUC would ordinarily accept?

- A full statement of the statistical analyses and power: The statistical analyses performed, how the power analysis was tailored to those analyses, a description of basic parameters such as effect sizes and standard deviations, and how they were estimated, etc., plus the results of the power analysis   13% (20/154)
- Full details of power: A power analysis statement with effect sizes, standard deviations, etc., along with the results of the power analysis, but without the full description of the analyses and some of the explanatory details in the first bullet above   12% (18/154)
- Simplified but adequate power: A basically correct, brief statement of a power analysis, without details but with enough information as to indicate competence in running it   46% (70/154)
- Any power analysis: If a power analysis is provided, the IACUC accepts it with little or no attempt to determine if it is properly done or to require specific details   10% (16/154)
- Statement of having been done: The statement that a power analysis has been done is sufficient, even if no details are provided   11% (17/154)
- Other: Six of these were Not Applicable, the remainder a mix of responses, mostly not informative   8% (13/154)

Surv. 5 Does your IACUC accept sample size justification in a statistical-type study that is based on a statement of previous experience of the PI, and nothing else?

- Yes   46% (72/156)
- No   51% (80/156)
- Other   3% (4/156)

Surv. 6 Does your IACUC accept sample size justification in a statistical-type study that is based on references to similar work in published literature?
- Yes   79% (120/152)
- No   18% (28/152)
- Other   3% (4/152)

Survey question 7 returned the result that a small percentage of committees will at least sometimes accept studies with substantial numbers of animals without any justification. Committees should review the regulatory requirements noted in 13:1. While justification need not be elaborate, it must be provided. A related question (not shown) asked whether IACUCs would accept protocols (otherwise unspecified) with no justification: did they see the number justification as "outside the purview of the IACUC"? About 4% said yes, 5% sometimes. Again, however, it is stressed that while power analyses are not required by law, justification of numbers *is* required. The answer should have been uniformly "No. We always expect the PI to provide a justification for the numbers used." Even pilot studies and other preliminary data collections require an explanation for why the requested numbers are optimal.

Surv. 7 Would your IACUC accept a protocol involving substantial numbers of animals, with neither a power analysis nor other serious attempts at justifications for sample size as listed in the preceding questions?

- Yes   5% (8/158)
- No   93% (147/158)
- Other   2 %(3/158)

Survey question 8 is the last one in the general justification of numbers set. It asked respondents about the more difficult situation in which research is partly exploratory, with ill-defined goals. Most committees are willing to approve the entire set of animals at the beginning, with varying levels of explanation required. However, about a quarter of the committees will not approve the entire set of animals under these circumstances. Only a small percentage will accept a justification based on the number of experiments that can be done per available time.

Surv. 8 Apart from pilot studies, in some research decisions about the number of animals needed later in the study can best be determined by analyzing the results of earlier parts of the study. How does your IACUC approach this problem?

- Not applicable   6% (10/160)

Of the remaining responses:

- We require an estimate of the number of animals needed based on the most likely outcome   54% (81/150)
- We approve the animals in sets, with only the immediately justifiable numbers being approved at the time of initial IACUC approval   27% (40/150)
- We allow the PI to "guesstimate" the animal numbers needed, without a justification for a particular likely outcome   7% (10/150)
- We allow the researcher to justify the numbers based largely on the number of experiments that can be carried out in the available time period   11% (17/150)
- Other   1% (2/150)

Attrition is a common problem. As shown in survey question 9, most IACUCs require some explanation or justification, and a substantial minority of committees require documentation.

Surv. 9 If a PI requests extra animals to account for attrition during experimentation, what degree of explanation or justification is required by your IACUC?

- Not applicable   12% (19/160)

Of the remaining responses:

- We almost always request an explanation and documentation to support the expected rate of attrition   41% (58/141)
- A statement of specific experience (e.g., "in our previous research we typically had *X*% attrition due to ...") but no additional documentation   38% (53/141)
- An explanation is only needed if the IACUC deems the attrition to be excessive   15% (21/141)
- No explanation needed; the researcher's attrition level is accepted as given   6% (8/141)
- Other   1% (1/141)

13:3 Relative to the number of animals used, are there different regulatory requirements for different species?

Reg. Definitions of *animal* relative to the PHS Policy and AWAR are provided in 12:1. U.S. Government Principle III, the *Guide* (p. 10), and the AWAR (§2.31,e,2) all require a justification for the appropriate species proposed for the animal use activity.

Opin. The definition of an animal varies with reference to which species are included under the purview of the IACUC but make *no* distinctions in terms of animal number justification. For example, there is no wording that distinguishes different phylogenetic levels of equally sentient animals in terms of animal number justification.

APHIS/AC Policy 12 (http://www.aphis.usda.gov/ac/polmanpdf.html), which is meant to help interpret the AWAR requirement for an alternatives search, clearly states that less sentient species (e.g., insects) should be used if possible. This guideline does not appear to have implications for decisions among sentient species, such as mammals and birds. Researchers sometimes speak of wanting to use the "lowest phylogenetic level" that will answer a question, but unless there is the possibility of using a species (such as an insect or worm) that is clearly less subject to pain than higher animals, there is nothing in the regulations to mandate this, nor to suggest that more justification for a particular number of subjects is required for "higher" species than for "lower" ones among sentient animals. (See 12:8.)

One distinction of interest involves the study of *endangered* species as covered under the Endangered Species Act of 1973 (P.L. 93-205, 87 Statute 884) and other relevant legal documents such as the Convention on International Trade in Endangered Species of Wild Flora and Fauna (CITES) treaty. (See www.cites.org for more information.) It seems likely that the total number of such animals to be studied, as well as the number to be euthanized, would be subject to stricter standards for endangered species than would otherwise prevail for other animals. Any IACUC dealing with such a situation will have to review all of the relevant documents.

Other than endangered species, there are few species-specific distinctions relative to justifying the number of animals proposed for use. From ethical, humanitarian, and regulatory standpoints, it is as important to justify the use of 100 mice properly as it is to justify the use of 100 dogs. Invertebrates are not included in the definition of an animal noted above, but it seems reasonable that certain apparently sentient and intelligent invertebrate groups (such as octopuses) be given the same consideration as vertebrates. Naturally, the *appropriate* species for the research must be used.

Is there a level of animal intelligence that renders killing animals less justifiable? Is it important to provide more justification for the utilization or euthanasia of ten apes compared to ten mice? There are no such distinctions in the regulations (at least in terms of justifying the number of animals used), but implicit in all animal number justifications is the idea that the knowledge gained must be adequate to offset the impact on the individual animals involved. Each IACUC will need to determine whether highly intelligent species require special consideration in maintaining this balance. The statistical analysis is not affected, but the decision as to whether the research should occur at all may be.

13:4 What are the key considerations in using a statistical power analysis to justify animal numbers?

Opin. There are several key considerations when a researcher wants to perform and present a fully usable and appropriate power analysis.

- The researcher should have a thorough understanding of the appropriate statistical analysis for the design or must be willing to seek out and collaborate with someone who does. Much basic science research that uses statistics involves several groups or conditions and often a number of time points. Such designs require advanced analytic techniques, such as the analysis of variance (ANOVA) and specific follow-up tests (such as the Tukey test). In some instances the major question involves the interaction of two independent variables, as tested in a two-way ANOVA. These analytic decisions have important implications for the power analysis. For example, a researcher with a complex design who estimates power merely for a two-group comparison may reach a substantially incorrect conclusion. Often, a thoughtful analyst can reduce a complex question to one for which sample sizes can be estimated with simpler methods, but this process requires a good understanding of the statistical issues.

- The design must be spelled out in sufficient detail for IACUC (or other) reviewers to understand it and to see how the power analysis is configured to match the analysis that will be employed.

- The researcher must present and justify the specific parameters and assumptions required by the power analysis. Typically, this will involve estimates of effect sizes (such as the mean difference between two groups) and estimates of variability, such as the standard deviation. Analytic methods have different requirements that must be determined before a power analysis can be run. Prior experience, the theory underlying the research, and pilot data may be useful.

13:5 What role can a biostatistician play in evaluating the justification of the number of animals requested?

Opin. Biostatisticians are trained to perform the kind of calculations used to determine sample size. They can advise the researcher how to analyze the data properly and can help to write the analysis and animal number sections of the protocol. To be most helpful, biostatisticians also should be familiar with the IACUC and its task. They should understand the audience that their comments will reach and understand enough of the science to be able to recognize what is central and what is not. Having a statistician either on the committee or available to help researchers could be very useful. Since the development of statistical methods and the performance of power analyses are both fairly lengthy tasks, IACUCs should not expect a statistician (whether a member of the committee or not) to perform these tasks routinely without making appropriate collaborative, and possibly financial, arrangements.

13:6 What exactly is a pilot study? How do pilot studies differ from other studies in animal number justification?

Opin. A pilot study is one involving a small number of animals (rarely more than ten) used to demonstrate that a technique can work or to estimate the variability in the data before performing a statistical analysis. A pilot study does not require the extent of sample size justification required by "full" studies. Nevertheless, a pilot study is part of an IACUC protocol and, therefore, does require a justification, even if the justification is little more than "This is the minimal number of animals needed to perform a pilot study." If pilot studies are to be exempted from all but the most lenient requirements for sample size justification, then common sense suggests that they be narrowly defined and limited to a few special cases. Every institution should develop and disseminate specific guidelines, or an SOP, concerning what will be accepted as a pilot study.

It is important to note that pilot studies are much more useful for estimating variability in the data than for determining the likely magnitude of effects. The latter requires a study much larger than most pilots. Furthermore, if a pilot study is used to estimate magnitudes of differences or to determine whether or not a particular manipulation is worth pursuing further, then the pilot animals should not be used in the succeeding full study (see 13:8). If pilot study animals serve only to estimate variability or to verify that a technique works in your laboratory (e.g., that you can actually get reliable results on a particular piece of equipment), it may be acceptable to include the pilot study animals in the full study.

13:7 Can a study using four groups of ten animals each be considered a pilot study?

Opin. In the author's opinion it is difficult to see how so many animals could be considered a pilot study or why a pilot study would require four groups. Variability can normally be adequately estimated with one group (or, at most, two groups) of animals. Viability of a technique also can typically be tested on one or two groups.

Every possible situation cannot be anticipated. For example, consider a researcher who has four technically difficult manipulations, each of which must be shown to work before a larger study can be undertaken utilizing all four. Conceivably, a pilot study with four groups could be justified here.

13:8 If animals are requested both for a pilot study and for the full study to follow, should the IACUC approve both at the same time?

Opin. Assuming the number of animals used in the pilot study is justified (see 13:6) and until the results of the pilot study are known, the IACUC should not approve the entire number of animals requested.

The pilot may show that the basic technique is flawed, and the whole study should not progress. In other instances, a pilot has as its purpose the estimation of variability in the data upon which a power analysis will be based. Until these estimates are in hand, any number of animals requested for the full study must be considered unjustified. Granting some kind of "conditional" approval pending the results of the pilot study sends the wrong message and may be confusing, yet many institutions require some preliminary report from the investigator before the full study is implemented. PIs could misunderstand this as in effective "approval but for a technicality," which may not be the intent. It is, in this author's opinion, best to grant approval only for the number of animals fully and initially justified. (See 9:39.)

As shown in the survey question, most committees use some sort of partial approval process; only a few give full approval to the entire request at the time of initial IACUC approval.

Surv. A PI requests animals for a pilot study and additional animals for the full study that will probably result from the pilot work. Assume that the number of animals required by the full study will be influenced by the results of the pilot study. How does your IACUC proceed?

- Approve all the animals at once   6% (10/162)
- Approve the pilot study, conditionally approve the rest,
  with some data to be provided to the IACUC before the
  remaining animals are released   36% (58/162)
- Approve only the pilot data, requiring the PI to submit
  a new request to get the remaining animals   37% (60/162)
- Decision varies, based mainly on factors involving
  the invasiveness, risk, and severity of the protocol   21% (34/162)

13:9 Can the IACUC demand that a pilot study be done?

Opin. The IACUC has the right and responsibility to require an adequate justification of the number of animals requested (see 13:1). If a researcher is unable to provide that, the IACUC may withhold its approval. If a pilot study can break the impasse, it may be an appropriate choice for both the IACUC and the investigator.

The IACUC, by exercising its responsibility to assure that the number of animals to be used is appropriate, may withhold approval of a project until a pilot study is completed. If a pilot study is agreed to, it requires IACUC approval (see 13:6–13:8). Actually, the IACUC may be able to demand a pilot study, but there is nothing in the AWAR or PHS Policy that states that it can. (See 9:39.) Since the IACUC will only grant full approval after the pilot study is completed (if it follows the preceding advice), it will necessarily need to receive a report on the pilot, with a revised request for the remaining animals.

Surv. Can your IACUC demand that a pilot study be done?

- Yes   69% (112/162)
- No   6% (9/162)
- We have no policy   24% (38/162)
- Other   2% (3/162)

13:10   Can an IACUC withhold approval of a protocol that requests fewer animals than the IACUC believes are needed for a valid study?

Opin.   Yes, if the researcher does not adequately justify the use of that number. The primary question is whether or not the sample size justification itself is adequate (see 13:1). If the justification is invalid or inappropriate, the researcher should be asked to improve it. If the researcher fails to do this acceptably, then approval of the protocol is withheld until the justification is satisfactory to the IACUC. The use of too few or too many animals can be equally deleterious. A researcher who requests fewer animals than one might have expected is not excused from showing how that number is sufficient for research purposes.

   As noted in the survey question, most respondents agreed with this interpretation. Many of the "other" responses could also broadly be classified as "yes," but with an additional step of request for further justification or a qualification that the option of withholding approval would depend on context.

Surv.   Would your IACUC withhold approval of a protocol that appears to be requesting too few animals?

- Yes   69% (111/161)
- No   24% (39/161)
- Other   7% (11/161)

13:11   When justifying the number of animals requested, should fetuses and neonates be counted as vertebrate animals?

Reg.   The AWAR and PHS Policy do not address this issue in detail. One useful indicator is the answer to a question on chick embryos in the *ILAR Journal*.[5] The question was "Are avian embryos covered by the PHS Policy, i.e., must their proposed use be reviewed and approved by institutional animal care and use committees (IACUCs)?" The reply was "PHS Policy is applicable to proposed activities that involve live vertebrate animals. While embryonal stages of avian species develop vertebrae at a stage in their development prior to hatching, OPRR [now OLAW] has interpreted 'live vertebrate animal' to apply to avians (e.g., chick embryos) only after hatching." This statement makes clear that hatched chicks *should be* counted. It would be reasonable to infer that the same logic would apply to neonate mammals. Thus it seems that born and hatched vertebrates should all be counted. The handling of fetuses per se is less clear (see opinion).

Opin.   Even fetuses should be counted in almost all cases, at least those advanced enough to feel pain or distress. This opinion is based on ethics, rather than the law and associated regulations, but behaving in a way that makes sense ethically is probably also good legal practice. The purpose of the IACUC is to help assure high standards of animal welfare in the institution. If a fetus can feel pain, then its welfare matters and it should become an IACUC concern. Nevertheless, it is difficult to develop comprehensive and systematic guidelines for preventing fetal pain, because so little is known about many species and their precise neural development. (See 8:11; 14:19; 16:36.)

If late-stage fetal or neonatal animals are directly the subjects of research studies, accounting for them follows fairly naturally from the sample size justification itself. The justification of animal numbers in such instances will be critically determined by the number of fetuses needed.

A survey question (data not shown) suggested that many IACUCs do not "count" fetal animals until they are born, and sometimes do not count even born animals until weaned. However, the intent of the respondents was not clear. Perhaps those committees that only count weaned animals are referring to breeding protocols in which animals are not in any way utilized for research until they reach adulthood. This author can understand why an IACUC might not want to count every fetus or neonate in a breeding situation when only those that survive to adulthood are used in experiments. Nevertheless, this author would still prefer to see all late-stage fetuses and born animals tallied and included. Suppose a researcher had a fragile strain of rat with 80% mortality rate in the postnatal period. An ethical analysis of the value of this research against the number of animals used would have to consider all of the animals produced, not just the 20% that survive to be manipulated. In addition, the possibility that those that died would have suffered from their disorder before succumbing to it would also be relevant, and would *only* be salient if the total number were reported.

The *ARENA/OLAW Institutional Animal Care and Use Committee Guidebook*[6] (pp. 130–133) section on breeding does not discuss fetuses but suggests that suckling animals be subdivided into those to be used in any way in the research versus those merely to be euthanized before weaning. Both groups are then counted and reported, as recommended here.

13:12 To reduce the number of animals used, how might an animal be used in more than one biomedical research, teaching, or testing protocol?

Opin. Any sacrificed animal should be utilized in as many studies as can benefit from the use of its tissues. It is appropriate to do so any time several protocols can use an animal without subjecting it to any additional pain or distress.

The real question concerns reuse of an animal in a painful or distressing way after it has already been used in this way once. Although animals are only occasionally subjected to a second survival surgery procedure, a second use in a nonpainful or nondistressful study can reduce the total number of animals involved in research in a particular institution. For example, after a survival surgery study in which animals need not be euthanized, a simple behavioral study might be appropriate.

13:13 How else can animal numbers be reduced?

Opin. It is worthwhile to consider ways that animal numbers can be reduced carefully. A table of possible methods for reducing the number of animals used is provided in the *ARENA/OLAW Guidebook*[6] (p. 98). Some of what is recommended is already standard practice for most researchers, such as standardized procedures; use of "healthy, genetically similar animals"; and good postoperative care to minimize losses. Other suggestions include sharing of tissue (as noted in 13:12), several terminal procedures per animal, appropriate choice of control groups, and maximization of the use of information with good statistical software.

Morton[7] describes some additional methods. For example, he notes that if a certain drug would be considered as too toxic for use if even one of six subjects had severe adverse reactions, it would make sense to test the six in sequence, then stop the study if one of them, in fact, had such a reaction. He notes that in certain kinds of research (such as testing antibiotics for efficacy in an animal model prior to use in a Phase I human study), there may be little purpose in a strict significance level (e.g., setting the cutoff for significance at $p = 0.01$) because any findings must still be replicated in humans.

Morton[7] notes that pilot studies can help to prevent problems and the data may be able to be used in the main study. Nevertheless, as noted in 13:6, one needs to be cautious with this concept. If pilot studies are used merely to test procedures and estimate variance, it is true that one can use the data in the full study. If, however, pilot studies are used to determine, for example, which drugs from a large set of drugs have an effect, one needs independent replication of the results with entirely new data. Anyone proposing to mix pilot data with full study data should consult a statistician.

13:14   What role can veterinarians and animal facility personnel play in reducing the number of animals used?

Opin.   The IACUC has a general responsibility to oversee the appropriate use of animals in biomedical research. It closely interacts with the AV and the animal care staff. Veterinarians and other animal facility personnel have responsibilities that generally include assuring that animals are obtained from high-quality vendors, reviewing the health status of incoming animals on the basis of a vendor's reports, assuring quarantine and testing of animals when appropriate, and facilitating the rederivation of animal strains when appropriate. They also are responsible for developing the entire program of preventive medicine and often oversee all animal care operations. The use of healthy, well-treated, and well-maintained animals helps decrease the numbers of animals needed by decreasing the need to repeat studies.

Veterinarians and animal facility personnel also can assist with preoperative, operative, and postoperative planning to help assure animal survival, further decreasing the numbers needed.

References

1. Silverman, J., Animal use for *in vitro* work: How much justification is enough? *Lab Anim.* (NY), 33(5), 15, 2004.
2. Doyle, R., Give some examples, *Lab Anim.* (NY), 33(5), 15, 2004.
3. Goad, M.E.P. and Block L., Justify more than numbers, *Lab Anim.* (NY), 33(5), 15, 2004.
4. Matthews, M., A little short, *Lab Anim.* (NY), 33(5), 16, 2004.
5. Division of Animal Welfare, Office for Protection from Research Risks, National Institutes of Health, The Public Health Service responds to commonly asked questions, *ILAR News*, 33(4), 38, 1991.
6. ARENA/OLAW Institutional Animal Care and Use Committee Guidebook, 2nd ed., National Institutes of Health, Bethesda, MD, 2002.
7. Morton, D.B., The importance of non-statistical design in refining animal experiments, *ANZCCART News*, VII(1), insert, March 1998.

14

Animal Acquisition and Disposition

Michael J. Huerkamp and David R. Archer

Introduction

Millions of animals are acquired and used for scientific research annually in the U.S. Because animal acquisitions for research and the eventual disposition of these animals are under the stringent regulation of federal law and the oversight of funding agencies, there is a lot at stake. If animals are not acquired and used in accordance with appropriate laws and standards, institutions risk punitive measures and/or public relations dilemmas. Complicating these issues are myriad details related to the potential sources of animals or research specimens (i.e., farms, commercial breeders, USDA Class B dealers, foreign import, pet shops, donations from citizens, transfer from other research projects, interstate movement, abattoirs), the type of research for which animals are acquired, the animal species to be used, stage of development/maturation, and eventual disposition (e.g., euthanasia, slaughter for food, donation to raptor rehabilitation program, adoption). After animals are acquired, the tracking and accounting of their use are arguably among the major administrative challenges at a research institution. Consequently, in order to act in a legal, fair, consistent, and rational manner, it is important for IACUCs to be cognizant of the many issues related to acquisition, records, tracking, and disposition of research animals.

14:1 What are the usual and ordinary sources of animals for research?

Opin. The most conventional way to obtain animals for research is to purchase them from outside the institution from a commercial vendor, licensed dealer, or farm. Within an institution, sources include breeding or stock colonies and those transferred from or exchanged with another research project. Other sources, usually requiring a certain level of justification, include pet shops, private pets (such as those used in veterinary clinical studies (see 8:31; 14:42), and those transferred from other research institutions. In certain geographic areas, dogs and cats may be available from animal control programs, but this availability is highly variable, depending on locality. Animals may also be studied in, or collected from, the wild.

In some instances, scientists may require tissues, rather than live animals, for research. As an alternative to purchasing animals to be immediately euthanized, scientists may obtain biological materials from slaughtered livestock, captured wildlife, harvested marine animals, or, perhaps, transfer from a colleague who has euthanized animals for another purpose.

It is important to understand that the breadth of species defined as *animal* varies with regulating body and does not necessarily adhere to the classic biological definition. PHS Policy (III,A) and the *Guide* (p. 1) include all vertebrates in their definition of animal, while the AWAR specifically excludes most mice and rats, birds bred for research, horses not used specifically for research purposes, and livestock or poultry used for food, fiber, or agricultural research (§1.1 Animal). (See 12:1.)

14:2 What are the legal requirements governing procurement of animals and tissues for research?

Reg.   The AWAR requires that animals used for biomedical research

- Be acquired by lawful means and in compliance with record-keeping requirements (§2.35)
- Be of an appropriate species (§2.31,e,2)
- Be tabulated in the form of approximate number of animals used (§2.36,b,5–§2.36,b,8) or to be used (§2.31,e,1)
- Be used in experiments that have the appropriate number of animals needed (§2.31,e,2)

For species used for food and fiber in agricultural research, the AWAR does not apply, but recommendations are given in the *Guide for the Care and Use of Agricultural Animals in Agricultural Research and Teaching*.[1] For research funded by entities of the PHS, the salient features of the AWAR are reinforced by the PHS Policy and the *Guide* (pp. 8–14).

The provisions in the *Guide*, including those that relate to justifying and tracking animal use, are also used by AAALAC in accrediting research institutions. While the AWAR excludes poikilothermic animals and typical laboratory rats and mice from these considerations, all vertebrates are covered under PHS Policy and, by extension, AAALAC. Consequently, in most research institutions in the U.S., federal regulations, AAALAC standards, and granting agencies require that virtually all vertebrate animals be acquired lawfully, used judiciously, and disposed of appropriately. Even non-PHS-funded research may be included in PHS requirements if an institution, via its Assurance statement to NIH/OLAW, indicates that it will include activities funded by sources other than PHS. Institutions may exclude non-PHS-funded or agricultural animal activities from its Assurance statement, yet still empower the IACUC to oversee the use of animals in those areas.

Also impacting on research animal acquisition and disposition are regulations and standards covering importation of animals and biologicals,[2] endangered species,[3,4] the safety of the American food supply,[5] the use of infectious agents (biohazards),[6] radioisotopes (usually state regulated), and chemicals and certain toxins.[7] Finally, beyond these bounds, an institution must also act and set policy in ways that serve to protect its good reputation.

14:3   Is a land grant college that uses animals only in agricultural research and training required to follow the AWAR, PHS Policy, or both?

Reg.   The AWAR (§1.1, Animals, §2.1,a,3,vi) specifically exclude livestock bought, sold, or used in agriculture production, agricultural research, or agricultural training. PHS Policy also does not normally cover animals acquired and studied for agricultural purposes; however, it does apply if the agricultural research and teaching component are part of a larger institution with an institutional Animal Welfare Assurance that states that all institutional entities will adhere to PHS Policy. This becomes a germane issue when PHS-funded research is done on institutional farms or other agriculture-related facilities.

Opin.   While it is possible to do so, and this option may be useful from an institutional cost accounting perspective, at this time it is only recommended that the IACUC be responsible for the number of animals used in agricultural research and training.[1] In order to preserve relations with the USDA and a good reputation and to obtain and maintain AAALAC accreditation, any institution doing agricultural research with food and fiber species should endorse the principles in the agricultural guide.[1] Adherence to standards in that document also promotes good science, enhances animal well-being, maintains credibility with the public, provides standardization of research conditions across institutions (and hence reduces variability), and ultimately protects the institutional privilege of conducting research with animals.

14:4   Since neither PHS Policy nor the AWAR address the research use of amphibians, reptiles, fish, and other ectotherms in any detail, should the use of these species in research be monitored by the IACUC?

Reg.   PHS Policy (III,A) does cover the use of all vertebrates in research including reptiles, amphibians, and fish.

Opin.   Although the PHS Policy is intentionally broad in scope and does not prescribe specifics about the care and use of any species, it assigns that task to the IACUC and allows for professional judgment. Many of the principles it advocates generally can be adapted to animal care and use programs for various kinds of amphibians, reptiles, and fishes. Recommendations for specific species are often available from organizations that have an interest in the appropriate care and use of these species in laboratory and field studies.[8–12] It is clear, however, that individual requirements for these three classes of vertebrates, which contain more than 28,000 species, cannot be addressed in a single set of guidelines. Consequently, NIH/OLAW recommends that the advice of experts be obtained to design and develop studies and suitable housing and care procedures for species not commonly used in research.[13]

14:5   For the purpose of IACUC review, are birds used in scientific research covered by the AWAR or PHS Policy?

Reg.   Birds bred for research are not covered by the AWA as a consequence of the Farm Security and Rural Investment Act of 2002 amending the definition of an animal in the AWA (AWAR, §1.1, Animal). (See 12:1.) Birds *not* bred for research are covered by the AWA; however, specific regulatory standards have not yet been published. Previously, all birds were excluded from the AWA.[14,15] PHS Policy covers the use of all vertebrates in research, and this category obviously includes birds (PHS Policy III,A).

Opin. As is the case for other vertebrate species, PHS Policy assigns oversight for care and use of birds to the IACUC and then allows for professional judgment. There are numerous scenarios in which birds may be used in research. Depending upon specific need, birds may be acquired from the wild or from domestic commercial sources. Commercial purveyors of birds may provide them for regulated (research) use or for nonresearch purposes. At our institution, purchased pigeons and trapped wild songbirds are used in neuroscience experiments. Canaries and zebra finches may be acquired from pet stores for similar use. House finches are captured and released in field studies of bacterial population dynamics. Domestic chickens, under some circumstances, have proved useful for antibody production, and quail may be used in cardiac research. Where avian species are not commonly used in research, NIH/OLAW recommends that expert advice be obtained to design and develop studies and suitable housing and care procedures.[13]

14:6   Are there any special legal requirements regarding accounting and tracking of specific species used in research?

Reg. The federal government specifically requires that detailed records be kept on individual dogs and cats acquired and used in research (AWAR §2.35,b,1–§2.35,b,7) and that annual use of regulated species be reported by pain/distress category (§2.36,b,5–§2.36,b,8). The species required to be reported are listed under Animal in §1.1 Definitions of the AWAR (see 14:1). Federal[3] and state laws generally require special permits for the use of endangered species in captivity or in field research. Specific record keeping can aid the researcher in proving that animals were born in captivity and in compliance with applicable laws.

Opin. Annual tracking of animal use is a potentially valuable management resource. Such data can be used to show trends in program size and changes in focus that may be useful in program justification and projections. Such information may also be requested by institutional administrators or public officials and may be helpful in clarifying facts about animal research for the public. Many institutions, ours, for example, require accounting of animal use in an annual report.

14:7   Should investigators be allowed to acquire, but not use, animals prior to IACUC approval of the protocol?

Opin. It is not generally advisable to allow advance purchase, capture, or acquisition by other means for several reasons. First, there may be unintentional use or temptation to use or to prepare to use the animals without appropriate authorization. The risk for this predicament may be increased given the "just-in-time" grant review and funding climate whereby IACUC approval of a grant is no longer a prerequisite for NIH peer review.[16] (See 8:22.) The conduct of any animal activities that have not been reviewed and approved by the IACUC is constituted by NIH/OLAW as a serious problem requiring reporting.[17] Second, if animals are acquired prior to consideration of a research proposal and the proposed use of the animals is subsequently rejected by the IACUC or the number of animals approved by the IACUC is reduced from the original request, this situation may present problems in managing the population and would constitute violation of legal requirements to use the minimal number of animals necessary (see 14:15). An exception to this might be fish maintained and studied in large numbers, those kept at marine or freshwater experiment stations for prospective use, and those trapped in the wild (see 14:25).

14:8   Is it necessary to have IACUC approval for animals that are just being maintained (no research being performed)?

Opin.   In addition to animals directly used in research, NIH/OLAW and APHIS/AC implicitly require that institutions establish mechanisms to monitor and document all animals acquired or produced.[13] This includes any animals that are kept for breeding purposes, nonhuman primates awaiting assignment in core pools at primate centers, dogs and cats in centralized institutional conditioning programs, and even animals held in abeyance in the case of a lapsed protocol. Without IACUC approval or at the expiration of a protocol, animal use is not valid. Continuation of animal activities beyond the expiration or use of animals without approval is a serious and reportable violation of PHS Policy.[18] In the case of an expired protocol, especially where the likelihood exists that it will be legally renewed or the animals will be transferred to another protocol, rather than the wasteful and internecine approach of animal depopulation, the IACUC should establish a mechanism for temporary oversight and management of these animals. Even in the case of fish caught and held for prospective use, a husbandry and holding protocol is appropriate.

14:9   What institutional entity should procure animals for research?

Opin.   One of the critical responsibilities of an IACUC is to assure that the institution lawfully procures appropriate animals for research. NIH/OLAW has deemed this to be best accomplished through the institution's animal resource program or other appropriately designated office.[19] Veterinarians involved in the animal resource program must have specialized training in laboratory animal medicine (AWAR §2.31,b,3,i; PHS Policy IV,A,3,b,1; *Guide*, p. 56) and usually have the expertise to evaluate animal sources and identify those that are appropriate for research given the institution, its resources, and its needs.

It is well within the jurisdiction of the IACUC, especially in consideration of its responsibility to ensure an adequate program of veterinary care and a safe work environment, to provide guidance to the animal resources program concerning the species, legal sources, genotypes, and microbiological background of animals acquired for research use.

In smaller academic or industrial programs with part-time or consultant veterinary staffs, the IACUC may well want or need to have considerable input into this process. Coordination of veterinary care, vendor approval, and a centralized purchasing system will aid the tracking of animal use as discussed in 14:6; 14:10; 14:11; 14:13; 14:14.

14:10   Should the IACUC become involved in approving an "acceptable" animal vendor?

Opin.   The loss of animals or research to preventable infectious diseases has ramifications for the institutional requirement to ensure that research is not unnecessarily duplicative (AWAR §2.31,d,1,iii; *Guide*, p. 10). Consequently, it is within the purview of the IACUC to work with the professional veterinary staff to identify acceptable animal health standards. In most instances, IACUCs defer to the professional judgment of the veterinary staff to identify vendors that meet appropriate health and legal standards. Often, this is the most effective means of addressing this obligation. Nevertheless, when rodent use is high, when certain animals are scarce, or when funds are inadequate, veterinarians may be under pressure to permit the entry of animals with suboptimal health status into the institution. In this situation, it may be worthwhile for the IACUC, as the de facto representative of the community of scientists, to work with the veterinary staff to develop and enforce standards that are in the best interests of the community of scientists within the institution.

14:11 What constitutes an acceptable animal vendor? What procedures can the IACUC or animal resources program use to assess vendor quality?

Opin. In general, animal vendors should have a practice of producing consistently healthy animals of a specific genotype and health status in compliance with applicable laws, statutes and regulations, and institutional needs. Under the AWAR (§2.40,a) vendors are required to have a program of veterinary care. Specifically, the AWAR require licensing of any of the following that supply warm-blooded animals, other than rats and mice, to research institutions:

- Pet shops (§2.1,a,3,i)
- Breeders selling more than 25 dogs or cats annually (§2.1,a,3,iv)
- Persons or commercial enterprises deriving more than $500 annually from such sales (§1.1, Dealer; §2.1,a,3,ii)
- Any person or entity (Class B Dealer) selling any dog or cat not born and raised on its premises (§2.1,a,3,iv).

Additionally, institutions should make a dedicated effort to ensure that all transactions involving animal procurement are done in a lawful manner (*Guide*, p. 57). Other considerations are whether the vendor has an adequate program of veterinary care and a history of consistently meeting consumer needs and expectations. Ideally, vendors should be AAALAC accredited or provide assurance that they meet standards for their industry (i.e., health quality, animal care, customer service) and applicable federal laws. For rodent vendors, the health status of the animals should be documented through a regular program of health surveillance and verified by a rational means that includes sampling technique, sampling strategy, appropriate diagnostic tests, sufficiently frequent testing, a reliable laboratory, and prompt reporting of changes in health status. In most instances, the evaluation and consideration of animal vendors are handled within the program of veterinary medical care, with IACUC oversight.

14:12 What precautions should the IACUC and animal resources program require if animals of unknown health status must be acquired?

Opin. Acquisition of animals of unknown health status should be limited to those instances when the animals are only available from one source. Nevertheless, this process still has several risks. The primary threat is the transmission of infectious diseases to pathogen-free animals. In addition to causing otherwise preventable pain and distress, research may be compromised or invalidated and therefore deemed unnecessarily duplicative (see 14:10). Consequently, the IACUC should ensure that the institution has the components of a program of veterinary medical care that provide for the stabilization, isolation, and health characterization of animals of an undefined health status (*Guide*, pp. 58–59).

Quarantine programs should be suitably long to permit the incubation stage of a disease to become clinically manifested or detected by culture, the presence of serum antibodies, pathogen DNA, or pathognomonic lesions. Where bacteriology or polymerase chain reaction (PCR) technology can be employed, results can often be obtained quickly; however, responsive antibody production may not be detectable for 1 to 3 weeks or more, depending on the amount and timing of the pathogen inoculum. Although investigators may feel a stringent quarantine and testing program is a hindrance to their research, data are readily compromised when using potentially infected animals.

An effective quarantine program should also provide for the rederivation of diseased animals by surgical or other appropriate means and other relevant services such as pathogen testing for cells and cell lines that may potentially affect the health of animals within the program.

14:13 Should the IACUC attempt to track the number of all animals acquired or only those used under approved protocols?

Opin. Although neither the PHS Policy nor the AWAR explicitly require an institutional mechanism to track animal usage by investigators under IACUC-approved activities, both require that proposals to the IACUC specify and include a rationale for the approximate number of animals proposed to be used (AWAR §2.31,e,1; §2.31,e,2; PHS Policy IV,D,1,b; *Guide*, p. 10). These provisions implicitly require that institutions establish mechanisms to monitor and document the number of animals acquired or produced and used in approved activities.[13] This includes any number of animals that are kept for breeding purposes or culled prior to research use and not subjected to any experimental manipulations.[20] Tracking is particularly important for institutions that maintain large breeding colonies of nonhuman primates where animals produced from this population may be held for lengthy periods before assignment to a research protocol. Institutions that use random source dogs and cats must maintain acquisition and disposition records in compliance with federal law (AWAR §2.35,b). These resources may also operate, depending on the institutional need, by maintaining pools of unassigned animals and allotting them to research protocols only after a suitable stabilization period. In some cases, however, tracking of animal numbers may be difficult or counterproductive (see 14:15).

14:14 How should numbers of animals acquired and used be tracked by the IACUC?

Reg. The AWAR (§2.36,b,5–§2.36,b,8) require that the common names and numbers of warm-blooded animals (other than rats and mice) used for teaching, research, experiments, or tests or held for all such purposes except applied production agriculture research be included in the annual report to APHIS/AC.

Opin. It is implicitly required that institutions establish mechanisms to monitor and document the number of animals acquired and used in approved activities (see 14:13).[13] The most effective approach combines a methodology for tabulating animal use with a policy regarding the accuracy or precision of the process. Methods of tabulation can be either manual or automated. The use of commercial animal ordering software and census tabulation by bar code scanning allows for most integrated animal resources programs to track, with relative ease, the number of animals purchased from commercial sources, imported into the institution through a quarantine program, or even bred on site. This process provides accurate figures with relatively little investment in time and personnel. When acquisitions are linked to IACUC approval numbers, investigators can be automatically informed when their use has reached a preset percentage (e.g., 80–90%) of the animals approved on that protocol and with a request to provide a specific justification if it is anticipated that the number of animals ultimately required will exceed the number approved.[13] Small institutions that use limited numbers of animals may find it most effective to maintain a hard-copy log of each IACUC-approved activity, merely subtracting the number of animals acquired for each order or bred and added to the census from the number approved, with verbal or written notification to the investigator as the number of animals approved is approached.[13]

With respect to the precision of the tabulation process, IACUCs in aggregate historically have approached this consideration with some lack of unity.[21] Some committees attempt to track all animal use with precise accuracy while others attempt to track use with approximate accuracy. More flexibility is often found in practices concerning rodents and ectotherms than those concerning nonrodents. The major reason for this diversity in accounting essentially relates to discrepancies in tracking and reporting requirements between the AWAR and PHS Policy. Species not regulated by the AWAR (e.g., rats, mice, birds, fish) are often accounted for by using approximate, rather than precise, figures (PHS Policy IV,D,1,a; *Guide,* p. 10). Even when a species may not be regulated under the AWAR, where the experimental use is highly invasive or has considerable potential for significant pain or distress, it may, at the discretion of the IACUC, be subjected to higher scrutiny and require precise justification and tracking.

Institutions that track and report the approximate number of certain species used in research may be in full compliance with the AWAR and PHS Policy. However, many internal and external factors (including financial management, public opinion, and the spirit of the law) support an accurate accounting of the animals used in research. Whatever approaches and processes an institution chooses, it must satisfy the PHS Policy requirement that the number of animals used be limited to the appropriate number necessary to obtain valid results.[13]

Surv. 1 Does your IACUC (or an administrative office) track the number of animals acquired for research with precise or approximate accuracy?

• Track all animal use with precise accuracy	99/171
• Track animal use with approximate accuracy	48/171
• No attempt to track	13/171
• Other	11/171

Surv. 2 In cases in which your IACUC tracks the number of animals acquired for research with only approximate accuracy, what species are involved? Check all applicable responses.

• Not applicable	71/135
• Guinea pigs	6/135

- Hamsters 4/135
- Rats 44/135
- Mice 50/135
- Fish 22/135
- Frogs 15/135
- Vertebrates used in field research 18/135
- Other (including rabbits, goats, sheep, donkeys, and cotton rats) 2/135

14:15 What factors prevent the precise prediction and accounting of the number of animals acquired and used for research?

Opin. Although precise accounting of animals is not always required (see 14:14), investigators must still satisfy the requirement that the number of animals used be limited to the appropriate number necessary to obtain valid results (AWAR §2.31,e,1–§2.31,e,2; PHS Policy IV,A,1,g; PHS Policy IV,D,1,a; U.S. Government Principles III). This requirement often produces conflicting pressures, on the one hand to justify animal numbers (including the use of statistical analysis and power calculations (see Chapter 13), on the other also to account for the inherent unpredictability of science. Foremost, it is fundamentally difficult to forecast animal requirements accurately for the 3- to 5-year funded lifetime of all but the smallest and simplest grants, especially given that results and their impact on subsequent research direction cannot typically be predicted. This difficulty has a virtually exponentially greater effect in sizable, active laboratories with multiple grants and numerous technicians, graduate students, and postdoctoral fellows carrying out various aspects of research. These conditions can often lead to discrepancies between the number of animals approved and the number necessary for the completion of the work. In addition to changes in research direction or increases in experimental complexity, failed experiments, unforeseen technical challenges, unexpected disease, and other complications may cause more animals to be used than originally expected (or requested). Given these circumstances, it is not reasonable to expect scientists to be able to project the exact number of animals needed to do the work. It is reasonable, however, to require a valid estimation and, where additional animals must be added to enable ongoing work to continue via the protocol modification process, to expect a logical explanation of why previous approximations were incorrect or why more animals are needed.

The real challenge in animal tracking is found in large breeding colonies of prolific animals such as ectotherms (e.g., fish), chickens, or rodents. Dealing with the sheer volume of animals produced can be a full-time job for animal caretakers and users, and attempts to count the animals accurately are often overwhelming. Activities involving the production of genetically manipulated animals are of low yield but require the production of large numbers of animals, the vast majority of which are unsuitable for research.[20,22] Avian and rodent breeding colony accounting may be facilitated if the populations are managed by a centralized animal resources program, but there are trade-offs in this practice such as insulation of researchers from their animals and risks of inequity between production and demand.

Additional challenges are presented when animals are used or collected in the field or bred in fisheries or when animals are transferred from one investigator to another. In large-scale pond and tank production systems, where fish are propagated or held for subsequent research projects or used in production aquaculture research, there may be tens to hundreds of thousands of fish involved.[23] In such studies, it is not ordinarily possible to count the precise number of animals used accurately and estimation of the number of fish acquired and maintained should be sufficient. Additionally, handling of fish, simply for purposes of counting, may be distressful and counterproductive to the maintenance of good health.[23] As such, IACUCs should consider whether the precise tabulation of fish is critical to the experiment itself or important to realistic minimization of fish pain or distress, or whether it constitutes an otherwise groundless exercise. In the context of protocol review involving fish, it is often reasonable for the IACUC to review the general number of animals involved, the rationale for that number, and justification for any vagueness.

While it is not outside the realm of possibility to count every single animal in an enormous population, the cost–benefit ratio is not conducive to the practice, and doing so arguably would have little impact on the humane care or use of animals. Although of less significance, receiving procedures may not account for extra animals included in a shipment to be counted against ceilings approved by the IACUC (see 14:21).

Additional accounting challenges include instances in which animals may be used in more than one study or animals are kept for several years for the same study. Animal use in more than one study usually is of one of two types. First are studies in which the animal is killed for a specific purpose but tissues taken for another study (discussed in section 14:30). Next are studies in which the longevity or rarity of the animal, the duration (e.g., aging) or innocuous nature of the primary study (e.g., behavioral), or the need to study disease pathogenesis or response to treatment over a lengthy period would allow for the animals to be used in long-term or justified sequential or serial experiments. The latter scenarios would usually involve nonhuman primates or higher vertebrates; thus it is especially important that each use of an animal is justified in an application and that multiple experiments do not jeopardize the quality of the data. For reporting purposes, animals should only be included once except when specific procedures are required to be reported. For purposes of annual reporting of species regulated by the AWAR, animals are to be counted only once, regardless of the number of studies in which they are used, and those used in more than one protocol should be counted in the most painful category.[24] When animals such as dogs, cats, or monkeys are studied for several years, they should be counted annually.[24]

In the context of the overall research program, scientists often fail to see precise accounting as valuable and thus do not commit precious human resources to the activity. Peer review groups of most funding agencies do not micromanage projects down to details such as animal numbers. Instead, most rely on the powerful influence of funding, an influence that is often underestimated by the public and largely ignored by the vocal opponents of research. IACUCs, in turn, may stringently enforce what has been conveyed as a loose regulatory standard for animal number approximation (AWAR §2.31,e,1–§2.31,e,2; PHS Policy IV,A,1,g; PHS Policy IV,D,1,a; U.S. Government Principles III). Sometimes, an "us and them" relationship can develop between investigators and IACUCs and can lead to less than optimal accounting of animal usage. Generally, IACUCs need to communicate their role to investigators in an effective way and be responsive to the sometimes rapidly changing needs of the investigators. An effective mechanism is the institutional training program (particularly for new investigators), supported by periodic workshops or communiqués to the established community of scientists explaining the relevant laws and the obligation of the investigators.

Newly arriving on the scene, especially for many larger programs, are techno-logical innovations that save labor and enhance process consistency, such as electronic census acquisition by bar code scanning. Scanning implementation will force institutions to address the issue of accounting for carryover animals as old protocols and associated cage cards expire, new protocols are initiated, and replacement cage cards must be produced. The identification or estimation of the number of carryover animals will be integral to the accuracy of this process and must be considered in the renewal along with any new animals to be added. Otherwise, investigators will run the risk of exhausting the number of animals approved by the IACUC before the anticipated conclusion of the project.

Despite all these considerations, the expectation of most organizations with research oversight responsibility is that there will be a good faith effort made in specifying and tracking the number of animals used.[13]

14:16 Should the number of animals generated by breeding colonies be tracked?

Reg.   The AWAR requires reporting of the species and number of regulated animals that are bred, conditioned, or held for use (§2.36,b,8).

Opin. Breeding animals for research purposes, stated simply, constitutes research. Consequently, animals produced by breeding colonies unarguably should be counted. The expectation of legal, regulatory, and accrediting bodies is that research facilities make a good faith effort at counting not only the number of progeny from these colonies that are used directly in research, but the total number of animals produced.[13,20] IACUCs consistently, and in overwhelming numbers, require the tabulation of rodents and other mammalian species produced from in-house colonies.[25] When a breeding colony is a centralized core or supports multiple projects (or even individual investigators), animal production should be tracked by a dedicated protocol. In the case of dogs and cats, federal law requires precise accounting at the time of weaning (AWAR §2.35,b; §2.38,g,3; §2.50,d). For most other species, progeny are likewise counted at weaning and debited against a numeric ceiling approved by the IACUC and include both those designated for use in research as well as those that will not be used.[20] Regrettably, the unavoidable production of unusable offspring is an integral part of the research enterprise.[20,22]

Surv. Does your IACUC require that animals produced from an in-house rodent breeding colony be accounted for in terms of the total number of animals bred?

- Yes   97/163
- No   22/163
- Not applicable to our IACUC   42/163
- Other   2/163

14:17 What information is useful and how should the IACUC address the acquisition of new research animals from breeding colonies?

Opin. The IACUC's role for oversight regarding breeding colonies includes ensuring that the need for a breeding colony has been established on the basis of scientific or animal welfare concerns; that the procedures for care, genotyping, and use employed in the breeding colony are consistent with the *Guide* and evaluated and approved by the IACUC on a regular basis (i.e., as part of both protocol review and the semiannual program review); and that there is a mechanism for tracking animals.[20] To review SOPs for breeding colonies, the IACUC will need information about colony management. These practices should be briefly described in the investigator's animal protocol and justification provided for departure from any standard institutional practices.

To reduce the potential for unnecessary animal wastage and assure the validity of data derived from experiments using animals from the colony, the IACUC should assess the specific mouse breeding colony management knowledge and skills of persons managing the colonies. Special emphasis should be placed on this individual's background in rodent genetics, breeding practices, genetic monitoring, and understanding of nomenclature.

Examples of other information that might prove useful to the committee include the typical number of breeders and number of young per cage, the breeding system (including number of females per male or continuous versus interrupted mating), the intended weaning age, methods for identification of individual animals, and even the breeding scheme (e.g., cross–intercross, backcrossing). This sort of information is especially important with polygamous or continuous mating systems when there is the prospect of large numbers of pups from several litters to be present in any cage. The estimated number of animals that are kept for breeding purposes and not subject to any experimental manipulations should be part of the animal protocol.[20]

Determining which animals to include in the estimated number of animals on an animal protocol can be especially daunting to the investigator and the IACUC in the absence of IACUC-developed guidelines. Studies involving genetic analysis are particularly animal intensive. It is possible for the investigator to estimate the number of animals required, but difficult for the IACUC to evaluate this estimate in the absence of experience.[20] A reasonable estimate is for the requirement of about 1200 mice to map a single gene with recessive inheritance and full penetrance derived from a nucleus of 10 to 12 monogamously mated mice.[20] Similar numbers of mice may be required for quantitative trait loci analysis using the F_2 progeny derived from an initial cross of four to six pairs. Using "speed" congenic technology, up to 750 mice may be necessary to derive a congenic strain providing the homozygous mutant is fertile.[20] Even in the straightforward case of established colonies in which all offspring consistently meet criteria to be suitable for research, the exact number of animals can only be approximated because it is impossible to predict the exact number and sex of offspring. After founder transgenic or "knock-out" mice have been identified, between 80 and 100 mice may be needed to maintain and characterize a line, assuming five breeder pairs per line, regular breeder replacement prior to senescence, no unusual infertility, and adequate numbers of weanlings for genotyping and phenotyping.[20]

In the preparation of a protocol, the estimated number of animals should clearly distinguish among breeders, young that cannot be used in experiments because they are of the wrong genotype or gender, and animals that will ultimately be subject to experimental manipulations. If a study requires fertilized one-cell eggs, embryos, or fetuses, the protocol should indicate the number of eggs, embryos, or fetuses that are required for proposed studies.[20] The estimated number of experimental animals may be limited to the number of female animals that are mated and euthanized or surgically manipulated to collect the required eggs, embryos, or fetuses. In this situation, males might be listed as breeders if they are not subject to any experimental manipulation. Likewise, when breeding colony animals will be used as sources of tissues, cells, or other biological materials, the total number to be used for experiments should be estimated on the basis of the typical amount of biological material needed for an experiment, the approximate number of animals necessary to yield the material, and the number of experiments projected.

The accounting of animal production can be done by animal care personnel recording births and weaning, but this is an intrusive and labor-intensive process often best done electronically. As a new cage of mice is populated with weanlings removed from a production cage, either the exact number of mice or an average (typically 3.25–3.5 mice/cage) can be deducted from the maximal number of mice approved from the protocol. As an alternative, investigators report to the designated office, at regular intervals, the number of animals born, weaned, or used in studies.[19] This report can be tallied against the numbers in the approved protocol.

14:18   Is there a percentage of unused breeding colony animals that are euthanized (without having been used as breeding stock or experimental subjects) that should be tolerated by the IACUC?

Opin.   Substantial numbers of animals may be required to maintain a breeding colony. For example, while trying to establish a breeding colony for a new mutant rodent or ectotherm model, the investigator will also be working to determine phenotype, to identify affected physiological system(s), and to define inheritance pattern in addition to breeding founder and subsequent generational animals. Under these circumstances, the production of unused animals that are subsequently euthanized is almost inescapable. This is a significant issue for commercial and core rodent production colonies that must operate at a production excess (approximately 30% or more) in order to meet unexpected increases in market demand or that produce animals at an excess when demand unexpectedly takes a downturn. A number of user factors such as newly funded projects, changing experimental initiatives, failure to plan, cessation of standing orders, and periods of relative research dormancy alone or collectively may have a dynamic impact on production. For animals from a production colony, matching supply with need is a moving target, and, therefore, production cannot reasonably be expected to match demand.

The case of creating genetically unique animals by using homologous or nonhomologous recombination presents many examples of animals that are produced and euthanized without further use. In the creation of a transgenic (nonhomologous insertion) mouse using the pronuclear microinjection technique, only 10–40% of the mice born will carry the transgene; thus, 60–90% cannot be used.[26] Although most transgenic founders will subsequently transmit the foreign gene in 50% of their offspring, approximately 20–30% are mosaic and transmit the gene at a lower frequency (i.e., 5–10%).[27] Consequently, in the breeding of transgenic founder animals anywhere from 50% to 95% of the offspring will not carry the gene of interest, will have no research value in many instances, and must be euthanized.

In the case of genetically unique mice or breeding of small colonies used by a single investigator and during periods when progeny may not be needed for experiments, mating must continue in order to perpetuate the colony. When this happens, the offspring that are not selected as breeders must be euthanized. While it is possible to freeze embryos for later rederivation, this is an emerging option and not a realistic contemporary approach for many investigators. In managing inbred animals by a single-line system, genetic divergence is prevented by euthanizing breeders, their progeny, and their grand-progeny that cannot be traced within a minimal number of generations to a common ancestor. As one might imagine, in a large production operation, involving multiple genotypes of animals, the euthanasia of large numbers of animals results.

For these reasons, it is recommended that IACUCs permit animals to be produced from breeding colonies in reasonable excess of projected or historical need.

14:19 Should unweaned animals (particularly rodents), or animals used or harvested *in utero,* be accounted for in the protocol review process?

Opin. If painful procedures are to be performed on animals, including those that are immature but at a stage of development able to perceive pain (i.e., late *in utero* or neonatal), it is clearly within the jurisdiction and responsibility of the IACUC to address this issue (*Guide,* p. 10). (See 8:11; 13:11; 16:36.) NIH/OLAW has stated that if suckling rodents will be subject to any manipulation, such as thymectomy, toe clip or ear notch for identification, tail tip excision for genotyping, or behavioral tests, the estimated number of manipulated sucklings must be included in the number of animals used.[20] The majority of IACUCs adhere to this doctrine. NIH/OLAW clarifies that institutions must appropriately monitor and document numbers of all animals acquired (through breeding or other means) and used in approved activities and not to exclude genetic culls or those produced in excess of need.[28]

Surv. At what age does your IACUC begin to count rodents for census purposes?

• Not applicable	33/161
• At birth or prior to weaning	58/161
• At or after weaning	47/161
• At weaning, except earlier when painful or invasive procedures are done	22/161
• Other	1/161

14:20 Should preweanling rodents be counted and debited from approved protocols especially if procedures are done, and at what age?

Opin. NIH/OLAW takes the position that if suckling rodents will be subject to any manipulation, such as surgery, permanent physical identification, biopsy for genotyping, or behavioral tests, the estimated number of manipulated sucklings must be included in the number of animals used.[20] While this expectation seems reasonable, it may be difficult to manage logistically and especially may add confusion to the process of monitoring and tracking animal use. Unfortunately, NIH/OLAW does not provide additional guidance in this regard. In particular, it is difficult to tabulate precise numbers of rodents born without disturbing them during rearing. The animal resources program's engaging in this process adds disruption and expense to the process without any apparent animal welfare benefit. While it is possible for the research staff to report these numbers to the animal resources program or IACUC office as they are generated by the manipulative event, a high degree of dedication to the process of all parties is required. After protocol review and depending upon the accuracy of information conveyed from the IACUC, there is a risk for the person responsible for tracking to mix, confuse, and interchange the total number of animals born, the subset of any used for a distressful manipulation, the accounting of weanlings, and any parental animals that may be purchased (see 14:19 and 14:21). At our institution the use of fetuses or animals up to the age of weaning is not tracked, but, rather, the adult breeding animals needed to produce the necessary number of offspring. However, if painful procedures are to be done on animals, including presumably young ones capable of pain perception or experiencing of distress, it is within the aegis of the IACUC to address this issue (*Guide*, p. 10).

14:21   A commercial animal vendor sends two more rats than were ordered. How should the two extra animals be accounted for in the IACUC's and PI's records?

Opin.   On one hand, it may not be necessary to tabulate the extra rodents given that the animals were not ordered, were received unexpectedly, and could not reasonably be considered by using statistical analysis or other means in the formulation of a protocol. Conversely, the animals do represent a windfall that could be used in manners or for purposes outside IACUC oversight. If the approved use involves highly invasive procedures, considerable pain or distress by design, or species of interest for any variety of reasons (e.g., regulated by the AWAR), the IACUC may be particularly interested in regulating any access to individual animals in excess of the IACUC-reviewed and approved figures. In the end, the housing density, census tabulation methodology, and protocol debiting procedures at the operative level in the animal research facility are likely to determine whether the extra animals are counted. For example, at our institution the rat census is done by counting each individual animal. Each rat has a corresponding individual cage card, and electronic census methods are used. Therefore, the extra rats would be debited from the protocol when new cage cards were made at the time of initial housing immediately after receipt. If the census is done by the cage, as for mice at our institution, extra mice in a shipment may or may not be counted. Capture of these animals for accounting purposes would hinge upon the research instructions for housing density and whether additional cage cards must be generated to accommodate all of the cages. In the context of the overall numbers of animals received by the typical investigator or acquired annually by the institution, these extra animals, included in shipments to buffer against any incidental mortality, are generally insignificant. The authors recommend that the IACUC work with the institutional animal resources program to develop a written policy and procedure regarding the disposition of animals received unexpectedly.

14:22   An IACUC approves the use of one litter of rodent pups (approximately ten pups) that will be born of one dam. What can the IACUC do to prevent a person from purchasing a total of 11 animals, rather than just the one dam?

Opin.   The most important considerations for the IACUC are to define when an animal is considered to be "acquired" and, in the context of protocol review and approval, to limit the approved number of animals to those that have been "acquired." Generally, this translates into the age or experimental conditions at which the institution's SOP dictate it is to be counted on the census. Under practical conditions and especially where the census is substantial, rodents are not tabulated until weaned. By extension, it is also critical for these definitions and expectations to be clearly conveyed to the animal purchasing party (commonly, a clerk affiliated with the animal resources program).

The NIH/OLAW mandate to document all animals produced, even if not used scientifically, including those culled and euthanatized before weaning, and whether or not painful procedures are done,[28] creates a possible conundrum. Institutions seemingly must commit resources to document every animal acquired and particularly all unweaned rodents, whether or not used in actual research. The alternative is to use human and technological resources rationally to apply constraints that in the end allow only the acquisition of the number of animals approved by the IACUC to conduct valid experiments. For example, when timed-pregnant female rats are purchased as the source of pups to be used in a study, to prevent excessive ordering it is necessary to segregate the number of sucklings projected for experimental use from the number of dams estimated to produce them, and to convey the appropriate information to the purchasing entity. If 1000 nursing rat pups of either sex have been justified for use in acute experiments and 100 outbred, pregnant female rats have been estimated to produce that number, the total number of rats approved to be *purchased* for the study would be 100 (the timed-pregnant females). If the IACUC was to approve 1100 rats (1000 + 100) and convey this to the purchasing entity (and even injudiciously to the investigator), this might create an error whereby up to 1100 timed-pregnant females could be ordered. The requirement to monitor, document, and scientifically justify the acquisition of all 1100 rats required by the study can be met in the context of the IACUC protocol review. After all, the pivotal regulatory requirements to approve a certain number of animals for a study are somewhat conflicting and ambiguous. The various charges are to consider the approximate (AWAR §2.31,e,1; PHS Policy IV,D,1,a), appropriate (AWAR §2.31,e,2; PHS Policy IV,D,1,b), or minimal (PHS Policy IV,A,1,g) number of animals needed. Arguably, this is not necessarily the precise number. As a facet of protocol review, and in an attempt to meet these requirements, it may be helpful to separate the number of adult animals to be acquired, those to be subsequently born at the institution, and, of those born, those to be used scientifically while still nursing (not ordinarily counted if rodents) and those ultimately to be weaned (customarily tabulated).

The authors caution institutions to weigh the pros and cons of tasking someone with the requirement of tabulating animals as they are born and debiting them promptly from an IACUC-imposed limitation. Such intrusive activity seemingly creates an animal welfare threat (e.g., litter cannibalism or abandonment) and may not provide much of a benefit beyond the rational and comprehensive consideration of animal numbers done in the context of IACUC review. Beyond these considerations, there is also an annoyance risk in this process for the investigator. For example, if the IACUC approves 100 rats and 100 adult rats are bought and debited, and some of the progeny are not only born, but also weaned and tabulated, then the scientist will be in a deficit situation that must be administratively reconciled. These undesirable predicaments can be obviated when the policies and procedures are clear and understood by all parties.

14:23 An investigator uses avian embryos in his research. What is the responsibility of the IACUC, if any, in overseeing this activity?

Opin. Although PHS Policy does not discuss considerations attendant to embryonated bird eggs (or mammalian fetuses, for that matter), nonvertebrate animals (chick embryos within eggs) have the potential to develop into hatched vertebrate animals. While embryonal states of avian species develop vertebrae at a stage in their development prior to hatching, NIH/OLAW has interpreted *live vertebrate animal* to apply to avians only after hatching.[29] However, the risk of eggs' hatching and producing chicks (requiring food, water, proper housing, and veterinary care and placing them under the purview of PHS Policy) dictates that IACUCs consider developing policies for different aged avian embryos, newly hatched birds, and the point at which bird embryos are considered vertebrate animals.[30] For chickens, the last 3 days of incubation (incubation days 18–21) represent the last stage of embryo development and coincide with the chick's drawing the yolk sac into the body and having sufficient pulmonary maturation to handle oxygen and carbon dioxide exchange.[31] During this period, some chicks may hatch normally and some prematurely hatched chicks may survive outside the egg with little additional care. (See 8:11; 16:37.)

14:24 Zebrafish are born with their yolk sac attached. At what stage of development are they considered vertebrate animals under the PHS Policy?

Opin.   Although fish are not regulated by the AWAR, as a rule of thumb, animals covered by the AWAR become regulated at birth from the maternal animal. Since NIH/OLAW has provided guidance for birds (see 14:23) in that they become regulated animals once they hatch from the egg,[29] it has been recommended that the same stage of development in fish be when the embryo has absorbed the yolk sac or begins to forage on its own.[23] This definition covers all reproductive strategies found in fish, including oviparous and viviparous species.[32] Although NIH/OLAW has been silent on this matter, the interpretation seems to be reasonable.

14:25 How, and with what precision, should animals be accounted for when used in field studies?

Reg.   The AWAR (§1.1, Field Study) grants an exemption to field studies in which free-ranging wild animals are observed in their natural environment and without the involvement of invasive procedures or intrusions that harm or materially alter behavior. Where federal funds are used to support field studies, PHS Policy and the U.S. Government Principles, as specified in the *Guide* (pp. 117–118), dictate that the approximate number of animals to be captured be addressed.

Opin.   Although the AWAR grants an exemption to observational field studies, few such studies can be done without interventions or some degree of invasiveness.[33] Given the variety of procedures that could be done in the field, the associated stress and risk of untoward outcomes associated with capture, and variation in researcher experience and skills, virtually any procedure, even simple handling, has the potential to cause harm. Because the tabulation of animal numbers revolves around the issue of invasiveness, it is important for the IACUC and involved researchers to address the specific procedures, potential refinements, and qualifications and training of personnel. After appropriate deliberation, it should be possible to define the relative invasiveness of the field study. If the study is determined to involve more than detached observation, the IACUC and investigator must address not only the acquisition of animals by capture, but any allowable rate of attrition related to capture and the experimental interventions. IACUCs also should be aware that the federal and state wildlife permitting processes are usually rigorous and involve an assessment of hazard of the procedures and overall risk to the population to be studied. However, this does not necessarily include the implications for individual animals. Investigators should define the potential number and species of targeted and untargeted species that might be captured. Additionally, the protocol should address, and the IACUC should consider, the merit of allowing investigators to use excess animal numbers. It is reasonable for the IACUC to review and approve a field study protocol with only a general description of animal numbers provided there are a rationale for that number and explanation for any vagueness.

Most critically, the IACUC may have to allow for some degree of ambiguity in the consideration of the number of animals captured or acquired. While in small-scale, well-defined experiments in the laboratory using only a few fish or birds or wild-caught rodents, it is possible to limit animal acquisition to a specific number, this is not ordinarily the situation in the field. For example, in the capture of fish, netting of birds, or live trapping of small mammals, IACUCs should appreciate that while investigators can define the number and identity of targeted species in the field required to meet their research objectives, it is not reasonable to expect them to know how many will actually be captured or caught. If captured targeted and accidental species are of such rarity as to be unlikely to be encountered, or if the captured animals are highly stressed and have a strong likelihood of dying upon release, some argue that it is irrational not to use them to advance the knowledge base to benefit society.[23] In the situation in which capture has an unexpected high yield, this serendipitous increase in animal numbers may provide a more robust database for evaluating specific hypotheses. Finally, the IACUC and field investigators should establish a mechanism to gain IACUC approval for protocol modification expeditiously from the field. It might be effective to permit investigators in these specific cases to submit an amendment with an appropriate justification post hoc for any additional numbers. Particularly if the research is not under the auspices of PHS Policy, IACUCs might consider allowing a degree of flexibility in utilizing unanticipated captures.

14:26 Should the IACUC allow for additional animals to be acquired for approved studies to compensate for animal losses due to random deaths, experimental error, husbandry-caused mortality, surgical complications, and the like?

Opin. It is reasonable to assume that not all animals that begin on a study will be alive at the end, as a result of causes other than those related to the experiment. In a breeding colony of healthy C57Bl/6J mice, for example, the mortality rate of adult mice in otherwise good health and not subjected to experimental manipulation has been documented at 0.13% per week.[34] At our institution, the tracking of mouse mortality by our animal resources program from 2001 to 2005 and for an approximate average daily census over that period of 35,000, the mortality rate on a weekly basis was at least 0.18% +/− 0.1% (mean +/− 1 standard deviation). This was for an uncharacterized and mixed population of inbred, outbred, mutant, and variously crossbred mice used for breeding, aging, invasive and noninvasive experimental purposes, and other applications. Separate from research factors, the husbandry program can be a source of unanticipated morbidity or mortality such as from cage flooding or dehydration if, for example, an automatic watering system fails. The risk of not allowing compensation for some degree of unplanned attrition is that data will be collected from a population with insufficient animals, statistical analysis will lose robustness, some opportunities to arrive at significant conclusions will be forfeited, and animals may be used wastefully. The jeopardy in allowing overly excessive animal numbers may be preventable animal pain and distress and undue loss of animals that may be undetected by the IACUC for a considerable period.

The important consideration for the IACUC is to determine a "normal" or "acceptable" rate of spontaneous mortality for the genotype and that associated with a given procedure or aspect of research. Excluding iatrogenic causes, attrition ordinarily from natural causes should not be very high and in the range of 5–7% annually for many populations of healthy, nonaged adult inbred mice[34] and substantially less for noninbred species higher on the phylogenetic scale. However, in situations in which the animal phenotype predisposes to a general lack of vigor or even death, and inexperienced persons are being trained in a new procedure, mortality above nominal rates may be observed. Especially with surgical models, IACUCs are often asked to allow for additional numbers of animals to account for technical mistakes, experimental error, and other complications.

Mortality related to surgery may be influenced by the genetic background of the animal, the anesthetic regimen, programs of intraoperative and postoperative care, and the invasiveness of the surgical procedure. High-risk procedures such as thoracotomy, bowel transplant, and three quarter and five sixth nephrectomy models will have ostensibly higher associated death rates than uncomplicated procedures such as embryo transfer, orchiectomy, or cranial implant. Mortality should not be very high in surgery, however, when well-trained and experienced persons use healthy animals and established procedures that meet contemporary veterinary standards including appropriate asepsis and postoperative care. In the absence of readily available published or internal data applicable to the model, a reasonable expectation for mortality in association with surgery in such cases might be 5% or less. When highly invasive, painful, or distressful experiments are planned, especially for species regulated by the AWAR, the IACUC may choose to exercise appropriate oversight by approving only the minimal quantity of animals rationalized as being necessary. Additional animals might only be allowed to be added by amendment, after IACUC review. Most IACUCs proceed with no guidelines regarding unforeseen losses. Instead, the rationale for extra animals is considered on a case-by-case basis in the context of protocol review. When IACUCs allow for a mortality "cushion" to be built into experiments, the general practice is to allow for 10% or lower mortality rate without additional review.

Given that it is reasonable for new technical employees (not to mention students engaged in research in academic institutions) to be learning procedures, the IACUC might prudently focus on training by initially allowing a number of animals to be used for nonsurvival training preparations rather than artificially increasing the sizes of experimental groups.

Surv. When considering animal usage, what allowance of animal numbers does your IACUC typically approve for unexpected deaths or related research problems?

- This has never arisen   16/169
- We have no policy; we handle each request on a case-by-case basis   77/169
- None; it should be accounted for in the experimental design   38/169
- 10% or less   31/169
- More than 10%   1/169
- Other   6/169

14:27 How can the IACUC prevent the excessive acquisition or breeding of animals beyond approved limits?

Opin. Many institutions have automated systems that will alert an appropriate individual when an investigator has reached a preset percentage (e.g., 80–90%) of the number of animals approved for a specific project and can prevent ordering of animals in excess of the number approved.[19] At our institution, the IACUC has instructed the animal resources program to notify investigators when 75% and 90% of the animals authorized on any approved protocol have been acquired, whether through purchase or breeding. Depending upon the pace and schedule of research, persons may elect to disregard the courtesy notice or act to modify the existing protocol to add animals. In extreme cases, when the authorized full allotment has been obtained, the PI is notified accordingly. The approach to this situation is fairly straightforward when animals are acquired by purchase. Animal ordering can be put in abeyance until the protocol has been revised accordingly to add new animals. The circumstances are more problematic when animals are produced to excess from an in-house breeding colony. In many instances, it may be not only impractical, but draconian, to cease all breeding immediately, euthanatize all animals that are in excess of the IACUC allowance, separate all breeders, and the like. It is incumbent upon the IACUC to articulate a clear policy and work closely and collegially with the investigators and animal resources program to nurture a culture of compliance. In the rare and unfortunate case of recalcitrance, the IACUC can resort to the options of protocol suspension or denial of access to facilities and notification of the proper regulatory authorities and funding entities.

14:28 Is IACUC approval needed for the use of animal tissues acquired from a slaughterhouse?

Reg. Ordinarily, neither the AWAR nor the PHS Policy protocol review requirements apply to dead animals in the context of research, including those obtained from abattoirs.[34] (See 8:7; 12:14.)

Opin. The use of shared tissues and slaughterhouse material is an effective application of the "three Rs" and is encouraged where scientifically appropriate.[35] IACUC approval is not required if the collection takes place post mortem and as a by-product of the commercial enterprise.[35] The use of dead animals or parts of animals in research, however, is an area for which a clear institutional policy can help prevent serious misunderstandings and possible compliance problems.[35] It is in the institution's best interest to ensure that issues such as occupational health and safety, potential disease transmission to colony animals, institutional and personal liability, and related matters are addressed. A few IACUCs delegate oversight for this to some combination of the AV, safety officer, and IACUC Chair or other official designee of the committee. Whatever the mechanism, it is recommended that the IACUC and investigators maintain appropriate documentation, and institutions tailor their policy to meet their needs.[35] If the samples are obtained ante mortem and are not incidental to slaughter (e.g., blood), or if they dictate the slaughter procedures in any way, then a protocol may be necessary, depending on the nature of the study.

PHS grant applicants using shared animal tissues or slaughterhouse materials are advised to specify the origins of the tissues when describing their proposed use in an application, especially if the "no" box is checked on the vertebrate animal block of the face page.[35] Otherwise, any reference to the use of animal tissues in the application is likely to trigger questions related to IACUC review. Therefore, an explanation in the application that the tissues will be harvested from dead animals at a slaughterhouse will help prevent complications in the peer review process.

Surv.   Does your IACUC review the use of animal tissues acquired from a slaughter-house?

- Not applicable to our IACUC   80/164
- No requirement or no policy   49/164
- Require formal IACUC approval   16/164
- Require IACUC approval via designated member review, superficial review, notice of intent, Memorandum of Understanding, or other means of notification   19/164

14:29   What should be required by the IACUC if a PI wishes to use blood or other body parts from an animal that was killed at a slaughterhouse, animal control institution, veterinary clinic or for reasons otherwise not related to research?

Opin.   It is recommended that IACUCs have a pro forma policy and procedure established for this sort of eventuality. Many IACUCs find it effective to have short, one-page forms for the purpose of specifically documenting animal tissue acquisition and use. A useful mechanism is to have a form for tissues from any source including abattoirs, intramural projects in which animals are euthanized and tissues are made available to others, or extramural transfers. Completed forms can be given a designated member review by the committee Chairperson or designee, acting under the auspices of the IACUC, although this is not required. (See 14:28.)

14:30   Is IACUC approval needed for use of animal tissues acquired from animals euthanized as part of an unrelated experiment? For example, without having an IACUC-approved protocol, can a scientist remove body parts immediately after euthanasia from an animal on a colleague's IACUC protocol?

Opin.   NIH/OLAW considers the use of shared tissues to be an effective application of reduction, refinement, and replacement and encourages this practice.[24] Assuming the animals and the method of euthanasia were covered under an approved IACUC protocol, IACUC approval is not needed.[35] Likewise, the harvest of tissues from dead animals itself is not considered research under the AWA and therefore is not a regulated activity.[35] Many institutions choose to require protocol review for all activities involving living or dead animals. This places the decision-making control at the IACUC level, and where the practice may best serve the interests of the institutions for a variety of regulatory and nonregulatory reasons (e.g., public relations, liability, occupational health and safety).[35] Committees that elect not to oversee this activity generally do so as long as the tissues collected do not alter the approved procedures in any way (including addition of extra animals), that the animals were used originally for the research of others (and not for the individual receiving the tissue), that hazard assessment showed no risk of injury or infection for personnel, and that the tissues were collected after the animal was dead. In these instances it is important for investigators and IACUCs to avoid the temptation to misuse the dead animal exemption to circumvent policy and regulation. For example, the idea that the sole determining issue is whether or not whole animals are involved (making oversight of tissue acquisition unnecessary since tissues do not constitute whole animals) is incorrect.[35] IACUCs should consider this issue and have a policy and practice. Some committees may abdicate this responsibility because the animal is dead or only parts are taken, but this has the potential for creating problems if safeguards or understandings are not in place to ensure personnel safety and oversight of the ultimate use(s) of the animal. In institutions where IACUC approval is required (see 14:28) the process often involves the use of a brief form and a designated member review by the IACUC Chair, AV, or other designee. In all cases, a "paper trail" is recommended to show that an institution has applied the appropriate standards to the acquisition, use, and disposition of animals.[35] Likewise, to prevent peer review complications, PHS grant application procedures should be assiduously observed.

Surv. Does your IACUC require its approval for the use of tissues acquired from animals euthanized (already dead) as part of an unrelated experiment?

- Not applicable to our IACUC   36/171
- Require approval   51/171
- Do not require approval   72/171
- Other   12/171

14:31 Are there reasons why IACUCs should review the acquisition of animal tissues acquired from animals euthanized as part of an unrelated experiment?

Opin. Assuming the animals and the method of euthanasia were covered under an approved protocol, IACUC approval is generally not needed.[35] However, approval is obviously necessary if the investigator receiving the tissue, in the course of meeting research needs, causes more animals to be used or different procedures to be performed on live animals (see 14:30). Review by the IACUC should also be considered when there is a risk of pathogen transmission from one facility to another within a research institution. For example, a number of rodent viruses may be transferred in tissues. Direct or indirect contact with such tissues may infect previously disease-free rodents in another animal research facility, invalidate the research in which the animals were used, and cause additional animals to be used unnecessarily as a replacement for those infected. Occupational health and safety issues also dictate IACUC consideration of tissue transfer and use. Some rodent pathogens, such as lymphocytic choriomeningitis virus, are transmissible to humans. Tissues from nonhuman primates, wild-caught animals, livestock, and unconditioned dogs or cats may also serve as a source of pathogens for humans. Additionally, there may be experiments done that could lead to public relations challenges for the institution, such as research using fetal tissue that may not be consistent with the institutional mission or philosophy, or research on tissues acquired from animals euthanized at an animal shelter.

14:32 Without having an IACUC-approved protocol, can a scientist remove body parts immediately before euthanasia from an anesthetized animal on a different scientist's IACUC protocol?

Reg. NIH/OLAW and APHIS/AC stipulate that any procedure involving antemortem manipulation of live animals must be subject to protocol review.[35]

Opin. In the context of the tissue-receiving investigator, it should be noted that NIH/OLAW has issued guidance that identifies a change, including addition of a species, as an example of a "significant change" requiring prior IACUC approval.[36,37] Likewise, IACUCs must have administrative mechanisms to ensure that all participating personnel are adequately trained and qualified, participate appropriately in the occupational health and safety program, and meet all other applicable criteria.[38] Acquisition of body parts from a live animal by a person not approved to be engaged in the procedure represents potentially serious noncompliance with PHS Policy and federal granting procedures. NIH/OLAW has indicated its determination that conducting unauthorized animal research activities is a serious and reportable violation.[36] This position is also consistent with the NIH Grants Policy Statement (Rev. 03/01) that identifies the substitution of one animal model for another (or presumably addition of an animal model where none existed previously) as an example of "change of scope" likely to require prior NIH (grants management officer) approval.[36] The signature block on the face page of the PHS form 398 grant application requires the institution to declare that it meets and will maintain compliance with all applicable terms and conditions of award. Consequently, IACUCs should have clearly articulated and well-known policies in this regard. When the situation of tissue sharing has been encountered by IACUCs, the majority require an IACUC review, but interestingly, a significant minority (20 of 164) are apparently not aware of this need. Given that prudent sharing of animals is encouraged as a means of reducing the number of animals used in research,[35] it may not be necessary for the second scientist to have an independent protocol. Nevertheless, the "piggybacker" should be added to the personnel list and meet credentialing qualifications, and the IACUC should specifically review any new or modified procedures for the existing protocol.

Surv. Does your IACUC require its approval for the use of tissues acquired from animals under anesthesia and about to be euthanized (i.e., not yet dead) as part of an unrelated experiment?

- Not applicable to our IACUC   35/164
- Require approval   105/164
- Do not require approval   20/164
- Other   4/164

14:33 Is IACUC approval needed for acquisition of animal tissues from abroad?

Opin. Investigators who collaborate with scientists in other countries may receive animal tissues from these foreign sources and may sometimes go abroad to collect samples from captive wild animals maintained in zoological collections and in research colonies. Just as animal-related activities supported by the PHS require review by the IACUC of the domestic awardee institution, foreign institutions that serve as performance sites must also have Assurances on file with NIH/OLAW.[13] NIH/OLAW considers institutions whose scientists are engaged in such collaborative work accountable for the animal-related activities from which they receive animals or animal parts. When a foreign institution holds a PHS Assurance, it also is expected to conduct the study in accordance with the applicable host nation's policies and regulations. In the specific case of sample collection, the review should take into account the species involved, the nature of the specimen, and the degree of invasiveness of the procedure and should give appropriate consideration to the use of anesthetics and analgesics. When samples are to be obtained directly by citizens of a foreign country for subsequent shipment, the recipient PHS-supported investigator should determine the proposed methods of collection and present that information to his or her IACUC for review. Prior to sample collection, and regardless of whether specimens are obtained by an awardee institution's investigator directly or by persons in a foreign country, NIH/OLAW strongly recommends that each awardee institution consult other agencies of the U.S. government concerning importation requirements.[13] Depending on the species involved and the nature of the specimen, the following may be of assistance: the U.S. Fish and Wildlife Service, Department of the Interior (for compliance with the International Convention on Trade in Endangered Species of Fauna and Flora [CITES]),[4] APHIS, USDA (regarding potential animal pathogens),[2] and Centers for Disease Control and Prevention (concerning importation of nonhuman primates and potential pathogens of human beings).[6]

14:34   Using PHS grant monies, a research institution purchases surgically modified mice from a commercial vendor. Must the vendor have an NIH/OLAW Assurance? (See 18:7.)

Reg.   The PHS Policy is applicable to all PHS-supported activities involving animals, whether the activities are performed at a PHS agency, an awardee institution, or another institution.[18]

Opin.   The determining issue is whether the surgery is conducted in response to a specific custom surgery request or whether the animals were previously surgically modified and available for sale "off the shelf" (e.g., from a catalog) before the PI's need was anticipated.[18] If an investigator requests that a specific custom surgical procedure or procedures be performed on an animal for use in activities funded by the PHS, then the organization or person who conducts the surgical procedure(s) is considered to operate from a performance site and must either have on file with NIH/OLAW an approved Animal Welfare Assurance or be included as a component of the applicant organization's Assurance.[18] If the animals are commercially available as surgically modified models, such as in the case of hypophysectomized rats or mice with implanted devices, then the supplier is not considered a performance site and is not required to file an Assurance with NIH/OLAW. It is noteworthy that some vendors may sell only certain models after receiving institutional assurance that their use has been approved.

14:35   Under what conditions can the IACUC permit the use of animals purchased from a retail pet store?

Reg.   The AWAR requires licensing of pet shops that supply warm-blooded animals, other than birds and domesticated laboratory rats and mice, for research purposes (§1.1, Retail Pet Store; §2.1,a,3,i).

Opin.   As a general rule, this is a practice that should be avoided. It may be acceptable for ectotherms, such as fish; for rare strains of rodents or rabbits; or for certain birds or reptiles when such animals cannot be acquired from a more traditional or conventional source. Where regulated species (see 14:1) are to be obtained from a pet shop, the institution has an obligation to obtain the informed consent of pet shop management and to procure the animals in a lawful manner (*Guide*, p. 57). The purchase of animals from pet shops should also fall under the auspices of the

IACUC because of the potential for abuse and the risk to the institutional reputation should this practice become known to the public. Animals that are obtained from pet stores may not be acclimated to a laboratory environment and are generally of unknown genetic and health backgrounds. With respect to the latter, the veterinary staff should be called upon to evaluate the health status of the animals and quarantine, isolate and otherwise manage these animals to protect other research colonies.

Surv.   Does your IACUC permit the acquisition of research animals from a pet store?

- Yes   8/160
- Yes, but only for species not regulated by the Animal Welfare
  Act regulations   11/160
- No, pet stores are prohibited as a source   85/160
- No, but only because this has not been encountered   45/160
- Other   11/160

14:36 Are there legal requirements that must be met for purchase of animals from a commercial farm?

Reg.   See 12:1 for the APHIS/AC and PHS Policy definition of an animal. Traditional farm animals, such as sheep, swine, or cattle, are regulated if used in research for purposes other than improving animal nutrition, breeding, husbandry, production efficiency, or improvement in the quality of food and fiber (AWAR §1.1, Animal).

Opin.   While APHIS/AC can regulate the sale of farm animals for research purposes, it has chosen not to do so. The only persons required to be licensed for selling farm animals to researchers are those who sell exclusively to researchers. Thus, the livestock producer or livestock marketer who sells an occasional animal is not required to be licensed and does not have to meet the AWAR transportation requirements, even if transporting animals to a research facility. However, the research facility becomes responsible at the point they assume custody of the animal. Therefore, if an employee of the research facility goes to a livestock market or farm and purchases livestock, then that person, as an agent of the research facility, must satisfy the AWAR transportation requirements (AWAR §3.136–§3.142) and the institution must satisfy health and husbandry standards (AWAR §3.125–§3.133).

14:37 Can animals be transferred between studies that have IACUC-approved protocols?

Opin.   Provided certain requirements are met, it is permissible to transfer animals between studies. Survey data from 2000[39] and this volume show this to be a universal and standard practice. Unrestricted trading of animals between research protocols, however, could lead to unauthorized use, defeat attempts at tracking animal utilization, put legal records pertaining to dogs and cats in jeopardy (AWAR §2.35,c,1–§2.35,c,2), and be potentially confusing to veterinarians who try to determine the ownership of animals requiring medical care. As such, transfer procedures should include informed IACUC consent for the transaction; approval by the IACUC of the use of the species in the research; a means to assure that multiple, unrelated major survival surgical procedures are prevented; and a mechanism to credit animals to one protocol and transfer them to another via a debit transaction. IACUCs may choose to discriminate in terms of the stringency of process control. For example, the process could be managed through an IACUC administrative function or by the animal resources program on behalf of the IACUC or deferred to the involved scientists. Important considerations in this regard might be whether or not the species is AWA regulated or the degree of invasiveness of the intended use. For those species that are not regulated, such as rats, mice, and fish, IACUCs might allow less restricted transfer. The technological innovation of census management by bar code scanning allows transfer procedures by crediting of the animals to the donor's account and debiting the exchanged animals from the total approved for the recipient collaborator.

Surv. Does your IACUC permit animals to be transferred between studies having IACUC-approved protocols?

- Not applicable to our IACUC   20/166
- Permit transfer   114/166
- Do not permit transfer   16/166
- Other   16/166

14:38 Under what circumstances or conditions may animals be transferred between studies and how can the IACUC track this activity?

Opin. At institutions where there are breeding colonies of animals supporting more than one research project, there must be a mechanism to transfer animals from the breeding colony to IACUC-approved research protocols. An example of the latter is an institution that has a core transgenic animal facility. Animals of a desired genotype are produced under one IACUC-approved activity (transgenic core) and are then transferred to the IACUC-approved activity of a specific scientist for research use. Other cases involve simple transfers in which one investigator receives a few breeders of one-of-a-kind genetically manipulated mice from the breeding colony of a colleague. Nonhuman primates used in operant conditioning or behavioral paradigms present another example. These animals are usually purchased from a commercial vendor and are quarantined in the animal research facility. During the quarantine phase (or shortly thereafter) the investigator may become aware of behavioral considerations that render the animal less appropriate for the intended research and more appropriate for use by a colleague. In this instance, the investigators may wish to exchange animals. The same may be true for dogs or cats obtained for use in research involving surgical procedures. In this circumstance, demeanor and size may lead scientists to trade animals between acute and chronic studies, or where the animals have anatomical or other attributes that are better suited for certain studies than others. Two investigators, one with expertise in a specific experimental manipulation and a second with animals of a unique genotype, may collaborate by transferring the animals from the original owner to the party doing the manipulation.

The most important consideration is that the IACUC employs a mechanism to ensure that the total number of animals used and the species of animal are appropriate in terms of the approved protocol, and that there are not multiple survival surgeries in unrelated projects (AWAR §2.31,d,1,x,A); *Guide*, p. 12). The IACUC should develop some guidelines for the transfer of animals between approved protocols if this is a circumstance that might occur repeatedly. An important consideration in setting guidelines is obtaining the surgical history of the animals from either the investigator or veterinary staff. This is necessary to prevent animals from being unintentionally subjected to multiple major survival surgical procedures, unless there is an approved exception (AWAR §2.31,d,1,x,A); *Guide*, p.12). The movement of animals from one protocol to another ideally should involve debiting or crediting the number of animals against those approved for the study. IACUCs often refer these activities to the animal resources program, which often has the staffing and computer resources to handle the task. Some IACUCs, however, find it advantageous to require the PI to inform the committee of the practice by memorandum or other mechanism.

Surv. Assume Protocol A allows for animals to be bred and used for Protocol B. Does your IACUC require tracking of the number of animals transferred to Protocol B?

- Not applicable to our IACUC   46/173
- Yes   113/173
- No   13/173
- Other   1/173

14:39 If a protocol is being renewed, how can the IACUC ensure that it is not unintentionally renewing the original number of animals requested?

Opin. This is an important issue from the perspective of preventing unnecessarily duplicative experiments and providing a justification of the total number of animals used in research. An example of this situation is useful for illustrative purposes.

An investigator is funded for a 5-year study involving 30 macaques. He purchases them over the first 3 years of the study and then submits a renewal after 3 years, requesting 30 macaques. How can the IACUC be sure that the 30 animals requested on the renewal are new monkeys, and not just a relisting of those already on the study and still in the investigator's colony?

As an extension of this problem, what if the investigator originally was approved for 30 animals but only acquired 20 by the time of the 3-year renewal? If ten animals are now requested in the renewal, how can the IACUC determine whether these are the remaining ten from the original approval, or ten more above the original request of 30 (for a total of 40)?

These problems can be addressed with careful formulation of the renewal and modification requests. IACUC review may be facilitated when requests are supported by progress reports of animal usage on the previous iteration. For an annual renewal (end of first or second year) the application should not include animal numbers, as the request has already been approved for the first 3 years. The annual accounting of animals for APHIS/AC should be monitored at the level of acquisition and then once annually for individual animals kept over multiple years.

Requests for additional animals before the end of the original protocol should be handled in a different manner than new or 3-year renewals and specifically indicate that additional animals are required. One mechanism to help eliminate the issues raised in the examples is to close out the original protocol regardless of the number of animals used or not used and to consider the continuation as an essentially new proposal, with a new set of experiments and requirement for justification of animal numbers. Obviously there are circumstances, as in the example, where further questions, such as the following, should be asked:

- Is this a new protocol or a 3-year renewal?
- If the protocol is a renewal, what is the previous protocol number?
- If the protocol is a renewal, are there *existing* animals that are already assigned to the study?
- If so, how many? (Do not include this number in your answer to the next question.)
- Justify the number and use of animals specific to this application.

Many IACUCs do not consider this issue with meticulous detail. Instead, they rely on faith that good science is not going to be repetitious or wasteful, that the vast majority of scientists can be trusted, that economic factors (as much as anything else) have a limiting effect on animal use and dictate judiciousness, and that the IACUC application will show that the research requires a certain number of animals and is novel and important.

When an animal resources program performs census management using electronic technology, it dictates downstream processes for the IACUC. Computerization of the practice typically requires that animals residing under an expiring protocol be managed in one of two ways at the point where the new protocol becomes active. Once the expiring protocol is deactivated, the animals are debited from the original number approved under the old protocol and considered to have been expended, *ipso facto*, as a result of termination. They then are immediately issued a new cage card (with new protocol and billing account numbers), considered as a new acquisition, and debited from the new IACUC protocol number. A second option is to transfer the animals from the expiring protocol to the new one, treating them as an internal receipt. To reduce confusion during consideration of the 3-year renewal, some IACUCs find it useful for the remaining animals to be added to the new protocol to be listed or described on a supplementary document separate from the renewal application.

In summary, special attention should be paid to the type of application (new or renewal), the numbers of animals used, the number requested, and the number to be carried over from the previous protocol. Although it is beyond the jurisdiction of individual IACUCs, in the opinion of the authors, many of these problems could be alleviated if periods covered by IACUC protocols and extramural grants could be matched in duration.

Surv. Assume a study is being renewed for an additional period. What actions does your IACUC take to assure that animals that were not used during the initial review period are considered part of the request for any additional animals needed? More than one response is possible.

- This question has not arisen to date   50/164
- We do not have a policy   17/164
- If there are any "leftover" animals from the initial approved protocol, they must be transferred to the reapproved protocol; they are counted as part of the number of animals approved for the renewed protocol   65/164

- Animals carried over are added to the total number of animals to be used via a separate line item; they are then debited from the grand total when the renewal is activated (such as when updated cage cards are made)   6/164
- The expired protocol is closed out and existing animals are deleted from the census; these can be added to the renewal via a special form but are debited from the total   7/164
- Carryover animals are specifically identified in the renewal; no additional consideration is given except that only the total number of new animals needed over the next 3 years is divulged to those who purchase the animals   9/164
- Investigators are encouraged to identify carryover animal numbers, but doing so is not enforced   3/164
- Other   7/164

14:40 What guidelines can the IACUC establish for the use of endangered species in research?

Opin. The use of endangered species in research must conform with all federal and state laws and international conventions to which the U.S. subscribes, including the Lacey Act, the Endangered Species Act, the Marine Mammal Protection Act, and the CITES treaty. It is desirable for IACUC guidelines to investigators to repeat this requirement.

14:41 Should institutional programs related to veterinary care (such as a rodent sentinel program, quarantine programs involving diagnostic sample collection from animals, necropsy service, rodent colonies for raptor rehabilitation programs, and therapeutic surgical interventions) be approved via the protocol review system?

Reg. The AWAR (§2.31,a) require that all animal activities for regulated species be reviewed by the IACUC. PHS Policy (IV,B,1) and the *Guide* (p. 9) state that the entire "program" must be reviewed by the IACUC.

Opin. PHS Policy applicability is not limited to research; it also includes all activities involving animals, such as testing and teaching. NIH/OLAW has determined that although animals used as sentinels, breeding stock, chronic donors of blood and blood products and for other like needs may not be part of specific research protocols, their use for these purposes contributes significantly to the institutional research program and constitutes activities involving animals.[36] Consequently, the IACUC must receive and approve protocols, have appropriate systems to monitor the use of animals prior to the commencement of such activities, and then perform reviews at the appropriate intervals.[36] The authors also interpret this to include rodent colonies used as a source of food for raptors in rehabilitation programs (see 8:35), animals used for training purposes, and those held on expired protocols. All clinical activities (e.g., diagnostic and therapeutic interventions) and the necropsy service involving existing research or instructional animals ordinarily are under the supervision of the AV, should be covered under the semiannual institutional program assessment, and do not require approval under the protocol review system. Nevertheless, some committees (fewer than 25% by survey) elect to oversee these activities via the protocol review process.

Surv.   Which of the following institutional programs related to veterinary care are reviewed by your IACUC and approved via the protocol review system? Check all appropriate responses. Values in parentheses show the percentage of programs managed by the protocol review system out of all institutions indicating the existence of the program. A total of 164 institutions responded to the survey question.

• Rodent sentinel program	111/130 (85%)
• Rodent quarantine program	83/122 (68%)
• Rodent surgical rederivation program	40/73 (55%)
• Necropsy service	38/101 (38%)
• Biological material quality assurance program (e.g., MAP testing of cell lines)	27/78 (35%)
• Veterinary therapeutic surgical interventions	36/164 (22%)
• Genetic quality control	24/74 (32%)
• Core rodent breeding service colonies	53/84 (63%)

14:42   Does the IACUC need to approve studies that use privately owned animals, such as those in a clinical trial in a school of veterinary medicine?

Reg.   The AWAR (§1.1) include all regulated species used, or intended for use, in research, teaching, and testing. The PHS Policy (III,A) includes any live vertebrate animals used or intended for use in research, research training, experimentation, or for related purposes (if the work is PHS funded or the institution voluntarily includes all animals under its Animal Welfare Assurance). Neither the AWA, AWAR, nor PHS Policy distinguishes between privately owned animals and those owned by the institution.

Opin.   If the study conditions do not meet the definitions of federal oversight described, it is strongly recommended for the protection of the institution that, at the least, an informed consent be obtained from the client as the representative of the animal (the subject). Ideally, IACUC approval should be obtained.

The law and regulations give the IACUC room to deliberate justifiable exemptions, particularly where the use is not covered in the AWAR or PHS Policy (see bulleted examples that follow). Ultimately this involves doing what is appropriate by all parties and documenting the exemption and its rationale. While one could argue that the IACUC specifically is not required to review the use of pets or other privately owned animals that may be involved in the research or related project, it is recommended that institutions have a written policy if federal oversight is not mandated. Further, as noted in the Regulatory comments to this question, when federal oversight is mandated, neither the AWAR nor PHS distinguishes privately owned animals from those owned by the institution. A number of factors are to be considered, including but not limited to the following:

• The specific type of work
• Species to be used
• The need to board or retain the animals for a period
• The nature and resources of the facility
• Involvement of federal funding

- Requirement of IACUC review by a nonfederal granting agency
- The content of the institutional Animal Welfare Assurance

Pets used in clinical trials are privately owned animals recruited from volunteers and, as such, usually remain the property of the owners, who maintain responsibility for their upkeep, housing, and transportation. However, once these animals appear on institutional property, as noted previously, rules should be established as to how the animals are managed and temporarily housed. Such rules should take into consideration facility operations, informed consent, institutional liability, public safety, and animal welfare. When client-owned animals are donated to the institution, the full range of compliance issues takes effect. At those veterinary schools where the IACUC does not review the research use of client-owned animals (e.g., does not have an NIH/OLAW Assurance), teaching hospitals typically have policies governing the conduct of research. Where the veterinary school has an Animal Welfare Assurance or the species to be used is covered by the AWAR or the Assurance of the parent university, the IACUC clearly has full oversight responsibility. In situations not covered by an Assurance or the AWAR, the proposed research may be reviewed by the department and then a hospital board or committee akin to an institutional review board for human subject experimentation. In practice, the overwhelming preponderance of IACUCs elect to oversee the experimental use of client-owned animals.

Surv. Does your IACUC approve studies using privately owned animals, such as those used in a clinical trial in a school of veterinary medicine?

• Not applicable to our IACUC	128/167
• Require approval via the regular IACUC application form	34/167
• No formal IACUC review, but specific written policy and procedure	1/167
• Do not require approval	2/167
• Other	2/167

14:43   An institution, a technical college, does not own or house animals, but, in the veterinary technology program, students encounter animals in classroom demonstrations of noninvasive procedures and during externships. Is the IACUC required to follow the AWAR, PHS Policy, or both, and is it required to monitor the acquisition and disposition of the animals?

Reg. The AWAR apply to the use of animals in teaching including those that might be encountered in a veterinary technology program such as dogs, cats, guinea pigs, hamsters, or rabbits (§1.1, Animals). Although not applicable where such animal use is supported by tuition or other institutional funds, NIH/OLAW has recommended for purposes of consistency and simplicity that all instructional use of animals be reviewed by the IACUC.[40]

Opin. The American Veterinary Medical Association requires that the instructional use of animals in accredited veterinary technology programs, regardless of funding source, species, or nature of use, be under the authority of an IACUC.[41] This logically extends even to instances when student-owned animals are used for the classroom demonstration of noninvasive procedures such as physical restraint for examination, vital signs recording, and positioning of conscious patients for radiography and that remain on the premises for less than 12 hours. Under this scenario at the teaching institution, it is within the purview of the IACUC to monitor and track the numbers of animals procured or participating, the frequency of use, and the eventual disposition of the animals. When students work with client-owned animals during an externship at a community veterinary private practice establishment for purposes of acquiring practical skills in specimen collection, medication administration, surgery assistance, or postoperative care, the institution has a stake in assuring that these experiences are meaningful and that the students will be appropriately supervised, mentored, and trained. Techniques learned in this environment will be carried into subsequent careers. While the level at which oversight occurs is a matter of institutional decision, it is logical to assign the responsibility to the IACUC. In this case, the IACUC should at least consider whether or not these experiences should be covered by protocols.

14:44   An institution's veterinarians or physicians occasionally are asked to provide veterinary services to a pet at no or reduced cost. How should the IACUC address these situations?

Opin.   Although research institutions have great resources and clinicians with unique skills, engaging in these activities is a legal "slippery slope" fraught with risk. There is great potential for both institutional and veterinary liability should something go awry.

All states have veterinary practice acts that regulate the practice of veterinary medicine in the state and protect veterinary private practice as a business. Many states permit veterinarians who specialize in laboratory animal medicine to practice without a license provided that certain conditions are met (e.g., licensed in at least one state, specialty board certification, limiting of activity to specialty area). When unlicensed specialists, including physicians, undercut fees or provide veterinary services for individuals who otherwise would seek care from private practitioners, the institution and veterinarian assume a legal risk.

Some institutions and individuals believe that they can protect themselves by rendering free service or requiring that clients sign consent forms. Nevertheless, the judicial system most likely would not enforce provisions of consent forms in which clients agree not to hold an institution or individual liable for the negligent treatment of an animal.[42] Nor will any consent by a client to have a veterinary procedure done by a nonveterinarian protect the nonveterinarian from prosecution for practicing veterinary medicine without a license.[42]

In some instances, veterinarians and institutions must be pragmatic. Acutely ill or dying animals on or near institutional property, such as stray cats in end-stage disease, dogs that have been hit by cars, or injured birds, may require humane intervention and cooperation with local animal control authorities. Whatever the reason, veterinarians are bound by oath to relieve animal suffering, but as noted previously, must also act in ways that protect research animals from those who have undefined health status. Consequently, institutions may want to develop policies that address situations in which emergencies involving animals that are not university property require prompt and humane intervention. The unwritten policy of the authors' institution is to stabilize the condition of animals that require critical care with the understanding that they will be transported by a county animal control officer immediately to a local veterinary hospital for subsequent care.

14:45   A laboratory animal facility is associated with a human hospital. What policy, if any, should the IACUC have relative to housing Seeing Eye dogs or other animals that might accompany patients admitted into the hospital?

Opin.   At the authors' institution, with a large medical center and several affiliated hospitals, Seeing Eye dogs and animals involved in pet-facilitated therapy programs are permitted in the wards and public areas of the hospital. These issues fall under the jurisdiction of hospital policy and practice and not under the oversight of the IACUC as the animals are privately owned and not used for research, testing, teaching, or training. Under the Americans with Disabilities Act, hospitals, as public accommodations, are expected to provide reasonable access to services and facilities (health care) for those who use service animals[43] but are not required to provide care or supervision while they are on site.[44] Likewise, at our medical center, pet-facilitated therapy is made available to patients through a relationship with a bonded, outside provider. The provider ensures that the privately owned animals participating in the program are of the appropriate age, size, and disposition; are acclimated to patients and the hospital environment; and are in apparent good health (including parasite control and full immunizations) as certified annually by a contract veterinarian.

14:46   What are the legal requirements regarding the disposition of research animals?

Opin.   Animals used in research may be disposed of by euthanasia, adoption, retirement (e.g., endangered species), return to production agriculture (food and fiber species), sale for slaughter, release to the wild, or transfer to another institution or research project. In most instances, animals used in research are euthanized. Considerations when disposing of these animals include the experimental history (e.g., method of euthanasia and exposure to chemicals, toxins, radiation, and drugs) and may involve burial in a landfill, incineration, alkaline hydrolysis with discharge into the sewage system, or rendering into animal feeds. In some instances, particularly in the case where radioisotopes have been given to an animal, the carcass must be held until the radioactivity decays to safe levels.

14:47   Can animals used in research, teaching, or testing be adopted as pets once they have been used in an IACUC-approved project? What is an appropriate mechanism for facilitating this process?

Opin.   Provided the animals are healthy and have not been disfigured or disabled by the research, adoption of animals that make suitable pets is appropriate. This practice, as have many others, may have legal ramifications. Guidelines for adoptions should be formally developed in consultation with legal counsel. In institutions where it is permitted, adoptions to a reliably suitable and permanent home occur on a limited basis after review and approval by the researcher, approval of the request by the IACUC, and confirmation of the good health of the animal by a veterinarian. There may be other restrictions, such as limiting adoptions to employees and students of the institution and not to the general public. Some public institutions do not permit adoption per se as adoption entails a change in ownership that involves divestiture of public property. However, at least one pragmatic institution skirts this issue by permitting animals to leave under the status of a long-term loan. Another institution prohibits outside adoptions but allows retired animals to remain in possession of the laboratory with care provided by staff. This leads to interesting questions of whether such "institutional pets" are covered in a protocol and how the program is funded. At the author's institution, healthy animals are released after the approval process and the completion of relevant records (i.e., USDA Veterinary Services Form 18-6: Record of Disposition of Dogs and Cats). Copies of all correspondence related to adoptions and the permanent medical records of adopted animals are maintained in the animal resources program office.

Surv. Does your IACUC permit adoption of animals as pets once they have been used in an IACUC-approved project?

• Not applicable	36/177
• Never discussed by our IACUC	18/177
• Yes, always	1/177
• Yes, under specified circumstances	71/177
• No	46/177
• Other	5/177

14:48 Should the IACUC permit farm animal species that have been used in teaching, testing, or research to be sold for slaughter? What guidelines would be useful to the IACUC in this regard? (See 18:8.)

Opin. Domestic livestock, usually raised for food or fiber, are also commonly used in research and training. In some instances, it may be appropriate to return live animals that have been used for research or training purposes to production agriculture or to render carcasses into livestock feed. However, if this is done, it should be done with caution and with full awareness of the potential for institutional and individual liability.

When animals used in research are subsequently slaughtered for human or animal food, the major concern is adulteration of the meat by residues (e.g., antibiotics, hormones, toxins, radioisotopes, carcinogens) or infectious agents that would render the meat or its food products unwholesome. The agencies that regulate slaughter are either USDA's Food Safety and Inspection Service (for federally inspected facilities) or state agencies for state inspected slaughter facilities. Endangerment of the nation's food supply by failure to adhere to food safety standards constitutes a felony in the U.S. that could be punishable by fines or sanctions to the institution or institutional officials, imprisonment of convicted parties, and revocation of the license of the AV. Given these potential legal ramifications, an institutional policy on the sale, slaughter, consumption, or rendering of research animals is strongly encouraged. The policy should be developed with appropriate legal counsel and should be consistent with the institution's Animal Welfare Assurance. Veterinary care of animals sent to market or slaughter and drug withdrawal periods must be consistent with the Animal Drug Use Clarification Act of 1994 (AMDUCA).[5] In the event that the animals are given a new animal drug as defined by the FDA,[45] no meat, eggs, or milk from those animals may be processed for human food without FDA approval. If animals are being slaughtered at the research facility for subsequent consumption, the method of euthanasia must be stated in the approved protocol and should be in compliance with the AWAR (§1.1, Euthanasia) and the American Veterinary Medical Association recommendations on euthanasia.[46] For animals that have had invasive surgical procedures, the general practice should be for the level of surgical asepsis and postoperative care to be consistent with current veterinary standards for the species. Where IACUCs allow this practice, Food and Drug Administration restrictions are adhered to and the animals must have no lingering evidence of manipulation such as the presence of a cannula or other permanent implant.

Surv. Does your IACUC permit farm animals that have been used for teaching, testing, or research to be sold for slaughter or released back to production agriculture?

- Not applicable   100/166
- Never discussed by our IACUC   9/166
- Yes, always   4/166
- Yes, under specified circumstances   12/166
- No   41/166
- Other   0/166

References

1. Committee to Revise the Guide for the Care and Use of Agricultural Animals in Agricultural Research and Teaching, *Guide for the Care and Use of Agricultural Animals in Agricultural Research and Teaching*, 1st rev. ed., Savoy, IL, 1999, p. 4.
2. Office of the Federal Register, Code of Federal Regulations, Title 7, Part 371 4(b)(2), Animal and Plant Health Inspection Service—Veterinary Services, Office of the Federal Register, Washington, D.C., 1989 (revised 1998).
3. Office of the Federal Register, Code of Federal Regulations, Title 50, Part 17, Section 17.22, Permits for Scientific Purposes, Enhancement of Propagation or Survival, or for Incidental Taking of Endangered Wildlife, Office of the Federal Register, Washington, D.C., 1975 (revised 1997).
4. Office of the Federal Register, Code of Federal Regulations, Title 50, Part 23, Endangered Species Convention, Office of the Federal Register, Washington, D.C., 1977 (revised 1997).
5. Office of the Federal Register, Code of Federal Regulations, Title 9, Part 530, Extralabel Drug Use in Animals, Office of the Federal Register, Washington, D.C., 1997 (revised 1998).
6. Office of the Federal Register, Code of Federal Regulations, Title 42, Part 71, Foreign Quarantine; Subpart F, Importations; Section 51, Dogs and Cats; Section 52, Turtles, Tortoises, and Terrapins; Section 53, Nonhuman Primates; Section 54, Etiologic Agents, Hosts and Vectors, Office of the Federal Register, Washington, D.C., 1985 (revised 1997).
7. Code of Federal Regulations, Title 29, Part 1910, Occupational Safety and Health Standards; Subpart Z, Toxic and Hazardous Substances; Section 1450(e), Chemical Hygiene Plan, Office of the Federal Register, Washington, D.C., 1990 (revised 1998).
8. Institute for Laboratory Animal Resources, National Academy of Sciences, Amphibians, in Guidelines for the Breeding, Care and Management of Laboratory Animals, Washington, D.C., 1974.
9. Institute for Laboratory Animal Resources, National Academy of Sciences, Recommendations for the Care of Amphibians and Reptiles in Academic Institutions, Washington, D.C., 1991.
10. Schaeffer, D.O., Kleinow, K.M., and Krulish, L. (Eds.), *The Care and Use of Amphibians, Reptiles, and Fish in Research*, Scientists Center for Animal Welfare, Greenbelt, MD, 1992.
11. American Society of Ichthyologists and Herpetologists, American Fisheries Society, American Institute of Fisheries Research Biologists, Guidelines for the use of fishes in field research, *Fish J.*, 13(2), 1, 1987.
12. American Society of Ichthyologists and Herpetologists, Herpetologists' League, and Society for the Study of Amphibians and Reptiles, Guidelines for the Use of Live Amphibians and Reptiles in Field Research, 2001. Available on the World Wide Web at: http://www.asih.org/pubs/herpcoll.html.

13. Potkay, S., et al., Frequently asked questions about the Public Health Service Policy on Humane Care and Use of Laboratory Animals, *Contemp. Topics Lab. Anim. Sci.,* 36(2), 47, 1997.
14. U.S. Department of Agriculture, Animal and Plant Health Inspection Service, Animal Care, Animal Care Policy Manual, Policy 26—Regulation of Agricultural Animals, November 17, 1998. Available on the World Wide Web at: http://www aphis usda gov/ac/policy/policy 26.pdf.
15. U.S. Department of Agriculture, Animal and Plant Health Inspection Service, Animal Care, Animal Care Policy Manual, Policy 29—Farm Animals Used for Nonagricultural Purposes, February 11, 2000. Available on the World Wide Web at: http://www.aphis.usda.gov/ac/policy/policy29.pdf.
16. National Institutes of Health, Department of Health and Human Services, Laboratory Animal Welfare: Change in PHS Policy on Humane Care and Use of Laboratory Animals, *Fed. Regist.,* 67(152), 51289, 2002.
17. Potkay, S., DeHaven, W.R., OLAW, and APHIS: Common areas of noncompliance, *Lab Anim.* (NY), 29(5), 32, 2000.
18. Wolff, A., et al., Frequently asked questions about the Public Health Service Policy on Humane Care and Use of Laboratory Animals, *Lab Anim.* (NY), 32(9), 33, 2003.
19. ARENA-OLAW, Institutional Animal Care and Use Committee Guidebook, 2nd ed., U.S. Government Printing Office, Washington, D.C., 2002, p. 153.
20. ARENA-OLAW, Institutional Animal Care and Use Committee Guidebook, 2nd ed., U.S. Government Printing Office, Washington, D.C., 2002, p. 130.
21. Huerkamp, M.J., and Archer, D., Animal acquisition and disposition, in *The IACUC Handbook,* Silverman, J., Suckow, M., and Murthy, S., Eds., CRC Press, Boca Raton, FL, 2000, p. 188.
22. Hogan, B., et al., *Manipulating the Mouse Embryo: A Laboratory Manual,* 2nd ed., Cold Spring Harbor Laboratory Press, Plainview, NY, 1994, p. 117.
23. Borski, R.J., and Hodson, R.G., Fish research and the Institutional Animal Care and Use Committee, *ILAR J.,* 44(4), 286, 2003.
24. U.S. Department of Agriculture, Animal and Plant Health Inspection Service, Animal Care, Animal Care Policy Manual, Policy 17—Annual Report for Research Facilities, March 17, 1999. Available on the World Wide Web at: http://www.aphis.usda.gov/ac/policy/policy 17.pdf.
25. Huerkamp, M.J., and Archer, D., Animal acquisition and disposition, in *The IACUC Handbook,* Silverman, J., Suckow, M., and Murthy, S., Eds., CRC Press, Boca Raton, FL, 2000, p. 189.
26. Hogan, B., et al., *Manipulating the Mouse Embryo: A Laboratory Manual,* 2nd ed., Cold Spring Harbor Laboratory Press, Plainview, NY, 1994, p. 120.
27. Wilkie, T.M., Brinster, R.L., and Palmiter, R.D., Germline and somatic mosaicism in transgenic mice, *Dev. Biol.,* 118, 9, 1986.
28. Wigglesworth, C., A word from OLAW, *Lab Anim.* (NY), 35(1), 16, 2006.
29. Division of Animal Welfare, Office for Protection from Research Risks, National Institutes of Health, The Public Health Service responds to commonly asked questions, *ILAR News,* 33(4), 68, 1991 Available on the World Wide Web at: http://grants2.nih.gov/grants/olaw/references/ilar91.htm#1.
30. Saif, Y.M., and Bacon, W.L., Protocol review: Simply stated, *Lab Anim.* (NY), 25(5), 22, 1996.
31. Swayne, D.E., Protocol review: Preventative measures, *Lab Anim.* (NY), 25(5), 21, 1996.
32. Williams, B., Wildlife research and the IACUC, *AWIC Bull.,* 10(1), 1999.
33. Mulcahy, D.M., Does the Animal Welfare Act apply to free-ranging animals? *ILAR J.,* 44(4), 252, 2003.
34. Reeb-Whitaker, C.K., et al., The impact of reduced frequency of cage changes on the health of mice housed in ventilated cages, *Lab Anim.* (NY), 35(1) 58, 2001.
35. Garnett, N.L., DeHaven, W.R., OPRR, and USDA, Animal care response on applicability of the animal welfare regulations and the PHS policy to dead animals and shared tissues, *Lab Anim.* (NY), 26(3), 21, 1997.
36. Potkay, S., et al., Frequently asked questions about the Public Health Service Policy on Humane Care and Use of Laboratory Animals, *Lab Anim.* (NY), 24(9), 25, 1995.
37. APHIS/AC, Research Manual Part 18. Available on the World Wide Web at: http://www.aphis.usda.gov/ac/researchmanual/18-4PROC.PDF.

38. National Institutes of Health, Department of Health and Human Services, Office of Extramural Research, Revised Guidance regarding IACUC Approval of Changes in Personnel Involved in Animal Activities, Notice NOT-OD-03-046, June 6, 2003. Available on the World Wide Web at: http://grants.nih.gov/grants/guide/notice-files/NOT-OD-03-046.html.

39. Huerkamp, M.J., and Archer, D., Animal acquisition and disposition, in *The IACUC Handbook,* Silverman, J., Suckow, M., and Murthy, S., Eds., CRC Press, Boca Raton, FL, 2000, p. 191.

40. ARENA-OLAW, Institutional Animal Care and Use Committee Guidebook, 2nd ed., U.S. Government Printing Office, Washington, D.C., 2002, p. 142.

41. American Veterinary Medical Association, Appendix B: Guidelines for the Use of Animals in Veterinary Technology Teaching Programs, in Accreditation Policies and Procedures of the AVMA Committee on Veterinary Technician Education and Activities (CVTEA), December 2005. Available on the World Wide Web at: http://www.avma.org/education/cvea/cvtea_appendix_b.pdf.

42. Tannenbaum, J., personal communication, 1998.

43. Office of the Federal Register, Code of Federal Regulations, Title 28, Part 36, Nondiscrimination on the Basis of Disability by Public Accommodations and in Commercial Facilities; Section 301(a), Eligibility Criteria, Office of the Federal Register, Washington, D.C., 1991 (revised 1998).

44. Office of the Federal Register, Code of Federal Regulations, Title 28, Part 36, Nondiscrimination on the Basis of Disability by Public Accommodations and in Commercial Facilities, Section 301(c)(1–2), Service Animals, Office of the Federal Register, Washington, D.C., 1991 (revised 1998).

45. Office of the Federal Register, Code of Federal Regulations, Title 21, Part 511, New Animal Drugs for Investigational Use, Office of the Federal Register, Washington, D.C., 1975 (revised 1998).

46. American Veterinary Medical Association, 2000 Report of the AVMA Panel on Euthanasia, *J. Am. Vet. Med. Assoc.,* 218, 669, 2001.

15

Animal Housing and Use Sites

Cynthia Gillett

Introduction

This chapter focuses on what are and are not acceptable housing or holding locations for animals used in research, teaching, and testing. The AWAR, *Guide*, and PHS Policy are the primary references utilized. There are few absolutes when considering animal housing and use sites. Professional judgment, institutional policy, and protocol goals and needs must all be considered when determining what is and is not an acceptable location for the care and use of animals.

IACUC Chairs from throughout the United States and Canada responded to a questionnaire about animal housing and use sites. Their responses are tabulated and summarized where appropriate to provide further insight into the breadth of policies currently in practice.

15:1 What is the definition of an animal facility?

Reg.   The AWAR (§1.1) have several definitions. A *housing facility* is any land, premise, shed, barn, building, trailer, or other structure housing or intended to house animals. Separate definitions also are provided for *indoor housing facility*, *outdoor housing facility*, *sheltered housing facility*, and *research facility*.

PHS Policy (III,B) defines an *animal facility* as "any and all buildings, rooms, areas, enclosures, or vehicles, including satellite facilities, used for animal confinement, transport, maintenance, breeding, or experiments inclusive of surgical manipulation."

The *Guide* (pp. 72–73) states that an animal facility consists of functional areas for animal housing, care, and sanitation; receipt, quarantine, and separation of animals, separation of species, or isolation of individual projects; and storage. Some facilities may encompass space for surgery, intensive care, necropsy, radiography, diet preparation, procedural space, clinical treatment, supply receiving, cage washing, carcass storage, facility administration, training, locker rooms, and break rooms.

Opin.   The author's working definition of an animal facility is an integrated concept that is inclusive of the many programmatic aspects that relate to animal housing, care, and use in an institutional setting. However, for the purposes of this chapter, the term *animal facility* shall refer to the primary housing location and support space for animals and animal studies overseen by the IACUC. It is important to note that the AWAR definition of a research facility broadly encompasses institutions or persons who use or intend to use live animals in research, tests, or experiments. Thus, a research facility, as defined in the AWAR, should not simply be thought of as those areas where animals are housed or used, but the institution as a whole.

15:2 What are the circumstances and length of time that animals may be removed from their primary housing site?

Reg. The AWAR (§2.31,c,2) require that any location where animals are housed for more than 12 hours be inspected by the IACUC at least once every 6 months. PHS Policy (III,B) states that "a satellite facility is any containment outside of a core facility or centrally designated or managed area in which animals are housed for more than 24 hours." Nevertheless, PHS Policy (II) requires compliance with the AWA where applicable. Therefore, the PHS Policy definition of a satellite facility and the duration of stay requirements may be applied to species that are not covered by the AWAR, such as typical laboratory rats and mice.[1] However, the more restrictive AWA regulation (§2.32,c,2) requiring IACUC semiannual inspection of locations where animals are housed for more than 12 hours must be applied to all AWAR-regulated species.

The *Guide* (p. 9) states that the IACUC is responsible for oversight of animal-activity areas; however, it does not define that term. It also states (p. 71) that "Animals should be housed in facilities dedicated to or assigned for that purpose and should not be housed in laboratories merely for convenience. If animals must be maintained in a laboratory area to satisfy a protocol, the area should be appropriate to house and care for the animals."

Opin. The author's interpretation of the term *animal-activity area*, as stated in the *Guide* (p. 9), is that it implies research procedural areas such as a laboratory or testing site. Neither the *Guide*, the AWAR, nor the PHS Policy prohibits removal of animals from the animal holding facility for procedures or for housing elsewhere, such as an investigator's laboratory or a satellite facility. However, doing so may trigger additional regulatory oversight responsibility of the IACUC or bring into play individual institutional requirements. Although one could apply different time standards on the basis of whether or not an AWAR-covered species is involved, applying the most restrictive criterion (12 hours) will result in compliance in all situations.

Surv. Under what circumstances and for how long does your IACUC allow animals to be removed from their primary housing site?

- Not applicable: no need to remove animals from their primary housing site — 33/185
- IACUC must approve any housing outside animal facility for any length of time — 50/185
- Up to 12 hours out of primary site permitted with no prior approval — 43/185
- Up to 24 hours out of primary site permitted with no prior approval — 13/185
- Up to 48 hours out of primary site permitted with no prior approval — 2/185

- No overnight out of primary area housing allowed without
  IACUC approval   24/185
- Never allowed   8/185
- Other   12/185

15:3   What are examples of acceptable circumstances for keeping animals outside the primary animal facility?

Opin.   The AWAR, PHS Policy, and *Guide*, while providing definitions and oversight rec-
           ommendations for animal facilities (including satellite facilities; see 15:1), do not
           provide examples of acceptable circumstances or rationale for keeping animals
           outside the primary animal facility.
               There are other considerations when housing animals outside a centrally man-
           aged facility. For reasons of disease control, some facilities have a policy stating
           that if animals leave the facility they may not be returned. This is particularly
           applicable to microbiologically defined (specific pathogen free) rodents and trans-
           genic mice. The method and route of transportation to the study area also must be
           monitored to minimize potential for disease transmission and to ensure conform-
           ance with transportation guidelines.
Surv.   What are some examples of acceptable circumstances and justification for keeping
           animals outside the primary animal facility? More than one response per respon-
           dent is possible.

- Not applicable   46/185
- Biohazard studies requiring specialized temporary
  containment   55/185
- Necessary equipment cannot be easily relocated to the
  animal facility   88/185
- Behavioral or stress testing that would be affected by daily
  transport from animal holding room   63/185
- Postoperative monitoring   43/185
- Insufficient procedure space in animal facility   30/185
- Continuous administration of test substances via highly
  technical, automated procedures (e.g., automated
  administration of intravenous drugs)   46/185
- Frequent, repeated observations are part of protocol   43/185
- Other   12/185

15:4   Are there regulatory considerations regarding the actual act of transportation that the IACUC must consider when animals are moved from their primary housing site to a secondary site?

Reg.   The AWAR (Part 3, Standards) have extensive transit regulations for each covered
           species. They include primary enclosures, primary conveyances, animal identifica-
           tion, food and water requirements, care in transit, handling, sanitation, ventilation,
           shelter, and temperature. The AWAR (§2.38,f) also cover the transit of animals as
           part of an adequate "handling" consideration. PHS Policy II requires compliance
           with the AWAR, where applicable.

The *Guide* (p. 57) states, "All transportation of animals . . . should be planned to minimize transit time and the risk of zoonoses, protect against environmental extremes, avoid overcrowding, provide food and water when indicated, and protect against physical trauma."

Opin. The AWAR on transport of animals apply to commercial transportation, but many of those requirements are common sense and designed to ensure the well-being of the transported animals. These regulations should be consulted when planning intrafacility and interfacility institutional animal transfers. While animal transport is not required to be directly reviewed and approved by the IACUC, transport activities should be reviewed during the semiannual program evaluation to ensure appropriate procedures are being followed. Animal transport issues also should be raised during protocol review when justification for off-site housing is being requested.

Surv. 1 Does your IACUC allow small animals (e.g., rodents) to be transported from one animal housing site to another by private vehicles?

• Not applicable	64/172
• No	48/172
• Yes with IACUC approval	42/172
• Yes, IACUC approval not necessary	9/172
• Other	9/172

Surv. 2 Does your IACUC allow larger animals (e.g., dogs, monkeys) to be transported from one animal housing site to another in private vehicles?

• Not applicable	97/162
• No	45/162
• Yes, with IACUC approval	13/162
• Yes, IACUC approval not necessary	3/162
• Other	4/162

15:5 Are there practical, nonregulatory considerations regarding the actual act of transportation that the IACUC must consider when animals are transported from their primary housing site?

Reg. The AWAR (§2.38,f,1) state that handling of all animals shall be done as expeditiously and carefully as possible in a manner than does not cause trauma, overheating, excessive cooling, behavioral stress, physical harm, or unnecessary discomfort.

Opin. The practical, nonregulatory considerations that can arise during animal transport tend to be situational and are best handled by discussions with the laboratory animal facility professional staff, rather than with the IACUC. Examples include situations such as whether or not a box of mice can be taken outside to transport them to a laboratory in another building; whether or not a cardboard box can be used to transport a rabbit to the laboratory; or which elevator can be used to transport animals to a laboratory.

The answers to these and similar issues depend on factors such as climate, time of year, facility design, and institutional policies. The IACUC, during the course of its semiannual facility inspection, may find evidence of unacceptable transportation practices (e.g., a soiled transport enclosure, inappropriate elevator or corridor use) that can then be addressed with the PIs and laboratory animal care staff.

15:6 If animals are being housed in a research laboratory setting, are there species differences to consider?

Reg.   The housing requirements for various species are addressed in Part 3 of the AWAR. These standards apply if the animals are approved by the IACUC to be "housed" in a laboratory. The IACUC may make exceptions to these standards if there is scientific justification to do so (AWAR §2.31,c,3).

Opin.   The need for laboratory housing should be protocol driven and not motivated by convenience or economics (*Guide*, p. 71). A comprehensive husbandry plan for the animals should be submitted to the IACUC. The IACUC should keep records of approved laboratory housing sites and perform, or delegate, periodic inspections of these areas when animals are present, in addition to the scheduled semiannual facility inspections.

Surv.   For which species, and where, has your IACUC approved housing outside the animal facility longer than 24 hours but still within your institution? Survey respondents indicated that decisions on laboratory housing were made on a case-by-case basis and that nonmammals (e.g., fish, birds) were more likely to be permitted to be housed in a laboratory setting. More than one response per respondent is possible.

- Not applicable   89/185
- Rodents in a laboratory   56/185
- Nonhuman primates in a laboratory   2/185
- Rabbits in a laboratory   6/185
- Nonmammals in a laboratory   24/185
- Other   14/185

15:7 May animals be housed at sites not belonging to an institution? Under what circumstances?

Reg.   The AWAR (§2.38,i) state,"If any research facility obtains prior approval of the APHIS AC Regional Director, it may arrange to have another person hold animals: *Provided* That:

(1) The other person agrees, in writing, to comply with the regulations in this part and the standards in part 3 of this subchapter, and to allow inspection of the premises by an APHIS official during business hours;

(2) The animals remain under the total control and responsibility of the research facility; and

(3) The Institutional Official agrees, in writing, that the other person or premises is a recognized animal site under its research facility registration."

APHIS/AC suggests that APHIS Form 7009 should be used for approved off-premises housing sites.

PHS Policy (IV,B) requires that awardee institutions and other participating institutions that receive PHS support have an approved Assurance on file with NIH/OLAW. NIH/OLAW has stated that vendors that supply custom antibodies to investigators must either have an Animal Welfare Assurance on file with NIH/OLAW or be included in the receiving research institution's Assurance statement.[2] Additionally, if AWAR-covered species, such as rabbits, are used to produce antibodies at the vendor's site or are in other research studies, then the vendor must be registered with APHIS/AC as a research facility. APHIS/AC Policy 10 clarifies the need for licensing or registration of a facility using covered species for antibody production.[3] (See 8:10; 19:15.) The *Guide* does not address this issue.

Opin.   Additional concerns include the need or desirability for the noninstitutional site either to be APHIS/AC registered, have an Assurance on file with NIH/OLAW, or be AAALAC accredited. These considerations are dependent on the primary institution's status and on the individual housing situation. Affiliate Veterans Administration Medical Center facilities are an example of facilities where housing animals at a noninstitutionally owned site might be routinely permitted. In the latter example, this should be listed as a "site" of the research institution.

Surv. 1   May animals belonging to your institution be housed at sites not belonging to your institution?

- Not applicable; the issue has not arisen   61/168
- No   22/168
- Yes, if approved by the IACUC   80/168
- Other   5/168

Surv. 2   Under what circumstances would your IACUC approve the housing of animals at sites not belonging to your institution? More than one response may be given.

- Not applicable; the issue has not arisen   69/146
- Space constraints   29/146
- Lack of specialized housing (e.g., farm animals, nonhuman primates)   49/146
- Collaborative research arrangements   91/146

15:8   Are there any circumstances under which animals may be housed in a person's home or on a private farm?

Reg.   Regulatory requirements are the same as for 15:7. The home or farm must be listed as a site of the institution.

Opin.   Prior approval by the IACUC for private housing is considered by surveyed respondents to be essential. When animals remain privately owned, as opposed to institutionally owned, they most often are housed in that person's home or farm (e.g., blood donor animals or veterinary clinical studies on companion and farm animals).

Surv.   Under what circumstances would your IACUC allow animals to be housed in a person's home or on a private farm?

- Not applicable; issue has not arisen   74/147
- None   53/147
- Certain animals may be housed at a private farm or home with IACUC approval   16/147
- Other   4/147

15:9   Do all wildlife studies need to be reviewed by the IACUC?

Reg.   The AWAR definition of *animal* (§1.1, Animal; see 12:1) includes most warm-blooded mammals; therefore, wild mammals are subject to IACUC protocol review. However, the regulations specifically exempt the IACUC from the requirement to review field studies that are conducted on free-living wild animals in their natural habitat and do not involve an invasive procedure, any harm to the animal, or any material alteration of the behavior of an animal under study (§2.31,d,1). PHS Policy II requires compliance with the AWA, as applicable, and (III,A) defines an *animal* as any live, vertebrate animal. Therefore, all vertebrate animals used in field studies fall under the PHS Policy. The *Guide* (p. 5) states that "biomedical and behavioral investigations occasionally involve observation or use of vertebrate animals under field conditions. Although some of the recommendations listed in this volume are not applicable to field conditions, the basic principles of humane care and use apply to the use of animals living in natural conditions."

Opin.   Clearly the IACUC must review wildlife studies conducted in the field on AWA-regulated species that consist of activities other than simple unobtrusive observation. PHS-sponsored activities involving vertebrate animals under field conditions must also be reviewed by the IACUC. It is the author's opinion that applying uniform policies to all animals involved in the institution's animal care and use program regardless of the type of research or teaching activity involved promotes consistent oversight.

15:10   If animals are wild-trapped, are there conditions the IACUC should place on that activity in terms of type of trap, frequency of checking of the trap, and euthanasia of injured or to-be-collected animals?

Reg.   The AWAR (§2.31,d,1) require the IACUC to review field studies involving an invasive procedure, any harm to an animal, or any material alteration of the behavior of an animal under study. AWAR §2.38,f,1 then applies: "Handling of all animals shall be done as expeditiously and carefully as possible in a manner that does not cause trauma, overheating, excessive cooling, behavioral stress, physical harm, or unnecessary discomfort." The IACUC may make exceptions to these standards if there is scientific justification to do so (AWAR §2.31,c,3).

The *Guide* (p. 5) states that "investigators conducting field studies with animals should assure their IACUC that collection of specimens or invasive procedures will comply with state and federal regulations and this *Guide*."

Opin.   To help ensure appropriate animal care and use, the IACUC is required to review field studies that involve animal trapping. Thus, protocols must provide sufficient detail on methods to be used (e.g., trapping, tagging, collaring, blood collections, euthanasia) and frequency of observations, and a contingency plan for animals hurt in the collection process. There are several references that provide advice on review of field studies.[4,5] One such book[4] reviews protocol aspects that IACUCs should assess for field research; it discusses species selection, site selection, and methodologies to be employed. (See 16:33.)

Surv.   What conditions does your IACUC place on field activities in terms of type of trap, frequency of checking of the trap, and euthanasia of injured or to-be-collected animals? More than one response is possible.

- Not applicable because there are no field studies   115/185
- Prior IACUC approval is required   39/185

- For small mammals: setting traps in evening and checking in the morning — 25/185
- For mist traps: constant observation and immediate checking upon capture — 22/185
- Frequent observation, daily at minimum — 36/185
- Food and water should be available, depending on timing — 17/185
- Shade or cover should be available — 20/185
- Trap is appropriate size for the animal — 28/185
- Traps are disabled when not in active use — 23/185
- Identification systems are atraumatic (i.e., minimize pain and discomfort and not likely to result in subsequent injury or increased predation) — 28/185
- Researcher safety (i.e., personnel made aware of potential zoonoses and preventive actions, properly trained in handling to minimize stress to animal and maximize personnel safety) — 33/185
- Euthanasia provisions (i.e., animals will be euthanized if required by study or as a result of accidental injury, and method meets field study guidelines) — 33/185
- Full description of trap function and use of least-stressful trap available — 23/185
- Permits if necessary — 35/185
- Case-by-case analysis — 29/185
- Other — 5/185

15:11 Should field sites be included in the semiannual IACUC inspections?

Reg. The AWAR (§2.31,c,2) state that "animal areas containing free-living wild animals in their natural habitat need not be included in such inspections." However, Animal Care's Research Facility Inspection Guide [http://www.aphis.usda.gov/ac/researchmanual/18-2FACI.PDF] states that "animal facilities which must be inspected include, but are not limited to: . . . field study areas where animals are confined."[6] Neither the *Guide* nor PHS Policy addresses the issue of inspecting field sites; however, PHS Policy II requires compliance with the AWA as applicable. Although the PHS Policy does not specifically discuss wild animals or field studies, the definition of *animal* in PHS Policy III,A applies to all live vertebrate animals. Therefore, field sites could be considered to be animal study areas by NIH/OLAW and require semiannual inspection by the IACUC.

Opin. It is the author's opinion that the question of whether field study sites are to be included in the IACUC semiannual inspection must be decided on a case-by-case basis. Strict interpretation of PHS Policy requires inclusion, but in practical terms, very few field studies are funded by the PHS. Certainly, field study sites that include housing enclosures should be inspected, as the animals involved are no longer truly free-living and are dependent on adequate monitoring and oversight. The IACUC should try to inspect such sites when they are actually in use. However, practical issues must also be considered. If the field site is geographically distant, sending an inspection team may not be feasible. Alternative review methods for such sites can include photographs, videos, and submission of comprehensive husbandry procedures.

Surv. 1 Does your IACUC include field sites in its semiannual IACUC inspections?

- Not applicable   107/158
- No   28/158
- Yes, if close by   9/158
- Yes, including the use of surrogate inspections if geographically distant   11/158
- Yes, if invasive procedures are used   3/158

Surv. 2 How does your IACUC inspect animal housing and use sites if they are at geographically distant institutions or field location? More than one response allowed.

- Not applicable   111/158
- The IACUC inspects all sites, whether near or distant   19/158
- If it is at a geographically distant institution, utilize that institution's IACUC   25/158
- If it is at a geographically distant field site, utilize nearby colleagues to inspect the site   6/158
- If it is at a geographically distant field site, utilize videos or similar means to evaluate the site   7/158
- If it is at a geographically distant field site, do not inspect   4/158
- Other   5/158

15:12 Are animals that are used in field studies included on the APHIS/AC annual report?

Reg. AWA-regulated species should be reported as noted in the AWA (Sect. 13,a,7,A) requirement for an annual report from research facilities. The AWAR (§2.36) also require an annual report. The AWAR (§2.36,b,4) do not exclude animals used in field studies, and the AWAR definition of an *animal* (§1.1) does not exclude wild animals (and, therefore, does include wild rodents). However, the AWAR (§2.31,d,1) specifically exempt from IACUC review field studies that are conducted on free-living wild animals in their natural habitat and do not involve an invasive procedure, any harm to the animal, or any material alteration of the behavior of an animal under study.

Opin. The answer to the preceding question seems straightforward because field studies tend to have less direct oversight by way of IACUC inspections, animal ordering, and so on. Including the animals on the APHIS/AC annual report was occasionally not done by some survey respondents. However, the institution and the IACUC should provide a mechanism for reporting (on the APHIS/AC annual report) the AWAR-regulated species used in field studies. For example, the IACUC can send each investigator who has an approved field study protocol an annual request for summary information on the number and species used during the reporting period (October 1 to September 30). These numbers can then be compiled for the institutional annual report, which is due before December 1 of each year.

Surv. Are animals used in field studies included on your APHIS/AC annual report?

- Not applicable 121/162
- Include AWAR-covered species used in field studies 26/162
- Include field study animals only if euthanized or more than momentarily restrained during the study 4/162
- Do not include field study animals in annual report 5/162
- Other 6/162

15:13 With reference to housing requirements, should the IACUC make an attempt to distinguish proactively between biomedical research and agricultural research when reviewing proposals that use large farm animals?

Reg. The definition of an *animal* in the AWAR (§1.1) excludes "farm animals, such as, but not limited to, livestock or poultry used or intended for use as food or fiber, or livestock or poultry used or intended for use for improving animal nutrition, breeding, management, or production efficiency, or for improving the quality of food or fiber." (See 12:1.)

The *Guide* (p. 4) states that "housing systems for farm animals used in biomedical research might or might not differ from those in agricultural research. Animals used in either biomedical or agricultural research can be housed in cages or stalls or in paddocks or pastures. . . . The protocol, rather than the category of research, should determine the setting (farm or laboratory)."

PHS Policy II does not make reference to a particular type of research. It states that it is applicable to all PHS-conducted or -supported activities involving animals. The HREA (Sect. 495a), on which the PHS Policy is based, does refer to "animals to be used in biomedical and behavioral research," a reference that, by inference, might be construed as not applicable to agricultural research. Nevertheless, it is prudent to follow PHS Policy II.

Opin. The AWAR language has become the working definition of agricultural research. The unstated converse is that farm animals used in studies whose goal is the advancement of biomedical science are regulated by the AWAR. Therefore, the practical reason why an IACUC might wish to distinguish between biomedical and agricultural research is that the latter need not be included in AWAR requirements such as annual reporting and search for alternatives and need not conform to recommendations in the *Guide* with particular reference to the space and physical plant recommendations that are focused on biomedical research animals. (See 15:14.) It is the author's opinion that applying uniform policies to all animals involved in the institution's animal care and use program regardless of the type of research or teaching activity involved promotes consistent oversight.

15:14 What criteria are used to distinguish biomedical from agricultural research? How are the standards different for biomedical research housing sites versus agricultural research housing sites?

Reg. (See 15:13.)

Opin. It is clear that farm animals such as sheep, cows, goats, swine, and horses used in biomedical research are covered by both the AWAR and the *Guide*. It is also clear that the same animals used for research on livestock production are not ordinarily covered by the AWAR. PHS-funded agricultural projects (which would be covered by the *Guide*) are very unlikely, given that the focus of PHS funding is biomedical research.

What is not clear are the circumstances under which farm animals used in biomedical research must be housed according to the *Guide*'s recommendations as compared to when they may be housed in more farmlike settings (such as paddocks, pastures, and barns). The *Guide* (p. 4) states that the housing system chosen should be IACUC protocol driven. This author interprets that to mean that farm animals used in studies in which the minimization of variables is vital to the study's outcome (e.g., transplant research) should be housed in the more environmentally controlled setting of a traditional laboratory animal facility. Conversely, farm animals intended for use in studies for which each animal's environment need not be finely controlled (e.g., antibody production) may be housed in farm-type settings.

Surv. 1 What criteria are used to distinguish biomedical from agricultural research? More than one response allowed.

- Not applicable because the institution does either all biomedical or all agricultural research   133/179
- If the end goal is to benefit human health, it is biomedical research   16/179
- If the end goal is to affect production practices, it is agricultural research   13/179
- The funding agency determines the category of research (e.g., NIH) funding means biomedical research, Pork Producers Council means agricultural research)   5/179
- The entire IACUC makes the determination   6/179
- An IACUC subcommittee makes the determination   3/179
- Other   3/179

Surv. 2 What standards does your IACUC use for biomedical research housing sites versus agricultural research housing sites? More than one response allowed.

- Biomedical housing largely follows the *Guide* and/or the AWAR   29/46
- Housing standards for agricultural research are those of a well-managed farm   10/46
- Housing standards for agricultural research are those in the *Guide for the Care and Use of Agricultural Animals in Agricultural Research*[7]   15/46
- Housing standards are protocol driven: biomedical research animals could be housed in a farm setting and agricultural research animals may need the controlled environment of biomedical research   14/46
- Other   2/46

Note: A few IACUCs have an agricultural animal subcommittee, or a separate IACUC for wildlife and agricultural research.

15:15 Is it acceptable to house different animal species in the same holding room?

Reg. AWAR adequate veterinary care standards (§2.33,b,1–§2.33b,2) require appropriate facilities for housing animals and appropriate methods for prevention of disease.

The AWAR (§3.33,b) prohibit the cohousing of hamsters or guinea pigs in the same primary enclosure. The *primary enclosure* (AWAR §1.1) is "any structure or device used to restrict an animal or animals to a limited amount of space, such as a room, pen, run, cage, compartment, pool, or hutch," and AWAR §3.58 state the same for rabbits. The AWAR (§3.33,c) require the separation of guinea pigs or hamsters that are under quarantine or treatment for a communicable disease from other guinea pigs, hamsters, or other susceptible species. The AWAR (§3.58,b) provide the same protection for rabbits. The AWAR (§3.7,d) also prohibit the cohabitation of dogs or cats with other species in the same primary enclosure unless they are compatible. Similarly, AWAR §3.133 (which covers animals other than dogs, cats, guinea pigs, hamsters, rabbits, nonhuman primates, and marine mammals) states that "animals housed in the same primary enclosure must be compatible. Animals shall not be housed near animals that interfere with their health or cause them discomfort."

PHS Policy is silent on the subject of mixing species, although PHS Policy II requires compliance with the AWAR, where applicable.

The *Guide* (p. 22) states that "the environment in which animals are maintained should be appropriate to the species, its life history, and its intended use." The *Guide* (p. 58) also states that "physical separation of animals by species is recommended to prevent interspecies disease transmission and to eliminate anxiety and possible physiologic and behavioral changes due to interspecies conflict. . . . In some instances, it might be acceptable to house different species in the same room, for example, if two species have a similar pathogen status and are behaviorally compatible."

Opin. Professional judgment usually dictates that mixing of certain species in the same room is inappropriate to their life histories (e.g., cats and dogs). The potential for disease and parasite transmission is an additional consideration regarding housing different species in the same room or enclosure. As examples of recommended separate housing by species, the *Guide* (p. 59) notes that nonhuman primates should be separated by geographical origin and rabbits should be housed separately from guinea pigs.

The NIH has an Intramural Position Paper on Housing Multiple Species of Large Laboratory Animals.[8] The content is directly applicable only to the NIH campuses; they are not recommendations promulgated by the *Guide* or any regulatory agency.

Surv. Which species do you house together in the same holding room? More than one response per respondent is possible.

- Not applicable: only one species used   16/185
- Species mixing in an animal holding room prohibited   83/185
- Rodent species, particularly rats and mice, housed in microcontainment-type cages   33/185
- Amphibians and reptiles (e.g., frogs and turtles) or multiple amphibian species   20/185
- Ungulates, such as goats and sheep   10/185
- Pigs and sheep, pigs and dogs   3/185
- Multiple macaque species (e.g., cynomolgus, rhesus)   3/185

- Baboons and macaques   1/185
- Whatever species the veterinary or animal care specialists
  say can be housed together   20/185

Note: Several respondents commented that mixing species is a veterinary, not an IACUC, decision. Institutions not responding to housing certain species together may have not done so because they do not house that species at all or they have not encountered a need to house species together.

References

1. Division of Animal Welfare, Office for Protection from Research Risks, National Institutes of Health, The Public Health Service responds to commonly asked questions, *ILAR News*, 33(4), 69, 1991. Available on the World Wide Web at: http://grants.nih.gov/grants/olaw/references/ilar91.htm#2.
2. Potkay, S., et al., Frequently asked questions about the Public Health Service Policy on Humane Care and Use of Laboratory Animals, *Lab Anim.* (NY), 24(9), 24, 1995. Also see OPRR Reports, #95-02, March 8, 1995. Available on the World Wide Web at: http://grants.nih.gov/grants/olaw/references/dc95-3.htm and http://grants.nih.gov/grants/olaw/references/laba95.htm#2.
3. U.S. Department of Agriculture, Animal and Plant Health Inspection Service, Policy 10, Licensing and Registration of Producers of Antibodies, Sera and/or Other Animal Parts and Pregnant Mare Urine (PMU), April 14, 1997. Available on the World Wide Web at: http://www.aphis.usda.gov/ac/policy/policy10.pdf.
4. U.S. Department of Health and Human Services, Public Health Service, National Institutes of Health, Institutional Animal Care and Use Committee Guidebook, NIH Publication No. 92-3415, Chapter B, 1992.
5. American Society of Mammalogists, Acceptable field methods in mammalogy: preliminary guidelines approved by the American Society of Mammalogists, *J. Mammal.*, 68(4), Suppl., 1, 1987. Available on the World Wide Web at: http://www.mammalsociety.org/committees/commanimalcareuse/98acuc*Guide*lines.pdf.
6. U.S. Department of Agriculture, Animal and Plant Health Inspection Service, Animal Care Research Manual/Research Facility Inspection Guide. Available on the World Wide Web at: http://www.aphis.usda.gov/ac/researchGuide.html.
7. Committee to Revise the Guide for the Care and Use of Agricultural Animals in Agricultural Research and Teaching, *Guide for the Care and Use of Agricultural Animals in Agricultural Research and Teaching*, 1st rev. ed., Federation of Animal Science Societies, Savoy, IL, 1999, p. 4. Available from the Federation of Animal Science Societies, 1111 N. Dunlap Ave., Savoy, IL 61874). Available on the World Wide Web at: http://iacuc.ufl.edu/2003%20Web%20Site/Animal%20Use%20*Guide*s/Ag_*Guide*/.
8. National Institutes of Health, Intramural Position Paper on Housing Multiple Species of Large Laboratory Animals. Available on the World Wide Web at: http://oacu.od.nih.gov/ARAC/HousingMultple.pdf.

16

Pain and Distress

Alicia Karas and Jerald Silverman*

Introduction

A great diversity of views are held concerning the acceptability of using animals in biomedical research, but many agree that it is desirable to reduce to a minimum any pain or distress associated with that research. This approach has been described most fully in Russell and Burch's classic text, *The Principles of Humane Experimental Technique*[1] and then popularized as the "3 Rs" of animal experimentation: reduction of the number of animals used, replacement of animals with nonsentient alternatives (or with human subjects), and refinement of experimental design to minimize pain and distress. Aside from its attraction on ethical grounds, the alleviation of unnecessary pain or distress also may improve the quality of scientific data obtained. It has become clear that the manner in which experimental animals are handled, housed, and fed has an impact on their physiological processes; examples relevant to development, toxicology, oncology, cardiovascular status, and other considerations abound in the literature.[2–6] Pain and stressors have been demonstrated to result in production of cytokines and activation of an endocrine state of catabolism, translating into an overall spectrum of effects such as decreased immune responses and healing rates, increases in tumor metastasis retention after surgery, and the development of chronic pain states.[7–11] Therefore, pain and distress may represent uncontrolled experimental variables that can have significant effects on research results. Reducing the magnitude of this variation can benefit both the welfare of the animals used and the scientific output from the project. At the very least, an understanding of the potential magnitude of the effects of pain and distress upon research data must be realized in order to help interpret results that may differ from laboratory to laboratory. It follows that a consideration of methods for minimizing or eliminating pain and distress should be of central concern to IACUCs for both ethical and scientific reasons.

In many instances we lack the scientific data necessary to make objective judgments concerning the significance of animal pain and distress. Often it is necessary to weigh the available evidence and try to strike a balance between a concern for animal welfare and the requirements of a particular research project. When attempting to achieve this balance in circumstances in which there are limited data, it is helpful to establish whether adoption of measures thought likely to improve animal well-being will interfere with a study or will simply require modest additional work of an investigator. When measures that may improve animal welfare can be implemented without compromising the scientific integrity

* The authors thank Paul Flecknell for his contribution to this chapter in the first edition of *The IACUC Handbook*.

of a project, such measures should be given strong consideration. In many instances there is insufficient scientific or empirical evidence that a given technique has the desired effect of relieving or preventing pain or distress. When this occurs it is worthwhile to attempt to determine whether additional effort or cost inputs do in fact lead to improvements in animal welfare.

16:1   What is the difference between stress and distress?

Opin.   Many attempts have been made to differentiate stress and distress,[12,13] and these vary in their approach. In general, stress is a normal biological event, whereas distress has some aversive or unpleasant component. Stress arises as a result of stressors in the animals' environment, for example, a change in environmental temperature. Normal homeostatic mechanisms are triggered and the animal adapts to the changes and restores its system to a normal state. These adaptive responses occur continuously and are a feature of all living organisms. As the effort required to adapt to stressors increases, a point may be reached where the animal fails to adapt fully, and this is considered by some authors to represent distress. Other opinions require that the animal becomes aware of the effort of its adaptive responses and perceives this effort as something it wishes to avoid, leading to a subjective state of "distress." Unless manifested by overt external physical signs or behaviors, states of distress may be difficult to recognize. It might be said that an aversive state has dimensions of time and magnitude and becomes "distress" when the product of the two exceeds a certain threshold. Thus, chronic exposure to a low aversive situation, such as barren housing, may not cause as much distress as weeks of increasing pain from tumor growth. However, the degree of aversiveness should be estimated and perhaps assigned a rank to guide considerations of refinement.[14]

Pain is one of a number of aversive states that can cause distress. Pain and other causes of distress might vary in their aversiveness between individual animals and between species. For example, being housed on wire cage bottoms might be more distressful for rodents that have injured footpads than for normal animals, and standing on a hot surface is likely to be more painful for a mouse than for a horse.

16:2   What is the difference between distress and suffering?

Opin.   Whether there is a difference between distress and suffering depends primarily upon how *distress* and *suffering* are defined. These definitions are still debated and, at present, IACUCs should determine which working definitions are acceptable to them. Useful discussions can be found in References 1, 15, and 16. (See 16:1.)

One possible definition is that the word *distress* is frequently used to encompass the state produced by exposure to stressors such as excess heat, cold, or lack of food (see 16:34). Those who advocate use of the term *suffering* usually require that the animal perceives the threat these stressors pose to its integrity and often include a temporal dimension to this emotion, requiring the state to last for some time before it is defined as suffering. *Distress*, as discussed in 16:1, can also range in magnitude. Therefore, suffering might be considered to be at the top of the aversiveness or maladaptation "distress" scale. However, unless a method of measuring or estimating distress becomes generally agreed upon, the two words can refer to either very different or very similar states.

16:3   What is the difference between pain and nociception?

Reg.   *Pain* is not directly defined in the AWAR, but a *painful procedure* (§1:1, Painful Procedure) is defined as one "that would reasonably be expected to cause more than slight or momentary pain or distress in a human to which that procedure is applied." The example provided is pain in excess of that caused by injections or other minor procedures. *Nociception* is not defined. Neither the PHS Policy nor the *Guide* provides definitions of *pain* or *nociception*, although the definition of a *painful procedure* in U.S. Government Principle IV (which is incorporated into the PHS Policy) and the *Guide* (p. 64) is similar to the AWAR definition.

Opin.   Nociception represents only the detection of certain stimuli by sensory nerves and the transmission of signals to the central nervous system. Pain is the subjective experience of nociception. Noxious input of a sufficient magnitude will cause depolarization of sensory nerve endings; examples of such stimuli types are heat or cold, pressure, and chemical (pH, irritant). The signal travels through the sensory nerve tracts to the spinal cord and potentially to the brain. Depending on the phylogenetic level of the species, the signal arrives at varying levels of the brain and there may be "processed" or "interpreted" as being painful. *Noxious stimuli* may or may not be of a type that would cause tissue damage, but in any case, a noxious stimulus has the potential to trigger both involuntary (e.g., reflex limb withdrawal, autonomic nervous responses) and voluntary (purposeful) aversive movements. If the overall interpretation of the stimulus is that it is painful, then an affective–motivational response occurs, involving prior experience and contextual information, and thus the *pain experience* depends upon many factors.

   Reflex motor and autonomic responses may occur in the unconscious or anesthetized animal in response to noxious stimuli; they are not typically defined as pain because the animal is "unaware" of the stimulus (the neurologic input–output situation is that of nociception). The administration of anesthetics to the point where both voluntary and involuntary movements are suppressed, and autonomic responses are blunted in response to a noxious input, is taken to be sufficient to eliminate the experience of pain for the duration of the anesthesia. Upon recovery to a conscious state it is assumed that at this point the animal is capable of "experiencing" or perceiving pain.

   Pain in humans is described as having both a sensory and emotional component. For example, the IASP[17] definition is as follows: "Pain is an unpleasant sensory and emotional experience associated with actual or potential tissue damage or described in terms of such damage." Defining animal pain is difficult because of uncertainties relating to the emotional (affective) component of pain in nonhumans. This has been avoided in some definitions by interpreting pain in relation to its effects on animal behavior. For example: "Pain in animals is an aversive sensory experience that elicits protective motor actions, results in learned avoidance, and may modify species-specific traits of behavior, including social behavior."[18] Nociception and pain are said to serve the organism; protecting it from damage. One neuroscientist's view of pain is that it evolved as part of the function of homeostasis (ensuring that the body regulates its own well-being).[19] Water and salt balance, for example, are regulated and involve internal physiologic compensations as well as behavioral drives. Similarly, pain is argued to have evolved to reflect the condition of the body arising from a tissue injury. Nevertheless, as a sensation, pain can vary dramatically. "The behavioral drive that we call pain usually matches the intensity of the sensory input but it can vary under different conditions, and can become intolerable or, alternatively, disappear, just as hunger or thirst."[19] Regardless of whether the reader agrees that animals experience emotions, for familiar animal species it can be argued that pain does exist: It has the behavior motivational component, it can vary depending on the circumstances of what else is happening at the time, and if the experience continues with sufficient magnitude, it can lead to adverse health consequences.

A useful summary of nociception and pain assessment and recognition is given in the guidelines of the Association of Veterinary Teachers and Research Workers.[20] Definitions, importance, recognition, and treatment goals for pain are also reviewed in the American College of Veterinary Anesthesiologists position paper on pain in animals.[21]

16:4   What should be the minimal expectations of the IACUC related to the relief of pain and distress?

Reg.   The IACUC, under the AWAR (§2.31,d,1,iv,A–§2.31,d,1,iv,C; §2.31,d,1,v), must assure that procedures causing more than slight pain or distress will be performed with appropriate sedatives, analgesics, or anesthetics unless certain specific criteria are met (e.g., IACUC-approved scientific justification to withhold analgesia). They must involve the AV or the AV's designee in their planning and not use paralytics without anesthesia. Animals experiencing severe or chronic pain or distress that is not alleviated must be euthanized during or at the end of the study. The AWAR (§2.31,d,1,ii) require consideration of alternatives to procedures that may cause more than momentary pain or distress to animals and require a written narrative description of the methods and sources used to determine that alternatives were not available. APHIS/AC Policy 12 states that a database search remains the most effective and efficient method for demonstrating compliance with this requirement. (See 16:5.)

The PHS Policy (IV,C,1,a–IV,C,1,c) is worded similarly to the AWAR relative to avoiding or minimizing pain and distress, the need to use appropriate means to alleviate pain and distress, and the need to euthanize animals during or at the end of a procedure if severe pain or distress cannot be alleviated. PHS Policy (IV,D,1,d) notes that applications and proposals to the PHS must contain a description of the procedures that are designed to assure that discomfort and injury to animals will be limited to that which is unpreventable in the conduct of scientifically valuable research, and the proper drugs will be used where indicated to minimize animal pain and discomfort. U.S. Government Principles IV to VI largely reiterate the AWAR and PHS Policy.

The *Guide* (p. 64), U.S. Government Principle IV, and the AWAR (§1.1, Painful Procedure) assume that procedures that cause pain in humans will probably cause pain in nonhuman animals. The *Guide* (p. 10) states that the IACUC should consider the use of appropriate sedation, analgesia, and anesthesia for animals. It also notes (p. 64) that the recognition of pain in different species is a key to its prevention or alleviation, and that professional judgment should be used to determine the appropriate analgesics or anesthetics. The *Guide* (p. 65) warns that some drugs, such as sedatives, anxiolytics, and neuromuscular-blocking agents, are not analgesics or anesthetics.

Opin.   Causing pain and distress is considered undesirable and forms the basis for much public disquiet about the use of animals in research. It is appropriate for an IACUC to require all protocols to be designed to prevent or minimize pain or distress. It also may be considered appropriate for an IACUC to require investigators to demonstrate that they have considered all appropriate alternatives in their use of animals.[22]

Each of the so-called 3 Rs can result in a reduction in pain and distress.[1] Alternatives that replace the use of sentient animals largely eliminate pain and distress since they use nonsentient alternatives (e.g., cadavers or tissue culture). Reduction alternatives (e.g., appropriate statistical methodology leading to a reduction in total animal use) result in fewer animals potentially experiencing pain and distress. Finally, refinement alternatives seek to reduce to a minimum the pain and distress experienced by those animals that are still necessary to use. Perhaps it is this last area that the IACUC will see as its greatest concern, and it should review not only the proposed use of analgesic and anesthetic drugs, but also the training and competency of those involved in the procedures. For example, an inexperienced investigator may handle an animal inappropriately and cause significant pain or distress that could have been prevented by proper training. It also is likely that a person skilled in surgical procedures will carry out a project using fewer animals (because of fewer technical failures), cause less postoperative pain (because of less tissue trauma), and effect a more rapid recovery (because of shorter anesthesia time) than an inexperienced or less competent investigator. The total amount of pain or stress from all aspects of animal use must also be considered.

For an IACUC to decide whether the PI has sufficiently considered pain and distress alternatives, all potential sources of painful and nonpainful distress should have been identified, and alleviation or prevention techniques sought. Although the regulations use broad terms such as *appropriate* (drugs) or *unavoidable* (discomfort), some methods to reduce pain or distress might be possible but extremely unfeasible (e.g., requiring years of surgical training prior to permitting a new surgeon to operate). In many instances it may be difficult or impossible to find a resource that offers advice on how to reduce pain or distress. Also, the treatment of distress caused by stressors other than pain (such as by providing additional bedding, more easily reachable food or water, fluid therapy, or other improvements in comfort and homeostasis) are frequently overlooked forms of refining animal use. This may occur because the wording of the AWAR can lead scientists to be primarily focused on analgesics or sedatives as refinements.

Finally, the IACUC should ensure that animals are observed appropriately for their species and condition at regular intervals, using generally agreed upon criteria (e.g., body condition scoring, food consumption, lameness, attitude) in both the normal state and during experimental conditions in which pain or distress is expected. Monitoring of animal well-being can ensure that measures to reduce pain or distress are utilized when needed. The frequency of observations should usually be greater when a significant abnormal state is anticipated. All of these factors should be integrated into the IACUC's expectations when assessing the efforts an investigator intends to make in order to reduce pain and distress.

16:5   What alternatives must be considered for protocols that will involve pain or distress? How can this be done?

Reg.   The AWAR (§2.31,d,1,ii) require the PI to consider alternatives to procedures that may cause more than momentary or slight pain or distress to the animal. The PI must provide a written description of the methods and sources used to determine that alternatives are not available. This requirement is clarified in APHIS/AC Policy 12, which provides examples of sources that can be used (e.g., Medline, Animal Welfare Information Center). Policy 12 states that the minimal written narrative should include the databases or other sources used, the date of the search, the years covered by the search, and the keywords or search strategy used.[23] More importantly, Policy

12 notes that reduction, replacement, and refinement (see 16:4), not solely animal replacement, are generally considered part of the entire concept of "alternative methods" and should be addressed as part of the search for alternatives.

The *Guide* (p. 10) states that the availability or appropriateness of the use of less invasive procedures, other species, isolated organ preparations, cell or tissue culture, or computer simulations should be considered.

U.S. Government Principle III, which is part of the PHS Policy, states mathematical models, computer simulation, and *in vitro* biological systems should be considered. (See 12:7–12:13.)

Opin. From 16:4 it can be seen that all protocols that have the potential to cause pain and distress should be reviewed with reference to reduction, replacement, and refinement. This requires significant effort of both the IACUC and the PI. PIs should at least explain, in some detail, how an animal is to be handled and justify their use of animals. They should indicate the literature sources they have reviewed to show that there is no practical alternative to animal use. From each step outlined in the protocol, handling or experimental details can be identified and used as keys to search for alternatives. If standard operating procedures can be developed by investigators at an institution, in concert with the veterinary experts, then uniform expectations are available to the IACUC. Deviations from standard operating procedures must be reviewed with respect to whether they will produce an acceptable level of well-being. As the number of databases on alternatives grows, reference to searches on these can be included on an IACUC application. (See 16:25.)

Investigators also should demonstrate to the IACUC that an appropriate experimental design that maximizes the information gained and minimizes the number of animals involved has been used. It is helpful to indicate whether expert advice from a statistician has been sought, particularly when the experimental design is complex. A useful practical approach to estimating numbers of animals required has been published.[22] Finally, a range of options are available for reducing pain and distress, and those appropriate to the investigation should be considered. For example, it might be possible to conduct a study entirely while the animal is under general anesthesia or reduce pain by using appropriate analgesics, or as technology expands, to use less invasive methods of collecting data. It also may be possible to limit the distress experienced by animals, or the duration of that distress, by requiring defined end points for a study (e.g., placing upper limits on total tumor burden in studies of carcinogenesis, or defining a set of criteria that will be used to determine when an animal should be removed from study and humanely killed).

Surv. Does your IACUC require a literature search for alternatives to procedures placed in USDA Category C (none or momentary pain or distress)?

- Not applicable as our work does not require
  literature searches   7/153
- We do not require a literature search for alternatives
  to those procedures, on the basis of their pain and distress
  classification   67/153
- We do require a literature search for alternatives
  to those procedures, on the basis of their pain and
  distress classification   78/153
- Other   1/153

16:6 Since the actual assessment of pain cannot be made until after a procedure is performed, is it appropriate for the IACUC to attempt to assess and categorize the potential for pain and distress when first reviewing a protocol?

Reg.   The AWA (Section 13,a,7,A) and the AWAR (§2.36) require an annual report for regulated species, encompassing animal use activity in testing, research, and teaching from October 1 through September 30 (the USDA fiscal year). The annual report (§2.36,b,5–§2.36,b,7) categorizes the number of animals used on the basis of the level of pain or distress they experienced. Category C (on APHIS/AC Annual Report form 7023) is used for no pain or distress and no use of pain-alleviating drugs, Category D is for pain or distress alleviated by drugs, and Category E is for pain or distress not alleviated by drugs. The PHS Policy has no requirement for the categorization of pain and distress. (See 16:3 for the definition of a *painful procedure*.)

Opin.   Empirically, many IACUCs use the APHIS/AC pain and distress categories for all animals, not only those regulated by the AWAR. Nevertheless, the AWA requirement applies to its regulated species only.

   The wording of AWAR §2.36,b,5–§2.35,b,7 is in the past tense, indicating that the information provided to APHIS/AC in the annual report should be based on the pain or distress actually experienced by the animal, not solely on an educated estimate made by the IACUC or the PI prior to the initiation of the study. While prospective estimates of the number of animals that will fall into each of the pain or distress categories can simplify the work of the IACUC, the reality is that research needs and experiences often change during the course of a study. If the IACUC uses a prospective system of assigning pain or distress categories, it is suggested that the IACUC have an established process that allows a PI (or the IACUC) to change the categorization of some or all animals, on the basis of animals' actual experience, retrospectively.

   Although the degree of pain cannot be determined for any particular individual animal until after a procedure has been performed, it is frequently possible to make an informed estimate of the likely degree of pain that may be caused. If the procedure, or a similar procedure, has been carried out previously at the institution, then this can be used to help predict the likely consequences. If the technique or species/strain is new to the institution, then colleagues at other institutions where the procedure has been performed should be consulted. If at all possible, individuals who actually performed or cared for animals should be asked for input. Extrapolations of pain experienced by humans undergoing similar procedures can act as a rough guideline. There should be few occasions when some indication cannot be obtained as to the likely pain and distress consequences. This initial assessment should certainly be sufficient to determine, for example, what level of analgesic use would be appropriate, and what type of aftercare the animal might require. In all instances, measures for the control of pain require monitoring of animals to ensure they are effective.

16:7 Can the IACUC approve protocols in which animals will experience pain or distress not relieved by the use of anesthetics or analgesics? Under what circumstances?

Reg.   If an investigator can justify, in writing, that withholding anesthesia or analgesia is required for scientific reasons and will only continue for the necessary period, then the IACUC can potentially approve such an activity (AWAR §2.31,d,1,iv,A). The PHS Policy (IV,C,1,b) has similar wording.

Opin. IACUCs may frequently be asked to approve protocols in which animals experience pain or distress that cannot be alleviated by the use of anesthetics or analgesics. For example, animals used in studies of arthritis or inflammatory conditions may experience pain, but analgesic administration may interfere with the protocol to such an extent that its use would invalidate the data obtained. Pain and distress also may occur in animal models involving neoplasia, chronic organ failure, infectious diseases, toxicity testing, and a wide range of other circumstances. In many of these studies, alleviation of pain and distress by pharmacological means may not be possible because of interactions between the drugs used and the research protocol. However, methods to reduce pain or disease-related disability, such as providing nutritional or environmental support, will act to reduce overall distress.

When pain or distress is unpreventable, IACUCs should ask the PI whether the potentially serious impact of pain or distress upon animal physiological processes will detrimentally influence the study's results. Pain and stress have been shown to cause metabolic and immunological alterations in animals. (See 16:1.) If a control group does not experience the equivalent amount of pain or stress of the experimental group, then results of certain types of studies might potentially be biased to a greater degree than if both groups were treated with analgesics. Another consideration is whether the benefit of the information or result outweighs the total degree of animal distress imposed by the study. Whenever there is unalleviated pain or distress, efforts must be made to reduce the number of animals used in such a study, replace the study with alternative techniques that do not cause pain, and refine the study design so that pain and distress are minimized. Although these basic tenets should be applied to all protocols, they clearly are of particular importance in circumstances in which significant pain and distress can be anticipated. (See 16:11; 16:13; 16:51.)

Surv. Does your IACUC place rats or mice in USDA pain or distress Category D (even though these categories are not required for rats and mice) if anesthesia is used during surgery but no postoperative analgesics are administered? Assume the surgery creates postoperative pain.

- Not applicable   41/149
- Yes, always   46/149
- Yes, sometimes   25/149
- No, those animals are placed in Category E   31/149
- Other   6/149

16:8 If the AV and a PI differ on the pain or distress potential for a particular procedure, is it the responsibility of the AV, IACUC, PI, or all to provide supporting evidence for their point of view?

Reg. With reference to surgery, APHIS/AC Policy 3 notes that all animal activity proposals must specify details concerning the relief of postoperative pain and distress. The specific details must be approved by the AV or his/her designee. Policy 3 also states that the AV retains the authority to change postoperative care as necessary to ensure the comfort of the animal and suggests that this change can be made without the approval of the IACUC. These policy statements are interesting in that they can be interpreted to mean that the AV, not the IACUC, can have final say over the provision of postoperative analgesia (other than the withholding of analgesia

Opin.
if scientifically justified). In contrast, the AWAR imply that the AV should only provide guidance to the PI relative to the use of anesthesia and analgesia (§2.32,b,4; §2.31,d,1,iii,B), while the IACUC maintains its responsibility to assure studies are performed with proper sedation, anesthesia, or analgesia (AWAR §2.31,d,1,iii,A). The PHS Policy itself is silent on the authority of the AV to change postoperative care; however, the *Guide* (p. 13) notes that the AV should give research personnel "advice that ensures humane needs are met and are compatible with scientific requirements." (See 16:9.)

Opin.
Veterinarians and veterinary care staff are professionals trained to understand species differences in behavior and husbandry needs. Whether or not specifically trained in pain medicine, they have the educational or practical background to be able to provide informed opinions. In some instances, the veterinarians, veterinary technicians, and animal care staff are in a position to observe animals more often than the PI's personnel. Some PIs also have expertise in animal husbandry or behavior or may have acquired substantial experience over the course of time. If the PI can offer evidence to the IACUC about the impact of a study on animal well-being, then this should be taken into account. However, as the science of animal welfare advances and other improvements in research refinement are published, veterinarians may be able to contribute new knowledge to the assessment of the potential for pain or distress and to suggest remedies for their alleviation. The IACUC's decision must incorporate input from all of those involved in animal care and experimentation.

Surv.
What information does your IACUC request if an investigator states that no postoperative analgesia can be used because it will interfere with the study goals or interpretation? More than one response is possible. [*Note:* To enhance clarity, the responses to this question were edited by the chapter authors. Affirmative responses to both the third *and* fourth bulleted items (*n* = 4) had their response changed to only the fourth bullet.]

- Not applicable   26/162
- We just accept the statement that analgesics cannot be used   16/162
- We ask for documentation, such as published references
  or laboratory records   90/162
- We ask for documentation, *and* we assign a person to observe
  the animals for a day or two and report back to the IACUC   29/162
- We only assign a person to observe the animals for a day or
  two and report back to the IACUC   0/162
- Other   1/162

16:9 A PI provides, in writing, a scientific rationale for withholding the provision of analgesia during and after a painful procedure. Is the IACUC obliged to accept this rationale?

Reg.
The AWAR (§2.31,d,1,iv,A) require the use of analgesia during a procedure that causes more than momentary pain or distress unless the PI provides the IACUC written scientific justification for not providing it. The PHS Policy (IV,C,1,b) has the same wording. The AWAR (§2.31,a) state that the IACUC cannot prescribe methods or set standards for the design, performance, or conduct of actual research or experimentation by a research facility. The HERA (Section 495,a,2,B) states that the guidelines established in the PHS Policy "shall not be construed to prescribe methods of

research." Both the AWAR (§2.31,c,6) and PHS Policy (IV,B,6) provide the IACUC with the authority to approve the use of animals in research or other animal activities.

Opin.   It has been argued that the scientific justification presented to the IACUC must have some substance; otherwise it is no better than no justification at all.[24] The same reference stated that if a logical justification could be provided, even if "reasonable minds may differ as to whether the justification suffices, the IACUC must defer to the investigator." In contrast, both APHIS/AC and NIH/OLAW responded that it is their interpretation "that the AWA and PHS Policy do not require an IACUC to approve a proposed project against its will."[25]

Any justification to withhold analgesia should be clearly reasoned and, whenever possible, documented by laboratory data or literature references. Generic statements, such as "Analgesia will interfere with the collection of useful data" may be correct, but the IACUC should not accept such statements at face value. (See 16:8.)

16:10   What are some typical criteria an IACUC or an investigator can use to determine whether an animal is in pain or distress?

Opin.   Most attempts to recognize pain or distress in animals rely on observing a combination of behavioral and physiological variables and looking for deviations from normality. For example, an animal in pain may change its spontaneous behavior so that it becomes less active, decreases grooming or nest building activity, and reduces its food and water consumption. It also may change its responses to handling, for example, by increased aggression. Conversely, some animals may become apathetic and unresponsive to handling. Body weight changes may occur in the face of acute or chronic pain or distress. The criteria that might be used have been discussed at length by various authors.[13,20,26] See reference 27 for an extended literature review.

The key point to note is that signs of pain and distress may be very subtle and their recognition almost always requires a detailed knowledge of the normal behavior of the animal species. Recognition also requires that sufficient time be allocated for observation of the animal, and, in some circumstances, it may be necessary to observe the animal's behavior in such a way that it is unaware of the presence of the observer (e.g., by using a video camera).

Spontaneous behavior can be studied in species that are not accustomed to being handled. Nevertheless, enriched housing may be necessary because full behavioral repertoires may not be evident in barren environments, even if the animal is not experiencing overt pain or distress. In socialized, easily handled species such as the dog or goat, palpation of surgical wounds may elicit evasive behaviors (flinching, glance at the site, vocalizations). If the administration of an analgesic or other therapy restores behavioral indices toward the normal, then unalleviated pain or distress may have been present.

16:11   An IACUC protocol indicates that a procedure is moderately painful to its canine subjects for approximately 1 hour, then never repeated. During that time, pain-relieving drugs cannot be used, but an alternative pain relief mechanism will be used. That is, calming music, a darkened room, and human petting of the animal will constantly occur. The technique has not been previously attempted. Should the IACUC consider this to be alleviated or unalleviated pain?

Reg.   See 16:3 for the definition of a *painful procedure* and 16:4 for related regulatory information. The AWAR (§2.36,b,7) specifically require painful procedures, for which appropriate anesthetic, analgesic, or tranquilizing drugs are not used, to be placed in Category E on the APHIS/AC annual report. The *Guide* (p. 65) notes that in addition to anesthetics, analgesics, and tranquilizers, nonpharmacological control of pain is often effective.

Opin. Practical experience in clinical veterinary practice has shown that many companion animals, especially dogs and cats, respond positively to human contact. For example, it is well accepted that good nursing, which includes stroking and verbal reassurances, can play a role in reducing pain and distress in the postoperative period. This is in accord with human clinical experience, in which the emotional state of the patient influences the degree of pain experienced and the analgesic dose required to alleviate that pain. Practical evidence that similar mechanisms can be involved in animal pain is provided by the changes in behavior of animals that are believed to be in pain postoperatively when they are given nursing attention. Such attention may silence an animal that is vocalizing, may trigger eating or drinking, or may result in a previously agitated animal's becoming calm and resting or sleeping. Obtaining a positive response to such contact depends critically upon the previous experience of the animal and the nursing skills of the personnel involved. If an animal has not been adequately socialized to accept (and welcome) human contact, then attempts to provide reassurance may be counterproductive and increase the animal's distress. If such a technique is to be considered as a replacement for conventional pain-alleviating techniques, then it is important that the IACUC require that the personnel involved along with their previous relationship with the animals used be specified. The personnel should have experience with the technique, and ideally the technique should be familiar to the animals (acclimation to restraint, devices, etc). It is important to carry out a pilot study, or to conduct the first few studies with veterinary oversight and intense monitoring, to ensure that the methods proposed to reduce pain or distress are adequate. Finally, it is appropriate to consider other, nonpharmacological methods of pain relief, such as application of heat or cold, acupuncture, or transcutaneous nerve stimulation. (See 16:12; 16:13.)

16:12   Should a decerebrated cat be placed in APHIS/AC pain/distress Category D (alleviated pain or distress) or E (unalleviated pain or distress)? (See 16:13.)

Reg. The AWA (Section 13,a,3,A) and the AWAR (§2.31,e,4; §2.36) refer to the use of drugs to alleviate pain. APHIS/AC Policy 11 implies the same by stating that individual animals experiencing pain/distress that is alleviated with anesthetics, analgesics, sedatives, and/or tranquilizers should be reported in column D of the APHIS/AC annual report, whereas animals for which needed anesthetics, analgesics, sedatives, and/or tranquilizers are withheld should be reported in column E. (See 16:6.) Neither the PHS Policy nor the *Guide* has any categorization of pain or distress; however, when the *Guide* or PHS Policy differs from the AWAR, PHS Policy requires absolute compliance with the AWAR for covered species (PHS Policy, IV,C,1).

Opin. An experimentally decerebrated animal cannot feel pain because of the inactivation of the cerebral cortex and thalamus. The decerebration is performed with traditional pharmacologic anesthesia, suggesting that USDA Category D is appropriate. However, once the effect of the anesthetic agent(s) has dissipated, the animal is still alive and one can argue that USDA Category E is appropriate since drugs are not used to alleviate any postoperative pain. In reality, the issue of pain is moot, since the animal cannot perceive any pain because the pain recognition centers of the brain have been inactivated or physically removed.

Thus, the animal is technically in Category E, but this status has no practical animal welfare meaning. The *Guide* (p. 65) recognizes that nonpharmacologic control of pain is often effective. The issue of decerebration has been discussed in the literature and it has been recommended that properly performed decerebration be considered the equivalent of continuous general anesthesia.[28]

16:13 Can mammals that have a cerebral cortex feel pain if made decerebrate?

Opin. When addressing this question it is important to differentiate between pain and nociception (see 16:3). Nociception is the response to damaging or potentially damaging stimuli, whereas pain has a subjective component and is interpreted by most humans (and other mammals) as an unpleasant experience.[13,20] It is widely accepted that the presence of a functioning forebrain is required for the perception (or more explicitly, *for the experience*) of pain, so decerebration should remove that capacity. Since thalamic structures are believed to play a role in pain perception, it is usually considered that removal of the forebrain, including the thalamic nuclei, ensures that pain cannot be perceived. The position in regard to decorticate animals is less clear. Comparison with humans leads one to presume that such animals have no capacity for pain perception. However, given the higher levels of organized behavior shown by decorticate or decerebrate animals of some species (e.g., rats), some caution is required in making these extrapolations. (See 16:12.)

16:14 Can nonmammalian vertebrates lacking or having only a primitive cerebral cortex (e.g., frogs) feel pain?

Opin. Since the capacity to experience the subjective, unpleasant component of pain in humans appears to rely on the presence of a functioning cerebral cortex, it is often assumed that animals with a less well developed cortex have less capacity to feel pain. Comparison of the frog with humans is simply a more extreme comparison than the more frequent extrapolation of a rat or mouse to humans. In both instances, we have little insight into the nature of the experience of pain in the animal, nor of its significance to the individual. We have adopted an approach that presumes an animal with a certain level of central nervous system development can experience pain, and research involving these species should be conducted in a way that reduces the likelihood of causing pain or distress.

It seems unlikely that frogs experience pain in the same manner as humans, but since we can demonstrate nociception in amphibians[29] and since their degree of cerebral development cannot be said to preclude the possibility of pain perception, we should assume that these species can experience pain. What is more problematic about the assumption that they experience pain is that pharmacologic alleviation of pain in nonmammalian species is poorly studied (although studies of nociception in amphibians and less cerebrally complex animals are available) and clinical signs of pain are not well characterized.

16:15 Can ketamine alone or with xylazine be considered adequate anesthesia for major surgery, such as abdominal surgery in laboratory rodents, rabbits, or cats?

Opin. Ketamine, when used alone, immobilizes most animals and in many species produces a significant degree of analgesia. It some instances its use as part of an anesthesia protocol may even help prevent postoperative pain.[30-32] The degree of analgesia and degree of immobilization vary considerably among species.[33] Ketamine is relatively ineffective in providing analgesia for the visceral organs,[34] and when used alone does not provide adequate anesthesia for abdominal surgery in rodents, rabbits, cats, or indeed any species except (possibly) nonhuman primates.[35,36] The degree of muscle relaxation is so poor that this often renders ketamine alone unsuitable even for superficial procedures. The addition of drugs with sedative effects such as xylazine, medetomidine, acepromazine, or diazepam greatly improves the quality of anesthesia. In general, combining ketamine with drugs that have analgesic and sedative properties (e.g., xylazine and medetomidine) produces more effective surgical anesthesia than combinations such as ketamine with acepromazine or ketamine with diazepam.[33] When combined with xylazine (or medetomidine), ketamine provides surgical anesthesia in most rodents, rabbits, and cats.[37,38] The degree of anesthesia provided by ketamine–medetomidine or ketamine–xylazine is usually sufficient for abdominal surgery; however, the duration of surgical plane of anesthesia varies with species and dose, and high doses may cause significant depression of homeostatic mechanisms.

16:16   Is chloralose considered to be an anesthetic or a hypnotic?

Opin.   Chloralose is a hypnotic, as are many other agents that are used to anesthetize animals (e.g., pentobarbital). It is often assumed, incorrectly, that because a drug is hypnotic it cannot be used to provide surgical anesthesia. The definition of *hypnosis* is "a condition of artificially induced sleep, or a state resembling sleep, resulting from moderate depression of the central nervous system from which the patient is readily aroused."[39] Although drugs classed as hypnotics produce this effect as the dosage is increased, progressively greater depression of the central nervous system (CNS) occurs, so that animals progress from sedation to hypnosis to general anesthesia. Some hypnotics cause such severe cardiovascular depression at the doses necessary to achieve general anesthesia that they cannot be used in this way without risking the death of the animal. Others, such as chloralose and the barbiturates, can be given at doses sufficient to produce general anesthesia in many species. Different species, and different strains of the same species, vary in their response, and in some circumstances the depth of anesthesia will be inadequate for surgery. Provided the anesthetic depth is assessed (e.g., by evoking a response to a painful stimulus such as a toe pinch), dose rates can be adjusted as necessary or additional analgesia can be provided by drugs such as morphine or fentanyl.[40]

Certain methods of administering chloralose may cause injury and subsequent pain or other morbidity during the postoperative period. Problems cited for chloralose include adynamic ileus and intestinal injury when administered intraperitoneally, phlebitis associated with extravascular accidental injection, and stressful recovery.[41] Chloralose is not a modern anesthetic agent. It gained favor and is still commonly used in certain types of studies (e.g., cardiovascular, respiratory and neurovascular physiology, neurotransmission) because of its reputation for creating less interference with the processes being studied than other anesthetic agents and because it can produce long-lasting immobility. However, in the past, chloralose was often compared to pentobarbital, and newer anesthetic techniques may provide acceptable hypnosis or anesthesia without the potential for pain or distress that chloralose has been cited to cause. In many types of studies, alternative modern anesthetic techniques, conscious trained animals, and decerebrate preparations have been shown to be equal or superior methods of immobilization.[42-46] There are still many instances in which use of chloralose, alone or in combination, is the most suitable technique for a measurement. Anesthetics have variable effects on experimental results, and an investigator should carefully scrutinize the literature for newer, less noxious anesthetics or conscious or decerebrate techniques, rather than relying on inflexible, traditional beliefs.[47,48] If, for example, an author uses modern anesthetics in one species and chloralose in another to measure the same parameters, it could be assumed the modern anesthetic would be appropriate for both species.[49] In addition, many experts recommend that chloralose not be used for survival surgery.[33,41,50]

16:17 Should the IACUC request special safety precautions when urethane is used as an anesthetic?

Opin. Urethane is mutagenic and carcinogenic,[51] and, if it is to be used as an anesthetic, appropriate precautions should be adopted to prevent safety hazards to personnel. The exact nature of the precautions may vary, but they generally mirror those required to ensure safe handling and use of carcinogens. The IACUC should consider utilizing the services of the institution's biosafety committee. (See 16:18 and Chapter 20.)

16:18 Should the use of urethane be allowed as anesthesia for recovery surgery?

Opin. Urethane has been used for the long-lasting immobility, anesthesia, and reported cardiovascular stability that it produces. Both the availability of newer injectable and inhalant anesthetic techniques and some rather disadvantageous features of urethane suggest that it should be used only with careful consideration of alternative methods.[42,43,46] Personnel safety is a major consideration, and exposure to urethane may cause neoplasia in some species (see 16:17). Hypertonic solutions may cause tissue damage and lead to adverse health consequences when it is given intraperitoneally and recovery can be very prolonged.[52]

16:19 Should the IACUC approve the use of a nonpharmaceutical grade drug for anesthesia or analgesia if pharmaceutical grade alternatives are available?

Reg. Neither the AWAR nor PHS Policy directly comments on the use of nonpharmaceutical grade drugs; however, the AWA (Section 13,a,3,A) states that adequate veterinary care with the appropriate use of analgesic or tranquilizing drugs is required. The AWAR (§2.33,a) also require adequate veterinary care. The HREA (Section 495,a,2,A) requires the appropriate use of analgesics and anesthetics. PHS Policy (IV,C,1,b) states that painful procedures will be performed with appropriate analgesia or anesthesia (unless otherwise justified). The *Guide* (p. 64) notes that professional judgment is required when selecting the analgesic or anesthetic that best meets clinical and humane requirements.

Opin. APHIS/AC Policy 3, which is meant to provide guidance on the AWAR, states, "Investigators are expected to use pharmaceutical-grade medications whenever they are available, even in acute procedures. Non-pharmaceutical-grade chemical compounds should only be used in regulated animals after specific review and approval by the IACUC for reasons such as scientific necessity or non-availability of an acceptable veterinary or human pharmaceutical-grade product. Cost saving alone is not an adequate justification for using nonpharmaceutical-grade compounds in regulated animals." PHS Policy (IV,C,1) requires compliance with the

AWA for regulated species. The preparation of nonpharmaceutical grade drug solutions can lead to inconsistent product quality and the potential for adverse effects, including tissue damage, pain, or lack of sufficient clinical effect. For this reason, their use may constitute less than adequate veterinary care.

Surv.   Does your IACUC allow investigators to use reagent grade drugs for anesthesia (such as tribromoethanol) if alternative (although not necessarily the same) pharmaceutical drugs are available?

• Not applicable	24/166
• Yes, an investigator can always use reagent grade drugs for anesthesia	28/166
• Yes, but the investigator must have IACUC approval to use reagent grade drugs	51/166
• No, investigators cannot use reagent grade drugs for anesthesia (unless the reagent grade drug itself is being studied)	61/166
• Other	2/166

16:20   The AWA, the AWAR, and the *Guide* have restrictions on performing multiple major survival surgeries on the same animal. (See 16:24.) Are there any such restrictions on performing multiple painful procedures on the same animal?

Reg.   There are no specific prohibitions on performing multiple painful procedures on the same animal, if approved by the IACUC. PHS Policy (IV,D,1,d) notes that applications and proposals to the PHS must contain a description of the procedures that are designed to assure that discomfort and injury to animals will be limited to that which is unpreventable in the conduct of scientifically valuable research, and that proper drugs will be used where indicated to minimize animal pain and discomfort. U.S. Government Principle IV speaks of avoiding or minimizing discomfort, distress, and pain when consistent with sound scientific practices. The goal of the *Guide* (p. 1) is to promote humane care for animals. The AWAR (§2.31,d,1,i) state that procedures involving animals will avoid or minimize discomfort, distress, and pain to the animals. The AWAR (§2.31,d,1,ii) also state that the PI must consider alternatives to procedures that may cause more than momentary or slight pain or distress. Further, the AWAR (§2.31,e,4) note that the procedures to be used should be designed to limit pain and discomfort to that which is unavoidable for the conduct of scientifically valuable research.

Opin.   It appears that the intent of the federal regulations restricting multiple major survival surgeries is to reduce the total amount of pain and distress caused to an individual animal and to prevent animals from being transferred at the end of one study to another study where the animal may be subjected again to major survival surgery. Therefore, it seems prudent that studies in which animals are potentially subjected to multiple painful procedures should be reviewed by IACUCs using a paradigm similar to that used for multiple major survival surgeries (AWAR §2.31,d,1,x; *Guide*, p. 12). (See 16:24.) If a procedure causes significant tissue damage (e.g., skin burns, osteoarthritis, or nerve damage) but is not a major surgery, there is still a potential that the animal is left with a chronic state in which pain or distress thresholds may be lowered and subsequent uses might result in more total pain than if another animal were used. On the other hand, a distinction should be made between *potentially* painful procedures and those that involve *unalleviated* pain. Techniques to reduce pain and distress include local and systemic analgesics, anesthetics, and operant conditioning/acclimation. Other modalities can be used alone or in combination and should be aggressively pursued if they will not specifically interfere with the study.

If a study requires that an animal be subjected to repeated painful or distressful manipulations, then the investigator should outline for the IACUC the impact, magnitude, and duration of the pain; its potential effects on experimental outcome; methods to monitor the animal's ability to adapt to or recover from the event; and reasons why the pain cannot be relieved, if this is proposed. (See 16:7.) Without such information any attempt to set a limit on the frequency with which such an event can be repeated is arbitrary. In the course of a particular study, repetition of procedures in which unalleviated pain is caused may be necessary. Allowing transfer of an animal to another study in which additional major painful procedures are performed may comply with the letter, but not the spirit, of the regulations. (See 16:23; 16:24.)

Surv. 1 If an animal is transferred from one study to another, does your IACUC have a means of determining whether prior major survival surgery was performed on that animal?

- Not applicable   47/157
- No   10/157
- Yes   96/157
- Other   4/157

Surv. 2 If an animal is transferred from one study to another, does your IACUC concern itself with the extent of pain or distress that the animal may have had in the original study?

- Not applicable   41/160
- No   5/160
- No, unless there was a reason to call it to the committee's attention   10/160
- Yes   102/160
- Other   2/160

16:21 Is it a good idea to have an institutionwide policy that some form of analgesia must be provided for any major survival surgical procedure?

Reg. The AWAR (§2.31,d,1,iv,A) state that procedures involving more than momentary pain or distress will be performed with appropriate sedatives, analgesics, or anesthetics unless their withholding is justified, in writing, by the PI. APHIS/AC Policy 3 states that all animal activity proposals involving surgery must provide specific details of pre- through postprocedural care and relief of pain and distress. U.S. Government Principle IV speaks of avoiding or minimizing discomfort, distress, and pain when consistent with sound scientific practices.

Opin. Analgesia, and often more than one approach to preventing pain, should be considered for any potentially painful survival surgical procedure, whether major or minor. (See 16:3–16:5, 16:10.) Published estimates of the degree of pain expected from various types of surgery[53,54] can be used to help decide on the amount, type, and combinations of analgesics to be used. In many cases, the amount of pain an animal will be expected to perceive is comparable to that felt by humans after the same surgery (AWAR, §1.1, Painful Procedure; U.S. Government Principle IV; *Guide*, p. 64). In order to get an idea of the incidence and magnitude of pain that a surgical procedure causes in humans, the authors find it instructive to consult the human surgical literature. Studies of analgesic interventions for a particular procedure (for example, hip replacement surgery) commonly include a description of the severity of pain typically caused by the surgery. The type of analgesia used should be related to the anticipated magnitude of potential for pain.

Surv.   Does your IACUC or institution have any specific requirements to use analgesia after a major survival surgical procedure? More than one response is possible.

- Not applicable   16/161
- No specific requirement; we evaluate the requirement for
  analgesia on a case-by-case basis   39/161
- Yes, we require some form of analgesia after any major
  survival surgical procedure unless the need to withhold
  analgesia is scientifically justified   102/161
- Yes, we require analgesia (unless withholding it is scientifically
  justified) but only for species covered by the Animal
  Welfare Act regulations   4/161

16:22   Is the use of an analgesic necessary after embryo transfer in mice?

Reg.   (See 16:4; 16:21.)
Opin.   The degree of pain caused by a flank incision and subsequent visceral manipulations will depend upon surgical finesse, the type of anesthetic agent used, and other factors. In the author's experience, mice lose body weight after bilateral flank ovariectomy, an indication that this surgery has an impact on them. Mouse strain and environmental factors (such as the season) have been shown to influence tests of experimental nociception in mice.[55] Some anesthetic agents (e.g., volatile inhalants, pentobarbital) do not provide any analgesia that lasts into the postoperative period, and some anesthetic agents (e.g., ketamine, xylazine) possess analgesic effects that may persist. The assessment of the typical amount of pain or abnormal conditions caused by reproductive tract manipulations in mice has not been specifically studied. (See 16:15.)

Surv.   Does your IACUC require the use of an analgesic after embryo transfer in mice?

- Not applicable   105/151
- We do not require the use of an analgesic   8/151
- We encourage but do not usually require an analgesic   14/151
- We usually require the use of an analgesic   22/151
- Other   2/151

16:23   If a simple injection, dietary manipulation, or other nonsurgical manipulation produces a permanent physical or physiological impairment, should the IACUC consider that a major operative procedure?

Reg.   The AWAR (§1.1, Major Operative Procedure) define a major operative procedure in two ways: first, in terms of a surgical intervention that penetrates and exposes a body cavity, and, second, in terms of any procedure that produces permanent impairment of physical or physiological functions. The *Guide* (p. 61) does the same, although it uses the term *substantial impairment* rather than *permanent impairment*. The *Guide* focuses on surgery as the cause of the impairment as compared to non-surgical procedures.

Opin.   (See 16:20.)

16:24   AWA-regulated animals arrive at an animal facility, already having undergone major surgery (under anesthesia) at the vendor's company. The animals healed well before arrival, and there was no apparent pain or distress upon arrival. There will be no painful or distressful procedures when used at the purchaser's institution. On the APHIS/AC annual report form (Form 7023), are these animals placed in column C (no pain or distress) or column D (pain or distress alleviated by drugs)?

Reg.   The AWAR (§2.31,d,1,x) and the *Guide* (p. 12) restrict multiple survival surgeries to those circumstances described in 18:6. (See 16:6 for general considerations of pain categories.)

Opin.   Vendors usually have their own APHIS/AC registration; in those circumstances the vendor's annual report states the appropriate pain or distress category for the animal while at the vendor's facility. If the vendor does not have an independent APHIS/AC registration, the receiving institution must arrange to have the vendor's animal use operations covered under the receiving institution's registration. How the animals in the study are categorized at the receiving institution should refer to the specific way that the animals are to be used there. In the scenario presented there was no apparent pain upon arrival. If there is no need from that point onward to administer analgesics or anesthetics to minimize pain or distress, then the study animals are placed in column C. An institution might choose to base its animal procurement decisions on whether or not a vendor claims to use analgesics at the time of the surgery.

16:25   When an IACUC reviews a literature search strategy for alternatives to painful or distressful procedures, must each potentially painful or distressful procedure be listed as a keyword in the search?

Reg.   (See 16:5.)

Opin.   (See 16:5.) For an investigator to consider adequately whether the proposed use of laboratory animals will be conducted by using currently available techniques (i.e., refined), a search of each potentially painful or distressful procedure should be done. This type of search is challenging but can be aided by use of Internet resources such as the University of California Davis Veterinary Medical Center for Alternatives site (available at http://www.vetmed.ucdavis.edu/Animal_Alternatives/databaseapproach.html) or the Animal Welfare Institute database for refinement (available at http://www.awionline.org/lab_animals/biblio/refine.htm). In addition to looking for nonsentient animal models (which may be found to be unsuitable), the PI is strongly urged to search for methods by which animal well-being can be monitored or improved. For example, researchers in unrelated disciplines may have developed behavioral assessment tools or otherwise refined animal use methods that can be applied to the new protocol as it is developed. Using the search term *alternatives* may unduly limit the number of useful hits in a literature search, whereas using terms such as *severity, assessment*, crossed with terms such as *model* and *animal*, and with *pain* or *illness*, may reveal established techniques for assessment of pain, disability, or humane end points. It should also be noted that such papers may not have as their primary focus the reduction of pain or distress, and that the methods section of the papers will have to be examined for the necessary details.

Surv.   When your IACUC reviews a literature search strategy for alternatives to painful or distressful procedures, must each potentially painful or distressful procedure be listed as a keyword in the search?

- Not applicable as our work does not require
  literature searches   8/143
- No, in general   49/143
- Yes, in general   86/143

16:26   Is a measured decrease in food or water consumption a reasonable indicator of pain in laboratory rodents?

Opin.   Numerous factors can influence food and water consumption in rodents and pain is only one of these. If a rodent reduces its food and water intake, pain should be included as a possible cause along with infectious disease processes, changes in environment, alteration in the husbandry regimen, and other factors. Nevertheless, in the postoperative period, immediate changes in food and water consumption seem especially useful as indicators of postoperative pain.[56,57] After surgery, rats and mice almost invariably show a small fall in body weight (between 5% and 15%, depending upon the nature of the surgical procedure) as a consequence of reduced food and water consumption. If analgesics are provided, this fall in consumption is reduced. Since both opioids (e.g., buprenorphine) and nonsteroidal antiinflammatory drugs (e.g., carprofen) have this effect, and since administration of these drugs to normal animals does not increase food and water consumption (i.e., they are not appetite stimulants per se), it seems reasonable to conclude that some of the reduction is due to postsurgical pain.[56,57] The degree of reduction ranges from 10% to 15% after minor procedures (e.g., skin incision) to 100%[58] after major invasive surgery.

   As rodents are often group-housed, measuring individual food and water consumption can be difficult. It is often more convenient to record body weight. It is important to note that many studies are carried out on growing animals, so establishing average weight gain for a few days before an operative procedure is necessary. Although this measure has been shown to be useful, it is essentially a retrospective index of the amount of pain or discomfort that was produced over the preceding 12 to 24 hours. Thus, significant weight loss may indicate that unalleviated pain or distress occurred, necessitating changes in future animal surgeries. Behavioral assessment, if carried out effectively, allows dose rates of analgesics to be adjusted, depending on the animals' responses. (See 16:27.)

16:27   What percentage of weight loss over what period might indicate that an animal is experiencing pain or distress?

Opin.   As discussed in 16:26, loss of weight does not necessarily indicate that an animal is experiencing pain and distress. Chronic weight loss, particularly if unexpected, should be investigated carefully. In many institutions, limits are placed on weight loss, on the basis of the assumption that a fall in weight must reflect some adverse occurrence and some degree of distress. Brief periods of inappetence will cause weight loss of 5–10% in rodents and, once weight loss exceeds 20%, general loss of body condition and fat or muscle mass becomes clinically apparent. Most guidelines are based on consensus, not on published data, and it is advisable to examine the issue on a case-by-case basis. It is important to establish why the weight loss is occurring. Is this due, for example, to lowered feed intake, increased metabolism, or decreased absorption of nutrients? Is the weight loss an unpreventable consequence of the research protocol, or could supplemental feeding, use of analgesics or antianxiety drugs, or other methods prevent or ameliorate it? It also is important to compare an animal's weight with that of an untreated, normally growing control. Finally, it is worth noting that adults of several species of laboratory animals when fed *ad libitum* become obese. This emphasizes the importance of adopting a reasoned, logical approach to interpreting the significance of weight loss in an animal.

Surv. If, on a particular protocol, your IACUC agrees to accept weight loss as one criterion that an animal may be in pain or distress, what percentage of loss is typically considered "significant"? Assume the weights are compared to those of untreated control animals.

• Not applicable	31/156
• Typically 5–10% after the normal recovery period from any procedure that was performed	15/156
• Typically 10–15% after the normal recovery period from any procedure that was performed	41/156
• Typically 15–20% after the normal recovery period from any procedure that was performed	43/156
• Over 20% after the normal recovery period from any procedure that was performed	19/156
• Other	7/156

16:28 What are useful guidelines for the IACUC to consider when the use of neuromuscular blocking (NMB) agents is requested? (See 16:29.)

Reg. The AWAR (§2.31,d,1,iv,C) state that neuromuscular blocking agents (paralytics) should not be used without anesthesia. The PHS Policy (U.S. Government Principle V) states that surgical or other painful procedures should not be performed on unanesthetized animals paralyzed by chemical agents. The *Guide* (p. 65) states that when these agents are used, it is recommended that the appropriate amount of anesthetic first be defined on the basis of results of a similar procedure that used the anesthetic without a blocking agent.

Opin. NMBs prevent voluntary muscle activity so that an animal can no longer respond to painful or other stimuli by moving. Since movement in response to a surgical stimulus is used to assess the adequacy of anesthesia, the use of NMBs as part of an anesthetic protocol requires special consideration. It is essential to preclude the very real possibility that the level of anesthesia is inadequate and the animal is aware of the surgical procedure being performed ("awareness"), which is potentially painful and distressful. Confusion commonly occurs over how to assess the appropriateness of the anesthetic level; when NMBs are used this confusion is compounded. In the management of anesthesia for human patients, estimates of "awareness" are based on postsurgical reports. Current reviews indicate that approximately 1 in 1000 patients can recall intraoperative events that occurred while he or she was under anesthesia.[59] Put into perspective, these human cases were anesthetized by anesthetists who had postgraduate professional training, whereas most animals are anesthetized by technicians or other individuals who have far less training.

Before examining what constraints might reasonably be placed on the use of these drugs, the IACUC should first establish why it is thought necessary to use them. In some instances, investigators may simply have taken a human anesthetic protocol and decided to use it in their animal model. This is often inappropriate, since the rationale for NMB use in humans often is not applicable to other animals.[60] Human (M.D.) surgeons are accustomed to operating under conditions of little or no muscle tone; veterinary surgeons are not. It is not necessary, for example, to use an NMB to allow assisted (mechanical) ventilation in nonhuman animals. Veterinary anesthesiology textbooks discuss mechanical ventilation during anesthesia, and these chapters do not mention use of neuromuscular blockers. In a discussion of respiratory management of anesthetized animals, three options for preventing spontaneous breathing are listed:

1. Hyperventilation
2. Deepening anesthetic levels
3. Use of NMB

Option 1 is listed as the easiest.[61] Thus, NMBs are not necessary for mechanical ventilation and the IACUC need not accept this as the sole reason for requiring NMBs. Neither are NMBs needed to produce sufficient muscle relaxation for the majority of surgical procedures. It is true that electrosurgery and cautery cause skin and muscle to twitch in the absence of NMBs, which can make the limb or head move slightly for brief periods. Nevertheless, this is only critical if the structures being accessed (e.g., brain, spinal cord) are sensitive to tiny movements. Under those conditions neuromuscular blockade is only necessary during specific portions of surgery (not, for example, for incision of skin and cranium). For the most part, patient movement can be prevented by ear bars and other securing measures. Certain intraocular surgeries may require a central eye position and an NMB may be needed to facilitate this. However, human intraocular surgery is performed on awake patients using a local anesthetic block of the orbit. Presumably, in the anesthetized animal, a skillful orbital local block will assure the central eye position as well as provide excellent preemptive analgesia and there will be no need for NMBs. If, however, NMBs are required, then the following points should be considered:

The PI should have experience with anesthesia and surgery in the species used and should be using a familiar anesthetic regimen that is effective for causing insensibility to pain and lack of awareness *on its own for that surgery* in the absence of an NMB.

The NMB should not be administered until after the anesthetic has reached a stable level and surgery has commenced. If practical, the NMB should be allowed to wear off periodically, so that somatic reflex responses can be assessed before additional doses of NMB are administered.

The heart rate and blood pressure should be monitored and elevations (15–20% or more) of either in response to painful stimuli indicate the need for additional anesthesia. It is important to note that this monitoring technique is not fully reliable and awareness in humans can occur without major changes in these variables.[62] With some anesthetic regimens (e.g., a volatile anesthetic such as isoflurane), monitoring the electroencephalogram can be of value, but this technique also can prove unreliable.[28] (See 18:1.)

If inhalant anesthesia is used, an end tidal agent monitor (which monitors the exhaled anesthetic gas concentration as an indication of the central nervous system concentration) is extremely useful to monitor and document that the animal's anesthetic level was in the range in which insensibility is typically present (1.5% for isoflurane).

The IACUC should scrutinize the anesthetic records of cases in which NMBs are used to determine whether there is documentation of all anesthetic techniques and indication that proper humane use of these agents occurred.

16:29   If an NMB is used along with an anesthetic, what might the IACUC request to help assure that the animal is anesthetized and not simply immobilized?

Opin.   Some common situations in which NMBs are used include neurophysiological and imaging studies. In many of these studies, a very light plane of anesthesia must be maintained to minimize interference between the anesthetic regimen and the study protocol. It is precisely in these circumstances that inadvertent production of inadequate anesthesia is most likely. Some investigators have suggested that after completion of surgery, the surgical wounds can be infiltrated with local anesthetic, general anesthesia discontinued, and the animal immobilized because of the effects of the neuromuscular blocking drug (although it must receive ventilatory support). This proposal raises two concerns. First, it is difficult to ensure the adequacy of local anesthetic blockade—either initially or later when the NMB has been administered. Second, even if the animal is pain free, it may become distressed if exposed to other stimuli and is unable to react because it is paralyzed. General anesthesia is used in animals not only to provide insensibility to pain but also to induce unconsciousness, which helps prevent the distress caused by physical restraint and experimental manipulations. A paralyzed, conscious animal, even if pain free, is likely to experience considerable distress. If NMBs are to be used during surgery or painful procedures, then the monitoring techniques that can be applied are those described in 16:28.

16:30   Is the use of a local anesthetic to perform a minor procedure considered to be alleviation of pain by anesthesia?

Reg.   (See 16:4.)

Opin.   Yes. Local anesthetics offer a valuable alternative to general anesthesia for workers carrying out a range of different procedures. In some species, the use of regional anesthesia produced by local anesthetics can be considered the method of choice (e.g., a paravertebral block to carry out abdominal surgery in cattle).[39,63] Provided the administration is carried out competently, pain can be prevented either by local infiltration of drugs such as bupivacaine, infiltration around nerve trunks, or epidural or intrathecal administration. Experience with both humans and other animals suggests that these techniques are useful, provided any distress caused by physical restraint or other (nonpainful) procedures can be controlled.[64-66] The duration of local anesthetic action varies with the agent and additional local anesthetic or other analgesic administration may be necessary to maintain alleviation of pain. In addition, the injection of a local anesthetic can cause momentary pain on injection. Techniques to reduce this pain involve the use of sedatives or tranquilizers (which by reducing anxiety or the level of consciousness can act to raise the pain threshold), buffering of acidic solutions, use of smallest needle possible, and other methods.

16:31 What records related to use of anesthetics and analgesics should be maintained and reviewed by the IACUC?

Reg. The AWAR (§2.35a; §2.35,f) require that all records and reports be maintained for at least 3 years. Those directly relating to proposed activities involving animals (i.e., those elements of research, testing, or teaching procedures that involve the care and use of animals [AWAR §1.1, Activity]) and proposed significant changes in ongoing activities reviewed and approved by the IACUC are to be maintained for at least 3 years. This includes records made during the activity and for at least 3 years after the completion of the activity. PHS Policy (IV,E,2) has the same requirement. APHIS/AC Policy 3 refers to maintaining animal health records for at least 1 year after the disposition or death of the animal; nevertheless, the policy notes that a longer retention period may be needed to comply with other applicable laws or policies (see 17:20).

Opin. Records of analgesic use should vary, depending upon the type and complexity of the procedure. As a minimum, the anesthetic drugs or analgesic drugs, dosage, time, route of administration, and effect should be recorded, together with details about the experimental animal (age, weight, sex, strain, etc.). If surgical procedures are to be carried out, the investigator should note the adequacy of anesthetic depth and the way this was assessed. If additional doses of drugs are given, the time and route of administration and the effect should be recorded.

The investigator also should record the duration of anesthesia, the recovery time to sternal recumbency, any morbidity, any unexpected adverse effects (e.g., vomiting), and when full recovery commenced (normal activity, feeding, drinking, etc.) and indicate how this assessment was made. As discussed in 16:28 and 16:29, it may be desirable to keep a continuous recording of heart rate and blood pressure, especially if neuromuscular blocking agents are used. These data permit a critical review of anesthetic practices and should be considered an essential part of any scientific protocol, because interactions between anesthetic complications and the study objectives can readily occur.

When using analgesics, the drug, dose, and route of administration should be recorded. Any additional doses given also should be noted. It may be considered helpful to link this information with the pre- and postprocedure observation of variables such as body weight and clinical appearance.

Although APHIS/AC Policy 3 states that medical records should be kept for at least 1 year, it is unclear whether that statement refers to all medical records related to a study or just those that arise from unanticipated clinical conditions (e.g., a surgical incision opening or self-inflicted trauma). In many instances it is difficult to discern whether a medical condition has emanated from study or nonstudy causes, and therefore it is prudent to maintain all medical records for at least 3 years after the euthanasia or other disposition of the animal.

Surv. 1 Does your IACUC require investigators to keep a postprocedural monitoring form for recording pain or distress signs in rats and mice?

- Not applicable   16/155
- Yes, always   58/155

- Yes, but only for USDA pain or distress Category E studies (even though rats and mice are not covered species)   15/155
- No   58/155
- Other   8/155

Surv. 2   Does your IACUC require investigators to keep a postprocedural monitoring form for recording pain or distress signs *in nonrodent* species?

- Not applicable   49/160
- Yes, always   74/160
- Yes, but only for USDA pain or distress Category E studies   11/160
- No   20/160
- Other   6/160

16:32   Should the use of anesthetics and analgesics be based on a strict dosage schedule or on varying dosages as determined by sound clinical judgment?

Reg.   The AWAR (§2.33,b,4) require the AV to guide the PI and other personnel in the appropriate use of anesthesia, analgesia, and immobilization. The AWAR (§2.33,b,5) give the AV additional authority, stating that adequate pre- and postprocedural care shall be in accordance with current established veterinary medical and nursing procedures. PHS Policy (IV,A,3,b,1) states that a veterinarian will have direct or delegated authority and responsibility for activities involving animals. The *Guide* (p. 56) notes that adequate veterinary care includes effective programs for anesthesia and analgesia, and the veterinarian must provide guidance to the researcher. Pages 64 and 65 of the *Guide* discuss pain, analgesia, and anesthesia. PHS Policy (IV,A,1) requires institutions to use the *Guide* as a basis for developing and implementing an animal use program.

Opin.   Anesthetics and analgesics should be given initially according to dose schedules. These are determined from the scientific literature but are often modified because of variations in response by different strains of animals. Administering a standard dose of anesthetic may produce the desired effect, but it also can cause animals to appear to be too deeply anesthetized or inadequately anesthetized. After assessing the response to an anesthetic for the particular strain, age, and sex of animal to be used in a study, a more appropriate dose schedule can be developed. The variation can be substantial. For example, when using the duration of unconsciousness as an indicator of anesthetic efficacy, doubling of sleep time can be observed in different strains of mice.[67]

It is relatively easy to adjust anesthetic drug doses to suit particular groups of animals, but more difficult to do this with analgesics because of our limited ability to assess postoperative pain. There is contention about whether a PI can propose to "give analgesics as needed" to treat pain. "As needed" dosing strategies are effective in humans, who can self-administer strong analgesics by means of patient-controlled analgesia devices, or by oral dosing, as they can determine when additional analgesics are needed and can titrate the dose accordingly. "As needed" analgesic dosing strategies in animals suffer from two major limitations: First, whether it will be possible at all to detect significant pain in the individual animal, and second, whether timing of the assessments will be frequent enough that lengthy periods of substantial pain are prevented. (See 16:57.) If pain assessment is not performed skillfully, then no amount of vigilance will suffice for preventing pain in that animal. For nonrodent mammalian species, the literature and experience with preemptive and multimodal analgesic techniques are rapidly expanding. Many authors recommend the administration of analgesics prior to incision as part of a balanced anesthetic technique, with the continuation of analgesia during the postoperative period. Fixed dose administration for 12–24 hours postoperatively, combined with monitoring for pain between doses, extending the duration of treatment for procedures that are felt to cause moderate or severe pain, and using modalities that overlap in terms of analgesic coverage are advocated in many current veterinary textbooks on pain management.[54,68] In our opinion, a problem with fixed dose schedules is that they may lead to insufficient pain relief. In fact, to be evaluated properly, an animal's pain should be assessed in the interval between doses and just prior to redosing in order to detect whether there are adequate duration and magnitude of effect. An argument for the use of a fixed dose schedule, with monitoring and ability to intervene between doses for the first 24–48 hours post surgery, is that this method rarely leads to complications of excessive analgesic side effects. It also has the benefit of assuring that if pain assessment skills are limited, the animal will be given at least a baseline of coverage for pain.

A general inadequacy of postoperative pain relief in humans, among other factors, was often ascribed to use of rigid dosing schedules with no inherent flexibility.[69] Individual animals will vary in their response to surgery. Therefore, dose rates obtained from the literature are a helpful starting point, but every attempt should be made to assess the adequacy of analgesia in each individual animal, in case a particular animal requires more analgesia than others. Even if retrospective measurements such as body weight (see 16:26; 16:27) are used, this permits variation of the "standard" dose in subsequent studies. In some species (dog, goat) complex behaviors, human–animal interactions, or animal–animal interactions are relatively well understood, and these behaviors can be used as baselines for postoperative observation to construct scales for the quantization of pain. Evaluation of pain in rodent and other "prey" species by observing behavior is more challenging: animals may be inactive because of photoperiod, observer presence, or pain. This is an area in need of study and educational efforts.

One major barrier to the provision of analgesia in the immediate postoperative period is that the dosing interval for opioid analgesics is shorter than the period animals are typically left without monitoring or treatment (e.g., overnight). This should be recognized, and if longer-duration techniques can be used (e.g., transdermal fentanyl patch) or the overlap of long-duration agents (e.g., nonsteroidal antiinflammatory drugs [NSAIDs]) is not sufficient for analgesic throughout the night, then overnight monitoring and treatment, or other scheduling modifications, will be necessary in order to meet regulatory and ethical requirements.

Surv. 1 Does your IACUC require overnight monitoring or analgesic administration for nonrodent species when a procedure has been performed and pain will be likely during the night?

- Not applicable   58/160
- Yes, if indicated by our veterinarian and approved
  by the IACUC   61/160
- Yes, under any conditions in which pain is expected
  to last overnight, if approved by the IACUC   30/160
- No, in most instances   10/160
- Other   1/160

Surv. 2 Does your IACUC require overnight monitoring or analgesic administration for rats or mice when a procedure has been performed and pain will be likely during the night?

• Not applicable	28/157
• Yes, if indicated by our veterinarian and approved by the IACUC	82/157
• Yes, under any conditions in which pain is expected to last overnight, if approved by the IACUC	19/157
• No, we do not require overnight monitoring or analgesic administration in most instances	26/157
• Other	2/157

16:33   If the use of anesthesia might impair the survival ability of an animal to be released in the wild as part of a field study, can the IACUC appropriately approve procedures that use either no anesthesia or anesthetic regimens that do not produce complete pain alleviation?

Reg.   Neither the AWAR nor the PHS Policy has pain relief exemptions for animals used in field studies. A field study is defined in the AWAR (§1.1) as any study conducted on free-living wild animals in their natural habitat. It excludes a study that involves an invasive procedure or harms or materially alters the behavior of an animal under study. The PHS Policy does not have a specific definition of a field study.

Opin.   The use of animals in field studies poses particular difficulties, since it is usually intended that the animal will survive the study and resume its normal activities. Anesthesia may impair this process. However, the introduction of reversible anesthetic regimens has greatly improved management of wildlife anesthesia. Similarly, in smaller species, use of modern inhalational agents (e.g., isoflurane) can result in very rapid recovery with few significant aftereffects. The use of potent inhalational agents such as halothane and isoflurane in simple induction chambers is not generally recommended, as dangerous concentrations of anesthetic are produced.[33] However, in field conditions at moderate to low environmental temperatures (<15°C), it is possible to use them, provided the animals are observed closely for signs of overdose. Finally, the use of local anesthetics can be considered as a means of minimizing an animal's experiencing of pain. If all these options have been considered and rejected as impracticable or ineffective, then an ethical judgment must be made as to whether the aims of the project outweigh the (presumably momentary) pain or distress caused by the required manipulations. Alternatively, if the project seeks only to obtain tissue samples from an animal, a judgment must be made as to whether the procedure should be allowed without adequate anesthesia and the animal allowed to recover, or the procedure carried out with anesthesia and the animal euthanized rather than released. (See 15:10; 16:9.)

16:34   What is the maximal practical length of time that the IACUC should allow a rat to be deprived of food or water? How do these suggestions change for a mouse, dog, nonhuman primate, or other common species?

Reg.   The AWAR (§2.38,f,2,ii) state that short-term withholding of food or water from animals is allowed when specified in an IACUC-approved activity that includes a description of monitoring procedures. APHIS/AC Policy 11[70] uses food or water deprivation beyond that necessary for normal presurgical preparation as an example of a procedure that may cause more than momentary or slight distress. PHS Policy (IV,A,1) requires institutions to follow the *Guide* (pp. 10, 12), which addresses food or fluid restrictions and notes the need for relevant objective information regarding the procedures and purpose of the study. The *Guide* specifically notes that restriction for research purposes should be scientifically justified, and a program should be established to monitor physiologic or behavioral indices, including criteria for temporary or permanent removal of an animal from the experimental protocol.

Opin.   Food and water deprivation can be carried out for many reasons, and attempting to establish maximal periods of deprivation without reference to the aims of a particular study is undesirable. For example, a project might be judged so important in terms of its potential benefits that very prolonged periods of food deprivation could be sanctioned if this were considered a necessity. As another example, the fasting period for a mouse may differ if insulin level is or is not to be measured during the construction of a glucose tolerance curve. At the extreme, an IACUC might consider whether any project could justify withdrawal of food until an animal dies as a result of this procedure.

Often, only relatively minor periods of food and water deprivation are required, for example, to ensure an empty stomach or gastrointestinal tract or to induce a catabolic state in the animal. In each study, it is important to determine the minimal period of deprivation needed to achieve the desired objective. Often 16- or 24-hour periods are chosen, as these are a convenient interval for removal of the food or water at the end of a working day, followed by use of an animal the following morning. In the case of small rodents, this period may be excessive and may induce unnecessarily severe effects because of the high metabolic rate of these smaller species. It has been shown that when presented with a limited quantity of food (e.g., half the amount normally consumed overnight), rats eat normally until the food is exhausted. Thus, by reducing the amount of food placed in their food hopper, effective food removal (e.g., from 3:00 A.M.) can be achieved. This is sufficient to produce an empty stomach.[71] Investigators should note, of course, that coprophagy occurs in these and other species, so complete food deprivation can only be achieved by combining fasting with the use of an anal cup to prevent ingestion of feces.[72]

A second aspect of the animal's normal biological processes also should be considered. Some species (e.g., rats) normally only feed in the dark phase of their photoperiod. If food is withdrawn overnight and a procedure carried out the next day, and if that procedure causes adverse effects, the rat may not eat the following night or the next day. As a result, 48 hours of fasting may have inadvertently been produced. This may have consequences for the particular study, as well as for the welfare of the animals concerned.

When dealing with larger species, longer periods of food withdrawal might be needed and, generally, are better tolerated than in smaller animals. If food is being withdrawn solely to reduce the risk of vomition during induction or recovery after anesthesia, a period of 8 to 16 hours is appropriate for dogs, cats, ferrets, or nonhuman primates, but unnecessary in rabbits and rodents as these latter species do not vomit. Withholding of food in pigs can be helpful in reducing the volume of gut contents for abdominal surgery, and in ruminants it may help reduce the incidence of rumenal tympany.[33]

16:35   Is it appropriate to use electric shock to stimulate animals to run or walk on a treadmill?

Reg.   PHS Policy (IV,C,1,a) requires that procedures with animals avoid or minimize discomfort, distress, and pain, consistently with sound research design. The AWAR (§2.31,d,1; §2.31,d,1,i) state that unless acceptable justification for a departure is presented in writing, procedures involving animals will avoid or minimize discomfort, distress, and pain to the animals. APHIS/AC Policy 11[70] cites inescapable noxious electrical shock as a procedure that may cause more than momentary pain or distress.

Opin.   When using any conditioning stimulus, it is desirable to use a reward system rather than a mild or moderate noxious stimulus such as electric shock. When examining such a proposal, an IACUC should require evidence to show that an aversive stimulus is the only technique that can be used to produce the required behavior in the animal. If it is concluded that aversive stimuli are the only practical conditioning stimuli, then the least aversive and potentially tissue-damaging stimulus used to motivate animals should be chosen (e.g., a puff of air, sound, or vibration) over one that is overtly pain producing.

16:36   Can a mammalian fetus feel pain? If so, at what age can a rodent fetus be presumed to feel pain? Should it be included as a vertebrate animal when the IACUC considers the number of animals requested for the study?

Opin.   The concern relating to fetal and neonatal pain has emerged largely from studies in human infants. Prior to the mid-1980s, many procedures were undertaken on human infants without effective anesthesia or analgesia. A series of studies indicated that human neonates can experience pain,[73,74] and investigations in rodents have shown not only that pain can (or at least responses to noxious stimuli) be demonstrated, but that these experiences produce long-term changes in the nervous system. The stage at which these abilities become functional during fetal development is still under debate, and case-by-case interpretation may be needed as the extent of mammalian neurophysiological development differs, depending on the species. In rodents and other species, anatomical studies suggest that nociceptive responsiveness is present in the second half of gestation; anesthesia is being advocated for human fetuses over the age of 20 weeks.[75–77] It is debatable whether this anatomical development translates into a capacity to experience pain.[78] Some authors have put forth the suggestion that consciousness begins at birth and that the experience of pain requires consciousness; others disagree.[74,75] However, since pain has a large subjective component, we have the same doubts about mature animals. It seems appropriate to assume that the mammalian fetus has some capacity to experience pain and adapt research protocols accordingly. Therefore, it is illogical to exclude the fetus when considering the number of animals to be approved in an IACUC submission. Of course, practical considerations should be noted in that an investigator cannot reasonably be expected to predict exactly how many fetal animals in a multiparous species may be present in each pregnant female. Useful reviews of pain in the fetus and neonate are given by Fitzgerald[79] and Narsinghani.[80] (See 8:11; 13:11; 14:19; 17:26.)

16:37   Can unhatched avian embryos be presumed to feel pain? Is IACUC approval needed for the use of avian embryos?

Reg.   NIH/OLAW has interpreted the term *live vertebrate animal* to apply to avians (e.g., chick embryos) only after hatching.[81] Birds bred specifically for research are not currently regulated under the AWAR.

Opin. Answering this question is difficult when dealing with avian species, since we
have a very limited knowledge of nociception and pain perception in these animals. Although it seems illogical to conclude that the capacity to experience pain
emerges at the instant of hatching, we cannot as yet determine the stage of development at which this capacity is sufficiently well developed to warrant concern.[82]
In the United Kingdom, arbitrary limits were set by legislation enacted in 1986, and
this pragmatic approach may be the best way forward for IACUCs: that is,
acknowledge that pain may occur, acknowledge our uncertainty as to the stage of
development at which this occurs, and set some initial guidelines that can be
reviewed as our understanding of avian neurobiology improves. This same
approach would allow studies on early embryonic stages to proceed without
IACUC approval, but would require approval once a particular developmental
stage has passed. (See 14:23; 16:36.)

16:38 Is hypothermia an acceptable form of anesthesia for fetal or neonatal homeothermic animals?

Opin. Hypothermia produces immobility and apparent insensibility in neonates and
fetuses and has become a well-established means of "anesthetizing" neonatal
rodents. At low temperatures nerve conduction slows and may be blocked, and
depression of body systems can produce unconsciousness. Nevertheless, to date
no convincing studies of neonatal central nervous system responses to noxious
stimuli during hypothermia have been carried out. In addition, it has been suggested that since rewarming from hypothermia is associated with pain in
humans,[83] the technique may be undesirable even if it produces a state of insensibility. Perhaps a more constructive approach is to examine the alternative anesthetic techniques available for use in neonates. Because volatile and injectable
anesthetics can be used to produce safe and effective anesthesia,[84,85] it seems reasonable for an IACUC to ask why it is necessary to use a questionable technique
when more acceptable alternatives are available. Investigators who have used all
of these techniques often find the use of conventional anesthesia more convenient,
especially for prolonged surgical procedures, since the neonates do not need to be
maintained on an ice pack for the duration of surgery.

Surv. Does your IACUC typically accept hypothermia as an acceptable form of anesthesia for fetal or neonatal homeothermic animals?

- Not applicable because we do not deal with fetal or neonatal
  homeotherms   59/152
- Yes, we typically accept hypothermia as an acceptable form
  of anesthesia   32/152
- We accept hypothermia only with justification as to why
  other anesthetics are inappropriate   32/152
- No, we do not accept hypothermia   25/152
- Other   4/152

16:39 Is hypothermia an acceptable form of anesthesia for poikilothermic animals?

Opin. Similar concerns to those described in 16:38 relate to the use of hypothermia as a
means of "anesthesia" in poikilotherms. As with mammalian neonates, well-established alternative anesthetic regimens are available and it, therefore, seems unnecessary to use a questionable technique.

16:40   A PI proposes to administer a small amount of nonirritating fluid daily for 1 month by intraperitoneal injection into hamsters. The injection is performed by a skilled person and the sites will be rotated. Is this APHIS/AC pain/distress Category C or Category E?

Reg.   (See 16:6.)

Opin.   The essential question is, Does the procedure cause momentary pain or distress, or does it cause more substantial pain? The investigator and IACUC are unlikely to be able to answer this question at the outset. Secondary questions include, On what basis can the fluid be deemed "nonirritating"? In other words, has histopathologic exam of the peritoneum been performed and does the animal indicate by behavioral cues that the effect of the injection procedure is transient? Also, what is the potential for complications—the risk of penetrating an abdominal organ, which would be associated with greater pain or illness? These questions can only be answered in a prospective manner; by observing of animals after a single and then after multiple days of injections, by body weight monitoring, and by comparisons to age-matched (and ideally sham-handled) animals. The degree to which a relatively minor noxious stimulus can lead to pain depends on whether inflammation occurs and on the "emotional" state of the animal. Following animals by daily monitoring of behavior and weight can aid in detecting the magnitude of the procedure's impact on them; use of operant conditioning (provision of a food reward) may aid in reducing some of the aversiveness by means of exerting a descending modulation of nociception to a less painful experience. Once this information is known, future categorization can be done with more certainty, and the initial study can be retrospectively characterized as C or E.

16:41   A PI proposes to administer a small amount of nonirritating fluid daily for 1 month by oral gavage to unanesthetized and untranquilized hamsters. The gavage is performed by a skilled person. Is this APHIS/AC pain/distress Category C or Category E?

Reg.   (See 16:6.)

Opin.   Oral gavage has the potential to cause irritation and inflammation of the esophagus, which might be expected to result in lower weight gain and food consumption. Body weight curves are likely to be very indicative of the prevalence of this pain. (See 16:40.)

16:42   An animal is given general anesthesia for several hours to permit a noninvasive imaging procedure to be performed. Should this type of restraint be considered to be a distressful procedure for which anesthesia was used to relieve the distress?

Reg.   See 16:6 for pertinent regulations that refer to the relief of either pain *or* distress. See 16:1–16:2 for definitions of *distress*.

Opin.   APHIS/AC Policy11[70] includes immobility as an example of a procedure that may lead to more than momentary or slight distress in a conscious animal. However, if the animal is anesthetized for the duration of the procedure, the potential distress has been alleviated by the appropriate use of a drug, as noted in the regulations. Assuming appropriate IACUC approval has been gained and no other experimental manipulations are involved, this would be reported as Category D on the APHIS/AC annual report for species covered by the AWA.

16:43 **An inhalation anesthetic is used to restrain an animal briefly for a painless procedure (such as an intranasal inoculation). Does this fall under APHIS/AC annual report Category C or D?**

Reg.   (See 16:6.)

Opin.   (See 16:42.) The primary advantage of using sedatives or anesthetics to immobilize animals for the performance of nonpainful procedures is that the drugs are used in lieu of more forceful or injurious physical restraint, which would likely be painful or distressful to the animal. Therefore, the IACUC must evaluate the rationale presented by the PI for using an anesthetic. For example, if the IACUC agrees with the PI that the extent of physical restraint needed to make an accurate intranasal injection is likely to distress the animal more than momentarily, then the use of an anesthetic (or perhaps a sedative) to reduce the need and consequences of that extreme restraint is appropriate. Alternately, the PI might argue that anesthesia is required to have the animal deeply inhale the intranasal inoculation and other means of assuring lung delivery would be more distressful or painful. With either explanation, the IACUC must agree with the stated need. Although using an anesthetic or sedative might be considered a refinement of the experimental technique, it does not change the fact that regulated animals that receive drugs for the purpose of alleviating pain or distress are required under the AWAR to be placed in Category D.

Surv.   If an anesthetic is used to restrain an animal briefly for a painless procedure (e.g., an intranasal inoculation), does your IACUC consider this as USDA pain and distress Category C or D?

- Not applicable   39/152
- We consider it Category C   89/152
- We consider it Category D   24/152

16:44 **What limits should the IACUC place on chronic restraint?**

Reg.   The PHS Policy (IV,C,1,a; IV,C,1,b; U.S. Government Principle IV) and the AWAR (§2.31,d,1,i; §2.31,d,1,ii) note that the IACUC should determine that a proposed activity prevents or minimizes discomfort and distress, and that alternatives to painful or distressful procedures have been considered. The *Guide* (p. 11) is more specific. It states that prolonged restraint (including chairing of nonhuman primates) should be avoided unless it is scientifically essential and is approved by the IACUC. It suggests the use of less restrictive systems, such as a tether system for nonhuman primates and stanchions for farm animals. Specific requirements for the use of restraint devices in nonhuman primates can be found in the AWAR (§3.81,d).

Opin.   Chronic restraint can be required for a variety of purposes but is usually needed when administering compounds or sampling body fluids via implanted catheters or obtaining continuous recording of physiological variables. The need for prolonged physical restraint has been reduced by the development of harness and swivel devices that allow an animal some degree of movement, and implantable telemetry devices or ambulatory infusion systems that allow complete freedom of movement. While these devices are refinements that may reduce the need for restraint and any accompanying pain and distress potential, there are some caveats. The weight of and surgical procedure required to insert telemetry implant devices can inflict a metabolic burden or postoperative pain. Tethers and metabolism cages allow freedom of movement but require animals to be isolated from conspecifics and kept in barren environments. Boredom and social isolation can also cause distress. The effects of environmental stressors on many experimental variables, including toxicity of substances, prenatal developmental biological processes, and cardiovascular, neurologic, and immunologic parameters have been well documented, and indeed, restraint methods are used as an experimental tool to study the effects of stress. [3–6,86–90]

When considering protocols that require restraint (or any of the refinements mentioned), an IACUC should question whether physical restraint is needed or whether an alternate approach can be adopted. This must be balanced by an appreciation that these alternate systems may still cause some pain, distress, or discomfort to an animal, both after implantation and during the earlier conditioning period. If a study can be completed with (for example) acclimated physical restraint of 1 or 2 hours, this might be preferable to the use of a tether system.

As the period of restraint increases, the balance between different systems changes. It is important to determine whether the animal can be readily trained to accept physical restraint. Measurement of stress sensitive indices, such as blood glucose concentration, heart rate, or behavioral stereotypy, suggests that animals can be acclimated or conditioned to accept restraint in slings or other restraint devices. [91,92] Problems arise, however, when physical restraint is carried out for prolonged periods in such a manner that the animal is unable to eat, drink, or carry out any normal behaviors, such as grooming.

The issue of restraint, therefore, is not simply one of "how long," but critically depends on the method used, the prior experience of the animal, and the availability of alternative, less stressful systems.

16:45 A research project with hamsters requires 0.5 ml of blood once a week for 6 months. Postorbital blood collection, under anesthesia, is approved by the IACUC. Even with a skilled technician, is it reasonable for the IACUC to assume pain or distress will occur over time from so many collections?

Reg. (See 16:4–16:6; 16:46.)

Opin. Abnormal behavior indicative of pain was studied in rats after a single postorbital puncture under anesthesia. Using intensive monitoring, the duration of the impact of the procedure appeared to be only about 5 hours. [93] We are not aware of any studies in other rodent species in which the impact of repeated postorbital bleeding was compared to that of repeated bleeds from other sites. Repeated tissue injury may cause altered threshold to pain sensitivity, and repeated sampling might cause increased pain over time. The severity of impact on an individual animal might be measured by alterations in body condition score, body weight, and of course, gross appearance of the eye. Alternating between the left and right eyes would reduce the number of times a single eye was sampled, lowering the overall potential for induction of a painful state.

16:46 Does postorbital blood collection in rodents normally require anesthesia, analgesia, or tranquilization?

Reg. Neither the AWAR nor the PHS Policy directly addresses postorbital blood collection. The *Guide* (p. 10) includes this procedure among those that may not have been previously encountered or may have the potential to cause pain or distress that cannot reliably be controlled.

Opin.   Postorbital (retroorbital) blood sampling in rodents can provide moderate quantities of blood quickly and easily and when carried out expertly seems to cause a
minimum of long-term complications. When it is carried out less competently, injuries to the globe can occur and these may be severe.[94] The procedure itself may be
no more painful than peripheral venipuncture (e.g., of the tail vein). Nevertheless,
many research units only carry out the technique with anesthesia, both to reduce
any pain and to minimize inadvertent injury should the animal struggle during the
procedure. The procedure is undoubtedly stressful, and perhaps a more appropriate question to ask is, Why is the retroorbital plexus to be used, rather than alternative techniques?[95] (See 16:5; 16:6; 16:45.)

Surv.   Does your IACUC require anesthesia or sedation for postorbital blood collection in
rodents? Check all appropriate responses.

- Not applicable   36/157
- Either anesthesia or sedation is required by our IACUC   58/157
- Only anesthesia is accepted by our IACUC   48/157
- Neither anesthesia nor sedation is normally required   7/157
- We require anesthesia or sedation for rats but not mice   3/157
- Other   5/157

16:47   Relative to pain or distress, how many episodes of blood collection by the postorbital sinus or postorbital plexus method should the IACUC reasonably allow on a single animal?

Opin.   A reasonable response to this question is for an IACUC to require a report on the
incidence of complications (corneal abrasions, retrobulbar hemorrhage or absces-
sation, damage to the globe, etc.) produced by the technique. Similar assessments
of sequelae to other methods of venipuncture can lead some IACUCs to conclude
that repeated orbital sinus puncture is acceptable, while others may conclude that
it should never be used. Very few studies have accurately assessed the incidence of
complications associated with this procedure, and all guidelines appear based on
a compromise between an investigator's wishes and an IACUC's unease at the
effects on the animal. The authors urge the IACUC to require follow-up (including
postmortem histologic evaluation of the globe and periorbital tissues) when allowing use of this technique. It then is possible to make an assessment based on data
rather than opinion. (See 16:6; 16:46; 19:19; 19:20.)

16:48   For performing procedures such as Southern blots or polymerase chain reactions, can the tail of a rat or mouse be clipped without causing more than momentary pain to the animal? If yes, what length of tail on mice?

Opin.   Removal of a small portion of the tail has become an established procedure in
many institutes that work with transgenic animals. There is little doubt that
removing the tip of a sensitive structure can cause pain. However, to date, no studies have been carried out to determine whether pain persists for more than a few
seconds or whether this is related to the length of tail removed. Different laboratories vary in their opinion as to whether the procedure requires anesthesia and
whether anesthesia should be local or general. One laboratory carries out the procedure with brief general anesthesia (with isoflurane) and provides postprocedure

analgesia (with carprofen). Rather than debate whether pain is produced and how much tail can be removed, it would be preferable to adopt the technique described by Meldgaard and associates[96] and sample saliva, thus replacing a procedure that is likely to cause at least momentary pain with one that causes none.

Surv. 1 Does your IACUC permit tail clipping as a means of obtaining tissue for DNA analysis in mice?

- Not applicable   51/158
- No, we do not permit tail clipping as a means of obtaining tissue   6/158
- Yes, we permit tail clipping up to 2–3 mm of tail   45/158
- Yes, we permit tail clipping up to 5 mm of tail   39/158
- Yes, we permit tail clipping up to 1 cm of tail   7/158
- Yes, we permit tail clipping up to >1 cm of tail   3/158
- Other   7/158

Surv. 2 If your IACUC permits tail clipping as a means of obtaining tissue for DNA analysis in mice, at what age do you allow this to be done *without* local or general anesthesia?

- Not applicable   69/150
- Less than 1 week old   9/150
- Less than 2 weeks old   14/150
- Less than 3 weeks old   24/150
- Less than 4 weeks old   21/150
- Other [primarily "no policy" responses]   13/150

Surv. 3 If your IACUC permits tail clipping as a means of obtaining tissue for DNA analysis in mice, is anesthesia required? Check all applicable responses.

- Not applicable as we do not do tail clipping of mice   55/153
- No anesthesia is required at any age   13/153
- We require local anesthesia at any age   14/153
- We require local anesthesia if the animal is more than [mean = 21.5; mode = 21; range = 7–42] days of age   26/153
- We require general anesthesia if the animal is more than [mean = 22.4; mode = 21; range = 14–35] days of age   32/153
- Other [primarily "no policy" responses]   13/153

Surv. 4 Does your IACUC require the use of alternatives to tail clipping for DNA analysis in mice?

- Not applicable as we do not do DNA analysis of this type in mice   54/150
- We do not require an alternative, but we encourage alternatives to be used   87/150
- We require an alternative (mark all alternatives used)   8/150

- Cheek scraping   2/8
- Saliva   2/8
- Ear punch   8/8
- Toe clip   0/8
- Hair analysis   3/8
- Other   0/8
  - Other   1/150

16:49 **What are the important issues and what are some general guidelines for the IACUC to consider relative to methods of animal identification?**

Reg.   The *Guide* (p. 46) states that toe clipping, as a method of identification of small rodents, should be used only when no other individual identification method is feasible and should be performed only on altricial neonates. Although the *Guide* lists other methods of identification (e.g., tattooing, ear notching) that can potentially be painful or distressful to an animal, only toe clipping is singled out. The AWAR (1:1, Painful Procedure) define a *painful procedure* as one that causes more than momentary pain or distress (e.g., as can occur from a needle prick). PHS Policy (IV,C,1,a; IV,C,1,b) has similar wording and the same intent. Identification requirements for dogs and cats are noted in the AWAR (§2.38,g). (See 16:50.)

Opin.   Accurate and reliable identification of laboratory animals is essential for most research projects. Methods should:

- Be reliable
- Cause minimal pain or distress to an animal
- Be simple to use
- Be standardized, so that the numbering system can be readily interpreted by all concerned

A further factor is the cost associated with the technique, which assumes greater significance when large numbers of animals must be identified. The IACUC should balance the consequences of failure of an identification system (loss of animals from a study and the need to repeat the investigation using additional animals) against the impact on the animal of the marking technique. These issues are often debated when an investigator wishes to use relatively low-cost (in economic terms) techniques such as ear punching or toe clipping, rather than more expensive methods such as microchip implantation. Aside from concerns that the former methods are less reliable (fighting can result in loss of further portions of the ear), both methods of physical marking result in a minor mutilation of the animal that may cause more than momentary pain. Checking the identification of an animal may cause stress from physical restraint. In contrast, the use of microchips appears to cause only momentary pain (although no well-controlled studies appear to have been carried out to support this assertion), and identification can be carried out more easily, often without the need to restrain the animal. An IACUC must weigh these issues and determine whether methods of identification that require removal of a piece of ear or a digit are acceptable, or whether less traumatic alternatives are preferred. As with many issues, limited published data are available and IACUC members must use their general biological knowledge to reach a conclusion.

16:50   Are there any circumstances under which the IACUC would allow a toe to be clipped as a means of animal identification or to obtain tissue for analysis by Southern blot or polymerase chain reaction?

Reg.   (See 16:49.)

Opin.   A similar approach to that of 16:49 must be taken to this question. Investigators should state why they consider it necessary to remove a piece of tissue when less invasive methods are generally available. These issues may seem relatively trivial when weighed alongside questions of end points in carcinogenesis studies or subjecting of animals to major survival surgery, but they should be addressed in a way that recognizes our general concern to reduce to a minimum the pain or distress caused to animals used in biomedical research. The pain and distress may be minor, but the numbers of animals involved can be considerable.

Surv. 1 Does your IACUC permit toe clipping of rats or mice as a means of animal identification?

- Not applicable as we do not use rats or mice   16/154
- No, we do not permit toe clipping as a means
  of identification   87/154
- Yes, with adequate justification   44/154
- Yes, with or without justification   4/154
- Other   3/154

Surv. 2 Does your IACUC permit toe clipping of rats or mice as a means of collecting tissue for DNA analysis?

- Not applicable as we do not use rats or mice   22/161
- No, we do not permit toe clipping for DNA analysis   89/161
- Yes, with adequate justification   37/161
- Yes, with or without justification   2/161
- Other   11/161

Surv. 3 If your IACUC permits toe clipping, how many toes are allowed to be clipped?

- Not applicable   103/147
- No more than one per foot   27/147
- No more than two per foot   12/147
- Other [primarily "no policy" responses]   5/147

Surv. 4 If your IACUC permits toe clipping, what rationale is acceptable, since the *Guide* suggests it be used only if other methods are not feasible?

- Not applicable   103/155
- A basic statement, such as "No other methods are available"   9/155
- A definitive statement, such as "We cannot use other methods
  because" (examples: needed to identify at recapture to prevent
  rehandling of animals; ear tags fell out; needed for pups
  <6 days old; needed to mark multiple animals in a litter)   41/155
- Other   2/155

16:51 What criteria can be used to determine whether an animal should be euthanized (or treated) because of the growth of an external or internal tumor?

Reg.   "Protocols should include criteria for initiating euthanasia, such as … tumor size, that will enable a prompt decision to be made by the veterinarian and the investigator to ensure that the endpoint is humane and the objective of the protocol is achieved" (*Guide*, p. 65).

Opin.   Determining end points in studies that involve production of neoplasia in animals is difficult, and there are no universal guidelines for all species and all tumors. The issue has been addressed thoroughly on two occasions (summarized in reference 97), and these reports provide a great deal of useful information and guidance. As with any issue of this type, before attempting to determine guidelines for termination, the PI should be asked to state the degree of tumor development that is required to meet the scientific objectives of the specific research protocol and whether the extent of tumor growth can be minimized. (See 17:37.)

      As in other types of research, termination at an early stage, resulting in loss of animals from a study, can result in the need to use additional animals. Termination late in a study may result in animals' experiencing unnecessary pain or distress. If animals die as a result of excessive tumor growth, this also can result in unnecessary pain and distress and in the loss of animals from the study since tissue or blood samples may not be available. Thus, there are compelling reasons to try to establish criteria for euthanasia.

      When formulating guidelines, it is critical that they should be unambiguous and clearly understood by all involved (the animal care staff as well as the investigator or AV) and that their application should be practical. Attempts to develop such criteria often rely on estimates of tumor mass, coupled with features such as ulceration or obvious necrosis of superficial tumors. These criteria are helpful, but it must be recognized that they may not relate to the degree of pain or distress experienced by the animal. As with postoperative pain, behavioral criteria may be more useful, but developing them is difficult and time-consuming, except when extreme changes in behavior (e.g., complete immobility) are used. Nevertheless, staff should be encouraged to refine the broad criteria adopted, and the IACUC should ask for feedback on the success or failure of the method used.

16:52 What justification should an IACUC expect if an investigator insists that the death of an animal or the death of 50% of her or his animals is the most appropriate end point for the study?

Reg.   Neither the AWAR nor the PHS Policy addresses the use of the lethal dose 50% (LD_{50}) test. The U.S. Food and Drug Administration does not require LD_{50} test data to establish levels of toxicity (Federal Register, 53: 39650–39651, 1988, and see http://iacuc.cwru.edu/policy/nihpolicies/iracld50.htm).

Opin.   There are two issues to consider here. First is the justification for the use of death as an end point, and second is the use of a 50% mortality criterion. The use of death as an end point has been regarded as essential in some investigations, but rapid developments are being made in this field and it is an area in which both investigators and the IACUC should make particular efforts to keep abreast of the current literature. The usual reason offered for selecting death as an end point is the difficulty of reliably differentiating animals that will die from those that will recover, despite their showing severe clinical signs of illness or toxicity. Surrogate humane end points have been proposed as a means to intervene when death can reliably be predicted from clinical signs or test results. In toxicity studies, death signifies a "cutoff," the dose was lethal, and by setting a 50% criterion, the toxic dose is "defined." In infectious disease, disease prevention, or therapeutic studies, death indicates a lethal infectious burden or failure of therapy. Often, these studies involve species for which detailed monitoring is not typically performed (e.g., rodents). If there is not a peracute onset of morbidity, animals gradually develop progressively more severe abnormalities, commencing with mild depression of normal activity and signs associated with lack of grooming activity (e.g., ruffled fur coat) and finally progressing to coma and death. When the clinical signs include subjectively distressing changes such as convulsions, weakness, cachexia, or severe dyspnea, there is pressure to euthanize an animal rather than allow further deterioration of its condition. As previously mentioned, the difficulty is the fear that some of these animals may recover and early euthanasia might invalidate a test result.

Several constructive suggestions have been proposed to reduce the need to use death as an end point in studies. There is also an alternative to using a strict 50% lethality criterion for acute toxicity studies that can reduce animal numbers used by 60–70%.[98] Furthermore, it is important to keep in mind that the results of a given test or exposure may be markedly affected by the handling or environment of the animals (see 16:43). This means that careful attention to laboratory conditions, training of personnel, and animal monitoring will play a vital role in experimental validity. Because rodents are typically not given a detailed health evaluation with the frequency that their rapid metabolic rate and comparatively shorter life span dictate, a conclusion that the agent, disease, or intervention caused death could be erroneous. Instead, the animals may have died as a result of weakness and inability to eat. Recognition of what is actually happening physiologically is missed when death is the only criterion evaluated; more detailed tracking of clinical signs or parameters might lead to better information, thereby increasing the discriminatory power of the experiment. Guidelines on the management of animals being used for the assessment of novel antibacterial or antifungal agents have been proposed by representatives of organizations involved in this type of investigation.[99]

In some circumstances, simple clinical indices such as the development of profound hypothermia can be used to predict death reliably. There remain many studies, however, in which no validated criteria have been devised. In these circumstances, an IACUC should carefully assess whether criteria can be developed during the progress of the study. In many instances, failure to develop these criteria may be due to infrequent or inadequate observations of the animals. At other times critical events may occur when personnel are not usually available, thereby precluding detailed observation. There seems no doubt that progress in refining end points has only occurred as a result of PIs carefully evaluating their own particular models. It may be that even after careful assessment, no progress is made and animals must be allowed to die if the aims of the study are not to be defeated. It is essential, however, for an IACUC to be proactive, require genuine attempts to develop nonlethal end points, and not allow investigators to use death simply because it is an unambiguous and easy criterion they can apply. Finally, the adoption of nonlethal end points can have beneficial effects on a study because they allow blood and tissue samples to be taken that may enable added data to be obtained.

16:53   A mammal is born with a genetic defect that leads to an abnormal but nonlethal health condition very soon after birth (e.g., an inability to use its hind limbs). It adapts as well as can be expected to that condition and requires no significant additional

levels of husbandry or veterinary care. **Should the IACUC consider this condition as nonpainful and nonstressful (Category C on the APHIS/AC annual report) or as unalleviated pain or distress (Category E on the APHIS/AC annual report)?**

Opin.   There are two possibilities for classifying such studies, and the IACUC should judge each individual circumstance as it arises. In the example, it is assumed that animals do not require additional levels of husbandry to supplement grooming of the affected limbs and there is also no risk of traumatic damage and self-inflicted wounds. Another example is congenital blindness or deafness. Although animals may have no apparent abnormality on clinical examination and require no special husbandry, one might consider that the lack of ability to move about or sensory deprivation causes an increase in the barrenness of the environment to that animal, and therefore a form of deprivation distress. On the other hand, one could argue that since we are unable to detect any apparent adverse effect, except by neurological examination, we should classify this as nonpainful or non-distressful, and therefore Category C. It also may be necessary to reconsider the classification as the animal ages and all aspects of the life cycle are taken into account.

16:54   In the example in 16:53, would the APHIS/AC classification change if this abnormality were caused by an experimental manipulation in an adult animal?

Opin.   The classification might well change if the defect were produced in an adult animal, since the animal's prior experience has some relevance. An animal that suddenly could no longer see or hear or move might require some time to adapt to the condition, and during this time the animal might be distressed by the defect, in contrast to an animal that had the defect from birth.

16:55   Cats are used to teach tracheal intubation techniques. They are anesthetized, recovered, and used again. What is a reasonable number of times, and at what intervals, might a cat be used for this purpose? How does the IACUC determine this?

Opin.   Not only are limits on frequency unhelpful, they make an assumption that somehow the repetition of a procedure can be added up to produce a "total score" of pain or distress. It is more appropriate to say that an animal can be used as often as required for such a procedure, provided it shows no resentment at the time of restraint for anesthetic induction; is treated appropriately for any pain or disability that might result from the procedure; continues to feed, drink, and grow normally; and does not develop clinical abnormalities. Also, the handling of the animal under anesthesia is critical, as the amount of trauma caused in one session can range from minimal to substantial. Animals should always be thoroughly examined for evidence of laryngeal, oral, or other trauma at the end of the procedure. Familiarizing an animal to the procedure and providing rewards for appropriate behavior and an enriched environment can lead to improved welfare of the cats (there can be no reason not to keep such animals in groups and in pens, rather than individual cages). This welfare enhancement could not easily be achieved when animals are used on a few occasions, euthanized, and a naive animal used for the next series of manipulations.

16:56 A drug to be used in a research project is known to cause tremors or mild seizures that occur once and last for approximately 2 minutes. Should the IACUC classify this as unalleviated distress? In this example, if seizures occur frequently, should the IACUC classify this as unalleviated distress?

Reg. The AWAR (§2.36,b,7) specifically require painful or distressful procedures for which appropriate anesthetic, analgesic, or tranquilizing drugs are not used to be placed in Category E on the APHIS/AC annual report.

Opin. Many drugs produce mild neurological side effects, and when these include seizures, often particular concern over the welfare of the animals being used arises. It is generally perceived that seizure activity can cause pain, particularly when marked muscle spasm occurs. In the case of seizures that cause a loss of consciousness, there is no evidence that the seizure itself is painful, although we must consider that postseizure muscle spasm can cause mild or moderate pain and that anxiety or fear may occur in animals cohoused with others who might react to the seizure episode. It is the opinion of APHIS/AC[100] that seizures such as this constitute unalleviated distress.

As with other areas of research using animals, we have no clear evidence regarding the degree of pain or distress caused by seizures. It could be argued that if the seizure is of a type that results in loss of consciousness and is not associated with potentially distressing aftereffects, then it should not be considered to produce unalleviated distress. If, in contrast, there are good reasons to believe that the animal remains aware during the seizure, then it should be classified as unalleviated distress. In summary, the IACUC should consider the type of seizure, try to determine the likelihood of persistence of consciousness, and consider the likelihood of postseizure effects (e.g., muscle pain after severe muscle spasms).

16:57 A PI performs a research procedure and states that he will euthanize an animal with a pentobarbital overdose as soon as any pain or distress is detected. Under those circumstances, would the IACUC place those animals in APHIS/AC annual report pain and distress Category C or D?

Reg. (See 16:6.) See 16:1–16:3 for definitions of *pain* and *distress*.

Opin. In this question, pentobarbital will be used for euthanasia, not anesthesia. The central issue is the timing of observations relative to the onset of pain or distress. If, for example, the animal is continuously observed until it shows the first signs indicative of *slight or momentary pain or distress* and then is immediately euthanized, the study should be placed in Category C. This is because Category C is reserved for animals that experience no more than slight or momentary pain or distress and the pain or distress in the example never progressed beyond the "slight or momentary" stage. The fact that euthanasia was performed does not change the APHIS/AC category to Category D (pain or distress alleviated by drugs) because we report to APHIS/AC the most severe category of pain or distress *actually experienced* by the animal. In this case, it was only slight or momentary (Category C).

In a real world situation animals are rarely observed with such rigor and (as another example) an animal that is observed only twice a day might experience 12–18 hours of *more than* slight pain before the pain is noted by an observer. If this happens, and the observer notes more than slight pain, Category E is appropriate because it is reserved for unalleviated pain or distress that is more than just slight or momentary.

As a final example, consider an animal that experiences no significant pain because an analgesic was used to alleviate what would otherwise be more than slight or momentary pain. That animal would be placed in Category D (pain or distress alleviated by the use of a drug). However, if the animal experienced a few hours of unalleviated, more than slight pain before the drug was administered, it would be placed in Category E because the drug was used *after* significant pain or distress occurred. That is, Category E was the most severe category of pain *actually experienced* by the animal. The fact that a drug was eventually used, or even the animal's being euthanized, does not "lower" the pain classification to Category D. As noted, the timing of euthanasia or drug administration is critical in determining the appropriate APHIS/AC category. (See 16:32.)

16:58   Should an IACUC allow cardiac puncture without prior anesthesia?

Reg.   Cardiac puncture as a means to effect euthanasia by exsanguination (unless there is prior anesthesia, stunning, or sedation) is considered unacceptable by the AVMA Panel on Euthanasia because of anxiety associated with extreme hypovolemia.[101] PHS Policy (IV,C,1,g) requires compliance with the AVMA Panel recommendations unless the IACUC approves a deviation. APHIS/AC Policy 3 states that the method of euthanasia must be consistent with the current Report of the AVMA Panel on Euthanasia. (See 17:2.)

There are no specific regulatory statements concerning cardiac puncture for purposes other than euthanasia, assuming the volume of blood removed is sufficiently small as not to cause hypovolemia. Personnel performing cardiac puncture must be appropriately trained (AWAR §2.31,d,1,viii; U.S. Government Principle VIII; PHS Policy IV,C,1,f).

Opin.   The authors know of no evidence to indicate that direct cardiac puncture is painful when used for clinical purposes (sampling of blood) or delivery of anesthetic euthanasia solution (pentobarbital). The entry point on the body surface is likely to be a factor—pericardial fluid removal (pericardiocentesis) is accomplished in patients by administering a local anesthetic intercostally. This is done to decrease the pain of inserting a needle as the intercostal nerves are easily impinged upon, causing marked pain. Clinically, pericardiocentesis is performed under echocardiographic and electrocardiographic guidance, to prevent accidental contact with critical structures inside the thorax. Rather than being sensitive to a needle prick, visceral structures (myocardium and pericardium) are innervated by sensory fibers that are responsive to distention, inflammation, or ischemia. Cardiac puncture can also be accomplished transdiaphragmatically. Insertion of a needle in this manner would be expected to cause no more pain than an intraperitoneal injection. Rapid removal of blood (exsanguination) might cause cardiac ischemia, which can be associated with moderate to severe pain. Removal of a small sample of blood may not be painful per se, but contact with a coronary artery can lead to pericardial hemorrhage and death. Thus, direct cardiac puncture might be useful only when other approaches to sampling blood do not exist. If the investigator proposes to use this technique in an unanesthetized or unsedated animal to sample blood, the anatomical approach and justification should address these concerns.

Surv.   Does your IACUC allow cardiac puncture without prior anesthesia?

- Not applicable as we do not do any cardiac puncture   24/158
- Yes, if requested by the investigator; no justification
  is required   10/158

- Yes, if requested by the investigator; scientific justification
  is required 17/158
- No 106/158
- Other 1/158

16:59 Is rabbit "hypnosis" a suitable form of restraint for painful procedures?

Opin. *Hypnosis* is one of many terms designating an immobile, unresponsive state that can be induced in numerous species, referred to as an *immobility response* (IR). IR can be triggered by various physical and psychological manipulations (predator threat, violent struggle, thoracic compression, restraint by neck skin "scruffing," supine position).[102] Reviews of reports in animals and a study of IR in rabbits indicate that between-individual responses are variable and may depend on conditioning or maturation.[103] Evidence exists for analgesia in some species (or an increased threshold to noxious stimulus, but animals can be roused from the IR state by stronger stimuli, and the duration of an IR is not predictable.[102] Commonly performed restraint techniques, such as application of a twitch to horses or restraint by neck skin pinch, can produce a degree of immobile tolerance to a brief stimulus, and thus can be considered preferable to more injurious or strenuous restraint. Nevertheless, because of the lack of predictability, induction of an IR in order to have an animal tolerate painful procedures lasting more than a few moments is not recommended. Also, those intending to use IR for momentary restraint must be prepared to accommodate a significant percentage of animals that do not develop the IR.

References

1. Russell, W.M.S., and Burch, R.L., *The Principles of Humane Experimental Technique*, Methuen, London, 1959.
2. Balcombe, J.P., Barnard, N.D., and Sandusky, C., Laboratory routines cause animal stress, *Contemp. Topics Lab. Anim. Sci.*, 43, 42, 2004.
3. Cory-Slechta, D.A., et al., Maternal stress modulates the effects of developmental lead exposure, *Environ. Health Perspect.*, 112, 717, 2004.
4. Damon, E.G., et al., Effect of acclimation to caging on nephrotoxic response of rats to uranium, *Lab. Anim. Sci.*, 36, 24, 1986.
5. Kerr, L.R., et al., Effects of social housing condition on chemotherapeutic efficacy in a Shionogi carcinoma (SC115) mouse tumor model: influences of temporal factors, tumor size, and tumor growth rate, *Psychosom. Med.*, 63, 973, 2001.
6. Sternberg, W.F., and Ridgway, C.G., Effects of gestational stress and neonatal handling on pain, analgesia and stress behavior in adult mice, *Physiol. Behav.*, 78, 375, 2003.
7. Beilin, B., et al., The effects of postoperative pain management on the immune response to surgery, *Anesth. Analg.*, 97, 822, 2003.
8. Ben-Eliyahu, S., et al., Evidence that stress and surgical interventions promote tumor development by suppressing natural killer cell activity, *Int. J. Cancer*, 80, 880, 1999.
9. Desborough, J.P., The stress response to trauma and surgery, *Br. J. Anaesth.*, 85, 109, 2000.
10. Kiecolt-Glaser, J.K., et al., Psychological influences on surgical recovery: perspectives from psychoneuroimmunology, *Am. Psychol.*, 53, 1209, 1998.
11. Perkins, F.M., and Kehlet, H., Chronic pain as an outcome of surgery: a review of predictive factors, *Anesthesiology*, 93, 1123, 2000.
12. Broom, D.M., and Johnson, K.G., *Stress and Animal Welfare*, Chapman & Hall, London, 1993.

13. National Research Council, Recognition of Pain and Distress in Laboratory Animals, National Academy Press, Washington, D.C., 1992.
14. Orlans, F.B., Review of experimental protocols: classifying animal harm and applying "refinements," *Lab. Anim. Sci.*, 37, 50, 1987.
15. Appleby, M.C., and Hughes, B.O., *Animal Welfare*, CAB International, Oxfordshire, U.K., 1997.
16. Australian Council on the Care of Animals in Research and Teaching, *Animal Pain: Ethical and Scientific Perspectives*, Kuchel, T.R., Rose, M., and Burrell, J., Eds., Australian Council on the Care of Animals in Research and Teaching, Adelaide, Australia, 1992.
17. International Association for the Study of Pain, Report of International Association for the Study of Pain Subcommittee on Taxonomy, *Pain*, 6, 249, 1979.
18. Zimmerman, M., Neurological concepts of pain, its assessment and therapy, in *Neurophysiological Correlates of Pain*, Bromm, B., Ed., Elsevier, Amsterdam, 1984, p. 15.
19. Craig, A.D., A new view of pain as a homeostatic emotion, *Trends Neurosci.*, 26, 303, 2003.
20. Association of Veterinary Teachers and Research Workers, Guidelines for the recognition and assessment of pain in animals, *Vet. Rec.*, 118, 334, 1986.
21. American College of Veterinary Anesthesiologists. Position paper on the treatment of pain in animals, *J. Am. Vet. Med. Assoc.*, 213, 628, 1998. Available on the World Wide Web at: http://www.acva.org/professional/Position/pain.htm.
22. Khamis, H.J., Statistics and the issue of animal numbers in research, *Contemp. Topics Lab. Anim. Sci.*, 36, 54, 1997.
23. U.S. Department of Agriculture, Animal and Plant Health Inspection Service, Animal Care Policy 12, Written Narrative for Alternatives to Painful Research. Available on the World Wide Web at: http//www.aphis.usda.gov/ac/policy12.html.
24. Francione, G.L., and Charlton, A.E., Bound by law, *Lab Anim. (NY)*, 27(3), 20, 1998.
25. Garnett, N.L., and DeHaven, W.R., A word from the government, *Lab Anim. (NY)*, 27(3), 21, 1998.
26. Morton, D.B., and Griffiths, P.H.M., Guidelines on the recognition of pain, distress and discomfort in experimental animals and an hypothesis for assessment, *Vet. Rec.*, 116, 431, 1985.
27. Flecknell, P.A., Advances in the assessment and alleviation of pain in laboratory and domestic animals, *J. Vet. Anesth.*, 21, 98, 1994.
28. Silverman, J., et al., Decerebrate mammalian preparations: unalleviated or fully alleviated pain? A review and opinion., *Contemp. Topics Lab. Anim. Sci.*, 44(4), 14, 2005.
29. Stevens, C.W., Alternatives to the use of mammals for pain research, *Life Sci.*, 50, 901, 1992.
30. Kehlet, H., Postoperative opioid sparing to hasten recovery, *Anesthesiology*, 102, 1083, 2005.
31. Vaculin, S., and Rokyta, R., Effects of anesthesia and nociceptive stimulation in an experimental model of brachial plexus avulsion, *Physiol. Res.*, 53, 209, 2004.
32. Weiskopf, R.B., Ketamine for perioperative pain management, *Anesthesiology*, 102, 211, 2005.
33. Flecknell, P.A., *Laboratory Animal Anesthesia*, 2nd ed., Academic Press, London, 1996.
34. Sawyer, D.C., Rech, R.H., and Durham, R.A., Does ketamine provide adequate visceral analgesia when used alone or in combination with acepromazine, diazepam, or butorphanol in cats? *J. Am. Anim. Hosp. Assoc.*, 29, 257, 1993.
35. Banknieder, A.R., et al., Comparison of ketamine with the combination of ketamine and xylazine for effective anesthesia in the rhesus monkey, *Lab. Anim. Sci.*, 28, 742, 1987.
36. Boschert, K., et al., Ketamine and its use in the pig: recommendations of the Consensus Meeting on Ketamine Anesthesia in Pigs, Bergen 1994, *Lab. Anim.*, 30, 209, 1996.
37. Flecknell, P.A., Medetomidine and atipamezole: potential uses in laboratory animals, *Lab Anim. (NY)*, 26(2), 21, 1997.
38. Green, C.J., et al., Ketamine alone and combined with diazepam or xylazine in laboratory animals: a 10-year experience, *Lab. Anim.*, 15, 163, 1981.
39. Thurmon, J.C., Tranquilli, W.J., and Benson, G.J., *Lumb and Jones' Veterinary Anesthesia*, 3rd ed., Williams & Wilkins, Baltimore, 1996.
40. Rubal, B., and Buchanan, C., Supplemental chloralose anesthesia in morphine premedicated dogs, *Lab. Anim. Sci.*, 36, 59, 1986.
41. Silverman, J., and Muir, W.W., A review of laboratory animal anesthesia with chloral hydrate and chloralose, *Lab. Anim. Sci.*, 43, 210, 1993.
42. Ching, M., Comparison of the effects of althesin, chloralose-urethane, urethane, and pentobarbital on mammalian physiologic responses, *Can. J. Physiol. Pharmacol.*, 62, 654, 1984.

43. Faber, J.E., Effects of althesin and urethan-chloralose on neurohumoral cardiovascular regulation, *Am. J. Physiol. Regul. Integr. Comp. Physiol.*, 256, 757, 1989.

44. Momosaki, S., et al., Rat-PET study without anesthesia: anesthetics modify the dopamine D1 receptor binding in the rat brain, *Synapse*, 54, 207, 2004.

45. Theodorsson, A., Holm, L., and Therodorsson, E., Modern anesthesia and preoperative monitoring methods reduce pre- and postoperative mortality during transient occlusion of the middle cerebral artery in rats, *Brain Res. Protocols*, 14, 181, 2005.

46. Yuan, C.S., Effects of chloralose-urethane anesthesia on single-axon reciprocal Ia IPSPs in the cat, *Neurosci. Lett.*, 94, 291, 1988.

47. Wagner, P.G., Eldridge, F.L., and Dowell, R.T., Anesthesia affects respiratory and sympathetic nerve activities differentially, *J. Autonom. Nerv. Syst.*, 36, 225, 1991.

48. Zaugg, M., et al., Differential effects of anesthetics on mitochondrial K_{ATP} channel activity and cardiomyocyte protection, *Anesthesiology*, 97, 15, 2002.

49. Savarese, J.J., et al., Preclinical pharmacology of GW280430A (AVA430A) in the rhesus monkey and in the cat, *Anesthesiology*, 100, 835, 2004.

50. Lukas,V.S., Animal use in toxicity evaluation, in *Anesthetic Toxicity*, Rice, S.A., and Fish, K.J., Eds., Raven Press, New York, 1994, chap. 2.

51. Field, K.J., and Lang, C.M., Hazards of urethane (ethyl carbamate): a review of the literature, *Lab. Anim.*, 22, 255, 1988.

52. Fish, R.E., Pharmacology of injectable anesthetics, in *Anesthesia and Analgesia in Laboratory Animals*, Kohn, D. et al., Eds., Academic Press, San Diego, 1997, 19.

53. Matthews, K.A., Pain assessment and general approach to management, in Management of Pain, Matthews, K.A., Ed., *Vet. Clin. North Am. Small Anim. Pract.*, 30, 929, 2000.

54. Gaynor, J.S. and Muir, W.W., Acute pain management, a case based approach, in *Handbook of Veterinary Pain Management*, Gaynor, J.S. and Muir, W.W., Eds., Mosby, St. Louis, 2002, p. 346.

55. Chesler, E.J., Identification and ranking of genetic and laboratory environment factors influencing a behavioral trait, thermal nociception, via computational analysis of a large data archive, *Neurosci. Biobehav. Rev.*, 26, 907, 2002.

56. Liles, J.H. and Flecknell, P.A., A comparison of the effects of buprenorphine, carprofen and flunixin following laparotomy in rats, *J. Vet. Pharm. Therapeut.*, 17, 284, 1993.

57. Liles, J.H. and Flecknell, P.A., The effects of surgical stimulus on the rat and the influence of analgesic treatment, *Br. Vet. J.*, 149, 515, 1993.

58. Karas, A.Z., Unpublished observations, 2005.

59. Sebel, P.S., et al., The incidence of awareness during anesthesia: a multicenter United States study, *Anesth. Analg.*, 99, 833, 2004.

60. Marsch, S.C.U. and Studer, W., Guidelines to the use of laboratory animals: what about neuromuscular blockers? (editorial), *Cardiovasc. Res.*, 42, 565, 1999.

61. Paddleford, R.R., Pulmonary dysfunction, in *Lumb and Jones Veterinary Anesthesia*, 3rd ed., Thurmon, J.C., Tranquilli, W.J., Benson, J.G., Eds.,Williams and Wilkins, Baltimore, 1996, p. 774.

62. Whelan, G., and Flecknell, P.A., The assessment of depth of anesthesia in animals and man, *Lab. Anim.*, 26, 153, 1992.

63. Hall, L.W., *Veterinary Anesthesia*, 9th ed., Balliere Tindall, London, 1991.

64. Branson, K.R., and Thurmon, J.C., Performing epidural anesthesia in swine, *Vet. Med.*, 85, 1345, 1990.

65. Kero, P., Thomasson, B., and Soppi, A.M., Spinal anesthesia in the rabbit, *Lab. Anim.*, 15, 347, 1981.

66. Valverde, A., et al., Comparison of the hemodynamic effects of halothane alone and halothane combined with epidurally administered morphine for anesthesia in ventilated dogs, *Am. J. Vet. Res.*, 52, 505, 1991.

67. Lovell, D.P., Variation in pentobarbitone sleeping time in mice. 1. Strain and sex differences, *Lab. Anim.*, 20, 85, 1986.

68. Pascoe, P.J., Perioperative pain management, in Management of Pain, Matthews, C., Ed., *Vet. Clin. North Am. Small Anim. Pract.*, 30, 917, 2000.

69. Smith, G., Postoperative pain, in *Quality of Care in Anesthetic Practice*, Lunn, J.N., Ed., McMillan Press, London, 1984, pp. 164–192.

70. U.S. Department of Agriculture, Animal and Plant Health Inspection Service, Animal Care Policy 11, Painful/Distressful Procedures. Available on the World Wide Web at: http//www.aphis.usda.gov/ac/policy11.html.

71. Vermeulen, J.K., et al., Food deprivation: common sense or nonsense? *Anim. Technol.*, 48, 45, 1997.

72. Waynforth, H.B., and Flecknell, P.A., *Experimental and Surgical Techniques in the Rat*, Academic Press, London, 1992.

73. Anand, K., Sippell, W.G., and Aynsley-Green, A., Randomised trial of fentanyl anesthesia in preterm babies undergoing surgery: effects on the stress response, *Lancet*, 1(8524), 243, 1987.

74. Anand, K., The biology of pain perception in newborn infants, *Adv. Pain Res. Ther.*, 15, 113, 1990.

75. Anand, K.J.S., and Maze, M., Fetuses, fentanyl and the stress response: signals from the beginnings of pain? *Anesthesiology*, 95, 823, 2001 (editorial).

76. Houfflin-Debarge, V., et al., Effects of nociceptive stimuli on the pulmonary circulation in the ovine fetus, *Am. J. Physiol. Regul. Integr. Comp. Physiol.*, 288, R547, 2005.

77. Senat, M.V., Fischer, C., and Ville, Y., Funipuncture for fetocide in late termination of pregnancy, *Prenatal Diagnosis*, 22, 354, 2002.

78. Benatar, D., and Benatar, M., A pain in the fetus: towards ending confusion about fetal pain, *Bioethics*, 15, 57, 2001.

79. Fitzgerald, M., Neurobiology of foetal and neonatal pain, in *Textbook of Pain*, 3rd ed., Wall, P., and Melzac, R., Eds., Churchill Livingstone, London, 1994.

80. Narsinghani, V., and Anand, K., Developmental neurobiology of pain in neonatal rats, *Lab Anim.* (NY), 29(9), 27, 2000.

81. Division of Animal Welfare, Office for Protection from Research Risks, National Institutes of Health, The Public Health Service responds to commonly asked questions, *ILAR News*, 33(4), 68, 1991.

82. Swayne, D.E., Protocol review: preventive measures, *Lab Anim.* (NY), 25(5), 22, 1996.

83. Morton, D., personal communication, 1998.

84. Danneman, P.J., and Mandrell, T.D., Evaluation of five agents/methods for anesthesia of neonatal rats, *Lab. Anim. Sci.*, 47, 4, 1997.

85. Park, C.M., et al., Improved techniques for successful neonatal rat surgery, *Lab. Anim. Sci.*, 42, 508, 1992.

86. Colomina, M.T., Albina, M.L., and Sanchez, D.J., Interactions in developmental toxicology: combined action of restraint stress, caffeine, and aspirin in pregnant mice, *Teratology*, 63, 144, 2001.

87. Cordero, M.I., Chronic restraint stress down-regulates amygdaloid expression of polysialylated neural cell adhesion molecule, *Neuroscience*, 133, 903, 2005.

88. Costa, A., Smeraldi, A., Tassorelli, C., Greco, R., and Nappi, G., Effects of acute and chronic restraint stress on nitroglycerine-induced hyperalgesia in rats, *Neurosci. Lett.*, 383, 7, 2005.

89. King-Herbert, A.P., et al., Effects of immobilization restraint on Syrian golden hamsters, *Lab. Anim. Sci.*, 47, 362, 1997.

90. Taylor, S.E., and Dorris, R.L., Modification of local anesthetic toxicity by vasoconstrictors, *Anesth. Prog.*, 36, 79, 1989.

91. Hassimoto, M., Harada, T., and Harada, T., Changes in hematology, biochemical values, and restraint ECG of rhesus monkeys (*Macaca mulatta*) following 6-month laboratory acclimation, *J. Med. Primatol.*, 33, 175, 2004.

92. Natelson, B.H., Hoffman, S.L., and Cagin, N.A., A role for environmental factors in the production of digitalis toxicity, *Pharmacol. Biochem. Behav.*, 12, 235, 1980.

93. Van Herck, H., et al., Orbital sinus blood sampling in rats: effects upon selected behavioral variables, *Lab. Anim.*, 34, 10, 2000.

94. Van Herck, H., et al., Orbital sinus blood sampling in rats as performed by different animal technicians: the influence of technique and expertise, *Lab. Anim.*, 32, 377, 1998.

95. BVA/FRAME/RSPCA/UFAW, Removal of blood from laboratory mammals and birds—first report of the BVA/FRAME/RSPCA/UFAW Joint Working Group on Refinement, *Lab. Anim.*, 27, 1, 1993.

96. Meldgaard, M., Bollen, P.J., and Finsen, B., Non-invasive method for sampling and extraction of mouse DNA for PCR, *Lab. Anim.*, 38(4), 413, 2004.

97. United Kingdom Coordinating Committee on Cancer Research, *UKCCCR Guidelines for the Welfare of Animals in Experimental Neoplasia*, 2nd ed., United Kingdom Coordinating Committee on Cancer Research, London, 1997.

98. National Institute of Environmental Health Sciences, The Revised Up-and-Down Procedure: A Test Method for Determining the Acute Oral Toxicity of Chemicals. Vol. I. NIH Publication 02-4501, 2001. Available on the World Wide Web at: http://iccvam.niehs.nih.gov/methods/udpdocs/udpfin/vol.1.pdf.

99. Rodent Protection Test Working Party, Guidelines for the welfare of animals in rodent protection tests: A report from the Rodent Protection Test Working Party, *Lab. Anim.*, 28(1), 13, 1994.

100. DeHaven, W.R., personal communication, 1999.

101. AVMA Panel on Euthanasia, 2000 Report of the AVMA Panel on Euthanasia, *J. Am. Vet. Med. Assoc.*, 218, 669, 2001.

102. Carli, G., Immobility responses and their behavioral and physiologic aspects, including pain: a review, in *Animal Pain*, Short, C.E., and Van Poznak, A., Eds., Churchill Livingstone, New York, 1992.

103. Danneman, P.J., et al., An evaluation of analgesia associated with the immobility response in laboratory rabbits, *Lab. Anim. Sci.*, 38(1), 51, 1988.

17

Euthanasia

Peggy J. Danneman

Introduction

This chapter addresses questions on the topic of euthanasia. The first two questions cover issues to consider when setting institutional policies on euthanasia. The remaining ones address dilemmas related to the use of specific agents and techniques (e.g., decapitation, exsanguination, pithing, carbon dioxide, alone or mixed with other gases). There is also a section on euthanasia of fetal and neonatal animals, a necessity that is frequently encountered in practice but is rarely addressed in the literature, and there are sections on determining the appropriate time for euthanasia (humane end points) and on effects on other animals from witnessing euthanasia.

17:1   What are some general guidelines for euthanasia?

Reg.   Euthanasia of experimental animals may be performed for many reasons, including procurement of tissues or blood as part of the experimental design. Animals also may be euthanized to alleviate otherwise untreatable pain or distress. Both the AWAR (§2.31,d,1,v) and PHS Policy (IV,C,1,c) require investigators to euthanize animals that would otherwise experience severe or chronic pain or distress that cannot be relieved by other means. In such instances, animals should be euthanized at the end of the procedure or, if appropriate, during the procedure. The *Guide* (pp. 65–66) further indicates that animal use protocols should include specific criteria for initiating euthanasia to "ensure that the endpoint is humane and the objective of the protocol is achieved."

Opin.   The term *euthanasia* is derived from Greek and means "easy death." According to definitions in the AWAR (§1.1, Euthanasia), the *Guide* (p. 65), and the 2000 Report of the AVMA Panel on Euthanasia,[1] euthanasia involves killing in a humane manner that causes rapid unconsciousness and death with little or no pain or distress. These goals are met by selecting an agent or technique that is appropriate for the situation, taking into account the species of animal and its age and temperament, the availability of appropriate equipment for restraint, and the environment in which the euthanasia will be performed. It also is important that the person performing the euthanasia be compassionate, gentle, fully trained in the technique, and technically proficient. Because other animals may be distressed by visual, auditory, or olfactory signals from the animal being euthanized, it has been considered preferable to perform the procedure in an area where other animals are not present.[1-3] However, some recent studies suggest that animals of at least some species (e.g., rats) are no more distressed by witnessing the death of conspecifics than they are by witnessing procedures such as routine cage changing.[4-7] For this reason, the 2000 Report of the AVMA Panel on Euthanasia[1] falls short of making a general recommendation to avoid having other animals present during euthanasia; instead, it advises against this practice only when dealing with "sensitive" species.

Aside from the all-important humane issues, three broad, overlapping criteria should be considered when selecting a method of euthanasia.

Regulatory: The method should be in compliance with the relevant regulations and guidelines noted. Deviations from these regulations and guidelines are permitted under certain circumstances and should be approved on the basis of written justification for scientific or medical reasons (*Guide*, p. 65; AWAR §2.31,d,1,xi; PHS Policy IV,C,1,g). It should be noted that some states and Canadian provinces also have specific laws related to euthanasia.

Human: The method should take into account personnel and management issues, including the qualifications and training of the personnel administering euthanasia; the need to minimize emotional distress in human participants and observers by proper attention to the aesthetic implications of the method; and attention to the health and safety of humans and other animals.

Scientific: The method should take into account the potential effect of the method on the scientific objectives of the research or postmortem evaluations. Any method of euthanasia should be viewed as having the potential to interfere with scientific objectives, and it is important to choose the method with care if any data will be collected from the animal after death.

Other criteria to consider are the reliability of the method in producing a rapid and humane death, the relative likelihood that an apparently dead animal might recover following disposal, the additional time or procedures that may be required to assure that an animal will not recover after disposal, the expense and availability of drugs or equipment, the potential for human drug abuse, and the ability to maintain euthanasia equipment in proper working order.[1,2,8]

17:2 The AVMA Panel on Euthanasia guidelines[1] are not recognized as law. Must the IACUC use them when determining the appropriateness of the proposed method of euthanasia?

Reg. It is true that the reports of the AVMA Panel on Euthanasia are presented as guidelines, not legal requirements. However, PHS Policy (IV,C,1,g) specifically states that the IACUC must determine that methods of euthanasia in a proposed research project are consistent with the AVMA Panel recommendations "unless a deviation is justified for scientific reasons in writing by the investigator." While this wording allows for deviation from AVMA Panel recommendations under specific circumstances, it clearly indicates that IACUCs are required to give the guidelines careful consideration when evaluating a proposed method of euthanasia. APHIS/AC Policy 3[9] states that "the method of euthanasia must be consistent with the current Report of the AVMA Panel on Euthanasia."

Opin. The recommendations of the AVMA Panel are supported by reports in the literature, but the panel's interpretation of those reports and final recommendations are not always consistent with the opinions of other experts. For example, there is considerable controversy in the laboratory animal science community regarding the humaneness of carbon dioxide asphyxiation and the most humane technique for performing euthanasia with carbon dioxide (see 17:3). Also, the Panel offers no guidelines regarding the acceptability of certain methods of euthanasia that are commonly used in research settings (e.g., fixative perfusion under deep anesthesia).

While approval of a method of euthanasia that is not consistent with the AVMA Panel guidelines is acceptable according to the *Guide* (p. 65) and PHS Policy (IV,C,1,g), the IACUC may do so only if the investigator has justified the method for scientific reasons; the *Guide* (p. 65) indicates that deviations from Panel guidelines also may be justified for medical reasons. APHIS/AC Policy 3[9] does not address whether, or how, an IACUC might approve a deviation from Panel guidelines. Neither PHS Policy nor APHIS/AC Policy 3 allows leeway for the IACUC simply to disagree with the interpretation and recommendations of the Panel (e.g., to decide, after careful consideration of the existing evidence and literature, that cervical dislocation performed by a skilled individual is a humane and fully acceptable method of euthanasia for small rodents).

Surv. Does your IACUC recommend or require compliance with the euthanasia recommendations of the American Veterinary Medical Association's (AVMA) Panel on Euthanasia?

- Not applicable; we do not euthanize animals covered by IACUC protocols   0/163
- We require following the AVMA Panel unless scientific justification is provided for an exception   104/163
- We always require following the AVMA Panel   50/163
- We recommend following the AVMA Panel, but we do not require it   7/163
- Other   2/163

17:3   What are the most appropriate conditions (e.g., concentration, flow rate) for euthanasia of adult rodents with carbon dioxide?

Opin. The literature contains numerous recommendations regarding the use of carbon dioxide (CO_2) for euthanasia. Nevertheless, there is no general agreement regarding the optimal CO_2 concentration or flow rate or whether the euthanasia chamber should be prefilled prior to introducing the animal. The 2000 AVMA Panel Report[1] recommends prefilling the chamber for species in which this has not been shown to cause distress; the report does not indicate which species might be distressed by prefilling of the chamber. Then, after the animal has been placed in the chamber, the gas flow should be sufficient to displace at least 20% of the chamber volume per minute. These recommendations are controversial. There is general agreement that death occurs more quickly if the chamber is prefilled. At least in part because animals collapse more quickly when exposed to initial high concentrations of CO_2, some observers argue that prefilling the chamber may reduce anxiety and struggling;[3] others suggest that pain and distress can be reduced if the chamber is not prefilled.[10–12] Some observers noted no differences in behavior of rats euthanized in prefilled versus nonprefilled chambers.[13,14]

Several sources suggest the use of low gas flow rates or other conditions that lead to a gradual increase in the concentration of CO_2.[3,10,12,15] Two reasons are cited for this recommendation:

- The turbulence and loud hissing associated with the rapid influx of gas into the chamber appear distressing to animals.
- Slowly rising carbon dioxide concentrations allow the animal to lose consciousness prior to asphyxiation and with less pain and distress.

There is evidence that animals experience significant pain and/or distress when exposed to high concentrations of carbon dioxide. [11,16–18] This is consistent with recent research indicating that carbon dioxide, which forms carbonic acid in a moist environment (e.g., the nasal mucosa), specifically excites small nerve fibers that transmit sensations of pain.[19,20] For this reason, it has been recommended that conscious animals should not be exposed to carbon dioxide concentrations greater than 70%.[11] On the other hand, it is difficult, if not impossible, to euthanize some animals (e.g., diving animals, such as mink) with carbon dioxide in concentrations less than 80%.[1,21] Because mink showed a strong aversion to the high concentrations needed to kill them, Cooper and colleagues[22] concluded that animals of this species cannot be euthanized humanely by using CO_2. Others have suggested that for humane euthanasia of any species CO_2 mixed with argon or nitrogen, or even 100% argon, may be preferable to 100% CO_2 administered in any manner.[18,23,24] (See also 17:22.)

What is clear at this time is that there is no consensus on the desirability of CO_2 as an agent for euthanasia, let alone the conditions under which it might best be administered. The best advice is that provided by NIH/OLAW in its recent clarification of PHS Policy regarding the use of CO_2 for euthanasia: that "IACUCs and veterinary staff should keep abreast of current peer-reviewed scientific literature and apply informed professional judgment to the design of institutional policies for CO_2 delivery systems and procedures."[25]

Surv. What conditions does your IACUC recommend for the euthanasia of adult rodents with carbon dioxide?

- Not applicable: we either do not use or do not recommend carbon dioxide — 22/162
- Nonprefilled chamber, 100% carbon dioxide used — 49/162
- Chamber prefilled with 100% carbon dioxide — 39/162
- Other combinations of carbon dioxide (e.g., 70%), flow rate, and chamber filling allowed — 30/162
- Never addressed chamber filling, concentration, or flow rate issues — 16/162
- Other — 6/162

17:4 Is the use of dry ice a satisfactory method of generating carbon dioxide to be used for euthanasia?

Opin. The most commonly used source of carbon dioxide (CO_2) for euthanasia is compressed CO_2 in cylinders, but dry ice has been viewed as an alternative source.[1–3] The 2000 Report of the AVMA Panel on Euthanasia states that dry ice is not an acceptable source of carbon dioxide for euthanasia because the inflow of gas into the chamber cannot be regulated precisely with this approach.[1] An additional problem with the use of dry ice is that animals are readily injured if they have direct contact with it. On the other hand, simple design features can be utilized to prevent the animal from having direct contact with the dry ice (e.g., by use of a wire mesh grid to elevate the animal or suspension of the dry ice in a compartment above the animal's head). Also, while it is true that the inflow of carbon dioxide into the chamber cannot be precisely regulated, the slow buildup of carbon dioxide that results from the use of dry ice may actually provide a more humane death than that which would result from the rapid introduction of high concentrations of carbon dioxide from compressed gas cylinders (see 17:3).

Surv. Does your IACUC consider dry ice to be a satisfactory method of generating carbon dioxide for euthanasia?

- Not applicable as we do not use carbon dioxide for euthanasia   27/161
- Satisfactory method, without any further IACUC-defined
  conditions   3/161
- Satisfactory method, with specific conditions defined by
  the IACUC   10/161
- Not usually permissible   94/161
- Other   27/161

17:5   For euthanasia with carbon dioxide, how many mice can be placed in the chamber at one time?

Opin. There are no regulatory guidelines regarding the number of mice that can be placed in the chamber beyond the admonition to prevent overcrowding.[1,25] For practical purposes, the density in the chamber must not exceed the point where the carbon dioxide (CO_2) can circulate freely and all animals will be equally exposed. Also, the density should be consistent with requirements for animal welfare. In the author's opinion, the density within the chamber should not exceed the point where every animal can stand with all four feet on the floor. It is preferable to load the animals into the chamber immediately prior to administration of euthanasia. However, if it will be necessary to hold animals in the chamber for any interval prior to euthanasia, density should be maintained below the point where the temperature, humidity, and levels of oxygen and carbon dioxide within the chamber would be adversely affected.

Surv. If your IACUC allows the use of carbon dioxide as a euthanasia agent for mice, what policies are established relative to the number of mice that can be put into a chamber at one time?

- Not applicable as we do not use mice in our work or
  we do not use carbon dioxide for euthanasia   25/164
- We have no policy or guidelines   30/164
- We allow euthanasia of only one animal at a time   11/164
- Our policy follows the space recommendations of the *Guide*
  relative to the size of the chamber versus the number of
  animals allowed in the chamber   50/164

- We allow euthanasia of up to five mice at a time in the
  same chamber   19/164
- We allow euthanasia of more than five mice at a time
  in the same chamber   16/164
- Other   13/164

17:6 Is carbon dioxide an appropriate agent for euthanasia of neonatal rodents?

Reg. The AVMA Panel on Euthanasia[1] recommends that inhalant agents, including carbon dioxide (CO_2), not be used as the sole method of euthanasia in animals less than 16 weeks of age, and the Canadian Council on Animal Care recommends that neonatal rodents euthanized with CO_2 be left in the chamber for a full half hour after all movements have ceased.[2]

Opin. Carbon dioxide is generally viewed as less than optimal for euthanasia of newborn animals of any species. Because they have been adapted to the comparatively low-oxygen, high-CO_2 environment of the fetus, neonates are typically more resistant to hypoxia and more capable of coping with low-oxygen, high-CO_2 environments than are older animals. [1-3,26] For this reason, it has been stated that CO_2 should not be used to euthanize newborn animals.[3] Other sources caution that while CO_2 can be used for euthanizing neonates, the concentration should be especially high and/or the exposure time should be prolonged.[1,2,27-29] Two recent studies examined the time required for neonatal mice of different ages to die after exposure to 100% CO_2. Klaunberg and associates found that neonates aged 1–14 days experienced cardiac arrest within an average of 4–5 minutes after initial exposure to CO_2.[28] However, since the youngest neonates (1–7 days of age) were sometimes able to recover if removed from the chamber after cardiac arrest, it was recommended that they be left in the chamber for several minutes after apparent death. Pritchett and coworkers found that time to death after initial exposure to CO_2 decreased from as long as 50 minutes on the day of birth to 5 minutes by 10 days of age.[29] On the basis of available data, it would be prudent to follow the recommendation that neonatal rodents euthanized with CO_2 be left in the chamber for a full half hour after all movements have ceased.[2]

Surv. 1 Does your IACUC consider carbon dioxide an appropriate agent for euthanasia of neonatal rodents? More than one response is possible.

- Not applicable   41/184
- Appropriate with prolonged exposure to carbon dioxide   47/184
- Appropriate when followed by physical method
  (e.g., decapitation)   69/184
- Permit its use only with scientific justification   12/184
- Not permitted for neonates under any circumstances   5/184
- Other   10/184

Surv. 2 If your IACUC allows the use of carbon dioxide for the euthanasia of mice, are there differing requirements for neonates and adults?

- Not applicable   38/165
- We have not addressed this issue   24/165

- We generally require a longer exposure time for neonates   67/165
- There are no differences in requirements   17/165
- Other   19/165

17:7   Is carbon dioxide an appropriate agent for euthanasia of rabbits?

Opin. Depending on the source, carbon dioxide is viewed as either a fully acceptable agent for euthanasia of rabbits[1,3,27,30] or as a less desirable, but still acceptable choice for this species.[2,27] Green[27] notes that it is a useful agent but causes apprehension in rabbits. In the author's opinion, carbon dioxide is an acceptable agent for rabbit euthanasia, but other methods are preferable. When rabbits are exposed to high concentrations of this gas, particularly at first, they tend to struggle and kick, and the potential for injuring themselves or an animal handler is significant. This is a common response of rabbits to any stimulus that causes pain or fear and is probably not indicative that rabbits experience more discomfort than other animals exposed to carbon dioxide. If scientific or other circumstances necessitate the use of carbon dioxide to euthanize a conscious rabbit, it is essential that the animal be properly restrained (e.g., in a commercial rabbit restrainer) to prevent it from kicking. The rabbit should never be placed loose in a euthanasia chamber, where the animal could injure itself if it were to panic during the procedure.

Surv. Does your IACUC consider carbon dioxide as an appropriate agent for rabbit euthanasia?

- Not applicable   68/160
- Fully acceptable   5/160
- Acceptable only with scientific justification   19/160
- Acceptable only if the rabbit anesthetized prior to exposure   7/160
- Not permitted   60/160
- Other   1/160

17:8   Which is preferable for euthanasia of adult mice, carbon dioxide or cervical dislocation?

Reg. The *Guide* (p. 66) states that in general, inhalant or noninhalant chemical agents are preferable to physical methods for euthanasia. On the other hand, the AVMA Panel report states that when performed properly by skilled individuals, physical methods may be more humane and practical than other methods of euthanasia.[1]

Opin. The reasoning behind the preceding recommendation in the *Guide* is not stated. There are many issues to consider when choosing a method of euthanasia (see 17:1), including the ability to ensure a rapid, humane death and possible adverse effects on scientific objectives. Many references document the adverse effects of carbon dioxide[11,30–35] on specific research goals, but under certain circumstances, cervical dislocation too could have confounding effects.[36] From a humane viewpoint, the case can be made that mice can be euthanized humanely or inhumanely by using either cervical dislocation or carbon dioxide. Carbon dioxide is generally viewed as an acceptable agent for euthanasia of adult mice.[1–3,8] There is evidence, however, that under certain conditions (e.g., high concentration and/or high flow rate), animals may experience pain or distress during carbon dioxide euthanasia (see 17:3).[3,10–12,16,17]

Cervical dislocation also is recognized as an acceptable method of euthanasia for adult mice by some,[3] but it is classified as "conditionally acceptable" in the 2000 AVMA Panel report[1] and is not listed as either a "most acceptable" or "acceptable" method by the Canadian Council on Animal Care.[2] There are two concerns associated with cervical dislocation:

The major concern is that, if it is performed incorrectly, cervical dislocation may result in painful injury rather than death.[1,2]

Electroencephalogram activity persists for up to 13 seconds after cervical dislocation[37] and has been interpreted as possible evidence that the animal may perceive pain for some period before death. However, other evidence, including data presented by the investigators who first documented the persistent brain activity, suggests that the animal should feel little or no pain if the technique is performed correctly.[2,3,37,38]

In summary, if the technical competence of the individual performing the euthanasia is questionable, the potential for pain or distress is greater with cervical dislocation than with carbon dioxide, and the latter is the preferable choice. However, if technical competence is not an issue, there is no clear basis for choosing one method over the other on either general scientific or humane grounds.

Surv.   Given the choice of carbon dioxide or cervical dislocation, which is generally preferred by your IACUC for euthanasia of adult mice?

• Not applicable	16/166
• I do not know	9/166
• Equally acceptable	18/166
• Carbon dioxide preferred	112/166
• Cervical dislocation preferred	5/166
• Other	6/166

17:9   What type of justification should the IACUC request if an investigator states that a rodent cannot receive any anesthetic or tranquilizer prior to decapitation or cervical dislocation?

Reg.   The 2000 AVMA report[1] states that personnel must be properly trained prior to euthanizing animals by decapitation or cervical dislocation and that untrained individuals should develop proficiency by practicing on dead or anesthetized animals. The need for a demonstrated high degree of technical proficiency is emphasized in the case of cervical dislocation, and the animal must be sedated or anesthetized prior to cervical dislocation if the technical proficiency of the operator is in question. Use of either cervical dislocation or decapitation must be justified on the basis of scientific requirements and must be approved by the IACUC. The Canadian Council on Animal Care recommends that "prior sedation or tranquilization should take place whenever possible" before use of any physical method.[2] However, in contrast to the 1993 report of the AVMA Panel,[39] the 2000 AVMA Panel report does not state that it is preferable for any animal to be sedated or unconscious prior to euthanasia using a physical method such as decapitation or cervical dislocation.

Opin.   As there are no current guidelines or regulations requiring sedation or anesthesia of an animal prior to decapitation by an experienced individual, it is the responsibility of the individual IACUC to determine whether, and under what circumstances, sedation or anesthesia might be mandatory in this situation. Performance of decapitation by an untrained individual is a different situation and both animal welfare and personnel safety would dictate against allowing an untrained person to perform decapitation on a fully conscious animal. A high level of technical proficiency is perhaps even more critical for assuring a humane death in the case of cervical dislocation. While the IACUC could allow a less technically competent individual to perform cervical dislocation or decapitation on a fully conscious rodent—provided that the investigator had justified the need for this deviation from the AVMA Panel recommendations in writing—it is difficult to imagine circumstances under which such a decision could be justified.

Surv.   What type of justification does your IACUC request if an investigator states that a rodent cannot receive any anesthetic or tranquilizer prior to decapitation or cervical dislocation?

• Not applicable	27/167
• General statement that anesthesia or tranquilization would (or might) interfere with the research	16/167
• Specific statement as to how the drugs will interfere with the research, but no references or data required	33/167
• Specific statement that includes one or more published references or laboratory data	87/167
• Other	4/167

17:10   If an investigator is granted IACUC approval to decapitate a rodent that will not be pretreated with an anesthetic or sedative, is the procedure considered to involve significant pain or distress without the use of drugs to alleviate them?

Reg.   There is no requirement under the PHS Policy or the AWAR to make a determination of "pain without anesthesia" for rodents. (See also 17:9.)

Opin.   The primary concern here is whether anesthesia or sedation is necessary to prevent pain or distress associated with decapitation. The AVMA Panel[1] notes that handling and restraint for this technique may be distressful to the animal and suggests that inexperienced personnel should practice any physical method on dead or anesthetized animals until they have gained sufficient expertise to decapitate fully conscious animals in a humane manner.

The brain electrical activity that persists for 13 to 14 seconds after decapitation has been another source of concern, but there is a significant body of evidence that this activity does not indicate a conscious awareness of pain.[2,3] Rather, it appears that loss of consciousness quite rapidly follows decapitation.[37,38,40] Provided that decapitation is performed properly by a skilled individual, there is no reason to believe that there is any greater pain or distress associated with this technique than with any other physical euthanasia method.

Surv.   If an investigator is granted approval by your IACUC to decapitate a rodent that will *not* be pretreated with an anesthetic or sedative, is the procedure considered to involve significant pain or distress without the use of drugs to alleviate them?

• Not applicable	43/160
• We consider it to be momentary pain or distress	44/160
• We consider it to be unalleviated significant pain or distress	9/160

- We consider it to be momentary pain or distress only if personnel performing procedure are skilled and the animal is accustomed to handling and restraint 60/160
- Other 4/160

17:11 Is sedation (e.g., with diazepam or acepromazine) an acceptable alternative to anesthesia for pretreatment of animals prior to decapitation or cervical dislocation?

Reg. (See 17:9.)

Opin. The AVMA Panel[1] notes that animals and personnel may be injured during the performance of any physical euthanasia procedure if the procedure is not performed correctly. As a result, they recommend that inexperienced personnel practice on dead or anesthetized animals until they have developed proficiency. Clearly, the goal is to render the animal completely unconscious during the practice period, and sedation will not have this effect. The AVMA Panel does not state that animals should be sedated or anesthetized prior to decapitation or cervical dislocation by an experienced operator. Most of the available data suggest that if these procedures are performed correctly, neither cervical dislocation nor decapitation is likely to cause significant pain (see 17:8; 17:10).[2,3,37,38,40] Therefore, there is no reason to view anesthesia as essential prior to decapitation or cervical dislocation performed by a skilled operator. Either sedation or no drug pretreatment would be acceptable in this situation. To protect both animal welfare and human safety, it is preferable to anesthetize any animal fully prior to decapitation by an unskilled operator. The AVMA Panel views cervical dislocation somewhat differently in that it recommends either anesthesia or sedation prior to the performance of this procedure by an operator who has not demonstrated proficiency in the technique.

Surv. Does your IACUC consider sedation (e.g., with diazepam or acepromazine) an acceptable alternative to anesthesia for pretreatment of animals prior to decapitation or cervical dislocation?

- Not applicable 49/161
- We have no policy on this matter 56/161
- Sedation is an acceptable alternative 27/161
- Sedation is acceptable, but only with scientific justification 11/161
- Sedation is not an acceptable alternative 17/161
- Other 1/161

17:12 Must a moribund mouse be anesthetized prior to euthanasia by decapitation or cervical dislocation?

Opin. Anesthesia is not required prior to decapitation or cervical dislocation of any animal provided that the person performing the procedure is proficient in the technique (see 17:9). However, the AVMA Panel[1] recommends that inexperienced personnel practice physical euthanasia procedures only on dead animals or animals that have been anesthetized, although the Panel also states that sedation is an acceptable alternative to anesthesia for cervical dislocation by an individual who has not demonstrated technical competency in the procedure. The concern is for both human safety, especially with decapitation, and animal welfare. Could an inexperienced operator perform these techniques as safely and humanely on a moribund mouse as on a dead or anesthetized mouse? The answer depends on the definition of *moribund*. One dictionary defines *moribund* as "being in the state of dying; approaching death,"[41] while another defines it as "in a dying state."[42] Neither of these definitions implies anything about the ability of the moribund animal to struggle—thereby posing a safety issue during decapitation by an untrained individual—or to experience pain or distress. In fact, in a clinical setting, the term could appropriately be applied to a fully conscious and responsive dying animal as well as an unconscious and unresponsive dying animal. It is best to assume that the moribund mouse would be capable of experiencing pain and distress, and to assign euthanasia by decapitation or cervical dislocation to a skilled operator. If for any reason the procedure *must* be performed by an inexperienced individual, the mouse should be anesthetized beforehand. (See 17:14.)

Surv. Does your IACUC require a moribund mouse to be anesthetized prior to euthanasia by decapitation or cervical dislocation?

• Not applicable	33/162
• The issue has never arisen	48/162
• We do not require anesthesia	23/162
• We require scientific justification for not using anesthesia	22/162
• We always require anesthesia	22/162
• We require anesthesia only if animal is still conscious	11/162
• Other	3/162

17:13 Is sedation (e.g., with diazepam or acepromazine) an acceptable alternative to anesthesia for pretreatment of animals prior to exsanguination?

Reg. The 2000 AVMA Panel on Euthanasia[1] states that exsanguination can be used to ensure death in animals that have been stunned or are otherwise unconscious. However, the Panel also states that exsanguination can be used to obtain blood products from animals that have been sedated, stunned, or anesthetized. PHS Policy (IV,c,1,g) and APHIS/AC Policy 3[9] require the IACUC to assure that proposed methods of euthanasia are consistent with the AVMA Panel report. PHS Policy further states that any deviation from the AVMA Panel recommendations must be "justified for scientific reasons in writing by the investigator."

Opin. As with decapitation and cervical dislocation (see 17:11), anesthesia is preferred to sedation only if it is essential for the animal be unconscious prior to the procedure. There is no reason to believe that there is significant pain associated with competently performed exsanguination through a cannula placed in a peripheral vein. However, there is concern that the extreme hypovolemia experienced by the exsanguinating animal prior to death is distressing.[1,3] Furthermore, there is disagreement on whether this distress can be adequately managed by using only a sedative. The Canadian Council on Animal Care states that "exsanguination is only acceptable as a euthanasia procedure if the animal is first rendered unconscious."[2] While the AVMA and Canadian guidelines differ on the use of sedation versus anesthesia prior to exsanguination through a peripheral vein, there is no question that an animal should be anesthetized prior to an exsanguination procedure that involves creation of a surgical incision.

Surv.   For your IACUC, is sedation (e.g., with diazepam or acepromazine) an acceptable alternative to anesthesia for pretreatment of animals prior to exsanguination?

- Not applicable   42/154
- We have no formal or informal policy   37/154
- Sedation is not an alternative to anesthesia   43/154
- We recommend anesthesia but permit sedative with
  scientific justification   25/154
- Sedation is fully acceptable   3/154
- Other   4/154

17:14 Must a moribund rabbit be anesthetized prior to euthanasia by exsanguination?

Opin.   In recognition of the anxiety that would presumably be experienced by a fully conscious animal experiencing extreme hypovolemia,[3] the 2000 AVMA Panel on Euthanasia[1] indicates that animals must be sedated, stunned, or anesthetized prior to collecting blood products by exsanguination. The Canadian Council on Animal Care[2] is less flexible in this regard, stating that exsanguination is acceptable as a method of euthanasia only if the animal is first rendered unconscious. The issue with a moribund rabbit is whether its physical condition would render it sufficiently insensitive to pain or distress that further sedation or anesthesia would be unnecessary. As discussed in 17:12, this cannot be determined only on the basis of the animals' being "moribund." If the dying animal is still fully conscious and responsive, exsanguinating it without prior sedation or anesthesia is not appropriate. Conversely, if the rabbit is already unconscious, the usefulness of a sedative or anesthetic is questionable, although some consideration should be given to the possibility of the animal's regaining consciousness during the procedure. The decision becomes more difficult if the rabbit is still conscious but exhibits diminished responsiveness. Such an animal could be considered to be in a mental state equivalent to "sedated" but not equivalent to "anesthetized." Further, even if an IACUC normally requires anesthesia of animals prior to exsanguination, there is an additional issue to consider in the case of a moribund animal. An animal near death might be unable to tolerate a general anesthetic and could very possibly die before or early in the exsanguination procedure, thereby severely limiting the amount of blood that could be collected. This could have a negative impact on the research if it is necessary to collect blood for use in the research project. In such a situation, the need to minimize distress in the rabbit might have to be balanced with the requirements of the research.

Surv.   Does your IACUC require that a moribund rabbit be anesthetized prior to euthanasia by exsanguination?

- Not applicable   67/164
- We have no formal or informal policy   28/164
- We require anesthesia   47/164
- We require anesthesia if the rabbit is still conscious   13/164
- We require anesthesia only if the procedure involves
  penetrating a body cavity, not for exsanguination from
  a peripheral vein   2/164

- We may not require anesthesia if anesthesia might result
  in animal's death prior to collection of the blood   5/164
- Other   2/164

17:15 Under what circumstances might an investigator be permitted to euthanize adult (large) rats or rabbits by cervical dislocation?

Opin.   It is generally acknowledged that it is physically difficult to perform manual cervical dislocation on larger, more heavily muscled animals. As a result, there is a high potential for severely injuring, but not killing, such an animal and this method is discouraged as a means of euthanizing larger rats (200 g or more) and rabbits (1 kg or more).[1-3] It is, therefore, especially important that scientific justification for such procedures be carefully reviewed by the IACUC. Nevertheless, an exception can be made if a person demonstrates proficiency in performing manual cervical dislocation on larger animals or if a mechanical dislocator is used.[1,2] As with cervical dislocation of smaller animals, it is important to assure that the person performing the procedure is properly trained and technically skilled, and that the use of this technique has been justified for scientific reasons by the investigator and approved by the IACUC.[1,2] Prior sedation or anesthesia of the animal should be given particularly careful consideration with larger animals, as the potential for improper performance of the technique, with resulting pain or distress, is much greater than with a mouse or small rat.

Surv.   Does your IACUC permit euthanasia of large rats (>200 g) or rabbits (>1 kg) by cervical dislocation? More than one response is possible. *Note:* Of the 160 people who responded to this question, 58 chose more than one response.

- Not applicable to our organization   33/160
- We have no policy   12/160
- We do not allow this form of euthanasia for large rats   88/160
- We allow this form of euthanasia for large rats
  by skilled persons   8/160
- We do not allow this form of euthanasia for large rabbits   55/160
- We allow this form of euthanasia for large (≥1 kg) rabbits
  by skilled persons   3/160
- We allow this euthanasia only if the rat or rabbit is first
  anesthetized   23/160
- Other   3/160

17:16 Is it acceptable to use scissors, rather than a guillotine, for decapitation of mice?

Opin.   The AVMA Panel[1] suggests the use of "sharp blades"—presumably scissors—for decapitation of neonatal rodents, but the use of scissors for decapitation of adult mice is not addressed by the Panel or in any of the U.S. regulations pertaining to euthanasia. The ILAR Committee on Pain and Distress in Laboratory Animals[3] states that decapitation of rodents may be accomplished by using a guillotine or heavy shears. In the author's opinion, humane euthanasia of adult mice using heavy, well-sharpened scissors is possible, provided that the operator is highly skilled in this technique. However, this procedure is exceedingly distasteful for most people, and this can lead all but the most resolute and experienced operator to perform the procedure less quickly and cleanly than is required for humane euthanasia. For this reason, use of scissors for decapitation is best limited to euthanasia of neonates or adult mice that have been heavily sedated or anesthetized. People who wish to use this approach to euthanize fully conscious adult mice should be especially skilled and experienced.

Surv. 1 Does your IACUC approve decapitation with a scissors as an acceptable method of euthanasia for adult mice?

• Not applicable	37/160
• Yes, if the animal is sedated or anesthetized	13/160
• Yes, even if the animal is not sedated or anesthetized	6/160
• No	100/160
• Other	4/160

Surv. 2 Does your IACUC approve decapitation with a scissors as an acceptable method of euthanasia for neonatal (1–2 days of age) rats and mice?

• Not applicable	47/155
• Yes, if the animal is sedated or anesthetized	44/155
• Yes, even if the animal is not sedated or anesthetized	31/155
• No	26/155
• Other	7/155

17:17 When reviewing a protocol that involves euthanasia of mice by cervical dislocation or decapitation, how should the IACUC determine that personnel performing these techniques are properly trained?

Reg. PHS Policy (IV,C,1,f), the AWAR (§2.31,d,1,viii), the *Guide* (p. 66), and the 2000 AVMA Panel report[1] emphasize that the personnel performing procedures on animals must be properly trained. In all instances, euthanasia procedures are explicitly or implicitly included in these requirements. (See Chapter 21.)

Opin. Although PHS Policy (IV,C,1,f) and the AWAR (§2.31,d,1,viii) specify that the IACUC is responsible for assuring that training requirements are met, neither of these documents defines who should provide the training or how the IACUC should determine that a particular individual has been properly trained. The 2000 AVMA Panel on Euthanasia[1] and Canadian Council on Animal Care[2] indicate that inexperienced persons should be trained and closely supervised by experienced individuals. The ILAR Committee on Pain and Distress in Laboratory Animals[3] is more specific in its recommendations, stating that the IACUC and AV should provide for the training and supervision of personnel who will perform euthanasia. This approach provides the IACUC with the highest degree of certainty that personnel are properly trained. Nevertheless, many committees and veterinarians do not have the resources to provide this kind of training for all individuals who require it. An alternative approach is to have the AV, institutional trainer, or an experienced member of the IACUC directly observe the skills of personnel who will perform euthanasia and make the determination that they either are proficient in the technique or require further training.

Surv.   When reviewing a protocol that involves euthanasia of mice by cervical dislocation or decapitation, how does your IACUC determine that personnel who will be performing these techniques are properly trained?

- Not applicable   30/168
- People must complete a training program or demonstrate skill to the satisfaction of the attending veterinarian or that person's designee   73/168
- We rely on the PI to provide assurance that personnel are properly trained   57/168
- Other   8/168

17:18   An investigator states that she has many years of experience euthanizing rats by stunning. She performs this by swinging the rat rapidly by the tail and hitting its head against the edge of a table. Should the IACUC accept stunning, if properly performed, as an acceptable form of euthanasia?

Opin.   Both the AVMA Panel on Euthanasia[1] and the Canadian Council on Animal Care[2] view stunning as unacceptable as the sole method of euthanasia. The concern is that although stunning may be an effective means for rendering an animal unconscious, it cannot be relied upon to cause death. It should, therefore, be followed by another technique that will cause death.

     Stunning by swinging the animal and hitting the back of its head on the edge of a table is an old method that is seldom addressed in the more recent literature. However, this technique has been described as a humane method of euthanasia or as a method of rendering animals unconscious prior to exsanguination or cervical dislocation in small animals.[12,27] Waynforth and Flecknell[12] emphasize that the procedure must be performed correctly and that training should involve deeply anesthetized animals. The need for training of personnel who stun animals by other methods is similarly emphasized by the AVMA Panel on Euthanasia and the Canadian Council on Animal Care.[1,2] Stunning by hitting the animal's head on the edge of a table requires a high level of concentration in addition to considerable technical proficiency. A small deviation in the force or angle of the blow can result in severe injury rather than instant death.

Surv.   Does your IACUC accept stunning of rodents, if properly performed, as an acceptable form of euthanasia? More than one answer is possible.

- We have no policy on stunning   66/171
- Not acceptable under any circumstances   76/171
- Acceptable only with scientific justification   6/171
- Acceptable with scientific justification and demonstration of proficiency   10/171
- Acceptable with IACUC approval and use of a second method to assure death   8/171
- Other   5/171

17:19   Is tribromoethanol (Avertin) acceptable as a euthanasia agent for mice?

Opin.   Tribromoethanol (formerly available as Avertin) has been used for decades as a rodent anesthetic. Despite significant concerns about postoperative complications associated with the use of this drug, it has gained popularity in recent years as a short-acting anesthetic for various procedures used in the production of transgenic mice.[27,43–45] It is not widely used as an agent for euthanasia. While it does not specifically address the use of tribromoethanol, ILAR's Committee on Pain and Distress in Laboratory Animals[3] cautions that most pharmacologic agents available for restraint, immobilization, analgesia, and anesthesia are not recommended for euthanasia because of practical considerations and their potential to cause convulsions prior to death. In the author's experience, tribromoethanol can be used for this purpose, but it will not reliably cause death unless used in doses that are considerably higher than those used for anesthesia. This requires administration of either a very large volume of the standard anesthetic preparation or a smaller volume of a more concentrated solution. With either approach, time to death may be prolonged. Other than the fact that tribromoethanol is not a controlled substance, there is little reason to consider it as an agent for euthanasia over other injectable agents such as pentobarbital.

Surv.   Does your IACUC approve the use of tribromoethanol (Avertin) as a euthanasia agent for mice?

- Not applicable   22/162
- The issue has not arisen   82/162
- Yes, when used with a second means of euthanasia
  (so-called double kill)   14/162
- Yes, even if it is the only means of euthanasia   7/162
- No   30/162
- Other   7/162

17:20   Is ethanol acceptable as a euthanasia agent for mice?

Opin.   Ethanol has been advocated as an immersion anesthetic for amphibians and invertebrates, and it is known to cause fatalities in these species if used in high concentrations and/or after prolonged exposure.[27,46] It is not widely used in mice, but some studies have shown that mice can be euthanized with 50–100% ethanol administered IP.[47,48] Some discomfort was noted with 100% ethanol, but there was no apparent discomfort with concentrations of 50–70%.

Surv.   Does your IACUC approve the use of ethanol as a euthanasia agent for mice?

- Not applicable   21/164
- The issue has not arisen   64/164
- Yes, when used with a second means of euthanasia
  (so-called double kill)   4/164
- Yes, even if it is the only means of euthanasia   0/164
- No   74/164
- Other   1/164

17:21   Is thoracic compression an ethical form of euthanasia for small rodents?

Opin.   The American Society of Mammalogists indicates that thoracic compression is one of the most commonly used methods for euthanizing small mammals in the field because it is quick and causes little pain.[49] The AVMA Panel on Euthanasia[1] agrees that this method is quick and causes no apparent pain but expresses some reservations about the degree of distress that this procedure may cause. As a method of euthanasia, thoracic compression can be compared to the use of neuromuscular blocking agents. With both approaches, the animal is rendered unable to breathe, leading to hypoxia and death. Because of the intense fear that the animal presumably experiences during the comparatively prolonged period between the onset of paralysis and death, the use of neuromuscular blocking agents to euthanize unanesthetized animals is widely and emphatically condemned.[1,2,27,50] The AVMA Panel views thoracic compression as a conditionally acceptable technique for the euthanasia of small- to medium-sized birds in the field when other methods of euthanasia cannot be used. The panel does not consider this technique appropriate for large or diving birds or in a laboratory setting.

Surv.   Does your IACUC consider thoracic compression an ethical form of euthanasia for small rodents? More than one response is possible. *Note:* Of the 158 people who responded to this question, 27 chose more than one response.

- Not applicable   29/158
- We have no formal or informal policy   30/158
- The issue has never arisen   85/158
- Absolutely unacceptable   39/158
- Acceptable, but only with scientific justification   3/158
- Acceptable (with or without scientific justification)   0/158
- Other   3/158

17:22   Is nitrogen or argon or mixtures of carbon dioxide with nitrogen or argon appropriate as an agent for the euthanasia of rodents or rabbits?

Opin.   Nitrogen and argon are inert gases that are available commercially in compressed gas cylinders. A mixture of argon (70%) and carbon dioxide (30%) is also commercially available as a mixture used in welding. These agents have been used to euthanize animals of several species, including dogs,[51] rats,[23] mink,[21] swine,[52,53] turkeys,[54] and chickens.[15,24,55,56] Although disturbing reactions, including convulsions, are seen in some species after induction of unconsciousness, only one study reported signs of distress in conscious animals exposed to either agent. Hornett and Haynes[23] reported that rats invariably showed signs of distress when euthanized with mixtures of nitrogen and air, whereas Leach and associates[18] noted that rats and mice appeared to find 100% argon less aversive than either carbon dioxide or a mixture of carbon dioxide and argon. Webster and Fletcher[24] also observed that 100% argon appeared to be less aversive to chickens than mixtures of argon plus carbon dioxide. The 2000 AVMA Panel on Euthanasia[1] views nitrogen and argon as conditionally acceptable agents that can be used to euthanize animals of various species, including rodents and rabbits. Because of the potential to cause distress—unlike carbon dioxide, neither argon nor nitrogen has any direct effect on the brain that causes narcosis prior to the onset of hypoxia—the panel recommends using these agents only when animals have been heavily sedated or anesthetized. The panel further states that flow rate should be sufficient to rapidly reduce the oxygen concentration in the chamber to <2%.

Surv. Has your IACUC approved argon gas, nitrogen gas, or mixtures of nitrogen or argon with carbon dioxide for the euthanasia of rodents or rabbits?

- Not applicable as we do not euthanize rodents or rabbits   14/167
- The issue has not arisen   83/167
- Yes   1/167
- No   68/167
- Other   1/167

17:23 An investigator wishes to euthanize monkeys by perfusion with formalin fixative with deep anesthesia. Is this acceptable?

Opin. Perfusion of anesthetized animals with fixative is a fairly common procedure in research settings, but it is not addressed as a technique for euthanasia by the AVMA Panel,[1] the Canadian Council on Animal Care,[2] or any of the U.S. regulations pertaining to animal welfare. The purpose of the procedure is to use the circulatory system of the living animal to distribute fixative through tissues. Cardiac arrest occurs as the heart becomes perfused with fixative. In the author's opinion, this is an acceptable form of euthanasia provided that the animal is deeply anesthetized prior to starting the perfusion procedure.

Surv. 1 Does your IACUC consider perfusion under anesthesia as an acceptable form of euthanasia?

- Not applicable   39/156
- Yes, with no further caveats   55/156
- Yes, but it must be followed by a physical method to ensure the death of the animal   44/156
- No   8/156
- Other   10/156

Surv. 2 If your IACUC permits perfusion as a means of euthanasia, are there any anesthetics that are *not* considered appropriate for this procedure?

- Not applicable   46/154
- We accept the use of any general anesthetic approved by our veterinarian   105/154
- We do not accept the following anesthetics: isoflurane   1/154
- Other   2/154

17:24 What agents or techniques are most appropriate for euthanasia of neonatal altricial rodents (e.g., rats and mice)?

Opin. There are few published guidelines that specifically address euthanasia of neonatal animals. The few specific recommendations that do exist are based on the fact that newborn animals are typically more resistant to hypoxia and more capable of coping with high environmental carbon dioxide than are older animals.[1-3] For this reason, it is often recommended either that inhalant agents (e.g., inhalant anesthetics, carbon dioxide, carbon monoxide) not be used to euthanize neonates[3] or that these agents be used only when the prolonged exposure times necessary to ensure death can be employed.[1,2,27-29] Klaunberg and colleagues[28] found that carbon dioxide was more effective than halothane for euthanasia of neonatal mice up to 7 days of age, and that carbon dioxide produced more rapid euthanasia of older preweanlings than either halothane or intraperitoneal pentobarbital.

When choosing a method of euthanasia for a neonate, it should be kept in mind that neonates, including neonatal altricial rodents, appear to be at least as capable of perceiving pain as adult animals.[57-61]

In the author's opinion, the most humane method of euthanasia for neonatal mice or rats ≤ 10 days of age is rapid decapitation with heavy, sharp scissors. Prior sedation or anesthesia is not necessary if performed properly (see 17:9). This procedure is aesthetically upsetting to many people, so carbon dioxide asphyxiation may be a more desirable alternative in some instances. Prolonged exposure of the pups to the gas is essential, as is verification of death prior to disposal.

Surv.   What agents or techniques does your IACUC consider acceptable for euthanasia of neonatal altricial rodents (e.g., rats and mice)? More than one response is possible. *Note:* Of the 155 people who responded to this question, 97 chose more than one response.

- Not applicable   44/155
- Anesthetic (inhalant or injectable) overdose   84/155
- Decapitation under anesthesia (other than hypothermia)   63/155
- Decapitation, typically no anesthesia needed   35/155
- Carbon dioxide alone   35/155
- Carbon dioxide followed by a physical method to assure death   66/155
- Hypothermia alone   10/155
- Hypothermia followed by a physical method to assure death   44/155
- Cervical dislocation   31/155
- Other   1/155

17:25 What agents or techniques are most appropriate for euthanasia of prenatal animals?

Opin.   There are no specific guidelines published for euthanasia of prenatal animals. Klaunberg and colleagues[28] found that none of the following methods used to euthanize pregnant female mice was effective in causing cardiac arrest in their fetuses (gestation days 14–20) within 20 minutes: carbon dioxide asphyxiation, barbiturate overdose, halothane overdose, potassium chloride injection (in dams anesthetized with halothane or isoflurane), or cervical dislocation (dam anesthetized or not anesthestized). On the basis of this study, it is recommended that late-term fetuses be euthanized individually by using a physical method of euthanasia; decapitation with heavy, sharp scissors would be quick and effective (see 17:24).

Surv.   What agents or techniques does your IACUC approve for euthanasia of prenatal animals? More than one answer is possible.

- Not applicable   66/152
- Decapitation with presumed residual anesthetic effect
  from anesthesia of mother   38/152

- Decapitation alone   24/152
- Intraperitoneal barbiturate   28/152
- Hypothermia followed by a physical method to assure death   30/152
- Anesthetic overdose, with or without a following physical
  method to assure death   33/152
- The method varies with the age of the fetus   32/152
- Other   3/152

Note: Of the 152 people who responded to this question, 50 chose more than one response.

17:26 Must neonates or preterm fetuses be anesthetized prior to decapitation or cervical dislocation?

Opin. There are no published guidelines that specifically address the issue of decapitation or cervical dislocation of neonates or preterm fetuses. In the absence of such specific guidelines and in recognition of the considerable data indicating that late-term fetuses and neonates are at least as capable of experiencing pain as are adults,[57-61] it is appropriate to follow the guidelines for euthanasia of adult animals by decapitation or cervical dislocation. These include use of an anesthetic or sedative only when the proficiency of the person performing the procedure is in question (see 8:11; 13:11; 14:19; 16:36; 17:9).[1]

Surv. Does your IACUC require that neonates or late gestation fetuses be anesthetized prior to euthanasia by decapitation or cervical dislocation?

- Not applicable or we have no policy   82/152
- We require some form of anesthesia for neonates only   14/152
- We require some form of anesthesia for late gestation fetuses
  and neonates   23/152
- We do not require some form of anesthesia for either   28/152
- Other   5/152

17:27 A mouse dam gives birth to a large litter. Two of the two pups never begin to breathe and two others breathe only when stimulated. A decision is made to abandon attempts to resuscitate these four animals. Must the deaths of these pups be ensured by some means?

Opin. It is generally recognized as essential that death be verified before disposal of an animal.[1-3,8] This can be accomplished by following up with another method designed to ensure death (e.g., pharmacologic agent, exsanguination, decapitation, or thoracotomy).[1,2] Alternatively, the animal can be examined by a person who is trained to recognize the cessation of vital signs in that species (e.g., absence of heartbeat, respiration, and reflex movements).[1-3,8] In the case of a neonatal mouse, follow-up by a physical method, such as decapitation with heavy sharp scissors (see 17:24), is preferable. Examination of such a tiny creature to verify cessation of vital signs would involve a particularly advanced level of expertise, and the ability of neonates to survive prolonged periods of hypoxia allows for the potential for recovery even in an apparently lifeless pup.

Surv.   Would your IACUC require that the deaths of newborn pups that never breathed or that breathed only when stimulated be ensured by a physical means (e.g., decapitation)?

• Not applicable	48/151
• We would probably require ensuring death by a physical means if it were likely that such a scenario would occur, but we have no standing policy	59/151
• We require that the presumed death of an animal be followed by a physical means of death (e.g., decapitation), and this policy would hold for these animals	22/151
• We would not require a second means of ensuring the death of the animals	16/151
• Other	6/151

17:28   What guidelines are there to assist an IACUC in determining whether a terminal procedure should be classified as euthanasia or nonsurvival surgery?

Opin.   There are no explicit guidelines for setting IACUC policy on this issue, other than admonitions that euthanasia should involve "rapid unconsciousness" and subsequent death[1,2] (AWAR §1.1, Euthanasia). The maximal interval between unconsciousness and death required to meet these definitions is not specified. In determining whether a particular procedure should be classified as euthanasia or nonsurvival surgery, the following considerations can be taken into account:

- Length of the procedure: longer procedures are more likely to qualify as nonsurvival surgery

- Invasiveness of the procedure: more invasive procedures, particularly those that require complex surgical manipulations, are more likely to qualify as nonsurvival surgery

- Purpose of the procedure: if the primary purpose of the procedure is to kill the animal, it is more likely to qualify as euthanasia; it is more appropriately classified as nonsurvival surgery if the death of the animal is a secondary objective or merely a consequence of the primary objective (e.g., removal of a vital organ for *in vitro* or *ex vivo* study)

Surv. 1   If a rat is deeply anesthetized and then perfused with fixative, does your IACUC considered this euthanasia or nonsurvival surgery?

• Not applicable	36/156
• Euthanasia	59/156
• Nonsurvival surgery	39/156
• Nonsurvival surgery if thoracotomy is involved, euthanasia if a peripheral vein is used	19/156
• Other	3/156

Surv. 2 Viable rat fetuses are removed from an anesthetized dam. The fetuses are then anesthetized and the fetuses and dam are decapitated. Does your IACUC consider this nonsurvival surgery or euthanasia for the dam? Is this nonsurvival surgery or euthanasia for the pups?

- Not applicable   58/152
- Euthanasia for dam and pups   35/152
- Nonsurvival surgery for dam, euthanasia for pups   51/152
- Both nonsurvival surgery and euthanasia for dam and pups   6/152
- Other   2/152

17:29 Is pithing an acceptable method for euthanasia of conscious frogs? Does it require justification?

Reg. Pithing of conscious frogs is classified as a "conditionally acceptable" method of euthanasia in the AVMA Panel report[1] and as an "acceptable," but not "most acceptable" method by the Canadian Council on Animal Care.[2]

Opin. Although the ILAR Committee on Pain and Distress in Laboratory Animals[3] states that "double pithing is an effective method of killing some poikilotherms," it does not list this procedure among its "General Recommendations for Euthanasia" of frogs. There is general agreement that pithing should involve destruction of both the spinal cord and brain ("double pithing") or destruction of the spinal cord followed by decapitation).[1-3] Pithing requires considerable technical proficiency and, if the person performing the technique is not skilled, the animal could experience considerable pain and suffering. The AVMA Panel recommends anesthetic overdose as a more suitable method than pithing for the euthanasia of frogs.

Surv. Does your IACUC consider pithing an acceptable method for euthanasia of conscious frogs? Does it require justification?

- Not applicable   80/166
- We have no formal or informal policy and the issue has
  never arisen   25/166
- Acceptable without further justification   8/166
- Acceptable, but only with scientific justification   25/166
- Acceptable, but only if the animal is anesthetized   24/166
- Acceptable for some types of frogs but not others   1/166
- Other   3/166

17:30 How should death be ensured prior to disposal of animals killed by carbon dioxide, stunning, or pithing?

Opin. It is generally recognized as essential that death be verified before disposal of an animal.[1-3,8] This is particularly important when animals are euthanized by techniques that can cause loss of consciousness or cessation of visible respiration well before death occurs. In some instances, animals that are merely unconscious may regain consciousness after disposal. Stunning, pithing, and carbon dioxide asphyxiation are all recognized as methods of euthanasia that may render an animal unconscious long before it dies. Therefore, it is particularly important when using these techniques that disposal not take place until death has been confirmed. This can be done by following up with another method designed to ensure death, such as pharmacologic agent, exsanguination, decapitation, or thoracotomy.[1,2] Alternatively, the animal can be examined by a person who is trained to recognize the cessation of vital signs in that species, for example, absence of heartbeat, respiration, and reflex movements.[1-3,8] It is important to note that cessation of respiration alone is not a reliable indicator of death, as continued cardiac function, with the potential for recovery, may persist after visible respiratory movements have stopped.[2,3]

Surv. 1 Prior to disposal of an animal that was euthanized by carbon dioxide, stunning, or pithing, how does your IACUC ensure that the animal is dead?

• Not applicable	19/159
• We have no policy	32/159
• Requires physical method follow up only after carbon dioxide	38/159
• Requires physical method follow up only after stunning or pithing	2/159
• Requires a physical method follow up after stunning, pithing, or carbon dioxide	43/159
• Other	25/159

Surv. 2 If your IACUC requires a physical method of assuring that an animal is dead after the use of carbon dioxide, stunning, or pithing, which physical methods are commonly approved (more than one response is possible). *Note:* Of the 152 people who responded to this question, 77 chose more than one response.

• Not applicable	34/152
• Thoracotomy	59/152
• Decapitation	63/152
• Cervical dislocation	63/152
• Exsanguination	48/152
• Removal of a vital organ	32/152
• Careful observation to assure no heat-beat, respiration, movement, etc.	52/152
• Other	6/152

17:31 What methods of euthanasia are acceptable in the field? Are there any differences between what is acceptable in the field and what is acceptable in the laboratory?

Opin. Techniques of euthanasia that are preferred in the laboratory also may be applicable in the field. However, the 2000 AVMA Panel on Euthanasia,[1] APHIS/AC Policy 3,[9] and Canadian Council on Animal Care[2] all note that there are situations in which it may be necessary to approach euthanasia in the field differently than in the laboratory. Different methods may be mandated because of feasibility, human safety, or animal safety. In more extreme circumstances, safety concerns may even mandate against the application of principles of euthanasia and require killing.[1] One technique that is viewed as unacceptable in the laboratory but that may be "conditionally acceptable" (AVMA Panel) or "most acceptable" (Canadian Council on

Animal Care) in the field is gunshot.[1,2] Acceptable application of this method requires that the operator be an expert marksman who is familiar with the proper, safe, and legal use of firearms. A firearm appropriate for the situation must be used, and the bullet must penetrate the brain of the animal in a location that will cause rapid loss of consciousness and subsequent death.

Other techniques that are acceptable for use in the field but not the laboratory include thoracic compression (see 17:21) and kill traps. The AVMA Panel considers the use of live traps—with subsequent euthanasia of the animal by an acceptable method—preferable to use of kill traps but recognizes that kill traps may be necessary for the collection and killing of free-ranging small animals when other methods are impossible or have failed. Careful attention to considerations such as size and placement of the trap, trigger type and sensitivity, and type of bait can help to assure the most humane outcome and guard against trapping of nontarget species. In addition, traps should be checked at least daily and any live, wounded animals should be killed quickly and humanely.

Surv.   What methods of euthanasia does your IACUC accept for field studies with small rodents? Check all responses that apply. *Note:* Of the 157 people who responded to this question, 23 chose more than one response.

- Not applicable   116/157
- Overdose of a parenteral anesthetic such as sodium pentobarbital   25/157
- Overdose of a gas anesthetic such as isoflurane   17/157
- Cervical dislocation   18/157
- Lethal traps   6/157
- Thoracic compression   3/157
- Drowning   2/157
- Decapitation with scissors   5/157
- Decapitation with guillotine   8/157
- Other   7/157

17:32   Sometimes animals are unintentionally injured during field studies. Is the investigator responsible for euthanizing these animals?

Opin.   Except for situations in which live, wounded animals are found in kill traps (see 17:31), the issue of how to deal with animals that are injured during field studies is not specifically dealt with in the guidelines or regulations. However, the AVMA Panel[1] emphasizes that situations in the field "do not in any way reduce or minimize the ethical obligation of the responsible individual to reduce pain and distress to the greatest extent possible during the taking of an animal's life." In the authors' opinion, investigators have an ethical obligation to kill severely injured animals provided that this can be done humanely and without endangering human safety or the safety of other animals.

Surv.   Does your IACUC usually have a field study provision that injured animals (e.g., from investigator trapping) be euthanized? Check all appropriate answers.

- Not applicable   108/161
- We have no policy   17/161

- We usually require such a provision   33/161
- Other   3/161

17:33 Is natural death an acceptable end point in biomedical research studies?

Reg. PHS Policy (IV,C,1,c) states that "animals that would otherwise suffer severe or chronic pain or distress that cannot be relieved should be painlessly killed at the end of the procedure or, if appropriate, during the procedure." Similar wording is included in the AWAR (§2.31,d,1,v).

Opin. The issue of humane end points has received considerable attention in recent years. The problem can be summed up as follows. If animals are terminated before scientific objectives are achieved, the value of the study is compromised and animal lives—not to mention time and resources—are wasted or used to poor end. However, if animals are allowed to suffer beyond the point where scientific objectives have been achieved, their suffering is unnecessary and the situation must be defined as inhumane. The ideal solution is to identify a point during the study when intervention can reduce pain or distress without adversely affecting scientific goals. Unfortunately, that may be easier said than done in situations in which experimental objectives require knowing with some degree of certainty that an animal would die in the absence of intervention. While many signs have been recognized as indicators of impending death, signs indicative of inevitable progression to death differ in different models, and the same signs that may predict impending death in one study may be seen in animals that will live for several more days or even eventually recover in another study.[57-65] For example, while hypothermia has proved to be an excellent predictor of impending death in several mouse infectious disease models, the degree of decline in body temperature that signals impending death is different in different models.[66-70] To prevent interference with scientific objectives, end points must be chosen with care, often based on data pertinent to the particular study in question. This may mean allowing animals to die a natural death (or at least progress to the moribund condition) as data are gathered that may identify an appropriate point when early euthanasia in later studies can enhance animal welfare without compromising scientific objectives. (See 17:34.)

Surv. If justification is provided, is "death as an end point" (i.e., not caused by euthanasia) potentially acceptable to your IACUC?

- Yes, potentially acceptable   93/163
- No, we do not accept death as an end point   49/163
- Other   21/163

Note: Many of those who chose "Other" as a response provided additional comments. The majority of these comments fell into two general categories: the issue has never arisen or death as an end point may be acceptable but only under exceptional circumstances (e.g., particularly outstanding justification or particularly extensive monitoring).

17:34 Is the moribund condition an acceptable end point in biomedical research studies?

Opin. As noted in 17:33, it should be a goal in any study to minimize animal suffering without compromising scientific objectives. For some studies this can be a challenge, as scientific objectives may be met only when it is clear that an animal will die or, in some cases, recover. In an effort to avoid "death as an end point" in such studies, the investigator or IACUC may identify the moribund condition as an appropriate point for euthanasia. The issue is whether intervening at this point is any more humane than allowing the animal to die a natural death. One dictionary defines *moribund* as "being in the state of dying; approaching death."[41] Another defines it as "in a dying state."[42] From a practical viewpoint, the term might be equally well applied to a fully conscious and responsive dying animal as to an unconscious and unresponsive dying animal. In the first case, intervention in the form of euthanasia would likely spare the dying animal additional suffering, whereas in the second case, it is unlikely that the animal would experience significant further suffering if allowed to progress to a natural death. ILAR's Committee on Pain and Distress in Laboratory Animals[3] cautions that unless euthanasia would interfere with experimental goals, animals should be killed before they become moribund.

Surv.   Is the moribund condition a potentially acceptable study end point for your IACUC?

- Yes, potentially acceptable with scientific justification   104/154
- Yes, potentially acceptable even without scientific justification   10/154
- No, we do not accept the moribund condition as a potentially acceptable study end point   32/154
- Other   8/154

17:35   In situations in which neither natural death nor the moribund condition is an acceptable end point, what clinical signs can be used to determine the appropriate time for euthanasia?

Reg.   (See 17:1 and 17:33.)

Opin.   Clinical signs that can be used to determine the appropriate time for euthanasia fall into two somewhat overlapping categories: (1) signs that indicate potential suffering in an animal in which effective treatments are contraindicated or have failed and (2) signs that are likely predictors of impending death. In general, the signs in the first category precede those in the second category and can be used to identify the appropriate time for euthanasia in studies in which it is not necessary to know precisely when an animal would die in the absence of intervention. For studies whose experimental objectives are met only when the likely time of death can be pinpointed with a fair degree of accuracy, only signs in the second category can be utilized without interfering with scientific goals (see 17:33).

    The following signs have been demonstrated to be useful predictors of impending death in rodents (most of the published information on humane end points is from rodent studies):[62–70]

- Weight loss
- Hypothermia
- Lethargy
- Ataxia
- Ruffled fur
- Hunched posture

- Labored breathing/cyanosis
- Abdominal or facial swelling
- Loss of the righting reflex
- Inability to ambulate
- Hind limb paralysis

Note, however, that the predictive value of these signs varies with the model (see 17:33). In general, progressive, sustained changes (e.g., in weight, body temperature, activity level) are better indicators of impending death than are acute changes, even in situations in which acute changes are quite substantial.

In situations when it is *not* essential to know when death will occur, and the goal is simply to prevent suffering that cannot be alleviated by other means, the following clinical signs may be used to identify an appropriate time for euthanasia:

- Species-specific signs of chronic pain (see Table 17.1)
- Rapid or excessive tumor growth (generally defined in terms of tumor size or weight)
- Ulceration of tumors or other tissues
- Chronically prolapsed organs
- Intractable diarrhea
- Abnormal discharge from any body orifice
- Persistent bleeding
- Large or deep skin ulcerations
- Persistent self-trauma
- Severe muscle wasting
- Lameness, paralysis, or other CNS signs that interfere with eating, drinking, or other normal activities
- Wasting
- Lethargy
- Failure to groom
- Abdominal distention
- Anorexia
- Persistent respiratory signs (wheezing, coughing, labored breathing, nasal discharge)
- Hypothermia

Surv. If a prescheduled time of euthanasia, "death as an end point," or the moribund condition is not an acceptable end point for a particular study, what clinical signs does your IACUC accept to determine the appropriate time for euthanasia? Check all appropriate responses.

- We do not euthanize any of our study animals	1/150
- If observable tumors are present, ulceration or defined size limits are used	112/150
- Weight loss, based on observations	63/150

TABLE 17.1

Species-Typical Signs of Severe or Chronic Pain in Laboratory Animals

Species	Signs of Severe or Chronic Pain/Distress
Mouse	Weight loss; dehydration; incontinence; soiled hair coat; eyes sunken, lids closed; wasting of muscles on back; sunken or distended abdomen; decreased vibrissal movements; unresponsive; separates from group; hunched posture; ataxia; circling; hypothermia; decreased vocalization
Rat	Eyes closed; poor skin tone; muscle wasting along back; dehydration; weight loss; incontinence; soiled hair coat; depressed/unresponsive; sunken or distended abdomen; self-mutilation; recumbent position with head tucked into abdomen; decreased vocalization; hypothermia
Syrian hamster	Loss of coat and body condition; increasing depression; extended daytime sleep periods; lateral recumbency; hypothermia; sores on lips, paws
Gerbil	Loss of weight and condition; sores on face; hair loss on tail
Guinea pig	Weight loss; hair loss; scaly skin; dehydration; decreased timidity; unresponsive; excessive salivation (oral problems); increased barbering; loss of righting reflex; decreased vocalization; hypothermia
Rabbit	Tooth grinding; apparent sleepiness; dehydration; weight loss; fecal staining; wasting of lower back muscles; decreased production of night feces; unresponsive
Nonhuman primate	Huddled or crouching posture, with hands folded over abdomen; clenching or grinding teeth; depression or increased restlessness; withdrawal from cage mates; increased (generally aggressive) attention from cage mates; anorexia; weight loss; decreased grooming
Dog	Unwillingness to move; crouching posture; depression or increased aggression; crying when handled or moved; increased restlessness
Cat	Hunched, crouching, or stretched posture; increased aggression, especially when palpated; anorexia; weight loss; vocalizing; wild escape behavior; unkempt appearance; pupillary dilation; stiff gait
Pig	Depression; unwillingness to move; attempts to hide; withdrawal from pen mates; anorexia
Sheep/goat	Rolling; frequently looking or kicking at abdomen; falling over; walking backward; rapid, shallow respirations; weight loss; tooth grinding; grunting; vocalization on handling (goats especially); rigidity; unwillingness to move
Bird	Eyelids partially closed; anorexia; ruffled, drooping, unkempt appearance; immobility when approached

From Danneman, P.J., Monitoring of analgesia, in *Anesthesia and Analgesia in Laboratory Animals*, Kohn, D.F., Wixson, S.K., White, W.J., and Benson, G.J., Eds., Academic Press, New York, 1997, p. 83. With permission.

- Weight loss, based on weighings   114/150
- Lack of activity or not using a limb(s)   104/150
- Ruffled fur   68/150
- Atypical vocalizations   64/150
- Difficulty breathing   110/150
- Lack of water consumption, as determined by dehydration   89/150
- Lack of water consumption, as determined by water
  measurement   39/150

- Infected wounds that are not responsive to veterinary treatment   96/150
- Progressively decreasing or increasing body temperature   67/150
- Atypical body discharges (ocular, nasal, rectal, or urinary)   88/150
- Other   20/150

Note: Of the 150 people who responded to this question, 133 chose more than one response.

17:36   If an animal is euthanized by a veterinarian for clinical reasons unrelated to the needs of the research, must the method of euthanasia be consistent with the method(s) approved during review of the protocol by the IACUC?

Opin.   There are no specific guidelines that address this question. In the author's opinion, the primary concern should be the welfare of the animal to be euthanized, although the objectives of the research should be taken into account also. If the method of euthanasia is critical to research objectives, it would be desirable to euthanize the animal in accordance with the approved protocol. However, if the method of euthanasia is not critical to research objectives, or if use of the method of euthanasia specified in the protocol is contraindicated by the animal's clinical condition, the method used should be the one that would provide the most humane death for the animal. In any case, the method chosen should be consistent with recommendations of the AVMA Panel on Euthanasia.[1]

Surv.   If an animal is to be euthanized by a veterinarian for clinical reasons unrelated to the needs of the research, does your IACUC require using the method(s) described in the approved protocol or may she or he use clinical judgment in the method to be used? More than one response is possible.

- We have not addressed this issue   38/181
- Veterinarians do not euthanize any of our animals   4/181
- The method specified in the protocol must be used unless the IACUC approves otherwise   22/181
- The veterinarian's clinical judgment can be used   109/181
- Other   8/181

17:37   Is it acceptable to euthanize animals in the presence of other animals of the same species?

Opin.   It has long been hypothesized that vocalizations, behaviors, or odors released by animals undergoing frightening or stressful experiences could cause distress in other animals, particularly animals of the same species. For this reason, it is widely recommended that other animals not be present during euthanasia, especially euthanasia of a conspecific.[1-3] There is some evidence that rats are aroused, and perhaps stressed, when exposed to conspecifics that are undergoing stressful procedures. De Laat and coworkers[71] observed that rats exposed to animals that had been stressed by transportation were more behaviorally aroused than rats exposed to animals that had not been transported. In a series of studies, Sharp and colleagues[4-7] observed that heart rate and/or blood pressure was elevated in rats that witnessed other rats' undergoing procedures such as decapitation, necropsy, routine cage change, restraint, vaginal lavage, or subcutaneous or tail vein injections. The animals also showed increases in heart rate and blood pressure when exposed to urine from stressed rats or dried rat blood. The magnitude of the cardiovascular changes observed in these studies was influenced by housing density and gender. In most cases, cardiovascular responses to witnessing decapitation were generally no greater than, and sometimes less than, the responses to witnessing procedures such as routine cage change. While these studies suggest that rats at least are stressed by witnessing a particularly violent euthanasia procedure, the evidence does not support the conclusion that witnessing euthanasia of conspecifics is any more stressful to other animals than is witnessing many other, more benign (at least from the human perspective) manipulations.

Surv. 1 Does your IACUC allow the euthanasia of animals in the presence of other animals of the same species?

• The issue has never been discussed	44/157
• We do not allow the euthanasia of animals in the presence of conspecifics	64/157
• We do allow the euthanasia of animals in the presence of conspecifics	24/157
• We do allow the euthanasia of animals in the presence of conspecifics, but only for these species (of the 14 people who chose this response, 6 listed more than one of the following species):	14/157
• Rats and mice	8/157
• Rodents	4/157
• Mice	2/157
• Amphibians	2/157
• Fish	2/157
• Rabbits	1/157
• Birds	1/157
• Other	11/157

Surv. 2 If your IACUC allows the euthanasia of animals in the presence of conspecifics, are there any practices intended to minimize fear or distress in the "witness" animals?

• Not applicable	97/150
• We do not have any special practices to minimize fear or distress in witness animals	28/150
• We use the following practices to minimize fear or distress in witness animals:	
• Visual ± auditory shields	5/150
• Animals to be euthanized taken to a different part of the room	4/150
• Animals to be euthanized taken to a different room	3/150
• Animals euthanized in cages	2/150
• "Witness" animals kept in cages	2/150

- Anesthesia of animal to be euthanized   1/150
- Highly trained personnel   1/150
- Not specified   2/150
- Other   5/150

Acknowledgments

The author would like to acknowledge with gratitude the assistance of Kathleen Schaefer-Keith and John Carr, both of whom spent long hours reviewing and recording survey responses for this chapter.

References

1. AVMA Panel on Euthanasia, 2000 Report of the AVMA Panel on Euthanasia, *J. Am. Vet. Med. Assoc.*, 218, 669, 2001.
2. Canadian Council on Animal Care, Euthanasia, in *Guide to the Care and Use of Experimental Animals*, Vol. 1, Olfert, E.D., Cross, B.M., and McWilliam, A.A., Eds., Canadian Council on Animal Care, Ottawa, 1993, 141.
3. National Research Council, Euthanasia, in Recognition and Alleviation of Pain and Distress in Laboratory Animals: A Report of the Institute of Laboratory Animal Resources Committee on Pain and Distress in Laboratory Animals, National Academy Press, Washington, D.C., 1992, p. 102.
4. Sharp, J., et al., Does witnessing experimental procedures produce stress in male rats? *Contemp. Topics Lab. Anim. Sci.*, 41, 8, 2002.
5. Sharp, J., et al., Are "by-stander" female Sprague-Dawley rats affected by experimental procedures? *Contemp. Topics Lab. Anim. Sci.*, 42, 19, 2003.
6. Sharp, J.L., Zammit, T.G., and Lawson, D.M., Stress-like responses to common procedures in rats: effect of the estrous cycle, *Contemp. Topics Lab. Anim. Sci.*, 41, 15, 2002.
7. Sharp, J.L., et al., Stress-like responses to common procedures in male rats housed alone or with other rats, *Contemp. Topics Lab. Anim. Sci.*, 41, 8, 2002.
8. National Research Council, Guide for the Care and Use of Laboratory Animals: A Report of the Institute of Laboratory Animal Resources Committee on Care and Use of Laboratory Animals. NIH Publication No. 86-23, U.S. Department of Health and Human Services, Washington, D.C., 1985, p. 65.
9. U.S. Department of Agriculture, Animal and Plant Health Inspection Service, Animal Care, Policy 3—Veterinary Care, April 14, 1997. Available on the World Wide Web at: http://www.aphis.usda.gov/ac/policy3.html.
10. Britt, D.P., The humaneness of carbon dioxide as an agent of euthanasia for laboratory rodents, in *Euthanasia of Unwanted, Injured, or Diseased Animals or for Educational or Scientific Purposes*, Universities Federation for Animal Welfare, Hertfordshire, U.K., 1986, p. 19.
11. Danneman, P.J., Stein, S., and Walshaw, S.O., Humane and practical implications of using carbon dioxide mixed with oxygen for anesthesia or euthanasia of rats, *Lab. Anim. Sci.*, 47, 376, 1997.
12. Waynforth, H.B., and Flecknell, P.A., Miscellaneous techniques, in *Experimental and Surgical Technique in the Rat*, Academic Press, London, 1992, p. 313.
13. Hewett, T.A., et al., A comparison of euthanasia methods in rats, using carbon dioxide in prefilled and fixed low rate filled chambers, *Lab. Anim. Sci.*, 43, 579, 1993.
14. Smith, W., and Harrap, S.B., Behavioural and cardiovascular responses of rats to euthanasia using carbon dioxide gas, *Lab. Anim.*, 31, 337, 1997.

15. Gerritzen, A., et al., On-farm euthanasia of broiler chickens: effects of different gas mixtures on behavior and brain activity, *Poult. Sci.*, 83, 1294, 2004.

16. Peppel, P., and Anton, F., Responses of rat medullary dorsal horn neurons following intranasal noxious chemical stimulation: effects of stimulus intensity, duration, and interstimulus interval, *J. Neurophysiol.*, 70, 2260, 1993.

17. Thurauf, N., et al., The mucosal potential elicited by noxious chemical stimuli with CO_2 in rats: is it a peripheral nociceptive event? *Neurosci. Lett.*, 128, 297, 1991.

18. Leach, M.C., et al., Aversion to gaseous euthanasia agents in rats and mice, *Comp. Med.*, 52, 249, 2002.

19. Steen, K.H., et al., Protons selectively induce lasting excitation and sensitization to mechanical stimulation of nociceptors in rat skin, *in vitro*, *J. Neurosci.*, 12, 86, 1992.

20. Thurauf, N., Ditterich, W., and Kobal, G., Different sensitivity of pain-related chemosensory potentials evoked by stimulation with CO_2, tooth pulp event–related potentials, and acoustic event–related potentials to the tranquilizer diazepam, *Br. J. Clin. Pharmacol.*, 38, 545, 1994.

21. Enggaard Hansen, N., Creutzberg, A., and Simonsen, H.B., Euthanasia of mink (*Mustela vison*) by means of carbon dioxide (CO_2), carbon monoxide (CO) and nitrogen (N_2), *Br. Vet. J.*, 147, 140, 1991.

22. Cooper, J., Mason, G., and Raj, M., Determination of the aversion of farmed mink (*Mustela vison*) to carbon dioxide, *Vet. Rec.*, 143, 359, 1998.

23. Hornett, T.D., and Haynes, A.P., Comparison of carbon dioxide/air mixture and nitrogen/air mixture for the euthanasia of rodents: design of a system for inhalation euthanasia, *Anim. Technol.*, 35, 93, 1984.

24. Webster, A.B., and Fletcher, D.L., Reactions of laying hens and broilers to different gases used for stunning poultry, *Poult. Sci.*, 80, 1371, 2001.

25. Shalev, M., OLAW clarifies PHS policy regarding use of carbon dioxide for euthanasia of small laboratory animals, *Lab Anim.* (NY), 31, 17, 2002.

26. Singer, D., Neonatal tolerance to hypoxia: a comparative-physiological approach, *Comp. Biochem. Physiol. A. Mol. Integr. Physiol.*, 123, 221, 1999.

27. Green, C.J., *Animal Anaesthesia*, Laboratory Animal Handbooks 8. Laboratory Animals Ltd., London, 1979, pp. 238, 240.

28. Klaunberg, B.A., et al., Euthanasia of mouse fetuses and neonates, *Contemp. Topics Lab. Anim. Sci.*, 43, 29, 2004.

29. Pritchett, K., et al., Euthanasia of neonatal mice with carbon dioxide, *Comp. Med.*, 55, 275, 2005.

30. Bivin, W.S., Basic biomethodology, in *The Biology of the Laboratory Rabbit*, Manning, P.J., Ringler, D.H., and Newcomer, C.E., Eds., Academic Press, San Diego, 1994, p. 72.

31. Fawell, J.K., Thomson, C., and Cooke, L., Respiratory artifact produced by carbon dioxide and pentobarbitone sodium euthanasia in rats, *Lab. Anim.*, 6, 321, 1972.

32. Howard, H.L., McLaughlin-Taylor, E., and Hill, R.L., The effect of mouse euthanasia technique on subsequent lymphocyte proliferation and cell mediated lympholysis assays, *Lab. Anim. Sci.*, 40, 510, 1990.

33. Pecaut, M.J., et al., Modification of immunologic and hematologic variables by method of CO_2 euthanasia, *Comp. Med.*, 50, 595, 2000.

34. Brooks, S.P., Lampi, B.J., and Bihun, C.G., The influence of euthanasia methods on rat liver metabolism, *Contemp. Topics Lab. Anim. Sci.*, 38, 19, 1999.

35. Malyapa, R.S., et al., DNA damage in rat brain cells after *in vivo* exposure to 2450 MHz electromagnetic radiation and various methods of euthanasia, *Radiat. Res.*, 149, 637, 1998.

36. Feldman, D.B., and Gupta, B.N., Histopathologic changes in laboratory animals resulting from various methods of euthanasia, *Lab. Anim. Sci.*, 26, 218, 1976.

37. Vanderwolf, C.H., et al., Neocortical and hippocampal electrical activity following decapitation in the rat, *Brain Res.*, 451, 340, 1988.

38. Derr, R.F., Pain perception in decapitated rat brain, *Life Sci.*, 49, 1399, 1991.

39. AVMA Panel on Euthanasia, 1993 Report of the AVMA Panel on Euthanasia, *J. Am. Vet. Med. Assoc.*, 202, 230, 1993.

40. Holson, R.R., Euthanasia by decapitation: evidence that this technique produces prompt, painless unconsciousness in laboratory rodents, *Neurotoxicol. Teratol.*, 14, 253, 1992.

41. *Merriam Webster's Collegiate Dictionary*, 10th Edition, Merriam-Webster, Springfield, MA, 1995.

42. *Dorland's Illustrated Medical Dictionary*, 26th Edition, W.B. Saunders, Philadelphia, 1981.

43. Wixson, S.K., and Smiler, K.L., Anesthesia and analgesia in rodents, in *Anesthesia and Analgesia in Laboratory Animals*, Kohn, D.F., Wixson, S.K., White, W.J., and Benson, G.J., Eds., Academic Press, New York, 1997, p. 165.

44. Papaioannou, V.E., and Fox, J.G., Use and efficacy of tribromoethanol anesthesia in the mouse, *Lab. Anim. Sci.*, 43, 189, 1993.

45. Zeller, W., et al., Adverse effects of tribromoethanol as used in the production of transgenic mice, *Lab. Anim.*, 32, 407, 1998.

46. Schaeffer, D.O., Anesthesia and analgesia in nontraditional laboratory animal species, in *Anesthesia and Analgesia in Laboratory Animals*, Kohn, D.F., Wixson, S.K., White, W.J., and Benson, G.J., Eds., Academic Press, New York, 1997, p. 337.

47. Lord, R., Jones, G.L., and Spencer, L., Ethanol euthanasia and its effect on the binding of antibody generated against an immunogenic peptide construct, *Res. Vet. Sci.*, 51, 164, 1991.

48. Lord, R., Use of ethanol for euthanasia of mice, *Aust. Vet. J.*, 66, 268, 1989.

49. Ad Hoc Committee on Acceptable Field Methods in Mammology, Acceptable field methods in mammology: preliminary guidelines approved by the American Society of Mammologists, *J. Mammol.*, 68, 1, 1987.

50. Breazile, J.E., and Kitchell, R.L., Euthanasia for laboratory animals, *Fed. Proc.*, 28, 1577, 1969.

51. Herin, R.A., Hall, P., and Fitch, J.W., Nitrogen inhalation as a method of euthanasia of dogs, *Am. J. Vet. Res.*, 39, 989, 1978.

52. Raj, A.B.M., and Gregory, N.G., Welfare implications of the gas stunning of pigs. 2. Stress of induction of anaesthesia, *Animal Welfare*, 5, 71, 1996.

53. Raj, A.B., Behaviour of pigs exposed to mixtures of gases and the time required to stun and kill them: welfare implications, *Vet. Rec.*, 144, 165, 1999.

54. Raj, M., and Gregory, N.G., An evaluation of humane gas stunning methods for turkeys, *Vet. Rec.*, 135(10), 222, 1994.

55. Raj, A.B.M., and Whittington, P.E., Euthanasia of day-old chicks with carbon dioxide and argon, *Vet. Rec.*, 136, 292, 1995.

56. Raj, A.B.M., Gregory, N.G., and Wotton, S.R., Changes in the somatosensory evoked potentials and spontaneous electroencephalogram of hens during stunning in argon-induced anoxia, *Br. J. Vet. Res.*, 147, 322, 1991.

57. Fitzgerald, M., Neurobiology of fetal and neonatal pain, in *Textbook of Pain*, Wall, P.D., and Melzack, R., Eds., Churchill Livingstone, London, 1994, p. 153.

58. Guy, E.R., and Abbott, F.V., The behavioral response to formalin in preweanling rats, *Pain*, 51, 81, 1992.

59. McLaughlin, C.R., and Dewey, W.L., A comparison of the antinociceptive effects of opioid agonists in neonatal and adult rats in phasic and tonic nociceptive tests, *Pharmacol. Biochem. Behav.*, 49, 1071, 1994.

60. Blass, E.M., Cramer, C.P., and Fanselow, M.S., The development of morphine-induced antinociception in neonatal rats: a comparison of forepaw, hindpaw, and tail retraction from a thermal stimulus, *Pharmacol. Biochem. Behav.*, 44, 643, 1993.

61. McLaughlin, C.R., et al., Tonic nociception in neonatal rats, *Pharmacol. Biochem. Behav.*, 36, 859, 1990.

62. Olfert, E.D., Godson, D., and Habermehl, M., Endpoints in infectious disease animal models, *ILAR J.*, 41, 99, 2000.

63. Krarup, A., et al., Evaluation of surrogate markers of impending death in the galactosamine-sensitized murine model of bacterial endotoxemia, *Lab. Anim. Sci.*, 49, 545, 1999.

64. Aldred, A.J., and Meckling-Gill, K.A., Determination of a humane endpoint in the L1210 model of murine leukemia, *Contemp. Topics Lab. Anim. Sci.*, 41, 24, 2002.

65. Anonymous, Guidelines for the welfare of animals in rodent protection tests: a report from the Rodent Protection Test Working Party, *Lab. Anim.*, 28, 13, 1994.

66. Stiles, B.G., et al., Correlation of temperature and toxicity in murine studies of staphylococcal enterotoxins and toxic shock syndrome toxin 1., *Infect. Immun.*, 67, 1521, 1999.

67. Wong, J.P., et al., Development of a murine hypothermia model for study of respiratory tract influenza virus infection, *Lab. Anim. Sci.*, 47, 143, 1997.

68. Soothill, J.S., Morton, D.B., and Ahmad, A., The HID50 (hypothermia-inducing dose 50): an alternative to the ld50 for measurement of bacterial virulence, *Int. J. Exp. Pathol.*, 73, 95, 1992.
69. Vlach, K.D., Boles, J.W., and Stiles, B.G., Telemetric evaluation of body temperature and physical activity as predictors of mortality in a murine model of staphylococcal enterotoxic shock, *Comp. Med.*, 50, 160, 2000.
70. Kort, W.J., et al., A microchip implant system as a method to determine body temperature of terminally ill rats and mice, *Lab. Anim.*, 32, 260, 1998.
71. de Laat, J.M., van Tintelen, G., and Beynen, A.C., Transportation of rats affects behaviour of non-transported rats in the absence of physical contact, *Z. Versuchstierkd.*, 32, 235, 1989 (preliminary communication).

18

Surgery

Lester L. Rolf, Jr.*

Introduction

The IACUC and AV have the responsibility to help assure humane care and use of animals on which surgery is performed and to help assure that individuals who perform that surgery are appropriately qualified and trained. This chapter addresses several major considerations of surgery using animals, such as anesthesia, aseptic technique, and the perioperative care associated with the conduct of survival and nonsurvival animal surgery in biomedical research and teaching environments. The first edition of this book included a survey that reviewed 12 IACUC protocol forms to identify those issues relevant to this chapter. The current edition includes results from a more extensive survey, which generated 160 responses. Where possible, results from these surveys have been combined; thus the number of respondents will be greater than 160 for some survey items.

18:1 What information regarding survival surgical procedures should be included by the investigator in the IACUC protocol? How detailed a description is necessary?

Reg.   There are numerous regulatory requirements that relate to survival surgical procedures. In the broadest sense, they revolve around adequate veterinary care (see Chapter 27; AWAR §2.33; PHS Policy IV,C,1,e). Specific areas of concern for the IACUC include the following:

- Use of appropriate anesthesia and analgesia (see Chapter 16)
- Use of aseptic technique (see 18:11–18:18)
- Occurrence of multiple survival surgeries (see 18:6–18:10)
- Qualifications of surgical personnel (see 18:19)
- Perioperative care (see 18:20–18:22)
- Whether major or minor surgery will be performed (see 18:3–18:5)
- Description of the procedures to be performed (AWAR (§2.31,d,2; §2.31,e,3); PHS Policy (IV,C,1; IV,D,1,c); compliance, as applicable, with the AWA required (PHS Policy, II; IV,C,1)

---

* The author thanks Kathleen L. Smiler for her contribution to this chapter. The author also thanks Elizabeth J. Dawe for her contribution to this chapter in the first edition of *The IACUC Handbook*.

Opin.   An investigator should provide sufficient surgical procedure detail in the IACUC protocol for the committee to confirm that acceptable surgical techniques are proposed and the AV can evaluate the perioperative care program pertaining to adequate veterinary care needed.[1] In the author's opinion, incision location, chronic instrumentation and implants (when applicable), and method of wound closure (including size and type of suture) should be included with the description of surgical procedures. In addition, the description of the method of anesthesia should include preanesthetic, anesthetic, and analgesic drugs and doses, routes, and frequency of administration. If neuromuscular blocking agents are to be used, the agent, dose, and administration route of the agonist and reversal agent should be listed. When neuromuscular blocking agents are used, it is appropriate to require the PI to document lack of pain response, including parameters such as heart rate, which are not affected by neuromuscular blockade, from the animal subjects during the procedure. This places the AV and IACUC in a much more enlightened position with regard to these study paradigms. (See 16:28.)

Surv.   What information should be included on IACUC protocol forms that have a separate section pertaining to survival surgical procedures? How detailed should the surgical procedure be? More than one response from each form was possible.

- Describe the surgical procedures in detail including the following:
    - Incision location and length   96/165
    - Organs involved   107/165
    - Instrumentation and implants   112/165
    - Type of suture material to be used   79/165
    - Method of skin or wound closure   94/165
    - Type of suture to be used for skin or wound closure   117/165
    - When (or if) sutures will be removed   97/160*
    - Major survival surgery is not performed   28/160*
- Regarding method of anesthesia during the surgical procedure:
    - List or include preanesthetic and anesthetic drugs, dose, route, frequency of administration   145/165
    - Provide supplemental dose of anesthetic drugs   1/5**
    - What is the duration of anesthesia?   108/165
    - Will neuromuscular blocking agents be administered?   95/165
    - Provide neuromuscular blocking agent dose and route of administration   91/165

     * Total surveyed is less than 165 because these two selections were not included in the survey used in the first edition of this book.

     ** This selection, identified by one of five respondents in the survey used in the first edition of this book, was not included as a selection in the second edition survey.

18:2   What information regarding *nonsurvival* surgical procedures should be included by the investigator in the IACUC protocol?

Opin.   As with survival surgical procedures (see 18:1), an investigator should provide sufficient detail in the IACUC protocol when describing nonsurvival surgical procedures

so that the committee can confirm that acceptable surgical techniques are proposed and the AV can evaluate the preoperative and intraoperative management
pertaining to adequate veterinary care. The method of anesthesia described
should include preanesthetic and anesthetic drugs, dosages, routes, and frequency
of administration. If neuromuscular blocking agents are to be used, the agent,
dose, and administration route of the agonist and reversal agent should be listed.
When neuromuscular blocking agents are used, it is appropriate to require the PI
to document lack of pain response, including parameters such as heart rate, which
are not affected by neuromuscular blockade, from the animal subjects during the
procedure. This places the AV program and IACUC in a much more enlightened
position with regard to these study paradigms. (See 16:28.) Some IACUCs choose
to be more restrictive than current regulations (i.e., *Guide*, p. 62; AWAR
§2.31,d,1,ix) with issues such as appropriate surgical attire, dedicated space
requirements, use of aseptic procedures, and use of expired medication and intravenous fluids. When this occurs, the IACUC should clearly define such requirements to the investigator.

18:3   How detailed a description is necessary in an IACUC protocol that contains *nonsurvival* surgical procedures?

Opin.   The description should be adequate to assure the IACUC that, in fact, normal procedures are not abbreviated and that conduct of the surgery will be done with the
same care and attention that is applied in survival procedures. These stipulations
are more easily met if the nonsurvival procedure is similar to survival procedures
performed as part of the same or other protocols.

   Some procedures require less detail than others, particularly those that do not
involve extensive intraoperative management. For example, a tissue harvest conducted alone or, perhaps, after some physiological measurements might only
require a "bare bones" surgical description, such as "Midline laparotomy is performed in surgical prepped and draped animals. The renal artery and veins are
ligated at the pelvis and the kidney extirpated. The animal is euthanized as
described in item. ..."

   More detailed description might be warranted when the intraoperative time
and care are likely to be substantial, such that failure to provide detailed attention
would more likely result in pain or distress to the animal. Examples include a series
of physiological measurements such as blood pressure, pressure volume curves of
the ventricular chambers, and transmural pressures, in which maintenance of vascular cannulas is involved. In such cases, the stability of the anesthetized animal is
required for reliable data. Also, because the potential for hemorrhage and the need
for volume replacement exist, the IACUC requires more detailed information.

   The method of anesthesia described should include preanesthetic and anesthetic drugs, dosages, routes, and frequency of administration. If neuromuscular
blocking agents are to be used, the drug, dosage, and route of administration
should be listed. (See 18:1; 18:2.)

Surv. 1 What information does your IACUC require in a protocol related to *nonsurvival* surgery?

   • Nonsurvival surgery is not performed   30/130
   • A bare-bones description of the surgical procedure   6/130
   • A description of the essential elements of the
     surgical procedure   63/130

- A detailed description of the surgical procedure   65/130
- Other   1/130

Surv. 2 What anesthesia information does your IACUC require for *nonsurvival* surgical procedures?

- Preanesthetic and anesthetic drugs   123/130
- The dose, route, frequency of administration of preanesthetic and anesthetic drugs   128/130
- The duration of anesthesia   96/130
- Whether neuromuscular blocking agents will be administered   88/130
- The dose and route of administration of neuromuscular blocking agents   85/130

18:4   Can the IACUC classify surgical procedures as minor versus major? By what criteria? How can this classification be useful to the IACUC when reviewing protocols?

Reg.   The AWAR (§1.1) define a *major operative procedure* as "any surgical intervention that penetrates and exposes a body cavity or any procedure that produces permanent impairment of physical or physiological functions." The *Guide* (p. 61) states that surgical procedures are categorized as major or minor. A *major survival surgery* is defined as surgery that "penetrates and exposes a body cavity or produces substantial impairment of physical or physiologic functions (such as laparotomy, thoracotomy, craniotomy, joint replacement, and limb amputation)." The *Guide* (p. 61) defines *minor survival surgery* as surgery that "does not expose a body cavity and causes little or no physical impairment (such as wound suturing; peripheral-vessel cannulation; such routine farm animal procedures as castration, dehorning, and repair of prolapses; and most procedures routinely done on an 'outpatient' basis in veterinary clinical practice)." PHS Policy (IV,A,1) requires institutions to use the *Guide* as a basis for their animal care and use program. (See 26:5.)

Opin. Classification of surgical procedures by an IACUC as major or minor is based on the preceding regulatory definitions. This classification is useful to an IACUC when reviewing a protocol because it is the basis for the IACUC's determination of the type of surgical facility and the degree of aseptic technique required (see 18:13). Major survival surgery on nonrodents must be performed only in facilities designed, operated, and maintained for that purpose. Minor survival surgery and all surgery on rodents do not require separate dedicated facilities; however, aseptic technique must be used (AWAR §2.31,d,1,ix; *Guide*, pp. 62, 78).[2] Further, for animals to be used in multiple major survival surgeries, their use must be scientifically justified in the protocol and approved by an IACUC (AWAR §2.31,d,1,x,A; *Guide*, p. 12). There is no regulatory limitation to multiple minor survival surgical procedures. Nevertheless, an IACUC should use professional judgment to limit the number of minor surgical procedures performed on an animal.[2]

18:5   Should laparoscopy be considered a major operative procedure (major survival surgery)?

Reg.   (See 18:4.) The AWAR (§1.1) define a *major operative procedure* as "any surgical intervention that penetrates and exposes a body cavity or any procedure that produces permanent impairment of physical or physiological functions." The *Guide* (p. 61) states that a major survival surgery is one that penetrates and exposes a body cavity or produces substantial impairment of physical or physiologic functions.

Opin.   The regulatory definition of major survival surgery is not presented as to the extent of entry or exposure of a specific body cavity. If the body cavity has been entered, a criterion has been met for defining the procedure as a major survival surgery. With laparoscopic procedures, there are usually at least two incision sites, one for the operative scope and one for the viewing scope. Depending upon the procedure, there may be more. The welfare of the animal patient should be the focus. While the laparoscopic technique minimizes trauma, it also places the animal at increased risk of hidden hemorrhage. Additionally, the repeated entry and withdrawal of the scopes in this technique and their awkward handling increase the likelihood of unnoticed contamination by surgeon or assistants. Further, laparoscopic equipment is more difficult to clean and sterilize, thereby increasing the risk of bacterial contamination, despite reduced wound size.

18:6   Under what circumstances can an individual animal be used in multiple survival surgical procedures?

Reg.   Although performing multiple major survival surgical procedures on an individual animal is discouraged, the *Guide* (p. 12) states that they may be permitted if scientifically justified by the investigator and approved by the IACUC. The AWAR (§2.31,d,1,x) and NIH/OLAW[3] permit multiple major survival operative procedures:

- When scientifically justified by the investigator in writing
- When needed as a routine veterinary procedure or to protect the health and well-being of an animal
- When other special circumstances are authorized by the Administrator, APHIS, USDA, on a case-by-case basis
- APHIS/AC Policy 14[4] further clarifies AWAR §2.31,d,1,x by stating that more than one major survival operative procedure is permitted on an individual animal:
  - When scientifically justified by the PI within one proposal and approved by the IACUC
  - When an animal that has an emergency major operative procedure required for proper veterinary care then may be used in a research proposal requiring a major survival operative procedure
  - When other circumstances are authorized by the Deputy Administrator (exemption requests are made to the appropriate Animal Care Regional Director, who forwards them to the Animal Care Assistant Deputy Administrator for review and recommendation to the Deputy Administrator; Annual IACUC evaluation of the exemption is required and must be included in the Annual Report (APHIS Form 7023) for consideration for renewal or continuation of the exemption)

Regulations that address multiple survival surgical procedures on an individual animal have exemption criteria where it is allowed. According to the AWA

(Section 13,a,3,D), exemptions are authorized by the Secretary of Agriculture. However, as specified by the AWAR (§2.31,d,1,x) and as noted by the NIH/OLAW,[3] requests are sent to the Administrator, APHIS, USDA. The Secretary of Agriculture has delegated this responsibility to the APHIS Administrator, who has further delegated this to the Deputy Administrator of Animal Care. According to APHIS/AC Policy 14,[4] requests are made to the appropriate Animal Care Regional Director, who forwards them to the Animal Care Assistant Deputy Administrator for review and recommendation to the Deputy Administrator.

Opin. Examples of special circumstances that may potentially justify performing multiple surgical procedures on an animal include procedures that are related components of a research project, involve conservation of scarce animal resources, or are needed for clinical reasons. The *Guide* (p. 12), APHIS/AC Policy 14,[4] and interpretation of PHS Policy[3] do not consider cost savings alone as an adequate justification for performing multiple major survival surgeries on a single animal. (See 18:9.)

18:7   An animal is surgically modified by the vendor to meet a research need (e.g., ovariectomy of a rat). Must that vendor have its own PHS Assurance statement if the animals are sold to an institution and used in a project supported by PHS funds? Should the IACUC obtain a copy of that Assurance statement? (See 14:34.)

Opin. Purchase of animals that have been surgically modified by a vendor and are available for general sale has not been formally addressed by NIH/OLAW. However, NIH/OLAW has published a response to a related scenario involving a commercial supplier that produces standard reagent antibodies by using its own resources and offers the antibodies for general sale as a catalog item. When antibodies are readily available and are considered off-the-shelf reagents, the supplier is not required to have an approved Animal Welfare Assurance on file with NIH/OLAW.[5] To apply the same principles, if an institution purchases an animal to be used in a PHS-supported study, and the animal has been surgically modified by the breeder in advance of sale, and the nonmodified animal is readily available as an "off-the-shelf" catalog item, then the breeder is not required to have his or her own Assurance on file with NIH/OLAW. On the other hand, if an investigator subgrants or subcontracts with a supplier to produce antibodies using antigens provided by or at the request of the investigator, the antibodies are considered "customized" and the supplier must file an Assurance with NIH/OLAW.[5] The institution must obtain the supplier's Assurance number by contacting the supplier or NIH/OLAW.[6] Again applying the same principles, if an investigator subcontracts with a breeder to modify an animal surgically specifically for use in a PHS-supported study, the breeder must file an Assurance with NIH/OLAW. In this author's opinion, the institution should obtain the breeder's Assurance number. Obtaining a copy of the Assurance statement would be a local decision made at the institutional level.

Another option is available to the vendor. The research facility that is requesting surgically altered animals, can, with approval of NIH/OLAW, include the vendor in its institutional Assurance. Only vendor animals on IACUC-approved protocols of the assuring institution would be covered. Further, the IACUC must inspect all of the vendor sites housing animals covered under their protocols and ensure that the person at the vendor's sites performing the proposed procedure is qualified to do so. This option becomes much less efficient and potentially more problematic as the number of institutions using this approach increases.

18:8   Should the IACUC permit a farm animal to recover from major surgery so that it can be sold for slaughter once it has fully recovered? (See 14:48.)

Opin.   The IACUC can permit or withhold permission for recovery from major surgery by citing applicable regulations. The AWAR (§2.31,d,1,ix) and the *Guide* (pp. 60–64) do not prohibit recovery from a first major surgery provided that all related requirements are met. However, PHS Policy (IV,C,1,a) states that "procedures with animals will avoid or minimize discomfort, distress, and pain to the animals, consistent with sound research design." This is similarly stated in the AWAR (§2.31,d,1,i). Assuming that the investigator's research or educational needs can be met with nonsurvival surgery, anesthesia without recovery supports the welfare interests of the animal by fully preventing potential pain and discomfort rather than *minimizing* potential postoperative pain with analgesics. If meat from the slaughtered animal was destined for human consumption, the IACUC has to assure that the animal received only FDA-approved medications and drug withdrawal periods were met. Ultimately, the IACUC's awareness of the public's expectations and perceptions regarding the use of animals in research and education should take precedence in making this decision. Use of animals in research is held to a higher ethical standard than many other uses. The potential negative public perception of returning farm animals to the food chain after use in research or education might be a basis for the IACUC's decision not to permit this practice.

18:9   Should the IACUC permit a hamster, ovariectomized as a customer service by the vendor, to undergo another major survival surgical procedure? For regulatory compliance purposes, should this be considered two major survival surgical procedures on the same animal?

Reg.   The AWAR (§2.31,d,x,A–§2.31,d,x,C) stipulate that no animal should be used in more than one survival major operative procedure unless justified for scientific reasons, required as part of the veterinary care of the animal, or otherwise exempted by the APHIS/AC Administrator. (See 18:6.) Further clarification is offered by APHIS/AC Policy 16,[6] which indicates that animals undergoing a major survival operative procedure by the vendor must be identified as such to prevent their use in another major survival operative procedure. The *Guide* (p. 12) also clearly identifies the need for scientific justification if more than one major survival surgery is to be performed on an animal. The *Agricultural Guide* (p. 22) is similarly specific with allowance given (1) when the surgeries are associated with the same protocol study and (2) can be scientifically justified. (See 18:18.)

Opin.   Ovariectomy, performed according to current veterinary practices, would have to be classified as a major surgery, since a body cavity has been entered, even when entered laparoscopically (see 18:5). Further, the removal of the ovaries would result in permanent physiological changes to the animal. On the surface it appears sophomoric to believe that the PI would order animals that were *not* needed for a particular study or in fact were destined to be used otherwise, perhaps without approval. For exceptions based on scientific merit, the onus of demonstrating scientific merit is on the PI and must be clearly enough connected to the scientific study to be approved by the IACUC.

18:10 A major surgical procedure is performed by a veterinarian for a spontaneous medical condition that is unexpected and unrelated to any ongoing research. Have conditions been established for prohibiting a subsequent survival major operative procedure unless scientifically justified and IACUC approval is given for "multiple major operative procedures from which the animal recovers"?

Opin.   This condition is specifically addressed by the AWAR (§2.31,d,1,x,B) and is clarified in APHIS/AC Policy 14,[4] which indicates that an animal that "has an emergency major operative procedure as part of proper veterinary care may still be used in a proposal that requires a major survival operative procedure." While not specifically stated, the assumption is that the emergency procedure was such that the procedure did not interfere with the potential use of the animal in a dedicated study. It is the responsibility of the AV and the PI to advise the IACUC on the medical and scientific suitability of these animals for further study. The *Guide* (p. 12) suggests that additional consideration should be given to the species availability or other unique features of the animal that might make it a valuable resource for additional studies involving major survival surgery.

18:11   If an investigator dedicates a part of a laboratory bench top for performing major survival surgery on a hamster, is this considered as being compliant with the AWAR?

Reg.   The AWAR (§1.1, Animal) specifically include hamsters as a regulated species. However, the AWAR (§2.31,d,1,ix) indicate that major operative procedures performed on rodents do not require dedicated facilities, though aseptic technique is required. The *Guide* (p. 78) indicates that dedicated space within a laboratory, managed in such a way as to minimize contamination from other activities in the room, is acceptable for rodent surgery.

Opin.   Conducting hamster survival surgical procedures on a laboratory bench is acceptable as long as there is no competing activity in the area at that time. Further, use of this area is compliant as long as conditions minimize contamination from other activities within the laboratory. The AWAR exception for rodents should not be interpreted to exempt the surgeon from the use of aseptic technique, nor from being appropriately garbed. Further, as it is a regulated species, adequate intraoperative and postoperative records of observations, analgesic or other medication administration, pain assessments, and notes on surgical healing should be maintained.

18:12   What is aseptic technique?

Reg.   The *Guide* (p. 62) states that aseptic technique includes:

- Preparation of the patient, with body hair removal and disinfection of the operative site
- Preparation of the surgeon, to include a surgical scrub of hands and arms and appropriate donning and wearing of surgical attire and sterile surgical gloves
- The use of sterilized instruments, supplies, and implanted materials
- The use of operative techniques to reduce the likelihood of infection

The *Guide* (p. 62) also states that "aseptic technique is used to reduce microbial contamination to the lowest possible practical level. No procedure, piece of equipment, or germicide alone can achieve that objective. Aseptic technique requires the understanding and cooperation of everyone who enters the operative suite. The contribution and importance of each practice varies with the procedure."

Opin.   By definition, *aseptic technique* is the performance of a surgical procedure in a manner to minimize exposure of the patient to pathogenic organisms.[7] (See 26:20.)

18:13   Is aseptic technique necessary for performing survival surgery in animals?

Reg.   The AWAR (§2.31,d,1,ix) require that all survival surgery will be performed on all regulated animals by using aseptic procedures, which include masks, caps, sterile surgical gowns, and gloves; sterile instruments; and aseptic technique. The *Guide* (p. 62) states that minor survival procedures may be "performed under less-stringent conditions than major procedures, but still require aseptic technique and instruments."

Opin.   An IACUC has a regulatory responsibility to assure use of acceptable standards of aseptic technique as stated. Semiannual inspections of areas where surgery is performed can help assure this standard. (See Chapter 26.) Performing survival surgery using aseptic technique also is necessary to achieve a satisfactory surgical outcome with reduced risk of infection. Aseptic technique is used to reduce microbial contamination of a surgical wound and exposed tissues to the lowest possible practical level (*Guide*, p. 62).[8] To use less than optimal aseptic technique potentially increases bacterial contamination and subsequent risk of infection, which can compromise an animal's health postoperatively. In addition, lack of clinical evidence of postoperative infection does not rule out clinically unapparent infection. A study performed in rats showed that infection can be clinically unapparent and yet cause adverse physiologic and behavioral responses that can affect research results and may not be recognized.[9]

18:14   Are the standards for aseptic technique different for rodents compared to other mammals?

Reg.   Standards for aseptic technique for rodents differ somewhat from those for nonrodents. The AWAR (§2.31,d,1,ix) and the *Guide* (p. 78) do not require a dedicated surgical facility for major survival rodent surgery as compared to nonrodent surgery. The AWAR (§2.31,d,1,ix) state that all survival surgery must be performed by aseptic procedures, including surgical gloves, masks, sterile instruments, and aseptic technique. The *Guide* (p. 63) identifies characteristics of rodent surgery, such as small incision sites, a one-person "surgical team," surgery performed on multiple animals at one sitting, and procedures of shorter duration, that make changes in standard aseptic techniques used in larger species necessary or desirable.[8,10]

Opin.   Modifications in standard aseptic techniques that are suitable for rodent surgery include use of one sterile instrument pack for up to five rodents incorporating techniques to maintain sterility between animals, and optional wearing of a mask with cap and sterile gown.[8,10] Although modifications in standard aseptic techniques may be necessary or desirable for rodents, the performance standards to prevent or minimize exposure of the patient to pathogenic organisms to reduce the likelihood of infection must be met and the well-being of the animals should not be compromised[10] (*Guide*, p. 61). For APHIS/AC-regulated rodents a mask is required.

A mask also seems required, as a legitimate component of the "surgical attire," if preparation of the surgeon is considered to include donning of a surgical mask. Nevertheless, use of a mask is not specifically mentioned in the *Guide* as part of the discussion of appropriate aseptic technique. (See 26:21.)

While the rationale for less stringent regulations governing rodent surgery appears obvious, there are reasons to be more restrictive, on an institutional basis, than required. Many laboratories work with both regulated and nonregulated species. The "lower-key" atmosphere of rodent surgery can spill over into surgical activities involving regulated species surgical work in these laboratories. If one is doing five sequential rabbit or guinea pig surgeries, five separate sterilized instrument packs will be required. It may not be unreasonable for an IACUC to have this same standard for rodents, even though it is not required by regulations. If a particular IACUC is more restrictive than the regulations, these additional requirements should be clearly defined so that the PI can perform studies in a compliant fashion.

18:15   Does aseptic technique need to be followed for nonsurvival surgery?

Reg.   APHIS/AC Policy 3[11] does not require aseptic technique or dedicated surgical facilities when performing nonsurvival surgery if the animal is not anesthetized long enough to show evidence of infection. The area to be used should be clean, free of clutter, and prepared by acceptable veterinary sanitation practices that would be used in a standard examination/treatment room. The *Guide* (p. 62) does not require using aseptic technique for nonsurvival surgery; however, "the surgical site should be clipped, the surgeon should wear gloves, and the instruments and surrounding area should be clean." If nonsurvival surgery is conducted in a dedicated "survival" surgery area, then aseptic technique should be used or the surgical room be returned to an appropriate level of cleanliness before it is used for major survival surgery (*Guide*, pp. 62–63).

Opin.   The extent to which an investigator should exceed the minimal regulatory requirements for nonsurvival surgery and apply any or all of the components of aseptic technique as described in 18:12 depends on the experimental protocol and the surgery being performed. Professional judgment is necessary to evaluate the probability of virulent bacterial contamination and subsequent host responses that would invalidate research results. Slattum and colleagues[12] reported a study demonstrating a need for aseptic technique when performing nonsurvival surgery. They found that gram-negative bacteremia and septic shock developed in dogs during nonsurvival cardiopulmonary studies performed without using aseptic technique. Laboratory-prepared nonsterile intravenous solutions were found to be contaminated with Gram-negative bacteria. Bacteremia and septic shock ceased to occur after initiating some components of aseptic technique such as using sterile commercial saline solution and other sterile intravenous injectables, disinfecting equipment and instruments, and using sterile gloves. (See 26:4.)

18:16   What level of aseptic technique should the IACUC require for field surgery involving wildlife?

Reg.   The AWAR (§2.31,d,1,ix) state that survival surgery conducted at field sites does not require dedicated facilities but must be performed using aseptic procedures, including surgical gloves, masks, sterile instruments, and aseptic technique. The *Institutional Animal Care and Use Committee Guidebook*[13] states that aseptic practices should be used when performing survival surgery on wildlife in field studies. The *Guide* (p. 61) recognizes that modification of standard aseptic and surgical techniques might be necessary when performing field surgery; however, animal well-being should not be compromised. When modifications are implemented, a thorough assessment of surgical outcomes should be performed to ensure that appropriate procedures are followed. Surgical outcome assessment may require other criteria in addition to clinical morbidity and mortality rates.

Opin.   An experimental animal surgical facility environment may be unnecessary when performing survival surgery on wildlife if it is not needed to improve animal well-being or surgical outcome as measured by lack of postoperative complications, improved surgical survival, or minimized pain and stress. Taking free-living wildlife to a dedicated facility to perform surgery could, in fact, be detrimental to the well-being and survival of the animals. Settings used in clinical veterinary practice may be suitable for field surgery involving wildlife.[2] Nevertheless, aseptic technique must be used to prevent or minimize exposure of the animal to pathogenic organisms. This includes preparation of the operative site (including hair removal and disinfection); surgical mask; sterile surgical gloves, instruments, supplies, and implanted materials; and use of operative techniques to reduce the likelihood of infection (*Guide*, p. 62). Steam sterilization or autoclaving is the preferred method of surgical instrument sterilization. However, chemical "sterilization" using liquid chemicals with appropriate contact time may be necessary in some field surgery settings. Ultimately, professional judgment must be used to optimize the circumstances for the environment where the surgery is performed and the aseptic techniques used.[2]

18:17   Would less than optimal aseptic technique be acceptable if a PI's records indicated excellent surgical success with a lack of postoperative infections?

Reg.   The AWAR (§2.31,d,1,ix) and the *Guide* (p. 62) require that all survival surgery on regulated species be performed by aseptic technique in accordance with professionally acceptable standards as described in 18:13 and defined in 18:12.

Opin.   No. In this author's opinion and that of another,[14] the performance of survival surgery using less than optimal aseptic technique indicates that although the IACUC and AV reviewed and approved the procedures, the review was not thorough enough or the investigator was inadequately trained in aseptic technique. The IACUC did not fulfill its responsibility to assure use of acceptable standards of aseptic technique.

   Aseptic technique is used to reduce microbial contamination of a surgical wound and exposed tissues to the lowest possible practical level[8] (*Guide*, p. 62). To use less than optimal aseptic technique potentially increases bacterial contamination and subsequent risk of infection which would compromise the animal's health. Lack of clinical evidence of postoperative infection does not rule out clinically unapparent infection. A study performed in rats showed that infection can be clinically unapparent and cause adverse physiologic and behavioral responses that can affect research results and may not be recognized.[9]

   Of course, it is hard to argue with success. Under some circumstances, less than optimal aseptic technique may be acceptable as long as the animal's health is not compromised and the procedures are approved by the IACUC. Although the IACUC should reconsider their approved aseptic techniques to meet current acceptable standards of proper veterinary care, the IACUC has chosen to focus on the product rather than the process.[14]

18:18   Should the IACUC demand that aseptic technique be used when surgery is performed on farm animals in the field (e.g., standing rumenotomy performed in a barn)?

Reg.   The AWAR (§2.31,d,1,ix), which apply to farm animals used in biomedical research, state that survival surgery conducted at field sites does not require dedicated facilities but must be performed using aseptic procedures, including masks, caps, sterile surgical gowns and gloves, sterile instruments, and aseptic technique. PHS Policy (IV,A,1) requires compliance with standards for survival surgery as outlined in the *Guide* when using farm animals in biomedical research. The *Guide* (p. 61) recognizes that modification of standard aseptic and surgical techniques might be necessary when performing field surgery; however, animal well-being should not be compromised. When modifications are implemented, a thorough assessment of surgical outcomes should be performed to ensure that appropriate procedures are followed. Surgical outcome assessment may require other criteria in addition to clinical morbidity and mortality.

When using farm animals in agricultural research, standards for survival surgery, as outlined in the *Guide for the Care and Use of Agricultural Animals in Agricultural Research and Teaching* (*Ag Guide*), should be applied.[15] The *Ag Guide* (p. 21) states that "major survival surgeries should be performed in facilities designed and prepared to accommodate surgery, and standard aseptic surgical procedures should be employed." Aseptic surgical procedures include use of cap, mask, gown, gloves, and sterile instruments as well as appropriate operative site preparation and draping. Minor survival surgical procedures that do not expose a body cavity and cause little or no physical impairment (e.g., wound suturing and peripheral vessel cannulation) may be performed under less stringent conditions if performed in accordance with standard veterinary practices. Therapeutic and emergency surgeries (e.g., cesarean section, bloat treatment, and displaced abomasum repair) are sometimes required in agricultural situations that are not conducive to rigid asepsis. However, every effort should be made to conduct minor and emergency survival surgeries in a sanitary and aseptic manner.

Opin.   When farm animals in the "field" require elective major survival surgery, it should be performed in facilities designed and maintained for surgery using appropriate aseptic procedures including cap, mask, sterile gown and surgical gloves, sterile instruments, and aseptic technique (see 18:12). Therefore, an elective rumenotomy should not be performed in a barn. Minor survival procedures may be performed under less stringent conditions than major procedures but require sterile instruments and aseptic technique. When therapeutic and emergency surgeries do not allow transport to dedicated surgical facilities, the facilities used should be clean and methods of aseptic technique used to prevent or minimize exposure of the animal to pathogenic organisms to reduce the likelihood of infection.

18:19   How can the IACUC assure that personnel are qualified to perform surgical procedures?

Reg.   The PHS Policy (IV,C,1,f) and AWAR (§2.31,d,1,viii) place responsibility with the IACUC to determine that personnel performing surgical procedures are qualified and trained in those procedures.

The *Guide* (p. 61) recognizes that personnel performing surgery on research animals have a wide range of educational backgrounds and might require various levels and kinds of training to ensure that good surgical technique is used.

To assist the AV and IACUC in developing appropriate training programs, the Academy of Surgical Research (ASR) developed and published training guidelines for research surgery commensurate with a person's formal education and training background.[16] Research institutions should also perform continuing and thorough assessments of surgical outcomes to ensure that appropriate procedures are followed. (See 21:13.)

Opin.   The goal is to assess surgical competency before IACUC approval of a protocol. The IACUC should review an individual's education, training, certification,[17] and experience for assessment of general surgical competency and qualifications to perform the specific surgical procedure. The ASR training guidelines noted previously[16] can be used when evaluating a person's educational background to determine what an individual may already know. For example, a physician trained in a surgical specialty is expected to be competent to perform surgery on animals within his area of surgical expertise. However, he or she may require training in interspecies variations of anatomy, anesthesia, analgesia, and postoperative care methods. The need for formal training can be waived if the surgeon participates in a multidisciplinary team approach to perform the specific surgical protocol and includes additional experienced personnel qualified to work with animals.[2,16] To assure a surgeon's competence, the IACUC may require that a laboratory animal veterinarian observe or assist with at least the first surgery. Assistance could continue with subsequent procedures until competency is achieved and a specific surgical procedure is predictable to the satisfaction of the veterinarian. Although there is no single credential to assure competency, the best credential is a documented record of previous successful performance of the proposed surgical procedure on the specified species, demonstrating minimal operative and postoperative complications.[18] With such documentation, there should be no need for additional training. After IACUC approval of a surgical protocol, continuing assessment of surgical outcomes should be performed. Participation and input from a laboratory animal veterinarian, animal care staff, surgical technicians, and the investigator are needed to assure ongoing use of appropriate procedures, to address complications and evaluate their rate of occurrence, and to initiate necessary corrective changes.

18:20   What is meant by perioperative care?

Opin.   Perioperative care encompasses all events associated with a surgical procedure.[2] A perioperative care program comprises three overlapping components:[2,19]

- Preoperative planning and management
- Intraoperative care
- Postoperative care

Detailed descriptions of the perioperative care program components have been published.[2,8,19] Preoperative planning should include:

- Identification of members of the multidisciplinary surgical team, all of whom should provide input into presurgical planning
- Roles and training needs of personnel
- Equipment and supplies required
- Facilities for conducting procedures

An anesthetic protocol should be developed, including anesthetic agents, techniques, and methods of anesthetic monitoring to be used. Planning of the surgical procedure should include aseptic techniques to be used (see 18:12) and assessment of indications for perioperative antibiotics. A postoperative care plan should be outlined and the location and facilities for postoperative recovery identified[2,8,19] (*Guide*, p. 61). Preoperative management should include[2,19] (*Guide*, p. 61):

- A preoperative animal-health assessment with a physical examination and laboratory examination if indicated
- A period of stabilization to a new environment for animals before undergoing surgical procedures
- Preoperative fasting of a specified duration if indicated for the species to be used
- Administration of preoperative medications or antibiotics

Components of intraoperative care include:[2]

- Monitoring of anesthetic level and vital organ function
- Provision of vital organ support such as parenteral fluid administration, supplemental oxygen, and maintenance of body temperature
- Proper surgical technique, which comprises:
  - Gentle tissue handling
  - Effective hemostasis
  - Maintenance of sufficient blood supply to tissues
  - Asepsis
  - Accurate tissue apposition
  - Proper use of surgical instruments
  - Appropriate use of monitoring equipment
  - Expeditious performance of the surgical procedure

The postoperative period can be divided into three overlapping phases: recovery from anesthesia, acute postoperative care, and long-term postoperative care.[2] Frequent assessment of thermoregulation and cardiovascular and respiratory function is required during anesthetic recovery and acute postoperative care. Additional care may include:

- Monitoring of the surgical incision
- Thermal support to combat hypothermia
- Parenteral fluid administration to maintain hydration
- Administration of analgesics for postoperative pain
- Administration of prophylactic antibiotics and other drugs

Long-term postoperative care after anesthetic recovery and adequate physiologic stabilization requires at least once-daily monitoring until sutures are removedl (usually at 10–14 days postoperatively) and any postoperative complications are resolved. Monitoring of vital signs, hydration, feed and fluid intake, feces and urine output, attitude and activity, surgical wound condition, body weight, and signs of postoperative pain and infection should be performed. Special diets, analgesics, bandaging, antibiotics, and other medications may be indicated. After suture removal, postoperative care required depends upon the species and surgical procedure. For example, chronic catheters or other partially exteriorized implants require ongoing monitoring and care. Monitoring of body weight should be scheduled throughout the postoperative period.[2]

18:21   What level of perioperative care is appropriate for rodents when compared to nonrodents?

Opin.   The general components of the perioperative care programs (see 18:20) are the same for rodents and nonrodents. Nevertheless, some elements of the programs differ. For rodent surgery, the surgical "team" may be reduced to one person, who serves as surgeon, anesthetist, surgical technician, and scrub nurse. When one person performs surgery on multiple animals at one sitting, as frequently occurs in rodent surgery, careful presurgical planning is required to assure availability of all supplies and equipment required to perform surgery and support necessary modifications in standard aseptic technique [8,10] (*Guide*, p. 63.) The preoperative animal-health assessment should include vendor-supplied colony health testing (e.g., serology) and a visual examination rather than a physical examination of each animal when received.[2] Serologic testing of sentinel animals in the facility or other health surveillance results also may be useful.

Although sophisticated methods for intraoperative monitoring of anesthetized rodents are available when scientifically required, such methods might not be practical or possible in many research situations. Simply observing chest wall movement to determine respiratory rate and palpating the apical pulse through the chest wall may be sufficient to assess cardiovascular and respiratory stability.[20] Procedures used in larger animal species for intraoperative vital organ support such as intravenous fluid therapy can be difficult to use in rodents and may require other routes of administration of fluids or other strategies for supporting circulating blood volume.[2]

The same intensity of monitoring and supportive care commonly provided for larger animals during recovery from anesthesia and acute postoperative care is not practical or feasible for rodents. In addition, performing surgery on multiple animals at one sitting, which frequently occurs, requires developing methods of supportive care for the anesthetic recovery of multiple animals simultaneously.[20]

18:22   How much detail should be provided in the IACUC protocol relative to perioperative care?

Opin.   In this author's opinion, the perioperative care detail that should be provided in the IACUC protocol includes the following:

- Names, qualifications, and responsibilities of participating personnel including the surgeon, the person who will be administering and monitoring anesthesia, and the person who will perform postoperative care
- Location where surgery will be done
- Duration of fasting, with rationale if greater than 24 hours
- Perioperative medications and antibiotics; dose, volume, route, and frequency of administration; duration of treatment

- Intraoperative monitoring and methods to be used to assess adequate anesthesia level
- Surgical anesthesia monitoring system if neuromuscular blocking agents are administered
- Aseptic techniques to be used
- Frequency of animal monitoring during anesthetic recovery and postoperative period
- Postoperative monitoring and care
- Indication of compliance with postoperative monitoring and care guidelines in the institution's IACUC handbook
- Indication of expected health changes or possible postoperative complications and description of methods of monitoring and care
- Postoperative analgesics to be given; agent, dose, route, and frequency of administration; a criteria for determining need for analgesics (see 18:31)

Surv. Which of the following items related to perioperative care are included on your IACUC protocol form (more than one answer is possible)?

Preoperative Period
- Surgeon's name and qualifications and/or names and qualifications of participating personnel — 121/175
- Name of person administering and monitoring anesthesia — 94/175
- Location where surgery will be done — 125/175
- Preoperative medications and antibiotics; dose, volume, route, and frequency of administration; duration of treatment; duration of fasting — 108/175

Intraoperative Period
- Description of intraoperative monitoring and methods to be used to assess adequate anesthesia — 113/175
- Descripton of monitoring of surgical anesthesia if neuromuscular blocking agents are administered — 82/175
- IACUC form guidelines or requirements for aseptic technique — 118/175

Postoperative Period
- Where animals will be recovered from anesthesia — 112/175
- Frequency of animal monitoring during anesthetic recovery — 116/175
- Frequency of monitoring during postoperative care — 108/175
- Person who will perform postoperative care — 101/175
- Description of postoperative monitoring and care — 112/175
- Indication of expected health changes or possible complications and description of methods of care — 97/175
- Postoperative analgesics to be used, including dosage, route, and frequency of administration — 114/175

18:23 Can expired pharmaceuticals be used for *survival surgery*?

Reg. APHIS/AC Policy 3[11] states that the use of expired medical materials such as pharmaceuticals "on regulated animals is not considered to be acceptable veterinary practice and does not constitute adequate veterinary care." Policy 3 specifies that expired anesthetics, analgesics, and emergency drugs cannot be used on any regulated animals for major survival operative procedures.

Opin. In this author's opinion, expired drugs and fluids of any kind should not be used in animals undergoing survival surgery. The use of expired anesthetics, analgesics, euthanasia, and emergency drugs should not be allowed regardless of the procedure intended. Loss of pharmacological integrity may lead to insufficient pain relief or other unexpected anesthetic problems. It is fundamental to proper veterinary care that the user have confidence in the administered drugs; this is best assured by use of nonexpired pharmaceuticals.

Surv. Does your IACUC consider expired pharmaceuticals to be acceptable for use in survival surgery?

- Not applicable   15/163
- Yes   2/163
- No   141/165

   Note: Difference in denominator in the last answer is due to incorporation of data from an earlier survey.

18:24 Is the use of expired pharmaceuticals acceptable for *nonsurvival surgery*?

Reg. APHIS/AC Policy 3 states that expired pharmaceuticals, other than anesthetics, analgesics, and emergency drugs, can be used if "their use does not adversely affect the animal's well-being or compromise the validity of the scientific study."[11] Facilities should have a written policy addressing the use of expired drugs in nonsurvival surgery or require that their intended use be described in the investigator's IACUC protocol submitted for approval.[11]

Opin. Without written documentation of safety and efficacy from the manufacturer and IACUC approval of an investigator's written request for their use in nonsurvival surgery, use of expired drugs and fluids should not be allowed for animals undergoing nonsurvival surgery.

Surv. Does your IACUC allow the use of expired pharmaceuticals in nonsurvival surgical procedures?

- No   102/158
- Yes, with IACUC-approved limitations   19/158
- Yes, with no limitations   11/158
- Not applicable   15/158
- Other   11/158

18:25 Should the IACUC request information on the type of suture materials that will be used?

Opin. The PI should describe the type of suture material and method of wound closure on the protocol form to assure that they are appropriate for the species and the surgical incision. To facilitate wound healing, the type of suture material used to close the skin should be chosen to minimize tissue reaction and potential for wound infection. For example, to produce a lesion on healthy skin, an inoculum of 10^7 or more *Staphylococcus aureus* organisms is required, but in the presence of a silk suture, the required inoculum is reduced to less than 10^3 bacteria.[21] The capillary action of braided suture material and the inherent difficulty in keeping a wound clean can combine to increase the probability of infection at the surgical site.[19] Appropriate wound closure using methods that bury sutures should be considered when animals may bite or pick at exposed skin sutures.

Surv. Which of the following details does your IACUC require in the protocol regarding wound closure and suture material?

• Incision location and length	103/160
• Organs involved	111/160
• Instrumentation used	77/160
• Type of suture material to be used	98/172*
• Method of skin or wound closure	117/172*
• Time sutures will be removed	97/160
• Major survival surgical procedures not done	28/160
• Other	18/160

Note: These figures incorporate data from the survey of the first edition of this text.

18:26 Should the IACUC request information about when sutures (or clips) will be removed, if that is necessary?

Opin. The appropriate time for suture, skin staple, or clip removal should be provided to investigators in an institutional IACUC guidebook, which should contain guidelines on postoperative care. Rather than request that the PI restate when sutures or clips will be removed, a written indication on the protocol form that the PI has read and will comply with the postoperative care guidelines provides adequate assurance that timely suture removal will be done. If written IACUC guidelines are not available to the PI, the IACUC should request that the PI state on the protocol form when sutures or clips will be removed.

18:27 Should the IACUC request information on how rodents or other small animals will be kept warm during and after surgery?

Opin. Because of the large surface area-to-body mass ratio of rodents and other small animals, heat loss resulting in hypothermia is likely to occur without adequate thermal support during surgery and recovery from anesthesia. Maintenance of normal body temperature significantly reduces anesthesia-related cardiovascular and respiratory disturbances (*Guide*, p. 63) and is often critical to the successful recovery of rodents from anesthesia.[20] Methods to combat hypothermia should be provided to investigators in an institutional IACUC guidebook, which should contain guidelines for rodent surgery and postoperative care. At the author's institution, perioperative observations and support are part and parcel of species-specific training, including with rodents. If the PI is expected to follow institutional guidelines with respect to body temperature maintenance, the protocol form could have an item in which the PI can indicate his or her intent to comply with these guidelines. This endorsement provides adequate assurance that appropriate thermal support will be provided. If written IACUC guidelines describing methods to combat hypothermia are not available to the PI, the IACUC should request that the PI state, on the protocol form, how rodents or other small animals will be kept warm during and after surgery.

18:28   Should the IACUC request information about the frequency of postoperative observations?

Opin.   The frequency of postoperative observation is determined by the nature of the surgical procedure and the stage of recovery.[2] Guidelines for minimal frequency of postoperative observation should be provided to investigators in an institutional IACUC guidebook included with guidelines for postoperative care. A written indication on the protocol form, by the investigator, that she or he has read and will comply with these guidelines provides adequate assurance that appropriate postoperative observation frequency will be carried out. However, if a surgical procedure necessitates more frequent observations than the guidelines recommend, the customized monitoring frequency and circumstances requiring monitoring should be described in the protocol. If there are no written IACUC guidelines available to the PI, the IACUC should request that the PI provide information about the frequency of postoperative observations on the protocol form.

18:29   When is an animal considered to be sufficiently recovered to return it to its home cage?

Opin.   Many biomedical scientists believe that an animal should remain in a recovery area until it has recovered from anesthesia, physiological parameters are adequately stabilized, and it has regained normal ambulatory and protective behaviors before being moved to its home cage.[2,22,23] The caveat to this is that application and practicality are at least species and size dependent. Physiological stability and normal ambulatory behavior might be very different for an animal that has undergone a thoracotomy versus one that has undergone arthroscopy. At the author's institution, continuous observation by caregivers is required until the animal can be extubated, the core temperature is 99.0° F, and the patient remains in sternal recumbency.

   Preemptive analgesia may cause a postoperative animal to appear lethargic. Unlike a tranquilized animal, however, one treated with narcotic or nonnarcotic analgesics will respond to nonpain stimulation (e.g., retching). The ability to stay sternal is associated with the ability to raise the head as needed and allows the animal to cope with postoperative vomition should it occur. The ability to raise the head is also an added safeguard when there is need to place water bowls that are shallow enough to prevent accidental immersion of the nostrils.

18:30   May animals that have undergone surgery be recovered in their home cage, or is there a need for a separate recovery area?

Reg.   The *Guide* (p. 79) states that "a postoperative-recovery area should provide the physical environment to support the needs of the animal during the period of anesthetic and immediate postsurgical recovery and should be so placed as to allow adequate observation of the animal during this period." The species and type of surgical procedure will dictate the type of caging and support equipment required, which should be designed to support physiologic functions, such as thermoregulation and respiration.

Opin.   During anesthetic recovery, an animal should be in an area that is warm, safe, quiet, comfortable, and appropriate for the needs of the individual animal and species.[24] Dedicated recovery rooms are recommended for species such as rabbits, cats, dogs, and swine.[23] A separate recovery area provides the necessary physical environment, ambient temperature control, and equipment for monitoring and supportive care to manage complications that may occur during recovery. Animals can be housed individually in appropriately sized cages designed to prevent injury to occupants. Individual housing also prevents potential trauma from cage mates. In addition, recovery areas can be located to facilitate the appropriate frequency of monitoring an animal by personnel.

Small ruminants that are too large to fit into recovery cages and large farm animals can recover in their own pen or stall if provisions are made for adequate observation, warmth, and animal safety.[2,23] The added advantage of conspecific smells, sounds, and visual contact may enhance the animal's recovery. Foam- or air-filled padding placed on the floor in secure, warm areas is also useful for these species. The additional space afforded by a pen or stall also allows for assistance with changing position and moving about larger species that may be unable to move themselves or begin thrashing about during recovery. Regardless of the location, the criteria noted in 18:29 also apply: monitoring of the animal until sternally recumbent, a core body temperature of at least 99.0°F, and extubation of the animal.

Rodents usually recover from anesthesia in the same laboratory where surgery is performed.[23] Provisions to support body temperature are frequently necessary as thermal support is often critical to the successful recovery of rodents from anesthesia.[20] Recovering rodents should be housed individually to prevent injury by cage mates. Although extensive monitoring may not be possible, the animals should be observed frequently until recovered from anesthesia and adequately stabilized before being returned to their home cages.

18:31   What records should investigators keep relative to surgery and the perioperative period?

Reg.   APHIS/AC Policy 3[11] requires appropriate postoperative record keeping in accordance with accepted veterinary procedures for covered species. The *Guide* (pp. 62–63) states that appropriate medical records should be maintained during and after the anesthetic recovery period.

Opin.   The specific information requested in medical records is detailed in 26:10 and 27:16. The *Institutional Animal Care and Use Committee Guidebook*[13] states that "an intra-operative anesthetic monitoring record should be kept and included with the surgeon's report as part of the animal's records. This record should be available to the personnel providing post-operative care. Post-operative records at a minimum should reflect that the animal was observed until it was extubated and had recovered the ability to stand. These should be supplemented by records evaluating the animal's recovery, administration of analgesics and antibiotics, basic vital signs, monitoring for infection, wound care, and other medical observations."

Preoperative records should include a health profile and medical and vaccination history (when applicable) provided by the vendor and subsequently supplemented by the clinical care staff during quarantine or with annual updates, as appropriate. Examples of additional information include vaccinations for dogs or cats, quarterly or semiannual blood work, and tuberculosis testing and serological profiles in nonhuman primates. A physical exam to assess an animal's health status should be performed. A preanesthetic evaluation should include records of body weight, body temperature, heart rate, and respiratory rate to establish baseline data, recognizing that these values might be influenced by preoperative or pre-anesthetic medications. The need for diagnostic radiographic tests and laboratory evaluation of blood, urine, and feces will depend on the animal species, research protocol, and health history of the animal or colony.[2] Preoperative administration of medications or antibiotics should be documented.

An anesthetic record should document anesthetic administration, monitoring of anesthetic depth (including reflexes), and frequent assessment of the physiologic status of an animal including body temperature and cardiovascular and respiratory function.[25] The extent of record keeping and monitoring sophistication depends upon the potential complications associated with the anesthetic regimen, the surgical procedure, and the animal species.[2] Minimal documentation should include vital signs, core body temperature, heart rate, and respiratory rate, recorded at an average interval of every 15 minutes, and the anesthetic adminis-tered, dose, route, and time of administration. The postoperative period can be divided into three overlapping phases: recovery from anesthesia, acute postopera-tive care, and long-term postoperative care.[2] The extent of record keeping and intensity of monitoring during the postoperative period depend upon the species and the surgical procedure. Generally, the most intensive and frequent monitoring is required during recovery from anesthesia, "when the animal is most vulnerable to the potentially adverse effects of anesthetic agents, hypothermia, and physio-logic disturbances secondary to surgery."[23] Frequent assessment of thermoregula-tion and cardiovascular and respiratory function should continue during anesthesia recovery and acute postoperative care until the animal is adequately stabilized. Vital signs, including body temperature, pulse rate and rhythm, respi-ratory rate and rhythm, and oxygen saturation, should be recorded approximately every 15 minutes during anesthesia recovery with less frequent monitoring possi-ble as an animal stabilizes. Documentation of clinical observations including mucous membrane color, capillary refill time, pain assessment, and surgical wound condition is recommended. Administration of analgesics, antibiotics, parenteral fluids, and other medications should be recorded and should include the drug name, dosage, route, and time given. A brief description of the surgery should be recorded with the postoperative care records or as a separate surgeon's report, with a notation about complications and/or unexpected need to deviate from the approved surgical plan.

During long-term postoperative care after anesthetic recovery and adequate physiologic stabilization, monitoring should continue at least once a day until suture removal (usually at 10–14 days postoperatively) and any postoperative complications are resolved. Vital signs, hydration, feed and fluid intake, feces and urine output, attitude and activity assessment, pain assessment, surgical wound condition and care, and monitoring for postoperative infection should be recorded. All postoperative care and medications administered as described with anesthesia recovery should be documented. After suture removal, frequency and parameters monitored depend on the species and surgical procedure. For example, chronic catheters or other partially exteriorized implants require documentation of ongo-ing monitoring and care. Monitoring of body weight should be scheduled through-out the postoperative period. (See 26:11–26:15.)

18:32 What is considered health record maintenance "consistent with current veterinary practice" and how are these records used?

Reg.   APHIS/AC Policy 3[11] requires appropriate postoperative record keeping in accordance with accepted veterinary procedures for covered species. (See 18:31.)

Opin. The animal health record describes a problem and observations, action plan, and resolution or lack thereof, whether alone or superimposed on other problems. It is a communication tool that allows for a variety of trained persons to interact with the patient in a meaningful way and have others on the care team recognize exactly what was done, why it was done, and why an anticipated outcome may or may not have been realized. It is a solid mechanism for communication between the clinical care staff and similarly trained physician/surgeons. In the research environment, especially with novel animal models, the reason for unexpected deaths can frequently be resolved by consulting a well-documented animal health record. The well-organized and documented animal health record serves to put everyone on the same page and is critical for good research. (See 26:10–26:15; 27:16–27:21.)

References

1. Silverman, J., Protocol review: whose responsibility is it? *Lab Anim.* (NY), 23(2), 22, 1994.
2. Brown, M.J., Pearson, P.T., and Tomson, F.N., Guidelines for animal surgery in research and teaching, *Am. J. Vet. Res.*, 54, 1544, 1993.
3. Potkay, S., et al., Frequently asked questions about the Public Health Service Policy on Humane Care and Use of Laboratory Animals, *Contemp. Topics Lab. Anim. Sci.*, 36(2), 47, 1997.
4. U.S. Department of Agriculture, Animal and Plant Health Inspection Service, Policy 14, Major survival surgery single vs. multiple procedures, April 14, 1997.
5. Potkay, S., et al., Frequently asked questions about the Public Health Service Policy on Humane Care and Use of Laboratory Animals, *Lab Anim.* (NY), 24(9), 24, 1995. Available on the World Wide Web at: http://www.aphis.usda.gov/ac/policy/policy14.pdf.
6. U.S. Department of Agriculture, Animal and Plant Health Inspection Service, Policy 16, Major survival surgery single vs. multiple procedures, April 14, 1997. Available on the World Wide Web at: http://www.aphis.usda.gov/ac/policy/policy16.pdf.
7. National Research Council, Survival surgery and postsurgical care, in Education and Training in the Care and Use of Laboratory Animals: A Guide for Developing Institutional Programs, Report of the Institute of Laboratory Animal Resources Committee on Educational Programs in Laboratory Animal Science, National Academy of Science Press, Washington, D.C., 1991, p. 61.
8. Cunliffe-Beamer, T.L., Applying principles of aseptic surgery to rodents, *AWIC Newslett.*, 4(2), 3, 1993.
9. Bradfield, J.F., et al., Behavioral and physiologic effects of inapparent wound infection in rats, *Lab. Anim. Sci.*, 42, 572, 1992.
10. Brown, M.J., Aseptic surgery for rodents, in *Rodents and Rabbits: Current Research Issues*, Niemi, S.M., Venable, J.S., and Guttman, H.N., Eds., Scientists Center for Animal Welfare, Bethesda, MD, 1994, 67.
11. U.S. Department of Agriculture, Animal and Plant Health Inspection Service, Policy 3, Veterinary Care, January 14, 2000. Available on the World Wide Web at: http://www.aphis.usda.gov/ac/policy/policy3.pdf.
12. Slattum, M.M., et al., Infusion-related sepsis in dogs undergoing acute cardiopulmonary surgery, *Lab. Anim. Sci.*, 41, 146, 1991.
13. Office of Laboratory Animal Welfare, National Institutes of Health, Institutional Animal Care and Use Committee Guidebook, 2nd ed., National Institutes of Health, Bethesda, MD, 2002, chap. C.3.g.
14. Banks, R., Protocol review: a question of technique, *Lab Anim.* (NY), 23(5), 22, 1994.

15. Federation of Animal Science Societies, *Guide for the Care and Use of Agricultural Animals in Agricultural Research and Teaching*, 1st rev. ed., Federation of Animal Science Societies, Savoy, IL, 1999, p. 21.

16. Academy of Surgical Research, Guidelines for training in surgical research in animals, *J. Invest. Surg.*, 2, 263, 1989.

17. Dennis, M.B., Surgical research specialist certification, in Proceedings of the Laboratory Animal Welfare Training Exchange, 1998 Meeting, St. Louis, August 20–21, 1998, p. 53.

18. Dennis, M.B., Surgical training and personnel qualifications, in *Research Animal Anesthesia, Analgesia and Surgery*, Smith, A.C., and Swindle, M.M., Eds., Scientists Center for Animal Welfare, Greenbelt, MD, 1994, p. 11.

19. Brown, M.J., and Schofield, J.C., Perioperative care, in *Essentials for Animal Research: A Primer for Research Personnel*, Bennett, B.T., Brown, M.J., and Schofield, J.C., Eds., National Agricultural Library, Washington, D.C., 1994, p. 79.

20. Wixson, S.K. and Smiler, K.L., Anesthesia and analgesia in rodents, in *Anesthesia and Analgesia in Laboratory Animals*, Kohn, D.F., Wixson, S.K., White, W.J., and Benson, G.J., Eds., Academic Press, New York, 1997, p. 165.

21. McCurnin, D.M., and Jones, R.L., Principles of surgical asepsis, in *Textbook of Small Animal Surgery*, 2nd ed., Slatter, D.H., Ed., W.B. Saunders, Philadelphia, 1993, p. 114.

22. White, W.J., and Blum, J.R., Design of surgical suites and postsurgical care units, in *Anesthesia and Analgesia in Laboratory Animals*, Kohn, D.F., Wixson, S.K., White, W.J., and Benson, G.J., Eds., Academic Press, New York, 1997, p. 149.

23. Smith, A.C., and Swindle, M.M., Post surgical care, in *Research Animal Anesthesia, Analgesia and Surgery*, Smith, A.C., and Swindle, M.M., Eds., Scientists Center for Animal Welfare, Greenbelt, MD, 1994, p. 167.

24. Haskins, S.C., and Eisele, P.H., Postoperative support and intensive care, in *Anesthesia and Analgesia in Laboratory Animals*, Kohn, D.F., Wixson, S.K., White, W.J., and Benson, G.J., Eds., Academic Press, New York, 1997, p. 379.

25. Brown, M.J., Principles of anesthesia and analgesia, in *Essentials for Animal Research: A Primer for Research Personnel*, Bennett, B.T., Brown, M.J., and Schofield, J.C., Eds., National Agricultural Library, Washington, D.C., 1994, p. 39.

19

Antigens, Antibodies, and Blood Collection

Harold F. Stills, Jr.

Introduction

Modern biologic research techniques often require the production of specific polyclonal and monoclonal antibodies as an essential component of the research protocol. Production of both polyclonal and monoclonal antibodies may require the immunization of animals, therein presenting the IACUC with the dilemma and duty of evaluating the immunization procedures and schedules with respect to potential animal pain and distress. Published guidelines from a number of sources are available to assist the IACUC in developing policies and procedures for animal use in antibody production.[1-6] Although *in vitro* alternatives to the use of live animals in the production of antibodies is often appropriate (see 19:16), this chapter primarily focuses on those circumstances in which the IACUC approves the use of live animals for antibody production.

Production of a high-quality antibody requires an appropriate stimulation of the immune system. The initial immune activating event involves processing of the antigen by antigen-presenting cells (APCs), which are primarily the macrophages and the dendritic cells, such as the Langerhans cells in the epidermis of the skin. Activation of these cells, along with the recruitment of additional cells for antibody production, is often enhanced by the addition of various biologically active compounds (primarily targeting toll-like receptors) in the adjuvant. Protecting the antigen from rapid degradation in the body and permitting the slow release of the antigen to the APCs are other functions often performed by the adjuvant. The end results of the use of an adjuvant are both an increase in the desirable antibody response and the production of a substantial inflammatory response with undesirable tissue destruction and potential pain and distress.

Of the available adjuvants in use today, none has proved more effective overall than Freund's complete adjuvant (FCA).[7,8] First described in 1942,[9,10] it was originally composed of paraffin oil, mannide monooleate, and killed mycobacteria. The presently available formulations of FCA, however, differ dramatically from the historic formulation in using a much purer and less toxic mineral oil and producing significantly fewer severe inflammatory lesions.[11,12] Injected into animals, FCA produces a chronic granulomatous inflammatory reaction at the injection site.[13] Granulomas often may be detected in draining lymph nodes, spleen, kidney, and other organs where microdroplets of the emulsion have been distributed by the vascular or lymphatic systems. Intradermal injections routinely produce ulcerations as early as 12 to 14 days post injection, which persist for 8 weeks or longer.[13] The microscopic lesions produced by injection of FCA are primarily those of granulomatous inflammation and focal necrosis.[13-18]

19:1   What type of justification should be required for the use of adjuvants in antibody production?

Reg.   (See 19:2.)

Opin.   The dramatic inflammatory reaction historically generated by the injection of the original formulation of Freund's complete adjuvant (FCA) has restricted its use to research antibody production. This has resulted in refinements in the original FCA composition designed to reduce the resulting inflammation, as well as the development of and efforts to encourage the use of other adjuvants that purport to produce similar antibody responses without the granulomatous lesions associated with FCA. Unfortunately, many of the alternative adjuvants selected to replace FCA produce severe inflammatory reactions.[14,16,19-21] All adjuvants, by nature of their mode of action, induce an inflammatory reaction. Since many of the compounds for which antibodies are required are poorly immunogenic, adjuvants are essential. The choice of any adjuvant should be based upon a sound scientific rationale that considers the specific immunogen, the species being immunized, and the desired antibody.[5,22]

Surv. 1 Does your IACUC require scientific justification for the use of Freund's complete adjuvant?

- Approve use without scientific justification   7/99
- Approve use with minimal scientific justification   31/99
- Approve use only with detailed scientific justification   61/99

Surv. 2 Does your IACUC require scientific justification for the use of an adjuvant other than Freund's complete adjuvant?

- Approve use without scientific justification   17/103
- Approve use with minimal scientific justification   46/103
- Approve use only with detailed scientific justification   40/103

19:2   Is the use of FCA a painful or distressful procedure?

Reg.   APHIS/AC Policy 11 lists the use of FCA under "examples of procedures that can be expected to cause more than momentary or slight pain."[23] That same policy goes on to state that FCA used for antibody production may cause perturbations ranging from momentary or slight pain to severe pain, depending on the product, procedure, and species. PHS Policy (U.S. Government Principle IV) and the *Guide* (p. 64) require the minimization of discomfort, distress, and pain in concert with good science.

Opin.   Although APHIS/AC considers the use of FCA as likely to be painful, other reports have failed to document any pain or distress in animals injected with FCA[16-18,24] or in tuberculin-negative humans.[25,26] In the author's experience, uncomplicated FCA–antigen emulsion intradermal and subcutaneous injections are not characterized by tenderness, irritation, or any other indication of distress in the animal. Feed consumption and body weights are unaffected by FCA–antigen injections.

      The choice of immunogen, regardless of the choice of adjuvant, may induce an autoimmune condition that is associated with pain and distress. The adjuvant-induced model of arthritis is a prime example and has even been recommended as a model of chronic clinical pain.[27] (See 20:23.)

Surv. What APHIS/AC pain classification does your IACUC assign to projects using Freund's complete adjuvant in antibody production?

- Category C (no pain or distress)   15/78
- Category D (alleviated pain or distress)   39/78
- Category E (unalleviated pain or distress)   24/78

19:3 Should the IACUC require different routes of antigen–adjuvant injection to have different APHIS/AC pain level classifications?

Opin. The type of inflammatory reaction produced by any adjuvant, including Freund's complete adjuvant (FCA) is not dependent upon the route of injection. Therefore, any differences in pain level classification result from the effect of the inflammatory reaction upon the tissues involved. For the common routes of injection for most adjuvants in nonrodents (intradermal, subcutaneous, and intramuscular), there is little evidence of pain or distress to the animal, even in cases when the pathologic lesions are significant. (See 19:11.) The intraperitoneal injection of FCA and other adjuvants in rodents, however, has been linked to the production of clinical signs indicative of pain and distress.[2,15,16,28–30]

Surv. Does your IACUC require different USDA pain/distress categories for different routes of antigen–adjuvant injection?

- No differences in classification   37/88
- There may be differences in USDA pain/distress category   51/88

19:4 Should the amount of killed mycobacteria in the Freund's complete adjuvant (FCA) be limited?

Opin. Formulations of FCA containing no more than 0.1 mg/ml of dry mycobacterial cell mass have been recommended as a method to reduce the associated inflammatory response.[13] Nevertheless, commercially available formulations of FCA generally range from 0.5 to 1.0 mg/ml of dry mycobacterial cell mass.

Surv. Does your IACUC limit the amount of mycobacteria in FCA, and if so, what is the limit?

- Limit 0.5 to 1.0 mg/ml   22/78
- No limit set   53/78
- Other limit set (<0.5 mg/ml)   3/78

19:5 What is the maximum volume of Freund's complete adjuvant (FCA)–antigen emulsion that the IACUC should permit for injection in any single site? What is the maximum total volume that should be permitted?

Opin. Recommended total injection volumes and injection volumes per site vary greatly. For the intradermal route of injection in the rabbit, recommendations vary from maximum volumes of 0.025 ml/site[2,5] to 0.1 ml/site in some laboratory immunology texts[31] and other reports.[32] Recommended subcutaneous injection volumes per site are generally higher, ranging from 0.25 ml/site[2] to 0.5 ml/site in the rabbit,[32,33] 0.2 ml/site in the guinea pig, and 0.1 ml/site in the mouse.[2] When separately listed and permitted (usually larger animals), intramuscular injection volumes are generally equal to the recommended subcutaneous volumes.

Surv.   What is the maximum volume of FCA–antigen that your IACUC approves for injection in any single site (intradermally, subcutaneously, and intramuscularly)? What is the maximum total volume of FCA–antigen that your institution approves for injection? (See Table 19.1.)

    • Not applicable   58/141

Of the remaining 83 applicable responses:

    • No IACUC policy for maximum single site volume injected   42/83
    • Established policy for maximum single site volume injected   41/83
    • No IACUC policy for maximum total volume injected   47/83
    • Established policy for maximum total volume injected   36/83

TABLE 19.1

Survey of Maximum Volume Permitted for FCA Injections
(Intradermal, Subcutaneous, and Intramuscular)

Species	Maximum Volume (ml) per Site		
	Mean	Median	Range
Rabbit	0.22	0.1	0.05–1.0
Mouse	0.12	0.1	0.01–0.5
Rat	0.16	0.1	0.05–0.5
Guinea pig	0.18	0.1	0.05–0.5
Goat/sheep	0.47	0.5	0.05–1.0

Species	Maximum Total Volume (ml) per Animal		
	Mean	Median	Range
Rabbit	1.26	1.0	0.1–10.0
Mouse	0.22	0.1	0.05–1.0
Rat	0.38	0.2	0.1–1.0
Guinea pig	0.49	0.4	0.1–1.2
Goat/sheep	0.92	1.0	0.1–2.0

19:6   Should the intradermal (ID) route of injecting Freund's complete adjuvant (FCA)–antigen emulsions be permitted by an IACUC?

Opin.   The dramatic inflammatory reaction created by the injection of FCA has led to a number of recommendations regarding injection routes. ID injection of FCA leads to skin ulcerations and the formation of large granulomas.[13–18] The *Canadian Council on Animal Care Guidelines on Antibody Production* restrict the ID use of FCA to only those instances that are absolutely necessary.[2] While the formation of ulcerative granulomas is not unusual with ID injections, granuloma formation is typical with FCA injection by any route. The localization of the granuloma lesion resulting from ID injections permits monitoring of the lesion's development while the ulceration tends to minimize the total tissue destruction. These advantages, coupled with the potentially increased immune response due to exposure of the dendritic Langerhans cells of the dermis, has led some authors to favor the ID route for antibody production and animal well-being, in spite of the aesthetically displeasing lesions.[4,17]

Surv.   Does your IACUC permit the intradermal injection of FCA–antigen emulsions?

- Intradermal injection permitted   52/67
- Intradermal injection not permitted   15/67

19:7   Should the subcutaneous (SC) route of injecting Freund's complete adjuvant (FCA)–antigen emulsions be permitted by an IACUC?

Opin.   The SC route of FCA–adjuvant emulsion injection has been recommended in the literature because of the ease of the injection technique[18] and the lack of the typical ulcerations commonly seen after intradermal injections. The extension of the granulomatous inflammation and formation of fistulous tracts are drawbacks of this route of injection.[13–18] The SC injection of FCA–antigen emulsions is generally considered an acceptable route of injection, especially in smaller animals, in which the intradermal route is technically more difficult.[2–6,17,22] Even when the SC route is used, use of multiple injection sites is recommended to maximize the presentation of the antigen to the immune system.

Surv.   Does your IACUC permit the subcutaneous injection of FCA–antigen emulsions?

- Yes, subcutaneous injection is permitted   70/75
- No, subcutaneous injection is not permitted   5/75

19:8   Should the intramuscular (IM) route of injecting Freund's complete adjuvant (FCA)–antigen emulsions be permitted by an IACUC?

Opin.   The IM route is the third most common route of injection in larger animals, after subcutaneous and intradermal injections. IM injections are generally effective for antibody production and have been recommended in larger animals (goats, sheep, horses, etc.).[2,6,32] Granuloma formation, muscle necrosis, and the formation of fistulous tracts have, however, all been reported[13–18] and the difficulty in evaluating lesions has led to cautions.[18,31] In addition, the potential of pain after injection into the closed IM space has led to cautions.[2,5] The extensive tissue destruction has caused this route to be discouraged for smaller animals.[2,6] Because of the extensive tissue destruction, the potential of pain after injection into a closed space, and the lack of any specific advantages of the IM route over either the intradermal or subcutaneous route, the IM route of injection is generally not recommended.

Surv.   Does your IACUC permit the intramuscular injection of FCA–antigen emulsions?

- Permit intramuscular injection   31/61
- Do not permit intramuscular injection   30/61

19:9   Should the intraperitoneal (IP) route of injecting Freund's complete adjuvant (FCA)–antigen emulsions be permitted by an IACUC?

Opin.   The IP route of injection is often used in rodents for polyclonal antibody production. While previously recommended by some as a route of FCA immunization, intraperitoneal injection of FCA is now strongly discouraged.[2,6] The formation of granulomas, fibrous adhesions, abdominal fluid distention, and the resulting clinical signs indicating pain and distress after intraperitoneal FCA injection have been described.[6,14,15,30,27,28,34] The use of intraperitoneal FCA in rats as a model of acute inflammation[30,35,36] is an additional indication of the potential for pain and distress. The potential complications of IP injection of FCA should be closely evaluated by all IACUC's prior to approving any project that proposes to use this route of injection. Scientific justification for the necessity of IP injection of FCA should be required. In cases of immunization for antibody production, there is little justification for the IP route over the intradermal or subcutaneous route of injection. (See 19:18.)

Surv. Does your IACUC permit the intraperitoneal injection of FCA–antigen emulsions?

- Permit intraperitoneal injections 46/64
- Do not permit intraperitoneal injections 18/64

19:10 Should the intravenous (IV) route of injecting Freund's complete adjuvant (FCA)–antigen emulsions be permitted by an IACUC?

Opin. The IV injection of FCA–antigen emulsions has been associated with adjuvant emboli in the lungs of rabbits and multiple pulmonary granulomas.[37] Several publications have stated that the IV route for FCA–antigen emulsion injections should not be permitted.[2,5,6,32] The IV injection of any non-water-miscible compound may result in the formation of emboli and have serious consequences. The author has not encountered any recommendation for or any acceptable justification for the IV injection of FCA–antigen emulsions for antibody production. If IV injection is desired (e.g., as in hybridoma formation), then the antigen should be injected in an aqueous form, with or without an aqueous soluble adjuvant.

Surv. Does your IACUC permit the intravenous injection of FCA–antigen emulsions?

- Permit intravenous injection 4/51
- Do not permit intravenous injection 47/51

19:11 Should footpad injections of Freund's complete adjuvant (FCA)–antigen emulsions be permitted by an IACUC?

Opin. As shown in the following, the footpad injection route for FCA–antigen emulsions in rabbits was prohibited by nearly all the institutions responding to the survey, corresponding to the recommendations in the literature.[2,4,6,32] The use of the footpad injection route in rodents is generally restricted to cases in which it is scientifically justified and then only in one foot. Even in instances in which a localized immune reaction is scientifically justified (e.g., lymphocyte isolation from the popliteal lymph node), justification for not using the dorsum of the foot instead of the footpad should be given. In this author's opinion, there are few, if any, acceptable scientific justifications for footpad injections in rodents.

Surv. Does your IACUC permit the footpad injection of FCA–antigen emulsions?

- Permit footpad injections in rabbits 4/42
- Permit footpad injections in rodents 36/56

19:12   What procedures should be required by the IACUC to minimize the potential for contamination of injection sites during the injection of adjuvant–antigens?

Opin.   Using aseptic procedures for antigen–adjuvant injections, including ensuring sterile preparation of the antigen–adjuvant mixture, clipping the injection site, and sterile scrubbing of the injection site with an antiseptic, has been recommended.[2,4,5,17,18] The injection of viable contaminating bacteria with an adjuvant–antigen mixture often results in the formation of an abscess with the influx of neutrophils, exudation, hyperemia, and pain.

Surv.   What procedures does your IACUC require for minimizing contamination of injection sites? More than one response per institution is possible.

- No required policy concerning injection site preparation   35/95
- Require sterile preparation of antigen–adjuvant mixture   47/95
- Require clipping of hair prior to injection   46/95
- Require surgical scrub to prepare injection site   40/95

19:13   What type of justification or information should be required by the IACUC for that committee to approve a schedule of antigen–adjuvant injections?

Opin.   The administration of booster injections prior to the optimal time not only increases the potential for pain and distress but also may be detrimental to the ultimate antibody level and affinity.[2,17,38–41] Animals mounting a robust immune response may develop an arthus reaction (an acute inflammatory and painful reaction) at the booster injection site. The antigen, quantity of antigen, adjuvant used, and injection route are all important factors in determining the optimal schedule for immunization and, optimally, the individual animal's antibody titer should be followed to determine when the antibody level has plateaued and booster injections are indicated.[42]

Surv. 1   Does your IACUC require investigators to justify the proposed interval between antigen–adjuvant injections?

- Require a justification   52/89
- Do not require a justification   37/89

Surv. 2   Does your IACUC require investigators to monitor serum titers before booster injections are given?

- Require monitoring   16/82
- Do not require monitoring   66/82

19:14   Should investigators be required to search for commercially available, acceptable antibodies?

Reg.   The AWAR (§2.31,d,1,ii; §2.31,d,1,iii) and the PHS Policy (IV,D,1,a) require that PIs provide assurances that, in the case of painful animal procedures, alternatives have been searched for and found unacceptable and that the proposed activities are not duplicative.

Opin.   The availability of a commercially available and acceptable antibody is a clear and definite example of an alternative to additional animal use and duplicative procedures. In fulfilling their responsibilities, IACUCs often require literature searches for alternative methods of antibody production but fail to require literature searches for the more obvious and more likely possibility, that an acceptable antibody is commercially available. The availability of both free and subscription Internet resources[43-45] with updated and current databases for locating commercially available antibodies makes the task reasonable.

Surv. 1   Does your IACUC require the investigator to search for commercially available antibodies and, if available, justify why these would not be acceptable?

- Require assurance that commercial antibodies are not available   53/89
- Require PI to search a standard reference   13/89
- No search required   23/89

Surv. 2   Does your IACUC require investigators to justify scientifically why commercially available antibodies are not acceptable for a study?

- Require scientific justification   62/92
- Do not require scientific justification   30/92

19:15   What oversight actions should an IACUC consider if an investigator has contracted with a commercial laboratory for the production of antibodies in animals?

Reg.   PHS Policy requires that any institution using animals with PHS funds, even if the funds are awarded by subcontracting or subgranting, must either have an approved Animal Welfare Assurance or be included as a component under the primary institution's Assurance. If both institutions have approved Assurances, it is only necessary that one of the institutions' IACUCs review and approve the activity. The subcontracting or subgranting institution, however, retains partial accountability for "providing effective oversight mechanisms to ensure compliance with the PHS Policy."[46]

Opin.   The contracting of antibody production is a difficult problem for many institutions. The availability of numerous commercial contracting companies along with the many alliances between peptide-generating and antibody-producing companies make institutional oversight a nightmare. IACUCs must be diligent in their attempts to identify potential situations and be proactive in the education of PIs as to the necessity of notifying and working with the IACUC when contracting antibody production from a commercial firm. (See 8:10.)

The IACUC's duty in reviewing the contracted facilities protocols and procedures for antibody production protocols is less well defined. Some IACUCs require a complete review of contracted protocols in addition to the review of the contracted facility. Others limit their requirements to the documentation of IACUC approval from the contracted PHS assured facility, while others do no review or have no requirements. Since the contracting institution retains partial accountability, it would seem prudent that IACUCs should review contracted antibody production protocols to assure that the procedures and policies involved adhere to those of the contracting institution.

Surv.   If an investigator proposes to contract with a commercial laboratory for the production of either a polyclonal or a monoclonal antibody using PHS funds, what type of review, if any, does your IACUC perform of the commercial laboratory?

- No review performed   25/71
- Check for PHS Assurance and USDA license   23/71

- Perform some level of protocol review and require approval of
  our IACUC, in addition to checking for PHS Assurance and
  USDA license   23/71

Twenty-two of the 87 institutions responding stated that they would rarely be aware of such subcontracting if and when it occurred. Of those 22, 6 are included in the preceding results while the other 16 indicated that they were not aware of any contracted antibody production at their institutions.

19:16   Should additional justifications be requested by the IACUC for the production of monoclonal antibodies using the mouse ascites method?

Reg.   NIH/OLAW issued a report directing that IACUCs "critically evaluate the proposed use of the mouse ascites method. . . . IACUC's must determine that (1) the proposed use is scientifically justified; (2) methods that avoid or minimize discomfort, distress, and pain (including *in vitro* methods) have been considered; and (3) the latter have been found unsuitable."[47]

Opin.   The development of practical and reasonably priced *in vitro* methods for hybridoma growth, coupled with evidence that the mouse ascites method causes discomfort, has led many IACUCs to require that *in vitro* methods be utilized. Published comparisons of the *in vitro* methods and the ascites method have shown the results to be highly dependent upon the specific hybridoma.[48] In the author's opinion, there are few justifications for the use of the ascites method for monoclonal antibody production. Nearly all hybridomas can be propagated *in vitro* and the levels of monoclonal antibody production can be increased by the use of specifically designed growth chambers, hollow-fiber bioreactor systems, and other means. In all instances, the burden of proof should be upon the PI to show that multiple *in vitro* methods have been attempted and were unsuccessful prior to the IACUC's approving the ascites method of monoclonal antibody production. (See 12:7–12:8.)

A report[49] commissioned by the NIH summarizes the findings of the National Academy of Sciences as to the scientific necessity for producing monoclonal antibodies by the mouse method, including ways to minimize any pain or distress that might be associated with that method. The report also discusses regulatory considerations for the mouse method and summarizes the current stage of development of tissue culture methods.

Surv. 1 Does your IACUC require additional justifications for using the mouse ascites method for monoclonal antibody production?

- Not applicable or do not perform   77/149
- Additional justifications needed   61/149
- No additional justifications needed   11/149

Surv. 2 What justifications does your IACUC accept for permitting the *in vivo* production of monoclonal antibodies by the mouse ascites method? Check all appropriate responses.

- Not applicable or do not perform   78/148
- PI requests; no additional justification needed   6/148
- *In vitro* production is too expensive   15/148
- PI unsuccessful with *in vitro* methods   52/148

- Literature indicates *in vitro* methods unsuccessful   51/148
- Molecular structure suggests *in vitro* production will fail   24/148
- Other justifications acceptable   6/148

19:17   Should the IACUC limit the number of allowable peritoneal taps for the collection of ascitic fluid?

Opin.   The process of removing the excessive peritoneal fluid from mice used in mono-clonal antibody production requires the use of a large-gauge needle and, often, some anesthesia. Many institutions limit the total number of peritoneal taps (with the last tap performed post mortem) while other institutions place limits on the animal's physical condition.[4] All hybridoma lines, based to a large degree upon the plasmacytoma fusion partner, behave somewhat differently when growing in the animal's peritoneal cavity. Certain hybridomas are extremely invasive, result-ing in bloody peritoneal fluid on the first tap. The tumor masses that hybridomas form in the peritoneal cavity may be either disseminated or solid and singular.[50] Likewise, the dynamics of monoclonal antibody production from specific hybri-domas also varies.[51] Absolute limitations on the number of taps that fail to recog-nize the differences between hybridoma lines are inappropriate; however, there is little reason for the maximal number of taps to exceed three.

Surv.   Does your IACUC limit the number of peritoneal taps permitted when using the mouse ascites methods for monoclonal antibody production and, if so, what is the limit?

- Not applicable or do not perform   76/143
- Limit to two or three taps   42/143
- Limit to five taps   1/143
- No absolute limits   17/143
- Only a single tap is permitted   7/143

19:18   Should the IACUC limit the amount of pristane or Freund's complete adjuvant (FCA) used to prime the peritoneal cavity for ascites production?

Opin.   Several authors have suggested that lowering the volume of pristane or FCA injected to prime the peritoneal cavity of mice may reduce potential pain and distress.[30,52] Scientific studies, however, have had contradictory findings; one study indicated that 0.5 ml of pristane was the most effective for ascites production,[53] while another study found no significant differences in ascites production after 0.1 or 0.5 ml of pristane.[54] Both FCA and pristane "prime" the peritoneal cavity for hybridoma growth by inducing a granulomatous inflammatory reaction (peritonitis). Since pain is a common sequel to peritonitis, IACUCs should require that the minimal effective quantity of FCA or pristane be used in those instances when the *in vivo* ascites method of monoclonal antibody production is warranted. (See 19:9.)

Surv.   Does your IACUC limit the volume of pristane or FCA used to prime the perito-neal cavity for mouse ascites production and, if so, what is the limit?

- Not applicable or do not perform   84/140
- Limit pristane or FCA to a maximal amount   40/140
- No limit   16/140

The maximal injected volume limits ranged from 0.01 ml to 1.0 ml with a median value of 0.25 ml and a mean value of 0.34 ml.

19:19 What is the limit for an acceptable volume of blood withdrawal from a single collection?

Opin. Numerous studies have associated severe hemodynamic changes and hemor- rhagic shock with blood losses greater than 30% of the total blood volume.[55,56] Smaller volume losses in rats (15–20% of the blood volume) reduce cardiac output by nearly 50%.[57,58] Overall, the recommendation of a maximal withdrawal of 15% of blood volume appears reasonable.[59]

Surv. What is the maximal blood volume withdrawal that your IACUC accepts as rea- sonable for a single withdrawal?

- No IACUC policy   39/112
- Maximal single blood withdrawal limit (percentage blood volume or body weight basis)   73/112

Withdrawal maximums varied from 0.8% to 15% (mean 4%, median 1.5%) of body weight and 2% to 25% (mean 14%, median 10%) of blood volume.

19:20 What clinical tests should be required if repeated blood sampling is a component of the protocol?

Reg. The AWAR (§2.33,b,2), the PHS Policy (IV,C,1,e), and the *Guide* (pp. 55–66) all require the provision of adequate veterinary care for the diagnosis, control, and treatment of diseases and injuries.

Opin. Standard hematology tests are often used to monitor the effects of multiple blood withdrawals upon an animal's well-being. The hematocrit and hemoglobin con- centration are commonly used because of the ease of these tests and the availability of rapid results. While these tests are generally excellent indicators of anemia in animals subjected to blood collections, they do have their limitations. The hemat- ocrit is unchanged immediately after blood withdrawal and underestimates blood loss for up to 72 hours after blood removal.[60] Hemoglobin concentration is even slower in responding to blood loss, and in human blood donors is lowest between 1 and 2 weeks after donation.[61]

Surv. Does your institution's IACUC require that clinical hematology tests be performed on animals when repeated blood sampling is a portion of the protocol?

- No routine testing necessary   45/103
- Require some clinical laboratory tests   16/103
- Rely on veterinary staff for monitoring   42/103

References

1. Grumpstrup-Scott, J., and Greenhouse, D.D., NIH intramural recommendations for the research use of complete Freund's adjuvant, *ILAR News*, 30, 9, 1988.
2. Canadian Council on Animal Care, CCAC Guidelines on Antibody Production, Canadian Council on Animal Care, Ottawa, 2002. Available on the World Wide Web at: http://www.ccac.ca/en/CCAC_Programs/Guidelines_Policies/GDLINES/Antibody/antibody.pdf.

3. United Kingdom Co-ordinating Committee on Cancer Research (UKCCCR), *Guidelines for the Welfare of Animals in Experimental Neoplasia,* 2nd ed., *Br. J. Cancer,* 77, 1, 1998.

4. Jackson, L.R., and Fox, J.G., Institutional policies and guidelines on adjuvants and antibody production, *ILAR J.,* 37, 141, 1995.

5. Leenaars, M., and Hendriksen, C.F.M., Critical steps in the production of polyclonal and monoclonal antibodies: evaluation and recommendations, *ILAR J.,* 46, 269, 2005.

6. Leenaars, P., et al., The production of polyclonal antibodies in laboratory animals, ECVAM Workshop Report 35, *Altern. Lab. Anim.,* 27, 79, 1999.

7. Munoz, J., Effect of bacteria and bacterial products on antibody response, *Adv. Immunol.,* 4, 397, 1964.

8. Altman, A., and Dixon, F.J., Immunomodifiers in vaccines, *Adv. Vet. Sci. Comp. Med.,* 33, 301, 1989.

9. Freund, J., and McDermott, D., Sensitization to horse serum by means of adjuvants, *Proc. Soc. Exp. Biol. Med.,* 49, 548, 1942.

10. Freund, J., The mode of action of immunologic adjuvants, *Adv. Tuber. Res.,* 7, 130, 1956.

11. Stewart-Tull, D., Freund-type mineral oil adjuvant emulsions, in *The Theory and Practical Application of Adjuvants,* Stewart-Tull, D.E.S., Ed., John Wiley & Sons, New York, 1995, p. 1.

12. Stewart-Tull, D., The future potential for the use of adjuvants in human vaccines, in *Biomedical Science and Technology: Recent Developments in the Pharmaceutical and Medical Sciences,* Hincal, A.A., and Ka, H.S., Eds., Plenum Press, New York, 1998, p. 129.

13. Broderson, J.R., A retrospective review of lesions associated with the use of Freund's adjuvant, *Lab. Anim. Sci.,* 39, 400, 1989.

14. Leenaars, P., et al., Evaluation of several adjuvants as alternatives to the use of Freund's adjuvant in rabbits, *Vet. Immunol. Immunopathol,* 40, 225, 1994.

15. Leenaars, P., et al., Comparison of adjuvants for immune potentiating properties and side effects in mice, *Vet. Immunol. Immunopathol.,* 48, 123, 1995.

16. Leenaars, P., et al., Assessment of side effects induced by injection of different adjuvant/ antigen combinations in rabbits and mice, *Lab. Anim.,* 32, 387, 1998.

17. Stills, H.F., Jr., Polyclonal antibody production, in *The Biology of the Laboratory Rabbit,* Manning, P.J., Ringler, D.H., and Newcomer, C.E., Eds., Academic Press, New York, 1994, p. 435.

18. Stills, H.F., Jr., and Bailey, M.Q., The use of Freund's complete adjuvant, *Lab Anim.* (NY), 20, 25, 1991.

19. Johnson, D.K., Adjuvant comparison in rabbits, in *Rodents and Rabbits: Current Research Issues,* Proceedings of a SCAW and WARDS conference, Niemi, S.M., Venable, J.S., and Guttman, H.N. Eds., Scientists Center for Animal Welfare, Washington, D.C., 1994, p. 77.

20. Johnson, D.K., Adjuvant comparison in rabbits, *Sci. Anim. Care,* 5(2), 2, 1994.

21. Hendriksen, C., and Hau, J., Production of polyclonal and monoclonal antibodies, in *Handbook of Laboratory Animal Science.* Vol. 1. *Essential Principles and Practices,* 2nd ed., Hau, J., and Van Hoosier, Jr., G.L., Eds., CRC Press, Boca Raton, FL, 2003, p. 391.

22. Stills, Jr., H.F., Adjuvants and antibody production: dispelling the myths associated with Freund's complete and other adjuvants, *ILAR J.,* 46, 280, 2005.

23. U.S. Department of Agriculture, Animal and Plant Health Inspection Service, Animal Care, Policy 11. Painful/Distressful Procedures, April 14, 1997. Available on the World Wide Web at: www.aphis.usda.gov/ac/policy11.html.

24. Smith, D.E., et al., The selection of an adjuvant emulsion for polyclonal antibody production using a low-molecular weight antigen in rabbits, *Lab. Anim. Sci.,* 42, 599, 1992.

25. Chapel, H.M., and August, P.J., Report of nine cases of accidental injury to man with Freund's complete adjuvant, *Clin. Exp. Immunol.,* 24, 538, 1995.

26. Hughes, L.E., Kearney, R., and Tully, M., A study in clinical cancer immunotherapy, *Cancer,* 26, 269, 1970.

27. Besson, J.M., and Guilbaud, G., Eds., *The Arthritic Rat as a Model of Clinical Pain?* Excerpta Medica, Amsterdam, 1988.

28. Toth, L.A., et al., An evaluation of distress following intraperitoneal immunization with Freund's adjuvant in mice, *Lab. Anim. Sci.,* 39, 122, 1989.

29. Lipman, N.S., et al., Comparison of immune response potentiation and *in vivo* inflammatory effects of Freund's and Ribi adjuvants in mice, *Lab. Anim. Sci.,* 42, 193, 1992.

30. Giffen, P.S., et al., Markers of experimental acute inflammation in the Wistar Han rat with particular reference to haptoglobin and C-reactive protein, *Arch. Toxicol.*, 77, 392, 2003.
31. Harlow, E., and Lane, D., *Adjuvants in Antibodies: A Laboratory Manual,* Cold Spring Harbor Laboratory, Cold Spring Harbor, NY, 1988, p. 96.
32. Amyx, H.L., Control of animal pain and distress in antibody production and infectious disease studies, *J. Am. Vet. Med. Assoc.*, 191, 1287, 1987.
33. Johnston, B.A., Eisen, H., and Fry, D., An evaluation of several adjuvant emulsion regimens for the production of polyclonal antisera in rabbits, *Lab. Anim. Sci.*, 41, 15, 1991.
34. Wanstrup, J., and Christensen, H.E., Granulomatous lesions in mice produced by Freund's adjuvant, *Acta Pathol. Microbiol. Scand.*, 63, 340, 1965.
35. Geisterfer, M., and Gauldie, J., Regulation of signal transducer, GP130 and the LIF receptor in acute inflammation *in vivo*, *Cytokine*, 8, 283, 1996.
36. Olivier, E., et al., A novel set of hepatic mRNAs preferentially expressed during an acute inflammation in rat represents mostly intracellular proteins, *Genomics*, 57, 352, 1999.
37. Brooks, R.D., Betz, R.D., and Moore, R.D., Injury and repair of the lung: response to intravenous Freund's adjuvant, *J. Pathol.*, 124, 205, 1978.
38. Herbert, W.J., The mode of action of mineral oil emulsion adjuvants on antibody production in mice, *Immunology*, 14, 301, 1968.
39. Hu, J.-G., Yokoyama, T., and Kitagawa, T., Studies on the optimal immunization schedule of experimental animals. IV. The optimal age and sex of mice, and the influence of booster injections, *Chem. Pharm. Bull.*, 382, 448, 1990.
40. Hu, J.-G., et al., Studies on the optimal immunization schedule of the mouse as an experimental animal: the effect of antigen dose and adjuvant type, *Chem. Pharm. Bull.*, 37, 3042, 1989.
41. Hu, J.-G., and Kitagawa, T., Studies on the optimal immunization schedule of experimental animals. VI. Antigen dose response of aluminum hydroxide–aided immunization and booster effect under low antigen dose, *Chem. Pharm. Bull.*, 38, 2775, 1990.
42. Hanley, W.C., Artwohl, J.E., and Bennett, B.T., Review of polyclonal antibody production procedures in mammals and poultry, *ILAR J.*, 37, 93, 1995.
43. Linscott's Directory of Immunological and Biological Reagents. Available on the World Wide Web at: http://www.linscottsdirectory.com/.
44. Weimer, R.V., Jr., Manufacturers' Specifications and Reference Synopsis (MSRS): Catalog of Primary Antibodies, MSRS/Aerie Corp., Birmingham, AL. Available on the World Wide Web at: http://www.antibodies-probes.com/.
45. The Internet resource for locating certain antibodies and suppliers of antibodies is available on the World Wide Web at: www.antibodyresource.com.
46. Potkay, S., et al., Frequently asked questions about the Public Health Service Policy on Humane Care and Use of Laboratory Animals, *Lab Anim.* (NY), 24, 24, 1995.
47. Ellis, G., and Garnett, N., OPRR Reports, Number 98-01, November 17, 1997. Available on the World Wide Web at: www.nih.gov/grants/oprr/dc98-01.htm.
48. Peterson, N.C., and Peavey, J.E., Comparison of *in vitro* monoclonal antibody production methods with an *in vivo* ascites production technique, *Contemp. Topics Lab. Anim. Sci.*, 37, 61, 1998.
49. Committee on Methods of Producing Monoclonal Antibodies, Monoclonal Antibody Production, National Academy Press, Washington, D.C., 1999. Available on the World Wide Web at: http://www.nap.edu/books/0309064473/html/.
50. Jackson, L.R., et al., Monoclonal antibody production in murine ascites. I. Clinical and pathologic features, *Lab. Anim. Sci.*, 49, 70, 1999.
51. Jackson, L.R., et al., Monoclonal antibody production in murine ascites. II. Production characteristics, *Lab. Anim. Sci.*, 49, 81, 1999.
52. McGuill, M.W., and Rowan, A.N., Refinement of monoclonal antibody production and animal well-being, *ILAR News*, 31(1), 7, 1989.
53. Brodeur, B.R., Tsang, P., and Larose, Y., Parameters affecting ascites tumour formation in mice and monoclonal antibody production, *J. Immunol. Methods*, 71, 265, 1984.
54. Hoogenraad, N.J., and Wraight, C.J., The effect of pristane on ascites tumor formation and monoclonal antibody production, *Methods Enzymol.*, 121, 381, 1986.

55. Noble, D., and Gregerson, M.I., Blood volume in clinical shock. II. The extent and cause of blood volume reduction in traumatic hemorrhage and burn shock, *J. Clin. Invest.*, 25, 172, 1946.
56. Williams, W.J., et al., *Hematology*, 3rd ed., McGraw-Hill, New York, 1983.
57. Saperstein, L.A., Saperstein, E.H., and Bredemeyer, A., Effect of hemorrhage on the cardiac output and its distribution in the rat, *Circ. Res.*, 8, 135, 1960.
58. Ploucha, J.M., and Fink, G.D., Hemodynamics of hemorrhage in the conscious rat and chicken, *Am. J. Physiol.*, 251, R486, 1986.
59. McGuill, M.W., and Rowan, A.N., Biological effects of blood loss: implications for sampling volumes and techniques, *ILAR News*, 31(4), 5, 1989.
60. Lee, G.R., The normocytic anemias, in *Wintobe's Clinical Hematology*, 9th ed., Lee, G.R., et al., Eds., Lea & Febiger, Philadelphia, 1993, p. 885.
61. Wadsworth, G.R., Recovery from acute hemorrhage in normal men and women, *J. Physiol.* (Lond), 129, 583, 1955.

20

Occupational Health and Safety

Robin Lyn Trundy and Susan Stein Cook*

Introduction

The use of animals for research purposes has always been directly or indirectly linked to human health concerns. Animal allergens, zoonotic diseases, and physical injuries are only some of the issues that need to be addressed in a general occupational health program associated with animal research. Although regulations and guidelines dealing with human health and protection are a valuable resource for establishing a comprehensive occupational health program, their scope and level of detail will always be limited. Fortunately, most current health and safety regulations are performance oriented, focusing on the goal (i.e., a safe work environment) rather than describing how to get there. This allows institutions to tailor their safety programs to protect employees in the most effective way.

General Occupational Health and Safety

20:1 What is the general responsibility of the IACUC toward ensuring a safe work environment for persons working with laboratory animals?

Reg. The need to protect the health and safety of employees involved in animal-based research is addressed in the *Guide* (p. 14) and the PHS Policy (IV,A,1,f). The PHS Policy (IV,A,1) requires institutions to use the *Guide* as the basis for their institutional program for activities involving animals. The *Guide* states that institutions must establish an occupational health and safety (OHS) program as part of the overall animal research program. The *Guide* (p. 14) specifically references the National Research Council publication *Occupational Health and Safety in the Care and Use of Research Animals*[1] to be used as a tool for establishing such a comprehensive OHS program. While the main responsibility of the OHS program lies with the institution, the IACUC is specifically charged with the oversight and evaluation of the animal care and use program. Since the OHS program needs to be part of the overall animal care and use program (*Guide*, p. 14), the IACUC also has to assume at least partial responsibility for the program.

Opin. The applicable referenced guidelines and regulations provide limited detail specifically addressing the IACUC's responsibility for occupational health and safety.

* The authors thank Robert J. Ceru and Kristin Erickson for their contributions to this chapter and to the first edition of *The IACUC Handbook*. The authors also thank Stefan Wagener for his contribution to this chapter in the first edition.

Therefore, the institution must manage and assign responsibilities covering the various programs required by the *Guide*. The quality of the occupational health and safety program will be significantly impacted by the quality of interaction of the IACUC with relevant institutional functions such as the research program, environmental health and safety, and occupational health services. Institutions are required to address OHS issues and assign responsibilities to various groups or individuals inside the institution for management and oversight of the different components. For example, occupational health requires the input of medical professionals, while physical, chemical, biological, and radiological hazards are best addressed by specific committees or safety professionals with expertise in these areas. Ideally, the IACUC assumes the role of a facilitator, interacting with the various programs, committees, and individuals to ensure the existence of a comprehensive OHS program and compliance with the *Guide*.

20:2 What is the general responsibility of the IACUC toward ensuring a safe work environment for persons having access to an animal facility but not working with laboratory animals (e.g., maintenance, clerical personnel, visitors)?

Reg. As stated in 20:1, the institution at large is primarily responsible for the establishment of an occupational health and safety program (OHS). The PHS Policy (IV,A,1,f) requires a health program for personnel who have frequent contact with animals, as well as all persons who work in laboratory animal facilities. According to the *Guide* (p. 14), the level and extent of personnel participation in the OHS program should be based on a risk assessment completed by OHS professionals. In addition, the National Research Council publication *Occupational Health and Safety in the Care and Use of Research Animals*,[1] as referenced by the *Guide*, suggests including not only animal caretakers, technicians, students, volunteers, investigators, and veterinarians, but also facility maintenance personnel, housekeepers, security, and other staff who must perform job duties in animal research environments.

Opin. Ideally, the IACUC facilitates the close interaction among institutional groups dealing with employee health and safety as it relates to animal research facilities. For example, the institution's environmental health and safety (EHS) office oversees the health and safety of maintenance personnel working in areas with known chemical and biological hazards. The institutional occupational health group and the medical professionals identify any necessary surveillance criteria. The assistance of these and other health and safety groups should be requested by the IACUC to perform risk assessments, project review, and health and safety recommendations.

20:3 Does the IACUC have a responsibility to ensure a safe working environment for research technicians, investigators, and students who use animals outside the animal facility?

Reg. The PHS Policy (IV,A,1,f) is applicable to all institutional programs involving animals, not only in animal facilities. Therefore, the occupational health and safety program (OHS) must address all risks to investigators, technicians, and students associated with PHS-supported animal activity, regardless of where it occurs.

Opin. The IACUC does not typically have direct responsibility or compliance oversight of OHS programs for research technicians, students, and investigators who work with animals outside (or inside) the animal facility. However, the IACUC is charged with the responsibility of reporting unsafe conditions and communicating openly with institutional health and safety professionals to foster a safe work environment and ensure safe project design.

General and specific regulatory health and safety requirements (i.e., OSHA regulations), however, do apply to worker protection inside and outside a facility. Hazards associated with animal research outside a facility can be of great concern. One example is field research involving wild animals. The incorrect use of equipment to trap or immobilize wild animals can pose a significant physical hazard. Zoonotic diseases, such as rabies or hantavirus pulmonary syndrome, can be life threatening. Electrical and chemical hazards can also be present. Prudent practices include risk assessment of all animal use protocols by the knowledgeable institutional safety groups or individuals to assure and maintain worker safety outside the facility.

20:4   Should the IACUC require occupational health examinations for those involved in animal care and use? If so, what should be the scope and extent of such a program in terms of personnel and diagnostic procedures?

Reg.   The *Guide* (pp. 14–19) outlines an occupational health program, as required in the PHS Policy (IV,A,1,f). This was summarized by NIH/OPRR[1] (now NIH/OLAW) as follows: "Basic elements of any health program, however, should provide: a pre-employment medical evaluation and history; immunization against tetanus; detailed training on how to perform required procedures safely; instruction in personal hygiene, zoonoses, and precautions for pregnant women and others at risk; protective clothing and devices; instruction in first aid procedures appropriate to potential hazards; and access to medical attention for the treatment of animal bites, scratches, allergies, and other job-related injuries or illnesses."

Opin.   The decision on a health examination is usually based on the type of work performed, the occupational risks involved, the species of animal, and the type and duration of animal contact required. All these factors are included in a project-specific risk assessment that needs to be performed in cooperation with health care professionals. Rather than requiring blanket health examinations, the IACUC must ensure that occupational health professionals are involved in tailoring the health assessment program for all personnel involved in animal care and use.

20:5   What circumstances may require additional medical surveillance activities beyond standard occupational health examinations for those who work with animals?

Reg.   Some Occupational Safety and Health Administration (OSHA) standards require specific medical surveillance activities if an employee is exposed to a hazard at levels or durations above those that are considered to be safe. Examples include the Occupational Noise Exposure Standard, the Respiratory Protection Standard, and the Bloodborne Pathogens Standard.

Under the Occupational Noise Exposure Standard,[2] employees who are exposed to noise levels at or above 85 decibels for an 8-hour time-weighted average (TWA) must have baseline and annual audiometric testing through the occupational health provider. These levels may be exceeded in some facilities, as a result of noise from animals or equipment.

The Respiratory Protection Standard[3] requires that any employee who is required to wear a respirator for protection against an airborne hazard on the job be medically evaluated before wearing a respirator; additional ongoing surveillance requirements are determined by the occupational health provider. This standard may apply to the use of certain chemical products in the research facility and for work with airborne infectious agents.

The Bloodborne Pathogens Standard[4] requires that employees who are exposed to certain human-derived materials (which include human cells under most circumstances) be included in the bloodborne pathogens exposure control program. These personnel must be offered the hepatitis B vaccination and receive specific postexposure services if an employee sustains an exposure to such materials through a splash to the eyes, nose, or mouth; absorption through broken skin; or a cut with a contaminated object. In the animal research environment, personnel who handle human cells during the introduction of these materials into an animal model or harvest these materials from the animal model are likely to be affected by these requirements.

Opin. In addition to the OSHA requirements, an institution should ensure that the occupational health–medical surveillance activities include additional components to protect personnel if infectious disease research is under way in the animal research environment or laboratories. The Centers for Disease Control and Prevention (CDC) publication *Biosafety in Microbiological and Biomedical Laboratories* is considered a de facto standard by many agencies and organizations. This document includes a specific appendix that addresses the need for an institutional policy on vaccinations for environments where an infectious disease risk is present. Additionally, the document provides specific vaccination and/or medical surveillance recommendations for work with certain organisms.[5] Examples include: poxviruses, cercopithecine herpesvirus 1 (herpesvirus simiae), retroviruses (including HIV and SIV), and arboviruses.

20:6 Should the IACUC establish safe working rules for an animal facility or laboratory?

Reg. The IACUC has no specific regulatory mandate for establishing safe working rules unless they pertain specifically to animal health and care and are based on the *Guide* or other relevant animal use and care regulations. (See 20:1.)

Opin. Safe working rules are a necessity for all facilities whether or not they are used for research with animals. Most safety rules and policies are hazard specific and based on regulatory requirements addressing those hazards. For example, occupational safety issues in laboratories working with chemical hazards are part of OSHA's Laboratory Safety Standard.[6] The chemical hygiene officer of the facility is the most likely person to establish OSHA-based safe working rules for the facility or laboratory if hazardous chemicals are involved. In the end, it will be the institution's responsibility to define the responsibilities clearly as well as to set the standards for health and safety.

20:7 What types of hazards are relevant to animal research settings?

Opin. While animal research settings involve some unique hazards related to working with the animals, such common workplace hazards as sharps injuries, burns, and falls also occur. In addition, the research might involve certain physical, chemical, or biological hazards. An overview of relevant physical, chemical, biological, and protocol-related hazards has been published.[1] In general, physical hazards in the animal research environment can include bites, scratches, and kicks; sprains and strains caused by moving of equipment; operational hazards involving electricity, machinery, and noise; or protocol-induced hazards such as radioactivity or the use of lasers. Chemical hazards can be directly related to a specific agent or procedure. Examples include chemicals used for processing tissues and cleaning or disinfecting research equipment. The accidental inhalation of waste anesthetic gases provides another route of chemical exposure. Research protocols might require the application of toxins, carcinogens, and other hazardous chemicals. Biological hazards, such as infectious pathogens, can be introduced through naturally occurring or experimentally infected research animals or include the application of recombinant DNA technology. Another significant hazard is manifested in the increasing development of animal-related allergies or occupation-related asthma.[1,7]

20:8   What sources of information can the IACUC access relative to hazards found in the animal research setting?

Opin.   The common hazards include chemical, physical, biological, mechanical, and environmental hazards and vary with type of research, geographic location, research species, facility design, and duration of exposure. The IACUC should rely on institutional health and safety officers and the AV to collect, collate, and share material from local, state, and federal agencies relevant to common hazards. (See 20:30.) Health and safety information and regulations may also be accessed via the World Wide Web. This often can be achieved by visiting such agency Web sites as the federal OSHA Web site (www.osha.gov) and performing a keyword search. Texts such as *Safety and Health on the Internet*[8] provide comprehensive health and safety-related listings.

The *Guide* (pp. 85–86) recommends a comprehensive listing of references on biological hazards, including CDC/NIH *Biosafety in Microbiological and Biomedical Laboratories.*[5]

20:9   What are the relevant institutional committees and departments that the IACUC should interface with relative to personnel safety?

Opin.   The IACUC has an oversight responsibility (see 20:1) related to personnel safety issues associated with animal protocols. To assure that personnel safety issues are effectively addressed as part of the review process, the IACUC must seek input and communicate with the committees and departments specifically charged with regulatory compliance, worker safety, and environmental health. These groups and individuals interact directly with agencies such as OSHA, the Nuclear Regulatory Commission (NRC), and the Centers for Disease Control and Prevention (CDC). These committees and departments include the following:

- Environmental Health & Safety (EHS)/Occupational Safety provides assistance relatied to common physical hazards, personal protective equipment assessments, and chemical use, including anesthetics and disinfectants.
- Biological Safety provides assistance related to biological containment practices, human-derived material use, safe handling of sharps, biological safety cabinet use, and biohazardous waste management.

- Radiation Safety provides assistance with radiation safety training, signage, personnel exposure monitoring, and area surveys for radioactive contamination. The scope of controls for radiological hazards is determined by the Institutional Radiation Safety Committee. When radiological hazards are used in conjunction with animals, the IACUC must maintain open communication with the Radiation Safety Committee, or a radiation safety officer, to assure that radiation safety compliance items have been addressed before IACUC approval is granted.

- Occupational Health provides medical screening of personnel who are enrolled in the occupational health program, as well as information on allergies and hygiene.

- A Chemical Hygiene Committee may not be established at an institution (see 20:19). If such a committee is not established, a chemical hygiene officer with oversight for animal research facilities should be included in the IACUC review for protocols involving hazardous chemicals, including anesthetics.

- An Institutional Biosafety Committee (IBC) is required for an institution if it receives any NIH funds for recombinant DNA research. This committee is charged with review of research involving recombinant DNA molecules (see 20:22), although many IBCs also review the use of biological hazards including potentially infectious agents and toxins. When recombinant DNA molecules are used in conjunction with animals, the IACUC and the IBC should work jointly to assure that all regulatory compliance items have been addressed before either committee approves a project.

Physical and Chemical Hazards

20:10   What is meant by a *physical hazard*?

Reg.   Physical hazards are those generally associated with:

- Noise
- Temperature
- Vibration
- Electricity
- Nonionizing and ionizing radiation (also considered a radiation hazard)
- Ultraviolet radiation
- Lasers
- Illumination
- Sharp objects

Occupational exposure limits and protective measures for some of these hazards are covered under specific OSHA standards:

- Occupational Noise Exposure[2]
- Nonionizing Radiation[9]
- Ionizing Radiation[10]
- Electrical[11,12]

In addition, a hazardous chemical is defined as a physical hazard if there is scientifically valid evidence that it is a combustible liquid, a compressed gas, explosive, flammable, organic peroxide, oxidizer, pyrophoric, unstable (reactive), or water-reactive.[13]

Opin. The IACUC should seek the assistance of occupational health and safety professionals to assess the potential physical hazards associated with animal protocols and facilities.

20:11 What are common physical hazards encountered in the research animal environment?

Opin. Physical hazards commonly encountered are associated with the operation and manipulation of machinery, equipment, or instruments. These hazards include heat exposure from autoclaves; injuries caused by sharp objects (e.g., cages, broken glass, scalpels); animal bites, scratches, and kicks; and back injuries.

Common hazards, such as noise, are often ignored and accepted as part of the animal facilities environment. Excessive noise can be produced by machinery and animals, especially pigs and dogs. Exposure to intense noise over time will result in hearing loss. If normal talking or phone conversation is not possible because of excessive noise, the noise level should be assessed by a safety professional. Electricity is of concern if electrical equipment is out of date, poorly maintained, or improperly grounded. Compliance with electrical code requirements is necessary, as is equipment maintenance.

20:12 Should physical methods of euthanasia be considered a physical risk?

Reg. Physical methods of euthanasia as defined by the American Veterinary Medical Association (AVMA)[14] include captive bolt, gunshot, cervical dislocation, decapitation, electrocution, microwave irradiation, exsanguination, stunning, and pithing. The last three are used in conjunction with other methods. According to the 2000 Report of the AVMA Panel on Euthanasia, "Given that most physical methods involve trauma, there is inherent risk for animals and human beings; therefore, extreme care and caution should be used."

PHS Policy (IV,C,1,g) requires euthanasia methods to be consistent with the AVMA recommendations unless a deviation for scientific reasons is justified, in writing, by the investigator. APHIS/AC Policy 3[15] requires that the method of euthanasia "must be consistent with the current Report of the AVMA Panel on Euthanasia."

Opin. As with all physical procedures and practices, the skill and expertise of the person performing the procedure determine the degree of risk. Prior to approval of such methods, the IACUC should evaluate the proficiency and level of training of individuals performing these procedures and be assured that the equipment is in good working order.

20:13 How can the IACUC evaluate physical hazard risk?

Opin. The IACUC does not have a specific mandate to evaluate physical hazards unless they are related to the animal as part of the research protocol. Potential physical hazards for animals and research personnel working with them are evaluated through routine inspections and protocol review. Since the IACUC is required to inspect the animal facilities regularly, these inspections should be used to observe practices and procedures involving potential physical hazards. The IACUC should involve institutional occupational health and safety professionals in the evaluation process as control of such hazards is within the purview of these individuals' responsibilities.

20:14   What is meant by a *chemical hazard*?

Reg.   A *hazardous chemical* is defined by OSHA as any chemical, chemical compound, or mixture of compounds that presents a physical or health hazard.[13] As noted in 20:10, a chemical is a physical hazard by OSHA definition if there is scientifically valid evidence that it is:

- A flammable or combustible liquid
- A compressed gas
- An organic peroxide
- An explosive, an oxidizer
- A pyrophoric
- An unstable material (reactive)
- A water-reactive material

A chemical is a health hazard by OSHA definition if there is statistically significant evidence based on at least one study conducted in accordance with established scientific principles that acute or chronic health effects may occur in exposed employees. These include allergens, embryotoxicants, carcinogens, toxic or highly toxic agents, reproductive toxicants, irritants, corrosives, sensitizers, hepatotoxins, nephrotoxins, neurotoxins, hematopoietic systems agents, and any agents that damage the lungs, skin, eyes, or mucous membranes.

20:15   What are some typical chemical hazards that can be found in a research environment where animals are used?

Opin.   Researchers commonly use a wide variety of chemicals. Toxic agents and carcinogens can be used to induce disease or study metabolic processes. Other chemicals are used in analytic techniques. Most laboratories have flammable solvents, corrosive liquids, and a variety of toxic chemicals. Cleaning materials, solvents, acids, disinfectants, and sanitizers, while outside the animal holding rooms, still can pose an exposure hazard if improperly stored or used.

20:16   What is a Material Safety Data Sheet (MSDS)?

Reg.   A MSDS is a document containing chemical hazard identification and safe handling information. It is prepared in accordance with the OSHA Hazard Communication Standard, also known as the "Right-to-Know" law.[13] Chemical manufacturers and distributors must provide the purchasers of hazardous chemicals with an appropriate MSDS for each hazardous chemical or product purchased.

Opin.   Copies of the MSDS for hazardous chemicals at a given worksite must be readily accessible to employees in that area. As a source of detailed information on hazards, they must be located close to workers and readily available during working hours.[13] MSDSs for hazardous chemicals used in the animal facility are required by law. Accessibility, however, does not necessarily require having copies available in each room or attached to each chemical. Central repositories (e.g., main office) can store all available MSDSs. Each area housing hazardous chemicals should post information identifying the location of MSDSs for that area. Many facilities use computer-based retrieval systems, loaded with software containing large numbers of MSDSs.

20:17   Should all approved anesthetic and euthanasia agents be considered hazardous?

Reg.   Approved anesthetic and euthanasia agents are considered hazardous if they meet one or more of the criteria identified for hazardous chemicals (see 20:14).

Opin.   The degree of hazard varies from agent to agent. Factors that determine the degree of hazard include the specific agent and its properties, exposure concentrations, work practices, engineering controls, personal protective equipment used, and route of administration.

The IACUC should rely on the AV, the chemical hygiene officer, and MSDS information (see 20:16) to address safety issues related to these agents. It is recommended that institutions have SOPs for use, storage, and disposal of anesthetic and euthanasia agents. Training in proper techniques involving inhalation and noninhalation agents should be provided.

20:18   How can an IACUC evaluate a chemical risk?

Opin.   An IACUC could partner with an industrial hygienist within its organization (or hired as a consultant) or any other health and safety professional with specific knowledge and experience (e.g., toxicologist) to review and evaluate chemical risks. Areas for review should include the following:

- Chemical agents used
- Engineering controls
- Special personal protective equipment required
- Route of excretion
- Precautions for handling animals exposed to specific chemicals
- Animal disposal
- Bedding disposal
- Cage decontamination
- Special precautions
- Storage and delivery procedures
- Level of training and experience of personnel

20:19   Is a separate chemical hazard committee necessary?

Reg.   The use of an institutional chemical hygiene committee is an option suggested by the Occupational Safety and Health Administration (OSHA) Laboratory Safety standard[6] to be established by the institution if appropriate. Currently, OSHA does not mandate a chemical hazard committee. However, if hazardous chemicals are used in the laboratory or animal facility, a chemical hygiene plan is required. This plan not only outlines all relevant safety procedures and practices as required by law, but also must designate personnel who are responsible for implementing the plan including the assignment of a chemical hygiene officer.

Opin. Establishment of a separate chemical hygiene committee is a recommended option in overseeing institutional compliance with all aspects of chemical safety. In addition, review of animal research projects involving hazardous chemicals should be done in cooperation with the chemical hygiene officer.

20:20 What institutional resources can the IACUC access to determine whether there is a suspected need for respiratory protection for an animal protocol?

Reg. The OSHA Respiratory Protection Standard requires the institution to develop and implement a written respiratory protection program with worksite-specific procedures and elements for required respirator use.[3] This program, and the use of respirators, are necessary if employees are exposed to air contaminated with harmful dusts, fogs, fumes, mists, gases, smokes, sprays, or vapors, and these contaminants cannot be controlled or eliminated by other means (e.g., engineering controls). Animal care–related examples of contaminants and their applications may include disinfection of animal rooms with aerosolized chemicals and project-specific feed preparations containing carcinogens. If an employer is required to have a respiratory protection program, the employer or institution must designate a program administrator who is qualified by appropriate training or experience to administer or oversee the respiratory protection program and conduct the required evaluations of program effectiveness.[3] This individual is the key resource for the IACUC for matters associated with respiratory protection.

Opin. Effective respiratory protection programs require a knowledgeable program administrator, an occupational health professional who can provide medical evaluations, and an understanding, at the institutional level, of the restrictions and requirements for use of respirators by employees. The IACUC should confer with the respiratory protection program administrator (usually a member of environmental health and safety or the occupational health department) whenever the need for respiratory protection is identified or suggested, as part of an animal protocol.

Biological and Radiological Hazards

20:21 What is the definition of a *biological hazard*?

Reg. Currently there are no regulatory-based definitions for biological hazards. A regulatory definition for bloodborne pathogens has been published.[4] Recombinant DNA molecules are defined in the *NIH Guidelines for Research Involving Recombinant DNA Molecules*.[16] Certain infectious materials are defined by the CDC,[17] USDA,[18,19] and Department of Transportation[20] regulations related to transfer, acquisition, and transportation.

Opin. The term *biological hazards* (or *biohazards*) refers commonly to agents or materials of biological origin that are potentially hazardous to humans, animals, or plants. Although there is no standardized definition for biohazards, institutions often include infectious agents, materials containing recombinant DNA molecules, toxins of biological origin, and human- or nonhuman primate–derived materials in the definition.

20:22   What is meant by *recombinant DNA molecules* and what are some examples?

Reg.   *Recombinant DNA molecules* are defined by the *NIH Guidelines*[16] as molecules that are constructed outside living cells by joining natural or recombinant DNA segments to DNA molecules that can replicate in a living cell, or molecules that result from the replication of those described.

Opin.   The ability to isolate, manipulate, and express genetic material has resulted in a new field of basic and applied research called *genetic engineering*. By using recombinant DNA techniques, genes can be isolated, expressed, and transferred, resulting in a multitude of possible applications. The development of transgenic and knockout animals (e.g., rodents) has proved valuable for studying a variety of diseases. Other applications involve the use of genetically modified infectious agents for vaccine development in animal models. Transgenic technology has also been used to alter certain aspects of the biochemistry, hormonal balance, and protein products in livestock.

20:23   Must an institution working with any biological hazards have a biological safety officer?

Reg.   The only agency currently requiring the position of a biological safety officer is the NIH as part of the requirements for recombinant DNA research.[16] As outlined in the *NIH Guidelines:* "The institution shall appoint a biological safety officer if it engages in large-scale research or production activities involving viable organisms containing recombinant DNA molecules." In addition, "the institution shall appoint a biological safety officer if it engages in recombinant DNA research at BL3 or BL4."

Opin.   Most institutions have gone beyond this basic requirement and have appointed a biological safety officer to oversee all aspects of biological safety. Biological safety officers commonly manage institutional programs addressing the OSHA requirements for bloodborne pathogens and compliance with CDC and USDA programs (i.e., Select Agents), develop biological safety operating procedures, evaluate the potential for zoonotic disease transmission, and provide training and education related to biosafety topics. To maintain a high standard of proficiency and education, certification and testing programs for biological safety professionals have been developed by the American Biological Safety Association (ABSA) and the National Registry of Microbiologists (NRM). The IACUC should consult the biological safety officer in all areas related to biosafety.

20:24   Do transgenic or knockout rodents that have been commercially purchased or received from another institution fall under the *NIH Guidelines for Research Involving Recombinant DNA Molecules*?

Reg.   As outlined in Appendix C-VI of the *NIH Guidelines*,[16] the purchase or transfer of transgenic rodents for experiments that require Biosafety Level 1 (BL1) containment are exempt from the *NIH Guidelines*.

Opin. The only aspect of transgenic rodent work that is currently exempt from the *NIH Guidelines* is the purchase or transfer of those animals that require BL1 laboratory containment only. Transgenic rodents generated at the institution are still covered by the *NIH Guidelines* and require, at a minimum, institutional biosafety committee notification simultaneous with the initiation of the work, even if the rodents require only BL1 containment. BL1 is the lowest level of physical containment and appropriate for well-characterized agents not known to cause disease, and of minimal potential hazard to laboratory personnel and the environment. Generally, combinations of laboratory practices, containment equipment, and facility design can be made to achieve different levels of physical containment, also referred to as *biosafety levels*. Four levels of physical containment (BL1 to BL4) are commonly used: from BL4, the highest level of physical containment, to BL1, the lowest. For a comprehensive overview of biosafety levels refer to the CDC and NIH documents previously mentioned.[5,16]

20:25 Should an IACUC member be a voting member of the institutional biosafety committee (IBC)?

Reg. There is currently no requirement that an IACUC member be a voting member of the IBC. However, the *NIH Guidelines*[16] require that one member of the IBC shall have expertise in animal containment principles if the institution is involved in recombinant DNA molecule research with animals, and those projects require IBC approval.

Opin. At smaller institutions, it may be advantageous to use an individual's expertise in a variety of functions. Institutions, therefore, might select an IACUC member with the appropriate expertise to be part of the local IBC. This double function, however, might require a significant time commitment by the individual.

20:26 Assume an agent transmissible to humans, such as a cercopithecine herpes virus 1 (B-virus), is known or suspected to be harbored by an animal, but the virus is not part of the proposed research. Should this protocol be reviewed and approved by the institutional biosafety committee (IBC) before receiving final IACUC approval?

Opin. In lieu of specific regulatory requirements, guidelines and recommendations for implementing a safety and health program specific for B-virus and nonhuman primates have been established by the CDC[5,21,22] and others.[23,24] None of these guidelines or recommendations requires a separate IBC review or approval of research involving animals considered to be infected with herpes B-virus prior to final IACUC approval. However, an institution that is working with or housing animals of the *Macaca* genus should have an established program to control this zoonotic hazard including training, specific SOPs, and exposure response procedures as recommended by the references cited. Development and oversight of the program should be verified as part of an IACUC review of protocols involving macaques or other Old World primates. The biological safety officer (BSO) or the institutional biosafety committee Chair (if there is no BSO) and the AV should be consulted as part of this verification.

    Herpes B-virus is one of the few examples for which specific guidelines are recommended to control a zoonotic disease hazard associated with a given species. When a protocol under review involves animals that are known, or strongly suspected, to be infected with agents transmissible to humans, the IACUC must consult the BSO or the IBC regarding applicable institutional programs or reviews related to biological safety. Research animals should be housed and handled in accordance with the animal biosafety level practices outlined in the CDC/NIH *Biosafety in Microbiological and Biomedical Laboratories.*[5]

20:27 What general guidelines can the IACUC follow to assure that biohazardous materials are being used safely in the animal facility and in individual laboratories?

Reg. The use of biohazardous materials, including recombinant DNA, is guided by a combination of laboratory practices and techniques, safety equipment, and special facilities commonly referred to as *biosafety levels*. One standard reference[5] contains information on four different biosafety levels specific for laboratories and vertebrate animals. The procedures and practices outlined in this publication have been widely accepted as a standard and are used as a basis for a comprehensive biosafety program. Although not federally mandated, compliance is expected by funding agencies including USDA, NIH, and the FDA. Similar information is outlined in Appendix G of the *NIH Guidelines*.[16] Nevertheless, the application of Appendix G is intended for recombinant DNA research involving animals that may not be maintained in enclosed caging systems that place an environmental barrier between the animals and the room environment because of size or growth requirements. When applicable, compliance is mandatory for institutions receiving NIH funding. Local and state agencies should be consulted for all waste-related issues pertaining to biohazardous materials, since specific regulations may exist for the treatment and disposal of biohazardous waste.

Opin. Because of the increasing complexity associated with the use and disposal of biohazardous materials, the IACUC needs to rely on the IBC or biological safety officers for advice, oversight, and compliance. One publication[5] can be used as a guidance document and should be available to all IACUC members. Additional biological safety informational resources are available through the American Biological Safety Association web site.[25] It is strongly recommended that the biological safety officer or institutional biosafety committee review and approve animal use protocols involving biological hazards prior to final IACUC approval. Projects may require specific protocols, safety procedures, safety equipment, inspection, waste disposal, and other considerations.

Specific practices and procedures (e.g., personal protective equipment, immunization) should be posted on animal room doors along with emergency information. All personnel involved should read and sign this information.

20:28 What generally accepted procedure should an IACUC require relative to recapping of needles in the animal facility?

Reg. Under recent changes to the OSHA Bloodborne Pathogens (BBP) Standard,[26] recapping of needles is generally prohibited and the use of sharps devices with safety features is required in the human patient care setting.

The OSHA BBP Standard applies to work settings where personnel are required to handle human-derived materials and may not apply to the majority of animal research settings. The standard applies to those personnel who are exposed to human-derived materials (e.g., tumor cells) during the introduction and harvest of such materials from research animals. However, the regulatory provision for the use of sharps devices with safety features does not extend to the use of sharps for tasks that do not involve delivery of human patient care. Even so, unnecessary recapping of needles in an occupational setting is a hazardous practice and could be cited under Section 5(a) of the Occupational Safety and Health Act, the OSHA General Duty Clause,[27] which states, "Each employer shall furnish to each of his employees employment and a place of employment which are free from recognized hazards that are causing or are likely to cause death or serious physical harm to his employees." In short, if a hazard that is present in the workplace can feasibly be eliminated and no actions are taken to do so, the employer may be cited for violation with the OSHA General Duty Clause.

Opin. Accidental sharp injuries involving animal fluids are a significant occupational hazard and have resulted in human infection.[28]

Prudent practice should require that needles are not recapped and are disposed of in approved sharps containers. In certain circumstances recapping might be necessary but should only be done when all nonrecapping alternatives, including the use of syringes with automated needle retraction function, cannot be employed. If necessary, training in these methods should be provided to investigators and staff. The biological safety officer should be consulted for assistance with establishing safe sharps handling and disposal practices in both laboratory and animal research environments.

20:29 What common personal protective devices should be required by the IACUC for personnel who enter a room housing macaque monkeys?

Reg. As a result of exposures with fatal outcome to cercopithecine herpes virus 1 (B-virus), the CDC and the National Institute of Occupational Safety and Health (NIOSH) have issued specific guidance documents on personal protective equipment.[21,22] The *Guide* (p. 17) specifically states that personnel exposed to nonhuman primates should be provided with such protective items as gloves, arm protectors, masks, and face shields.

Opin. The selection of appropriate personal protective equipment (PPE) should be based on the following process:

- Identification of the most likely hazards to be encountered
- Assessment of the risk and any adverse effects caused by unprotected exposure
- Identification of all other control measures available and feasible to be used instead of personal protective equipment
- Performance characteristics for the required protection (e.g., splash or impact protection)
- Need for decontamination (e.g., reuse or disposal)
- Assessment of any constraints that might negatively influence the use of PPE (e.g., vision, dexterity)

PPE, such as gowns, aprons, gloves, and eye or face protection, is considered the last line of defense. It should only be relied on when the specific hazard cannot be removed or contained with any other control measures. The primary hazard posed by macaques is a potential herpes-B virus exposure through infectious body fluids, bites, or scratches caused by the animal or contaminated objects. The PPE selected should provide protection against the physical hazards as well as fluid exposure. Primary routes of entry are the mucous membranes, which require appropriate splash protection in the form of splash goggles and masks (for the nose and mouth). The use of face shields provides additional face protection. Other routes of entry include parenteral exposure through cuts, breaks in the skin, and needle sticks. PPE selection should address these risks and be based on the task performed. Heavy-duty reinforced leather gloves are appropriate for restraining animals, but inappropriate for handling syringes. In the latter case, double gloving is recommended, since it facilitates the safe changing of gloves (due to contamination) without compromising skin protection. Depending on the task performed, and in accordance with local and state regulations, PPE may have to be disposed of as biohazardous waste. It is highly recommended to use CDC and NIOSH guidelines[21,22] as the basis for a comprehensive B-virus protection program.

20:30   What role does the AV have in evaluating the impact of biological hazards in the animal facility?

Opin.   Biological hazards, in general, or specific infectious agents can have a significant effect on the health of animals. Disease prevention needs to be an essential component of a comprehensive veterinary care program as identified in the *Guide* (p. 57) and is the responsibility of the AV, who is certified or has training or experience in laboratory animal science and medicine involving the species being used. It is in the institution's best interest to support the AV in all aspects of a disease prevention program. There is a special need for close collaboration among the IACUC, the AV, and the biological safety officer when a research project involves the use of naturally occurring pathogens, experimentally induced infectious disease, or wild caught domestic and exotic species. The AV and biological safety officer can assess the potential impact and highlight any concerns for the IACUC.

20:31   Many adventitious agents, such as mouse hepatitis virus, can be transmitted via cell lines or tumors. Is it necessary or advisable for an IACUC to request documentation that cell lines, tumors, or nonsterile biologic fluids are free of such agents before a protocol receives final IACUC approval?

Opin.   The *Guide* (p. 60) states that transplantable tumors, hybridomas, cell lines, and other biologic materials can be sources of murine viruses that can contaminate rodents and that appropriate testing should be considered to detect these agents in biological materials. Introduction of adventitious agents into an animal colony can alter research results, cause disease and death of animals, and waste animal life and time. Some agents have zoonotic potential under certain conditions. Therefore, it is strongly advised that the IACUC request documentation that such agents are not introduced into the animal facility via cell lines or tumors. Responsibility for oversight of this effort is frequently assigned to the AV as part of the veterinary care program.

20:32   What is a select agent?

Reg.   Infectious agents and toxins that are considered by the Department of Health and Human Services (DHHS) or the USDA as having the potential to pose substantial harm or a severe threat to human, animal, or plant health or plant products are regulated as select agents. Under the DHHS and USDA regulations,[17,18] anyone who wishes to transfer or possess any quantity of a select agent pathogen must successfully complete an intensive registration and approval process through the appropriate regulatory agency. This process includes a security clearance component for all who will have access to the select agent. Possession of select agent–listed toxins does not require this intensive registration process with the federal agencies if specified quantities of toxin are not exceeded by the individual principal investigator.

Opin.   All researchers (and research-related committees such as the IACUC) should be aware of which agents and toxins are currently regulated as select agents. Current lists, including specific exemptions, are available through the applicable federal agency web sites. Institutions should be vigilant in implementing policies to assure that select agents are not unknowingly introduced to, and possessed by, the institution. The biological safety officer and chemical hygiene officer can contribute significantly in this regard.

20:33   How does the use of a select agent in an animal study impact the IACUC review process?

Opin.   The review process should not change for protocols involving animals. However, this process should include direct communication with the PI and the Responsible Official for the select agents. Unique access restrictions and emergency response requirements for select agent work can be best clarified by these individuals. The Responsible Official is the individual who is charged by the institution with the authority and responsibility of ensuring compliance with Department of Health and Human Services and/or USDA select agent regulations. In some instances, this individual will be the biological safety officer. In all instances, the biological safety officer will be affiliated with select agent regulatory compliance activities. Animal studies involving the use of a select agent should also be reviewed by the institutional biosafety committee.

20:34   What is meant by a *radiological hazard*?

Reg.   According to Title 10, Code of Federal Regulations,[29] the definition is as follows: "Radiation (ionizing radiation) means alpha particles, beta particles, gamma rays, x-rays, neutrons, high-speed electrons, high-speed protons, and other particles capable of producing ions." Radiation may be machine produced (x-ray machines, accelerators), by-product materials (e.g., radioisotopes produced by a reactor) or naturally occurring atoms that emit radiation, such as uranium, radium, or other naturally radioactive atoms.

Opin.   The level of hazard is determined by the amount of radiation, type of radiation, chemical form, method of use (procedures and protocols), and other factors. Radiological hazards must be assessed carefully by a qualified radiation protection professional. Many radiation safety issues are more public relations and regulatory compliance than real risk problems.

20:35   What are typical sources of radiological hazards in the research animal environment?

Opin.   Typical sources of ionizing radiation in the animal research environment are radioactive materials administered as part of a research protocol or machine-produced radiation, all of which can be used for treatment, diagnostic, or research tools. Some examples are H-3, C-14, In-111, Cr-51, I-131, and Tc-99$_m$. These radioisotopes may be administered to animals as radiolabeled antibiotics, chemical toxins, and blood flow tracers for trauma and injury studies. X- or neutron radiation may be administered to animals to study brain or other physiological characteristics, to treat cancers, or to pursue other areas of research or treatment. New technologies are now in use, including positron emission tomography, in which unique radiation risks are present. The use of such radiologic technology tools in animal studies can present exposure hazards for personnel handling these animals during and after exposure. Careful review and oversight by the radiation safety staff are needed.

20:36   Is external beam radiation such as X-irradiation considered a radiation hazard of concern to an IACUC?

Opin.   External beam radiation effects should be considered a radiation hazard. Local and state regulations may have specific requirements for exposure limits to animals or limits for machine-produced radiation and the effects on animals. Therapeutic or research uses involving high doses of X-radiation that potentially can cause somatic effects should be considered similar to other procedures that cause pain and suffering for animals. The IACUC should communicate with the radiation safety department to assure that protocols involving X-irradiation, as well other forms of radiologic technologies, are properly planned, documented and monitored.

    The use of external beam radiation in the animal research environment is normally limited to therapeutic purposes (diagnostic and treatment). Nevertheless, it should be included in the overall animal care program oversight of the IACUC, as it can be a hazard. If a research protocol requires the use of X-radiation, that use requires IACUC review and approval. Therapeutic or research uses involving high doses of X-radiation that potentially can cause somatic effects should be considered similar to other procedures causing pain and suffering for animals.

    All human health aspects related to external beam radiation must be addressed by institutional safety professionals in the radiation safety and occupational health areas. This area is usually beyond the IACUC's expertise and scope.

20:37   How can an IACUC evaluate radiological hazard risk?

Reg.   The risk potential is regulated by federal, state, and local regulations for radiation. Agencies include the U.S. Nuclear Regulatory Commission, the Department of Energy, state regulatory agencies, and agreement states (states that have entered into an effective agreement with a federal regulatory agency and have become the regulatory control for the given area in that state).

Opin.   The IACUC should only evaluate the effects of animal experiments involving radiation. The radiation safety officer and the institutional radiation safety committee must assess radiation risk, determine precautions and management practices, and document the radiation protection evaluation. They also must inspect and assure safety and compliance with the regulations. In certain cases, precautions must be taken by animal care staff. The instructions for precautions are often posted on the door of the room or on the cages of the animals for which the precautions are necessary.

Animal care staff who will handle or assist with the management of animals that have been administered radioactive materials or who will handle or manage the radioactive waste or bedding must be trained in radiation safety at their institution.

20:38 What kind of documentation relative to an investigator's approved radioisotope use may be provided to the IACUC from a radiation safety office?

Reg. Records are mandated for the following radiation safety–related activities:

- Radiation licensing and permits
- Approvals
- Ordering and receipt of radioisotopes
- Training
- Inspections (both by researchers and safety managers)
- Protocols and procedures
- Radiation monitoring instrument certification
- Transportation
- Waste management
- Bioassays
- Other aspects of radiation use, including decommissioning of the use locations when done

Each institution must have the required licenses and permits, which, in turn, may mandate further requirements. Records for all of the radiation uses, from beginning to end, must be maintained for review by the radiation safety office.

Opin. The safe use of radiation and radioactive materials entails a very comprehensive and strict program of safety and compliance performed by qualified radiation safety professionals. IACUCs should initiate and maintain a close and friendly working relationship with the radiation safety managers at their institutions and defer to their findings.

References

1. National Research Council, Occupational Health and Safety in the Care and Use of Research Animals, National Academy Press, Washington, D.C., 1997.
2. Office of the Federal Register, Code of Federal Regulations, Title 29, Part 1910.95, Occupational Noise Exposure, Washington, D.C.
3. Office of the Federal Register, Code of Federal Regulations, Title 29, Part 1910.134, Respiratory Protection, Washington, D.C.
4. Office of the Federal Register, Code of Federal Regulations, Title 29, Part 1910.1030, Bloodborne Pathogens, Washington, D.C.
5. Centers for Disease Control and Prevention and National Institutes of Health, Biosafety in Microbiological and Biomedical Laboratories, 4th ed., U.S. Government Printing Office, Washington, D.C., 1999.
6. Office of the Federal Register, Code of Federal Regulations, Title 29, Part 1910.1450, Occupational Exposure to Hazardous Chemicals in Laboratories, Washington, D.C.
7. National Institute of Occupational Health Alert, Preventing Asthma in Animal Handlers, U.S. Department of Health and Human Services, Publication No. 97-116, 1998.

8. Stuart, R.B., and Moore, C., Safety and Health on the Internet, Government Institutes, Rockville, MD, 1999.
9. Office of the Federal Register, Code of Federal Regulations, Title 29, Part 1910.97, Nonionizing Radiation, Washington, D.C.
10. Office of the Federal Register, Code of Federal Regulations, Title 29, Part 1910.1096, Ionizing Radiation, Washington, D.C.
11. Office of the Federal Register, Code of Federal Regulations, Title 29, Part 1910 Subpart S, Electrical, Washington, D.C.
12. Office of the Federal Register, Code of Federal Regulations, Title 29, Part 1910.301–1910.399, Electrical, Washington, D.C.
13. Office of the Federal Register, Code of Federal Regulations, Title 29, Part 1910.1200, Hazard Communication, Washington, D.C.
14. American Veterinary Medical Association, 2000 Report of the AVMA Panel on Euthanasia, *J. Am. Vet. Med. Assoc.*, 218, 669, 2001.
15. U.S. Department of Agriculture, Animal and Plant Health Inspection Service, Policy 3, Veterinary Care, April 14, 1997. Available on the World Wide Web at: http://www.aphis.usda.gov/ac/policy/policy3.pdf.
16. National Institutes of Health, Guidelines for Research Involving Recombinant DNA Molecules, 66 FR 64052, December 11, 2001.
17. Office of the Federal Register, Code of Federal Regulations, Title 42, Parts 72 and 73, Possession, Use, and Transfer of Select Agents and Toxins, Final Rule, Washington, D.C.
18. Office of the Federal Register, Code of Federal Regulations, Title 7, Part 331, and Title 9, Part 121, Agricultural Bioterrorism Protection Act of 2002, Possession, Use and Transfer of Biological Agents and Toxins, Final Rule, Washington, D.C.
19. Office of the Federal Register, Code of Federal Regulations, Title 9, Animal and Animal Products, Parts 92–95, 122, Washington, D.C.
20. Office of the Federal Register, Code of Federal Regulations, Title 49, Parts 171–180, Hazardous Materials Regulations, Washington, D.C.
21. Centers for Disease Control and Prevention, Guidelines for the prevention of herpesvirus simiae (B-virus) infection in monkey handlers, *MMWR*, 36, 680, 687, 1987.
22. National Institute for Occupational Safety and Health, Health Hazard Evaluation Report: 98-0061-2687, Yerkes Primate Research Center, Lawrenceville, GA, April 1998.
23. Holmes, G.P., et al., Guidelines for the prevention and treatment of B-virus infections in exposed persons, *CID*, 20, 412, 1995.
24. Institute for Laboratory Animal Research, Occupational Health and Safety in the Care and Use of Nonhuman Primates, National Academy Press, Washington, D.C., 2003.
25. American Biological Safety Association, Mundelein, IL 60060. Available on the World Wide Web at: http://www.absa.org.
26. Office of the Federal Register, HR 5178, Needlestick Safety and Prevention Act, Washington, D.C.
27. Office of the Federal Register, 29 USC 654, Section 5, OSH Act of 1970, Washington, D.C.
28. Miller, C.D., Songer, J.R., and Sullivan, J.F., A twenty-five year review of laboratory acquired human infections at the National Animal Disease Center, *AIHA J.*, 48, 271, 1987.
29. Office of the Federal Register, Code of Federal Regulations, Title 10, Part 20.1003, Standards for Protection against Radiation, Washington, D.C.

21

Personnel Training

Howard G. Rush

Introduction

Personnel training is a subject that has gained increasing attention in the years since revision of the AWAR, the PHS Policy, and the *Guide*. All three documents specifically require institutions to provide training for personnel engaged in animal research, although the specific recommendations in these three documents differ. The mechanisms whereby institutions provide such training, the identification of individuals to provide the training, and the content of training courses have become the subject of much discussion among professionals in research administration and in laboratory animal science and medicine. Many symposia and seminars have been sponsored by national professional organizations to address these issues, and a great deal of resource material has become available. Numerous methods have arisen to provide training, which range from one-on-one sessions to online tutorials. In response to the demand for increased training, many institutions have created specific positions for trainers as part of the IACUC staff. IACUCs increasingly utilize training and retraining as a means to achieve compliance, especially in response to episodes of noncompliance encountered with research personnel.

A survey was conducted by the editors and authors of this book. One hundred fifty-seven institutions responded to questions on personnel training. In this chapter, the survey results are expressed as the number of responses per number of respondents. Thus, if more than one answer was provided, the sum of all responses will be greater than the number of respondents. If no answer was provided by some respondents, the number of respondents will be less than 157.

21:1 Is there any requirement in either the AWAR or the PHS Policy for general training in laboratory animal care and use?

Reg. The AWAR (§2.32,c,2; §2.32,c,4; §2.32,c,5) include requirements for general training in laboratory animal care and use. General training includes information on:

- Research methods that limit the utilization of animals or minimize animal distress
- Institutional procedures for reporting deficiencies in animal care and treatment

- Services such as the National Agricultural Library and the National Library of Medicine that can be used to acquire information on appropriate methods of animal care and use, alternatives to the use of live animals in research, the intent and requirements of the AWA, and the use of which can prevent duplication of research involving animals

The PHS Policy does not specify the content of training programs to the same degree as the AWAR; nevertheless, PHS Policy (IV,C,1,f) requires the IACUC, in its review of protocols, to determine that personnel conducting procedures are appropriately qualified and trained. In addition, the PHS Policy (IV,A,1,g) requires the institution to include in its NIH/OLAW Assurance a "synopsis of [the] training or instruction in the humane practice of animal care and use, as well as training or instruction in research or testing methods that minimize the number of animals required to obtain valid results and minimize animal distress, offered to scientists, animal technicians, and other personnel involved in animal care, treatment, or use."

The *Guide* (pp. 13–14) further specifies that individuals who care for or use animals should be properly trained and that the institution has a responsibility for providing either formal or on-the-job training for personnel. Training of animal care personnel is necessary to implement an effective animal care and use program and to foster humane animal care and use. Investigators and other personnel who perform surgery, administer anesthesia, or perform other manipulations must be appropriately trained to accomplish these tasks in a humane and scientifically acceptable manner.

Opin. The prevailing attitude among personnel engaged in animal research is that animals should not be subjected to unnecessary pain or distress. Universally, animal care and use personnel want to perform research that uses animals humanely. Experience has demonstrated that utilization of animal care and research personnel who are well trained will ultimately reduce animal pain and distress in research because well-trained personnel can perform animal research techniques with greater skill and fewer adverse outcomes for the animals used. Consequently, it is an institutional imperative to provide training to animal care and research personnel to ensure humane care for animals used in research.

21:2 Is training in animal care and use mandatory for personnel who will be working with animals?

Opin. Training for animal care and research personnel is certainly desirable in order to ensure the highest level of humane care. However, from a regulatory standpoint, the AWAR does not actually stipulate that training be mandatory for personnel working with animals. The AWAR (§2.32) state that personnel must be qualified to perform their duties and that training be made available to ensure that this requirement be fulfilled. By extension, if someone were already qualified to perform his or her duties by previous training or experience, no further training would be required. It is interesting if not surprising, then, that training is mandatory at most institutions. The reasons for this are not clear from the results of the survey. One possibility is that compliance personnel at institutions believe that the only way compliance can be achieved is through a regulatory approach. This is unfortunate because it emphasizes the "cops" side of the "cops versus colleagues" dilemma that laboratory animal care and IACUC personnel encounter in ensuring compliance with regulations and standards. Approximately half of the institutions provided only basic information in the mandatory training sessions. At most of the institutions there was little or no resistance to attending training (see 21:34).

Surv. At your institution, is there mandatory training in animal care and use for personnel who will be working with animals?

- Not applicable as we have no training program, mandatory or not mandatory   4/154
- Yes, and it is the same for all persons   77/154
- Yes, but it varies with the needs of the person   65/154
- No   5/154
- Other   7/154

21:3   Is there any requirement in either the AWAR or the PHS Policy for additional training in the care and use of a particular species?

Reg. The AWAR (§2.32,c,1,i; §2.32,c,1,ii; §2.32,c,3) require training on the basic needs of the relevant species, proper handling and care for various species used by the facility, and the proper use of anesthetics, analgesics, and tranquilizers for any animals used by the facility.

The PHS Policy does not specify any type of species-specific training, but it does require that Assured institutions use the *Guide* as a basis for their animal care and use program (PHS Policy IV,A,1).

The *Guide* (p. 14) states that research personnel who perform experimental manipulations on animals, including anesthesia and surgery, must have adequate training or experience to perform those procedures humanely. In addition, species-specific training for animal care personnel is implied in the general training statement in the *Guide* (p. 13), as noted earlier.

Opin. There is a vast diversity of species used in biomedical research. Although rats and mice account for more than 90% of the animals used in research, numerous other species have been utilized and will continue to be utilized, depending on the suitability of particular models for the research being conducted. It is impossible for any one person to be familiar with all species that are or might be used in research. Therefore, institutions must provide animal care and research personnel with access to training when experiments are planned using new species.

21:4   Do the AWAR or the PHS Policy state whether or not training is required for specific research procedures (e.g., performing an arterial cut-down)?

Reg. Procedure-specific training specified in the AWAR (§2.32,c,1,iii; §2.32,c,1,iv) includes training on pre- and postprocedural care of animals and aseptic surgical methods and procedures. The PHS Policy does not require procedure-specific training, and, as noted in 21:3, the *Guide* (p. 14) implies that procedure-specific training may be necessary to ensure that research personnel who perform experimental manipulations on animals can conduct their research humanely.

Opin. Innumerable animal research techniques are described in the scientific literature for use in the many disciplines in biomedical research. It is impossible for any one person to be skilled in all techniques commonly used in animal research. Training personnel should be prepared to teach common animal research techniques to individuals who conduct animal research. In addition, they should identify individuals at their institution who work in research laboratories and have unique skills in performing specific research procedures with animals. These individuals should be cultivated as ancillary training staff who can be called upon to help train personnel when training requests for their particular skills are received. Engaging research personnel to participate in the training effort can be a fruitful and rewarding means to expand the training provided at an institution while conserving fiscal resources.

21:5   **In addition to the training topics noted in 21:2 and 21:3, what other information is useful to be included as part of a training and education program?**

Opin.   Training programs for animal care and use personnel can be regarded in a tiered manner for optimal delivery of training. All personnel engaged in research with animals and all personnel who provide animal care services should participate in an introductory training session that includes information on federal laws and national standards, institutional policies and procedures, the institutional animal care and use program, institutional occupational health programs, reporting of concerns about animal care and use, the politics and ethics of animal experimentation, and alternatives to animal experimentation. A second tier or level includes species-specific training for personnel. It is directed at increasing their familiarity with the species they will be utilizing or for which they will be caring. Such training can include information on the biological characteristics and care of a given species and basic research techniques such as restraint, injection techniques, and euthanasia. A third tier includes technique-specific training for personnel to improve their skill level for more advanced research techniques such as surgery, anesthesia, and even specific surgical procedures (e.g., vascular catheter implantation). Personnel not directly engaged in animal care and use need not participate in the second and third tiers, but all personnel participate in the first tier. Additionally, personnel who have specific care or research duties also participate in training that is directed at improving their knowledge of and skill level in the species with which they work. These types of core modules have been described.[1]

Surv.   Beyond the basic APHIS/AC regulations and PHS Policy requirements, what other information might your IACUC consider as part of a training and education program? More than one answer is possible.

• Institutional policies and procedures	138/152
• Organizational structure of the institutional animal care and use program	93/152
• The politics and ethics of animal experimentation	87/152
• Alternatives to animal use	92/152
• Pain and distress and its alleviation	108/152
• Acceptable euthanasia methods	118/152
• An introduction to surgery	58/152
• Occupational health program	99/152
• Record keeping requirements	93/152
• Reporting of concerns about animal care and use	131/152
• Basic care and use of common laboratory animals	117/152

- Informational resources on animal care and use   102/152
- Hands-on training in basic research techniques   116/152
- Other   11/152

21:6 Who is responsible for assuring that research and animal care personnel working with animals are adequately trained?

Reg. According to the AWAR (§2.32,a), the research facility (see definition in 15:1) is responsible for ensuring that research personnel are qualified to perform their duties. To accomplish this, the research facility must make training available to personnel engaged in animal care and use. The HREA (Section 495,c,1,b) and the PHS Policy (IV,A,1,g) require the awardee institution to assure that all personnel involved in animal care, treatment, or use have training available to them. Both the AWAR (§2.31,d,1,viii) and the PHS Policy (IV,C,1,f) specify that the IACUC, in its review of protocols, should determine that personnel are appropriately qualified and trained. The *Guide* states that the institution is responsible for providing training to animal care personnel (*Guide*, p. 13). (See 21:7.)

Opin. Regardless of the letter of the law or the PHS Policy, the responsibility for assuring that personnel are adequately trained must be shared by the institutional officers, the IACUC, the animal care and veterinary staff, and the PI. Overall, it is an institutional responsibility to ensure that animal care and research personnel are adequately trained to care for research animals and conduct animal research procedures. The institution must be willing to commit the personnel and financial resources to accomplish this objective. Nevertheless, in practical terms, the responsibility for assuring that training is available lies with the IACUC. The IACUC is able to identify the personnel who should be trained and coordinate the resources and activities necessary to provide the training. The responsibility for actually providing the training usually rests upon the veterinary staff or the staff that supports the IACUC. These are the individuals who possess the scientific, clinical, and technical skills that the animal care and research staff need to perform their duties. Finally, the PI has a responsibility to convey to her or his staff the importance of receiving proper training in order to conduct humanely the animal studies in which they will be participating. The PI's attitude toward training sets the tone for laboratory personnel with regard to their participation in the institution's training program. These views are consistent with the survey results (listed later), which indicate that, at most institutions, the PI and the IACUC share the responsibility for assuring adequate training of research personnel while the responsibility for providing training to animal care personnel rests with the veterinarian and animal care supervisors.

Surv. 1 At your institution, who is responsible for assuring that research personnel working with animals are adequately trained in pertinent animal care and use procedures? Select all that apply.

- Responsibility of principal investigator and IACUC   91/155
- Responsibility of principal investigator and veterinarian   27/155
- Responsibility of principal investigator   38/155
- Responsibility of the IACUC   23/155
- Responsibility of veterinarian or animal facility supervisors   50/155
- Other   11/155

Note: The category "Other" included variations on the preceding entries plus designated training program personnel, compliance personnel, study directors, and managers.

Surv. 2 At your institution, who is responsible for assuring that animal care personnel working with animals are adequately trained in pertinent animal care and use procedures? Select all that apply.

• Responsibility of principal investigator and IACUC	37/155
• Responsibility of principal investigator and veterinarian	17/155
• Responsibility of principal investigator	23/155
• Responsibility of the IACUC	40/155
• Responsibility of veterinarian or animal facility supervisors	107/155
• Other	12/155

Note: The category "Other" included variations on the preceding listings plus designated training program personnel, compliance personnel, study directors, the institutional official, and managers.

21:7 Who is typically involved in training persons for the appropriate care and use of laboratory animals?

Opin. The qualifications of trainers can vary considerably between institutions and are likely to be affected by many factors. Among these are the size of the institution, its mission (academic, industrial, etc.), the organization of the IACUC and animal care services, the physical resources, and the animal population and species maintained (*Guide*, p. 13). A wide variety of individuals with differing qualifications are recruited to provide training, including veterinarians, veterinary technicians, laboratory animal technicians and technologists, and research personnel. It is interesting to note from the survey response that follows that at the majority of institutions the veterinarian or research personnel were most often involved in providing training on the appropriate care and use of animals. Only at 19% of the institutions did the IACUC staff actually provide the training. This is in contrast to the author's institution, where one member of the IACUC staff is identified as the trainer with responsibility for training research personnel on appropriate care and use. Our experience has been that specialization among the IACUC staff members (i.e., training, inspection, protocol review) is a more efficient method at a large institution because it provides a specific person for the investigators.

Surv. At your institution, who actually trains people in appropriate care and use of laboratory animals? More than one answer is possible.

• The veterinarian(s)	102/155
• The veterinary technician(s)	67/155
• Research personnel	97/155
• The IACUC staff	30/155
• Not applicable as we do not train people in care and use of lab animals	1/155
• Other	43/155

Note: The category "Other" included a wide variety of responses such as operations manager, compliance manager, animal facility manager, laboratory animal technologist, supervisor, designated training program personnel, principal investigator, online tutorials, audiovisual material, and offsite training.

21:8   What are some effective means of instruction in the proper care and use of laboratory animals?

Opin.   By far, the most effective means of instruction for animal care and use training is the individual or small group hands-on training session, sometimes termed a "wet lab." This setting provides the optimal conditions for student–teacher interaction and is excellent for fostering skill development. This type of training session is ideal for teaching animal research techniques such as handling, restraint, anesthesia, and surgery. On the other hand, the wet lab is not an efficient use of time or personnel when it is necessary to present introductory material such as federal laws and national standards, institutional policies and procedures, the institutional animal care and use program, and institutional occupational health programs. For this type of material, slide presentations, videos, written material, and even computer-based training are more appropriate and efficient. This assertion is borne out by the results of the following survey, which demonstrate a range of training strategies at the surveyed institutions.

Surv.   How does your institution instruct people in the proper care and use of laboratory animals? More than one answer is possible.

- No instruction   0/156
- Lectures   82/156
- Individual or small group hands-on training sessions   134/156
- Slide or PowerPoint presentations   74/156
- Written material (handouts, newsletters, and institutional manuals on animal research)   108/156
- Videos   59/156
- Computer-based training (either auto-tutorials or Web-based)   82/156
- Outside, for-profit training company   4/156
- Other   10/156

Note: The category "Other" included variations on the answers plus on the job training.

21:9   What are some reference sources that can be used to develop a general training program for animal care and use?

Opin.   Individuals charged with developing training programs should have access to the following resources:

- Office of the Federal Register, Code of Federal Regulations, Title 9 (Animals and Animal Products), Subchapter A (Animal Welfare), Washington, D.C., 1985
- U.S. Department of Health and Human Services, Public Health Service, Public Health Service Policy on Humane Care and Use of Laboratory Animals, Washington, D.C., 2002

- Institute of Laboratory Animal Resources, Committee on Educational Programs in Laboratory Animal Science, Education and Training in the Care and Use of Laboratory Animals: A Guide for Developing Institutional Programs, National Academy Press, Washington, D.C., 1991

- Institute of Laboratory Animal Resources, Committee to Revise the Guide for the Care and Use of Laboratory Animals, Guide for the Care and Use of Laboratory Animals, 7th ed., National Academy Press, Washington, D.C., 1996

- Interagency Research Animal Committee, U.S. Government Principles for Utilization and Care of Vertebrate Animals Used in Testing, Research, and Training, Federal Register, Washington, D.C., May 20, 1985

- Applied Research Ethics National Association (ARENA), National Institutes of Health, Office of Laboratory Animal Welfare (OLAW), Institutional Animal Care and Use Committee Guidebook, 2nd ed., Bethesda, MD, 2002

- Bennett, B.T., Brown, M.J., and Schofield, J.C., *Essentials for Animal Research: A Primer for Research Personnel*, National Agricultural Library, Beltsville, MD, 1990

- Committee on Guidelines for the Use of Animals in Neuroscience and Behavioral Research, Guidelines for the Care and Use of Mammals in Neuroscience and Behavioral Research, National Academy Press, Washington, D.C., 2003

- Lawson, P.T., *Assistant Laboratory Animal Technician Training Manual*, American Association for Laboratory Animal Science, Memphis, TN, 2004

- Lawson, P.T., *Laboratory Animal Technician Training Manual*, American Association for Laboratory Animal Science, Memphis, TN, 2004

- Lawson, P.T., *Laboratory Animal Technologist Training Manual*, American Association for Laboratory Animal Science, Memphis, TN, 2003

- Rollin, B.E., and Kesel, M.L., *The Experimental Animal in Biomedical Research, Volume I: A Survey of Scientific and Ethical Issues for Investigators*, CRC Press, Boca Raton, FL, 1990

- Russell, W.M.S., and Burch, R.L., *The Principles of Humane Experimental Techniques*, Methuen & Co., London, 1959 (reprinted as a special edition in 1992 by the Universities Federation for Animal Welfare)

- Federation of European Laboratory Animal Science Association's Working Group on Education, FELASA Recommendations on the Education and Training of Persons Working with Laboratory Animals: Categories A and C, *Lab. Anim.*, 29, 121, 1995

- Institute of Laboratory Animal Resources, Committee on Occupational Safety and Health in Research Animal Facilities, Occupational Health and Safety in the Care and Use of Research Animals, National Academy Press, Washington, D.C., 1997

- American Veterinary Medical Association, 2000 Report of the AVMA Panel on Euthanasia, *J. Am. Vet. Med. Assoc.*, 218, 669, 2001

- Laboratory Animal Welfare Training Exchange (available on the World Wide Web at: http://www.lawte.org)

- ResearchTraining.Org (available on the World Wide Web at: http://www.researchtraining.org/)

- IACUC.Org (available on the World Wide Web at: http://www.iacuc.org/)

21:10 What are some reference sources for training on the care and handling of a particular species?

Opin. The following references cover a wide variety of species and topics:

- Fox, J.G., Anderson, L.C., Loew, F.M., and Quimby, F.W., *Laboratory Animal Medicine*, 2nd ed., Academic Press, New York, 2002

- Poole, T., *The UFAW Handbook on the Care and Management of Laboratory Animals, Volume 1: Terrestrial Vertebrates*, 7th ed., Blackwell Publishing, Ames, IA, 1999

- Poole, T., *The UFAW Handbook on the Care and Management of Laboratory Animals, Volume 2: Amphibious and Aquatic Vertebrates and Advanced Invertebrates*, 7th ed., Blackwell Publishing, Ames, IA, 1999

- Olfert, E.D., Cross, B.M., and McWilliam, S.S., *Guide to the Care and Use of Experimental Animals, Vol. I*, 2nd ed., Canadian Council on Animal Care, Ottawa, 1993

- Canadian Council on Animal Care, *Guide to the Care and Use of Experimental Animals, Vol. II*, Canadian Council on Animal Care, Ottawa, 1984

- Hedrich, H., *The Laboratory Mouse (Handbook of Experimental Animals)*, Academic Press, New York, 2004

- Rollin, B.E., and Kesel, M.L., *The Experimental Animal in Biomedical Research. Volume II: Care, Husbandry, and Well-Being, An Overview by Species*, CRC Press, Boca Raton, FL, 1995

- Quesenberry, K.E., and Carpenter, J.W., *Ferrets, Rabbits, and Rodents: Clinical Medicine and Surgery*, 2nd ed., W.B. Saunders, Philadelphia, 2003

- Hrapkiewicz, K., Medina, L., and Holmes, D.D., *Clinical Laboratory Animal Medicine: An Introduction*, 2nd ed., Iowa State University Press, Ames, 1998

- Harkness, J.E., and Wagner, J.E., *The Biology and Medicine of Rabbits and Rodents*, 4th ed., Williams and Wilkins, Baltimore, 1995

- Wagner, J.E., and Manning, P.J., *The Biology of the Guinea Pig*, Academic Press, New York, 1976

- Van Hoosier, G.L., Jr., and McPherson, C.W., *Laboratory Hamsters*, Academic Press, New York, 1987

- Manning, P.J., Ringler, D.H., and Newcomer, C.E., *The Biology of the Laboratory Rabbit*, 2nd ed., Academic Press, New York, 1994

- Foster, H.L., Small, J.D., and Fox, J.G., *The Mouse in Biomedical Research. Vol. I. History, Genetics, and Wild Mice*, Academic Press, New York, 1981

- Foster, H.L., Small, J.D., and Fox, J.G., *The Mouse in Biomedical Research. Vol. II. Diseases*, Academic Press, New York, 1982

- Foster, H.L., Small, J.D., and Fox, J.G., *The Mouse in Biomedical Research. Vol. III. Normative Biology, Immunology, and Husbandry*, Academic Press, New York, 1983

- Foster, H.L., Small, J.D., and Fox, J.G., *The Mouse in Biomedical Research. Vol. IV. Experimental Biology and Oncology*, Academic Press, New York, 1982

- Bennett, B.T., Abee, C.R., and Hendrickson, R., *Nonhuman Primates in Biomedical Research: Biology and Management*, Academic Press, San Diego, 1995

- Suckow, M.A., Franklin, C.L., and Weisbroth, S.H., *The Laboratory Rat*, Academic Press, San Diego, 2006

- Hau, J., and Van Hoosier, G.L., *Handbook of Laboratory Animal Science, Vol. 1. Essential Principles and Practices*, CRC Press, Boca Raton, 2002

- Laber-Laird, K., Swindle, M.M., and Flecknell, P., *Handbook of Rodent and Rabbit Medicine*, Pergamon Press, Oxford, 1996

- Suckow, M.A., and Douglas, F., *The Laboratory Rabbit*, CRC Press, Boca Raton, FL, 1997

- Sharp, P.E., and LaRegina, M.C., *The Laboratory Rat*, CRC Press, Boca Raton, FL, 1998

- Suckow, M.A., Danneman, P., and Brayton, C. *The Laboratory Mouse*, CRC Press, Boca Raton, FL, 2000

- Field, K.J., and Sibold, A.L., *The Laboratory Hamster and Gerbil*, CRC Press, Boca Raton, FL, 1998

21:11   What are some reference sources for training personnel in anesthesiology, perioperative care, and aseptic surgery?

Opin.   The following sources are helpful:

- Block, S.S., *Disinfection, Sterilization, and Preservation*, 5th ed., Lippincott, Williams & Wilkins, Philadelphia, 2000

- Waynforth, H.B., and Flecknell, P.A., *Experimental and Surgical Technique in the Rat*, 2nd ed., Academic Press, London, 1992

- Adams, R.J., Techniques of experimentation, in *Laboratory Animal Medicine*, Fox, J.G., Anderson, L.C., Loew, F.M., and Quimby, F.W., Eds., Academic Press, Orlando, 2002, chap. 23

- Swindle, M.M., Vogler, G.A., Fulton, L.K., Marini, R.P., and Popilskis, S., Preanesthesia, anesthesia, analgesia, and euthanasia, in *Laboratory Animal Medicine*, Fox, J.G., Anderson, L.C., Loew, F.M., and Quimby, F.W., Eds., Academic Press, Orlando, 2002, chap. 22

- Cunliffe-Beamer, T., Biomethodology and surgical techniques, in *The Mouse in Biomedical Research*, Vol. III, Foster, H.L., Small, J.D., Fox, J.G., Eds., Academic Press, New York, 1983, chap. 18

- Vogler, G.A., Anesthesia and analgesia, in *The Laboratory Rat*, Suckow, M.A., Franklin, C.L., and Weisbroth, S.H., Eds., Academic Press, San Diego, 2006, chap. 19

- Markowitz, J., Archibald, J., and Downie, H.G., *Experimental Surgery*, Williams and Wilkins, Baltimore, 1964

- Lumley, J.S.P., et al., *Essentials of Experimental Surgery*, Butterworth and Co., London, 1990

- Schwartz, A., Experimental surgery, in *The Experimental Animal in Biomedical Research*, Vol. I: *A Survey of Scientific and Ethical Issues for Investigators*, CRC Press, Boca Raton, FL, 1990

- Cunliffe-Beamer, T.L., Applying principles of aseptic surgery to rodents, *Anim. Welfare Inform. Cent. Newsl.*, 4, 3, 1993

- Academy of Surgical Research, Guidelines for training in surgical research in animals, *J. Invest. Surg.*, 2, 263, 1989
- Lang, C.M., *Animal Physiologic Surgery*, 2nd ed., Springer-Verlag, New York, 1982
- Berg, J., Sterilization, in *Textbook of Small Animal Surgery*, 2nd ed., Slatter, D., Ed., W.B. Saunders, Philadelphia, 1993, chap. 11
- Brown, M.J., and Schofield, J.C., Perioperative care, in *Essentials for Animal Research: A Primer for Research Personnel*, Bennett, B.T., Brown, M.J., and Schofield, J.C., Eds., National Agricultural Library, Washington, D.C., 1994
- Schofield, J.C., Principles of aseptic technique, in *Essentials for Animal Research: A Primer for Research Personnel*, Bennett, B.T., Brown, M.J., and Schofield, J.C., Eds., National Agricultural Library, Washington, D.C., 1994
- Brown, M.J., Pearson, P.T., and Tomson, F.N., Guidelines for animal surgery in research and teaching, *Am. J. Vet. Res.*, 54, 1544, 1993
- Cunliffe-Beamer, T.L., Surgical techniques, in *Guidelines for the Well-Being of Rodents in Research*, Guttman, H.N., Ed., Scientists Center for Animal Welfare, Bethesda, MD, 1990
- Smith, A.C., and Swindle, M.M., *Research Animal Anesthesia, Analgesia, and Surgery*, Scientists Center for Animal Welfare, Greenbelt, MD, 1994
- Kohn, D.F., Wixson, S.K., White, W.J., and Benson, G.J., *Anesthesia and Analgesia in Laboratory Animals*, Academic Press, San Diego, 1997
- Flecknell, P.A., *Laboratory Animal Anesthesia: A Practical Introduction for Research Workers and Technicians*, 2nd ed., Academic Press, London, 1996
- Institute of Laboratory Animal Resources Committee on Pain and Distress in Laboratory Animals, Recognition and Alleviation of Pain and Distress in Laboratory Animals, National Academy Press, Washington, D.C., 1992

21:12   **What are some reference sources that can be used to develop a training program for use of hazardous agents in animal research? (See Chapter 20.)**

Opin.   Useful reference sources include the following:

- Centers for Disease Control and Prevention and National Institutes of Health, Biosafety in Microbiological and Biomedical Laboratories, 4th ed., U.S. Government Printing Office, Washington, D.C., 1999
- Centers for Disease Control and Prevention and National Institutes of Health, Primary Containment for Biohazards: Selection, Installation and Use of Biological Safety Cabinets, U.S. Government Printing Office, Washington, D.C., 1995
- Office of the Federal Register, Code of Federal Regulations, Title 10, Part 20, Standards for Protection against Radiation, Washington, D.C., 1984
- Office of the Federal Register, Code of Federal Regulations, Title 29, Part 1910, Occupational Safety and Health Standards; Subpart G, Occupation Health and Environmental Control, and Subpart Z, Toxic and Hazardous Substances, Washington, D.C., 1984

- Office of the Federal Register, Code of Federal Regulations, Title 29, Part 1910,Occupational Safety and Health Standards, Subpart I, Personal Protective Equipment, Washington, D.C., 1984
- Clark, J.M., Planning for safety: biological and chemical hazards, *Lab. Anim.*, 22(7), 33, 1993
- Department of Health and Human Services, National Institutes of Health, Guidelines for research involving recombinant DNA molecules, Federal Register, 59:34496 (amended 59 FR 40170, 60 FR 20762, 61 FR 10004, 62 FR 4782, 62 FR 53335, 62 FR 56196, 62 FR 59032, 63 FR 8052, 63 FR 26018, 64 FR 25361, 65 FR 60328, 66 FR 1146, 66 FR 64051, 66 FR 57970, 66 FR 64052, 66 FR 57970)
- Holmes, G.P., et al., Guidelines for the prevention and treatment of B-virus infections in exposed persons, *Clin. Infect. Dis.*, 20, 421, 1995
- Committee on Hazardous Biological Substances in the Laboratory, Biosafety in the Laboratory: Prudent Practices for Handling and Disposal of Infectious Materials, National Academy Press, Washington, D.C., 1989
- Richmond, J.Y., et al., What's hot in animal biosafety?, *ILAR J.*, 44(1), 20, 2003
- Thomann, W.R., Chemical safety in animal care, use, and research, *ILAR J.*, 44(1), 13, 2003

21:13 How can training versus qualifications be effectively evaluated by the IACUC?

Reg.   (See 21:1–21:3.) The *Guide* (p. 14) states that investigators and other research personnel engaged in animal research must be qualified through training or experience in order to conduct the research humanely.

Opin.   In order for the IACUC to evaluate personnel training and qualifications, investigators should provide the IACUC with specific information on their prior experience and training with the species and procedures proposed in their protocol as well as the experience and training of their staff members. The IACUC then should determine, in the course of reviewing the investigator's protocol, whether the research personnel have sufficient training or experience to conduct the proposed procedures. For example, an investigator may be trained as a human surgeon but, without specific training or experience in animal surgery, may not be qualified to perform surgical procedures in animals. Such an individual may need additional training on species-specific anatomy, physiology, behavior, anesthesia, and analgesia. If the prior training or experience of the research staff is deemed adequate by the IACUC, the protocol, once approved, may be initiated. However, if the IACUC determines that the prior training or experience of research personnel is inadequate, the IACUC should request or require the research staff to receive additional instruction before proceeding with their studies. The requirement for additional training might even be made a condition of approval. That is, the IACUC could place a stipulation on the investigator's protocol approval that the studies could not proceed until the staff had received the necessary training. Another interesting approach is for the IACUC to approve the protocol with a stipulation that the procedures be observed by a veterinarian or another staff member the first few times they are attempted. If the observations indicated that additional training were needed, it would be provided by the observer at that time or scheduled in the future before additional animals were used.

When semiannual inspections of animal care and use facilities are performed, the IACUC also can gain some insight into the adequacy of training by interviewing research personnel during these inspections. This approach not only identifies inadequacies in training after protocol approval has taken place, but also can be useful as an audit of the adequacy of the review process in identifying personnel in need of training.

21:14   How, and to what extent, should training and education efforts be documented?

Opin.   Neither the PHS Policy nor the AWAR specifically require that training records be maintained. The common view (see 21:14 survey) that it is necessary to maintain training records may stem from the institution's responsibility to make training available and the IACUC's responsibility to ensure that personnel are trained and qualified. Logically, it is difficult to meet these obligations without a mechanism to track the training and experience of individuals engaged in animal research. The most common methods of documentation were filing the sign-in sheets from training classes and maintaining an electronic database of personnel training.

Surv. 1 How, and to what extent, does your institution document training and education efforts? More than one answer is possible.

- We do not document our training and education efforts   14/156
- Files containing the sign-in sheets from training classes   104/156
- Electronic database of personnel training records   86/156
- Other   20/156

Note: The category "Other" included a wide variety of responses such as filing of training and certification records, issuing of training certificates, records from online training, attendance certificates, and Good Laboratory Practice training records.

Surv. 2 What records do you keep relative to training and education of animal use personnel? More than one response is possible.

- Not applicable as we do not keep such records   13/154
- Person's name and organizational affiliation   105/154
- Inventory of prior experience   70/154
- Names of classes or training sessions   109/154
- Dates of classes or training sessions   121/154
- Names of trainers   70/154
- Other   13/154

Note: The category "Other" included a wide variety of responses such as copies of certificates, documentation of Web-based training, a record of proficiency level for skills learned, and copies of online training documentation.

21:15   What documentation of training activities must be provided to the IACUC?

Opin.   The PHS Policy and the AWAR do not specifically require reports on training activities. However, given the increased reliance of IACUCs on the use of training to manage compliance issues, it would be prudent to provide training reports to the IACUC at regular intervals. Even if training information is incorporated into the protocol, summary reports could be very useful in evaluating training effort and staffing levels.

Surv. What documentation of training activities must be provided to the IACUC? Select all that apply.

- We do not provide reports of training activities to our IACUC   28/149
- We provide the IACUC with a summary report on training activities in the semiannual report to the Institutional Official   38/149
- We provide reports on training activities only upon request from the IACUC   38/149
- We update the IACUC at its regular meetings   41/149
- Other   20/149

Note: The category "Other" included a wide variety of responses. In a significant number of instances the response indicated that the training records are incorporated into the protocol form being reviewed by the committee.

21:16 Should the IACUC require any form of testing subsequent to training?

Opin. Current regulations do not require that individuals receiving training be tested to evaluate their mastery of the subject matter. Most of the institutions surveyed did not test after training, but a small yet significant number of institutions considered such testing necessary. The need for posttraining testing should be carefully evaluated by the IACUC from the perspective of the institutional culture and the value added to the program.

Surv. Does your institution have any form of testing subsequent to training?

- Yes   55/156
- No   93/156
- Other   8/156

21:17 Should an IACUC require evidence of technical proficiency for individuals conducting procedures involving animals?

Opin. Individuals conducting animal research should provide the IACUC with some evidence of technical proficiency. In general, a signed statement of assurance should be sufficient to indicate that personnel conducting research procedures have the experience to perform the procedures or will seek training. Requiring the IACUC or training staff to observe and document proficiency for all individuals would place an undue burden on them especially at a large institution, where hundreds or even thousands of individuals would have to be evaluated. At some institutions where individual testing of competency is feasible, such a program might be warranted, but there is no requirement to evaluated proficiency in this manner.

Surv. Does your IACUC require some form of evidence of a person's technical proficiency when performing a procedure with animals?

- Yes   116/150
- No   26/150
- Other   8/150

21:18   After an individual has completed basic training, should the IACUC establish technical proficiency and skill of that individual?

Opin   There is no regulatory requirement for the IACUC to establish proficiency standards. On the other hand, without some system to determine proficiency, it may be difficult for the IACUC to ensure adequate training of personnel.

Surv.   In the opinion of your IACUC, is it important that the IACUC establish proficiency standards for those performing animal-related tasks beyond basic training?

- This has never been addressed by our IACUC   77/152
- Our IACUC believes that there is a need for proficiency
  standards that extend beyond basic training   64/152
- Our IACUC does not believe that there is a need for
  proficiency standards that extend beyond basic training   7/152
- Other   4/152

21:19   If an IACUC requires individuals to establish technical proficiency with respect to procedures involving animals, what sort of evidence or documentation is needed?

Opin.   (See 21:20.) A statement of assurance would be an acceptable method of documentation.

Surv.   If your IACUC requires some form of evidence of a person's proficiency when performing a procedure with animals, what sort of evidence is acceptable? Check all that apply.

- Not applicable to our IACUC   19/146
- Written statement or check-off box that implies proficiency is
  present   73/146
- Hands-on testing (or IACUC observation) of basic skills only
  (e.g., animal handling, aseptic technique)   19/146
- Hands-on testing (or IACUC observation) of all skills required
  in the protocol   38/146
- Other   15/146

Note: The category "Other" included a wide variety of responses such as observation at the time procedures were first performed (as opposed to prior to beginning actual experiments), official statement of proficiency, observation by study director or PI (as opposed to veterinary or IACUC staff), and hands-on testing of selected procedures.

21:20   How might an IACUC go about establishing the technical proficiency of individuals?

Opin.   Although it is important that research personnel be proficient in the tasks that they perform, it may be impractical for them to demonstrate proficiency to the IACUC.

Because at most institutions the responsibility for training animal use personnel is shared by the IACUC and the investigator, it may be sufficient for the investigator to provide written assurance of proficiency to the IACUC. (See 21:19.)

Surv. What mechanism(s) does your IACUC use to assure proficiency in specific methods and procedures? More than one response is possible.

• We have no assurance mechanism	24/151
• We have a formal hands-on testing program	27/151
• We have a formal written test that is used for this purpose	6/151
• We accept a written statement from the principal investigator that personnel are proficient	73/151
• If necessary, we review data on mortality and related parameters	35/151
• We do "site visits" to watch what is actually happening during experimentation, testing, or teaching	64/151
• Other	18/151

Note: The category "Other" included a wide variety of responses such as evaluation of animal handling, procedural, surgical, and euthanasia skills on an individual basis; direct observation and evaluation by the veterinarian; and written statements of assurance.

The previous four questions indicate that most institutions require some evidence of proficiency for technical procedures to be performed but, in most cases, this requirement was satisfied by a written statement of proficiency provided by the investigator. In a small but significant number of instances proficiency had to be demonstrated in hands-on testing or via observation by the veterinary or IACUC staff.

21:21 Can an investigator offer experience with procedures in one species (including humans) as sufficient evidence of qualification to perform the same procedure in a different species?

Opin. This question is a practical application of the principles stated in 21:12. There are no universal guidelines that can be applied across the board to determine whether an individual has sufficient training or experience to perform a proposed procedure. Thus, the type and duration of experience that are acceptable at one institution may be unacceptable at another. Ideally, the IACUC should take into account a variety of factors in order to make this decision, including prior experience, previous species-specific and procedure-specific training, the difficulty and complexity of the proposed procedures, the possible and probable adverse consequences to the animals, and even the past interactions of the investigator and the IACUC. The assessment of this latter issue might include objective and subjective information, such as prior problems with investigator compliance, the number of animal health problems encountered with previous animal research protocols, IACUC staff appraisals of the competence and skill levels of research personnel, and complaints from other staff members against the investigator or his or her staff members.

Surv. Can an investigator offer your IACUC experience with procedures in one species (including humans) as sufficient evidence of his or her qualification to perform the same procedure in a different species?

- The issue has never arisen and we have no policy   45/149
- Yes, but we still must evaluate each person's training and
  experience on a case-by-case basis   69/149
- No, experience on the appropriate species is always mandatory   31/149
- Other   4/149

21:22   Should an IACUC approve a protocol with novel procedures based on the surgical or other procedural experience of the investigator or other personnel involved?

Reg.   Although the PHS Policy and the AWAR are mute on this matter, the *Guide* provides some insight into this issue. If novel procedures are encountered in protocols under review, the *Guide* suggests that the IACUC seek pertinent information on the possible effects on the animals from the literature as well as experienced animal care and use personnel (*Guide*, p. 10).

Opin.   In the absence of sufficient information, the IACUC can require the investigator to perform pilot studies in order to evaluate the effects on the animals (*Guide*, p. 10). Similar to what was suggested in 21:12, the IACUC can approve the protocol as a pilot study or might place a stipulation or contingency on the protocol that the procedures be observed by a veterinarian or other qualified staff member to ensure that the novel procedures are performed in a humane and painless manner.

Surv.   Would your IACUC approve a protocol with novel procedures on the basis of general surgical or other procedural experience of personnel involved?

- We would likely approve a protocol with novel procedures
  on the basis of the surgical or other procedural experience
  of the investigator or other personnel involved   37/149
- We would be likely to approve such a protocol, but
  require the veterinary staff to observe one or more
  procedures and report to IACUC   102/149
- We would not approve such a protocol   1/149
- Other   9/149

21:23   Must students or short-term employees (e.g., summer help) fulfill the same training requirements that the IACUC requires of investigators and their staff?

Reg.   See 21:1, as the AWAR and the PHS Policy do not distinguish between different types of employees or students.

Opin.   The IACUC should determine that all research personnel conducting animal studies are appropriately qualified and trained to perform their duties. Ideally this includes full and part-time employees, students, and visiting scientists. There is no requirement that all personnel receive identical training, only that the training qualifies them to perform their duties. Thus, tiered training programs (see 21:5) might be valuable when differing training needs exist. From a practical standpoint, it is generally easier to identify full-time employees at institutions because orientation training for their position would likely include exposure to the IACUC and animal care unit. Individuals in the other categories can easily be unaccounted and may only be identified after they have begun their duties. This may be because others in the PI's laboratory remember to send them for the training, or because they are recognized by the animal care staff as new to the facility. Regardless, it is important for the trainers to foster good communication with laboratory personnel so that new employees are referred promptly to the IACUC for training.

Surv. Does your IACUC require that students or short-term employees (e.g., summer help) fulfill the same training requirements as investigators and their staff?

• Not applicable	14/150
• Always receive the same training as full-time research personnel	102/150
• Receive the same training as full-time personnel, but only if they can be identified as short-term personnel	14/150
• Receive less or no training	13/150
• Other	7/150

21:24 Should the IACUC require students using live animals to participate in the training program for animal users?

Opin. Students working in laboratories should receive the same training as all other research personnel. Students who utilize animals in classroom settings should receive abbreviated training specific to the procedures being conducted. This may be provided by the IACUC staff or by the instructor.

Surv. If your institution has courses in which students use live animals for any reason or length of time, are they required to participate in an animal use training program? More than one response is possible.

• Not applicable	76/152
• Student training is voluntary on the part of the student	3/152
• Our own students do not have to partcipate in any prior training	4/152
• Our own students have to take training similar to that provided to animal care personnel	13/152
• Our own students have to take training similar to that provided to research personnel	37/152
• Summer or other temporary students do not have to participate in any prior training	2/152
• Summer or other temporary students participate in the same training as our own students	28/152
• Training of our own students includes	40/152
• Classroom instruction	40/40
• Hands-on instruction	37/40
• Training of summer or other temporary students includes	35/152
• Classroom instruction	28/35
• Hands-on instruction	27/35
• Other	9/152

21:25 Can a high school student perform animal research if he or she works under the guidance of an investigator with an IACUC-approved protocol?

Opin. It has become increasingly common for high school students to work in animal research settings as volunteers, students, or employees. There is no inherent reason that a high school student cannot perform research procedures using animals if he or she is under the guidance of an investigator with approval to conduct the research. High school students should receive whatever training is necessary for them to perform their duties. Their training should be the same as that provided to college-level students and regular employees of the institution.

Surv. At your institution, can a high school student perform animal research if he or she works under the guidance of an investigator with an IACUC-approved protocol?

- Not applicable as we do not have high school students   48/150
- This has not been addressed by our IACUC   33/150
- Yes   53/150
- No   11/150
- Other   5/150

Note: The category "Other" included a variety of responses such as the requirement that the individual be enrolled as a student, be an employee, or participate in a special program; that he or she meet all training requirements; or that he or she be 18 years old or older.

21:26 Is there any lower age limit below which the IACUC should disallow student participation in animal care and research?

Opin. There is no specific reason to place an arbitrary age limit on student participation in animal care and use. However, it is certainly uncommon for students in elementary and middle school to volunteer or seek employment in research settings. IACUCs are advised to treat any such request on a case-by-case basis. Any student, regardless of age, must be adequately supervised by an investigator with approval to conduct the animal studies. Other factors such as labor laws and institutional policies on employment and volunteers potentially may limit the participation of younger students in animal research.

Surv. At your institution, is there any age limit below which the IACUC does not allow student participation in animal care and research?

- Not applicable   32/150
- No policy   85/150
- Yes, not allowed to participate below the age of   35/150
  - 18   18/35
  - 16   10/35
  - Other   7/35

21:27 In terms of animal care and use, who is ultimately responsible for the activities of a student?

Opin. The PI who has approval for animal use bears responsibility for any and all personnel working in his or her laboratory, including students. In the course of review and approval of her or his protocol, the PI should provide assurance to the IACUC that all personnel working with animals will be adequately trained to perform their duties. The investigator must understand that this responsibility extends to students working in the laboratory.

Surv. At your institution, who is ultimately responsible for the activities of a student who participates in a research project?

- Not applicable   30/150
- Principal investigator   102/150
- IACUC   6/150
- Principal investigator and IACUC   8/150
- Other   4/150

21:28 Is there any regulation or policy requiring periodic retraining or recertification of personnel working with animals?

Opin. There is no statement in the PHS Policy or AWAR that requires periodic retraining or recertification of animal care and use personnel. The requirements for recertification are generally determined by a certifying professional organization or agency that governs each specialty area and should not be required by an IACUC. That is, a state board of veterinary examiners determines whether a veterinarian should be recertified at periodic intervals. Thus, the IACUC cannot impose recertification on an individual when no professional recertification program exists. On the other hand, periodic retraining may be utilized to improve the understanding and skills of the animal care and use staff, but, again, there is no externally imposed requirement to do so. Similarly, institutions can develop in-house certification programs along with recertification requirements, but the author is not aware of any that have been implemented.

Surv. At your institution, is there an IACUC policy requiring periodic retraining or recertification of personnel working with animals?

- No policy   57/151
- We have such a policy   41/151
- Retraining required only if required by IACUC for animal welfare or regulatory compliance reasons   46/151
- Other   7/151

Routine retraining as an institutional policy was not common at surveyed institutions. Retraining as imposed by the IACUC for animal welfare or compliance reasons was as likely to be institutional practice. Roughly an equal number of institutions had no policy on retraining.

21:29   Should training be continuing and ongoing?

Reg.   The PHS Policy and the AWAR do not address the frequency or nature of training. The *Guide* recommends that animal care and use personnel participate regularly in continuing education activities (*Guide*, p. 13).

Opin.   As noted in 21:23, retraining should be encouraged, but there is no regulatory mandate that requires personnel to be retrained. However, the biomedical research environment is constantly changing. New methods and techniques are being described, new animal models are being identified, and new management approaches are being implemented at a rapid pace. Ideally then, all personnel engaged in animal care or use should participate in some form of continuing education or training in order to stay abreast of changes in their areas of interest. Such participation can occur through the existing institutional training program or through local, regional, and national meetings, conferences, and workshops.

Surv.   Should training in animal care and use be continuing and ongoing?

- Our IACUC believes that training should be continuing and ongoing   91/153
- Our IACUC has not addressed this issue   57/153
- Our IACUC believes it is not necessary for training to be continuing and ongoing   1/153
- Other   4/153

21:30   What are some effective ways to provide and document continuing education?

Opin.   Continuing education is a wise investment for institutions because it improves personnel engagement in the institution's success and improves employee morale. Continuing education opportunities can be provided by existing staff (trainers, supervisors, veterinarians, etc.), by institutional human resource and development personnel, by outside trainers, or by off-site training sessions at local, regional, or national meetings. Although continuing education and retraining are not specifically required, it is generally accepted that if such programs are implemented, then adequate records must be maintained.

Surv.   What are some effective ways for your IACUC or institution to provide and document continuing education? Check all appropriate responses.

- We do not provide continuing education   28/153
- We provide continuing education but do not document it   12/153
- Animal care and research personnel are notified and encouraged to attend various upcoming training sessions, off-site meetings, and special educational opportunities   94/153
- We document on-site training sessions   90/153
- We document off-site training sessions   58/153
- We review the stated credentials on protocol forms   44/153
- Other   5/153

At many institutions, animal care and research personnel are notified of and encouraged to attend upcoming training sessions, off-site meetings, and special educational opportunities. If personnel attend any on-site training sessions, their participation is generally recorded in the institution's regular training records. On the other hand, considerably fewer respondents record participation in off-site training programs. Some indicated that the credentials of all personnel were rereviewed every 3 years when the protocol was resubmitted to the IACUC.

21:31   Does an investigator who predominantly writes grants and protocols but has little hands-on animal contact need to be trained and qualified in the same way and to the same extent as an animal caretaker or a research technician?

Opin.   Investigators who spend most of their time writing grants and venture into the laboratory less often still have a responsibility to ensure that their personnel are trained and that they have the same level of understanding of current animal care standards. It is because of this responsibility that this type of investigator is obligated to participate in at least introductory level training that reviews federal laws and national standards, institutional policies and procedures, the institutional animal care and use program, institutional occupational health programs, reporting of concerns about animal care and use, and the politics and ethics of animal experimentation. The training should impress upon them their responsibility for the personnel working under them to ensure that the individuals who actually conduct the animal procedures are appropriately trained. At most institutions (see later discusssion), these investigators are required to take the same level of training as animal caretakers and research technicians.

Surv.   At your institution, does an investigator who mainly writes grants and protocols and has little hands-on animal contact need to be trained and qualified in the same way and to the same extent as an animal caretaker or a research technician?

• Not applicable	30/154
• Those investigators require the same level of training	78/154
• Those investigators require only basic introductory level training	39/154
• Other	7/154

21:32   Should an investigator who will not be involved in any hands-on animal contact be required to undergo any training?

Opin.   Investigators who will not be conducting procedures on animals themselves should still be required to take basic introductory training in order to become familiar with federal, state, and institutional policies and procedures. As supervisors of research personnel who will be performing animal research procedures, they have responsibility to convey the importance of utilizing the most humane methods in animal research and of regulatory compliance.

Surv.   At your institution, if a PI will have absolutely no hands-on animal contact at all, does she or he need to have any animal use training of any kind?

• Not applicable	30/154
• Does not require any training at all	12/154
• Requires only basic introductory level training	55/154

- Requires the same level of training as those working directly
  with animals   53/154
- Other   7/154

21:33   If a person has successfully completed an animal care and use training course at a different institution, should the IACUC at a new institution require that she retake a similar course at that institution or otherwise demonstrate her capabilities?

Opin.   As the IACUC examines the credentials of animal use personnel in the course of conducting its review of proposed activities, any training that has occurred at another institution should be taken into account. In order to do this, the IACUC may request the individual to provide detailed information on the nature of training at the previous institution. The amount of "credit" awarded by the IACUC will likely vary with the institution's policies and the content of the training at the previous institution. Regardless of the nature of the training at the previous institution, the individual should take any orientation training (see 21:5) offered in order to acquire institution-specific information that would not be acquired in any other way.

Surv.   If a person has successfully completed an animal care and use training course at a different institution, does your IACUC require that he or she retake a similar course at your institution or otherwise demonstrate his/her capabilities?

- Not applicable   14/154
- Must take at least basic orientation training at our institution   87/154
- Original institution's training program is sufficient   26/154
- If the animal work is the same, we accept the training
  from another institution; if the animal work changes, he or
  she must take our course   17/154
- Other   10/154

21:34   What type of information might the IACUC consider as essential to basic required training?

Opin.   Basic training will vary somewhat from one institution to another, but certain common elements will most assuredly be incorporated. These include historical perspectives on animal use, the current animal research climate, animal care and use regulations and policies, institutional policies and procedures, the institutional animal care and use application, the institutional occupational health and safety program, and the procedures for registering concerns about animal use. Some institutions may also include species-specific information on the basics of normal and abnormal behavior, signs of illness, and research techniques.

Surv. 1   If your IACUC has mandatory training of some sort, is it limited to basic information (e.g., institutional policies and procedures, laws and standards, protocol submission, occupational health and safety program availability, how to register concerns)?

- Not applicable   10/153
- Yes, for the most part it is limited to the type of basic
  information described in this question   73/153

- No, the training exceeds the type of basic information
  described in this question 69/153
- Other 1/153

Surv. 2 If your training program is mandatory, do you find resistance from investigators to send staff to this training?

- Not applicable 22/153
- We find little or no resistance 106/153
- We find moderate or heavy resistance from investigators
  based on the following (check all appropriate boxes): 24/153
  - Time commitments from courses or lab work interfere
    with training 20/24
  - They prefer to train their own staff in their own way
    of doing things 14/24
  - They do not trust other trainers to do a good job 2/24
  - They believe their needs are specialized and cannot be
    taught by others 6/24
  - Other 2/24
- Other 1/153

21:35 If a representative of a commercial company is to demonstrate the proper use of a new surgical product (e.g., for intestinal anastomosis) on a live animal, how can the IACUC assure that the representative is properly trained in the use of the product on the species to be used?

Opin. This circumstance would most often be encountered at large academic medical centers with well-developed continuing medical education programs. Such programs are important for providing training to faculty and local practitioners on new techniques and equipment. When institutions encounter such requests, it is important to develop mechanisms to manage them effectively rather than forbid them outright so as not to inhibit education and training.

Surv. At your institution, if a representative of a commercial company is to demonstrate the proper use of a new surgical product (e.g., for intestinal anastomosis) on a live animal, how does your IACUC assure that the representative is properly trained in the use of the product on the species to be used?

- Not applicable as we do not allow this type of demonstration
  or it is not pertinent to our institution 64/152
- This issue has never been addressed by our IACUC 46/152
- The person would have to provide our IACUC with acceptable
  documentation of skill and experience on the species to
  be used 24/152
- Our IACUC would accept a statement from the principal
  investigator that the person is skilled and has experience
  on the species to be used 15/152
- Other 3/152

Note: The most common response in the category "Other" was that the AV would be present for the procedures and would determine whether the representative was adequately trained.

21:36   Should personnel be trained before a research project begins, or is it acceptable for them to train during the course of the project via observation, assisting a fully trained person, and so on?

Opin.   The AWAR and PHS Policy do not provide guidance on when or how training is provided. Thus, any combination of prior training and hands-on training would be acceptable. Some types of training might be so specialized that it can only be provided by other members of the research staff in a mentor–trainee setting. It would always be wise to provide training in a stepwise fashion, starting with the most basic techniques and progressing to the most complex and specialized.

Surv.   At your institution, must personnel be adequately trained before a research project begins, or can they be trained during the course of a project via observation, assisting a fully trained person, and so forth?

- In most instances, the person must be fully trained before
  the project begins   36/155
- In most instances, the person must have the basic skills needed
  before the project begins, but more detailed skills can be gained
  during the course of the project   87/155
- In most instances, all hands-on training can be gained during
  the course of the project   30/155
- Other   2/155

21:37   What information is useful for training of IACUC members?

Opin.   The AWAR do not explicitly identify IACUC members as needing training, but it can be inferred from §2.32,a, which states that all individuals involved in animal care, treatment, and use be qualified to perform their duties. IACUC members, in order to perform their duties, also must have sufficient knowledge or experience, which may be provided by training. First and foremost, IACUC members should receive training on animal research laws and standards and on institutional policies and procedures. This training may cover many of the topics listed in the following (standards for housing and care, euthanasia standards, pain and distress, etc.). Training on routine and minor procedures (handling, restraint, injections, etc.) might be valuable for some members who have limited experience with animals, but training on complex procedures such as surgery is probably not warranted. In addition to material provided by the institution, a variety of online resources are available such as the American Association for Laboratory Animal Science Learning Library and IACUC.org. Also, attendance at IACUC 101 and Public Responsibility in Medicine and Research (PRIM&R) may be helpful. PRIM&R is a national organization that promotes ethical standards in research. They sponsor training for professionals in human and animal research protection. IACUC 101 is an educational workshop sponsored by NIH/OLAW that provides training to IACUC members, administrators, veterinarians, animal care staff, researchers, regulatory personnel, and compliance officers.

Surv. When training new IACUC members, what topics are covered? Check all appropriate responses.

- We do not have any training program for new IACUC members   18/150
- We train new members via their observations of ongoing procedures, but there is no formal program   48/150
- We cover animal use justification   70/150
- We cover what constitutes an appropriate literature search, if applicable   59/150
- We cover animal number justification   69/150
- We cover means of alleviating pain and distress   69/150
- We cover euthanasia techniques   70/150
- We cover animal enrichment and/or physiologic well-being considerations   52/150
- We cover means of conducting semiannual inspections   75/150
- We cover the general policies of the IACUC   89/150
- We cover the minimal training and skill qualifications of those using animals   54/150
- We cover means of reporting and responding to complaints   70/150
- We cover expectations for animal housing and general care   61/150
- Other   30/150

Note: The category "Other" included a variety of responses. Some approaches that were listed include being mentored by a senior IACUC member, using training courses on the Veterans Administration Web site, sending new members to IACUC 101, having additional reading material, and reviewing sample protocols.

21:38 Are any special approaches useful for training of the unaffiliated member for IACUC service?

Opin. The AWAR and the PHS Policy do not specifically address training of IACUC members; however, the *Guide* (p. 9) notes that "it is the institution's responsibility to provide suitable orientation, background materials, access to appropriate resources, and, if necessary, specific training to assist IACUC members in understanding and evaluating issues brought before the committee." It is advisable to provide such training to all IACUC members, not just the unaffiliated member.

Surv. 1 Do you use any training methods for the unaffiliated member of your IACUC that you do not use for other members?

- Yes   13/146
- No   130/146
- Other   3/146

Surv. 2 What mechanism(s) does your IACUC use to assure that the unaffiliated member is provided adequate training to become an active participant of the IACUC? Check all that apply.

• All of our IACUC members receive the same training	95/151
• The unaffiliated member is given additional training in the form of laboratory tours, animal facility tours, additional reading materials, and other types	52/151
• The unaffiliated member is sent to special meetings (e.g., IACUC 101)	42/151
• Other	6/151

21:39 Is it practical for the IACUC to rigidly enforce attendance for mandatory training courses?

Opin. Enforcement of attendance at mandatory training sessions must be viewed in the context of the institutional culture. If the IACUC is touted as a regulatory enforcement body, then noncompliance will likely not be tolerated. On the other hand, if the IACUC is the facilitator of research, then investigator cooperation will likely be high and compliance will not be a problem.

Surv. How flexible is your IACUC regarding attendance for mandatory training courses? More than one response is possible.

• Not applicable as we do not have mandatory training courses	55/146
• We count "full attendance" if a person attends more than half of the session	11/146
• We allow people to attend if the class has just started	26/146
• We allow for "makeup" sessions if a person cannot attend a regularly scheduled session	61/146
• Other	17/146

Note: The category "Other" included a variety of responses. The most common alternative approach that permitted flexibility was that training was provided online.

Reference

1. Institute of Laboratory Animal Resources, Committee on Educational Programs in Laboratory Animal Science, Education and Training in the Care and Use of Laboratory Animals: A Guide for Developing Institutional Programs, National Academy Press, Washington, D.C., 1991

22

Confidential and Proprietary Information

Marilyn J. Chimes, Laure Bachich Ergin, and Kathryn A. Donohue*

Introduction

The purpose of this chapter is to provide a basic framework for understanding how the concepts of confidential information and proprietary information relate to IACUC activities. In providing institutional review and approval for research activities using animals, IACUCs may review information considered confidential or proprietary by the researchers. IACUCs at government institutions, such as public (state agency) universities, are subject to federal and state laws for open meetings and Freedom of Information acts for "public records"; however, even among such government IACUCs, the application of these access-to-information laws varies. Some IACUCs have open meetings, and, therefore, whatever information is discussed by the committee becomes public information. Conversely, others have closed meetings and do not have this automatic release of information into the public domain. Similarly, state laws on access to public records may lead to radically different results, as some jurisdictions require release and other jurisdictions find an exemption that preserves confidentiality.

In reading the court decisions cited in the following sections, it is important to keep in mind that a court's decision is based on the set of facts before it. Thus, a different factual situation might lead to a different legal decision. This is one reason why courts within the same jurisdiction may appear to reach contradictory decisions. It also is important to remember that a published decision provides binding precedent only within the court's jurisdiction and only as to the issue addressed. Of course, a broadly and well-written decision might provide persuasive authority in other jurisdictions and to other related issues. Thus, for example, a decision by the Vermont Supreme Court, the state's highest court, on the applicability of that state's public records act to an IACUC is binding only within Vermont. The Vermont decision, however, might provide persuasive reasoning that would guide judges in other states when determining the applicability of their state public records laws to an IACUC in their state. Readers are encouraged to check with their institution's general counsel or state attorney general for specific laws and interpretations in their particular state.

* The authors thank Sally Thieme Sanford and Lisa A. Vincler for their contribution to this chapter in the first edition of *The IACUC Handbook*.

22:1 What are the sources of law that can apply to an IACUC's activities?

Reg. IACUCs must comply with the AWA (7 U.S.C. §2131 et seq.), particularly AWA §2143 (setting forth functions of the IACUC) and §2157 (prohibiting release of confidential information by members of the IACUC), and the federal regulations promulgated thereunder, 9 C.F.R. §1.1 et seq. (AWAR), particularly AWAR §2.31 (detailing the IACUC's functions). If research involving animals is funded by a federal agency, federal laws specific to the funding agency will also apply to the IACUC. See, for example, 42 U.S.C. §289d (setting forth requirements of an IACUC at institutions receiving funds from the NIH). Additionally, federal and state laws not specific to animal-related activities can apply to an IACUC's activities, such as laws pertaining to patents or trade secrets, laws requiring public disclosure of records or meetings open to the public, and laws protecting personal privacy. Occasionally, laws designed to protect wildlife, employees, or the environment also can have an impact on an IACUC's activities. (See 2:1 and 2:2 for specific laws requiring the formation of an IACUC.)

Opin. Many sources of law can bear on an IACUC's activities. They include federal and state constitutions, federal and state statutes, federal and state regulations, and federal and state case law.

Constitutions set out the fundamental, broad legal principles that govern the nation or the state that enacted them. The principle of academic freedom, for example, is often considered to be grounded in the First Amendment to the U.S. Constitution. State constitutions are sometimes found to be more protective of certain rights (such as the right of privacy) than is the federal Constitution, even where the wording in the two documents is similar.

Statutes are laws enacted by the federal Congress, a state legislature, or, in some states, by a vote of the people. In a particular state, both federal laws and the laws of that state apply. Laws are codified, or collected, in books that are divided into titles or chapters for different categories of laws. The AWA, for example, is codified in the U.S. Code at Title 7, Section 2131 et seq. The HREA is codified in various places within the U.S. Code, most particularly with respect to animal research at Title 42, Section 289d. Citations to codified law properly include a year in parentheses; this is the year the code collection was published, not the year the statute was enacted. Code collections are generally not published every year, although a supplement with only those sections that have changed might be published after each legislative session. Having the publication date is helpful to ensure that the cited statute includes any recent amendments.

Regulations are rules enacted by a federal or state agency pursuant to authority granted in a statute. As with statutes, these rules are collected in books that are divided into titles or chapters, typically by topic. Pursuant to the AWA, for example, the USDA has enacted various rules that are found in the Code of Federal Regulations. Various administrative procedures govern the adoption of regulations. Typically, for example, the agency is required to publish a notice of proposed rule making and to provide an opportunity for comment. Federal notices of this sort are published in the Federal Register; states have similar publications for the rule-making activities of their agencies.

Case law sets out judges' written interpretations of constitutions, statutes, regulations, and other sources of law on the basis of the way the law applies to a particular situation. The three primary types of U.S. judges are federal judges, state judges, and administrative law judges; each type of judge has jurisdiction over specified types of cases, and sometimes the jurisdictions overlap. Only published decisions of appellate-level judges (those who review trial-level decisions) can be cited as binding precedent in other lawsuits. Although it may seem as though there are innumerable books of appellate-level decisions, comparatively few lawsuits result in this kind of decision. The vast majority of filed lawsuits settle prior to trial; only some that go to trial are appealed, and only a few of those that are appealed result in a decision written for publication. The citation to a case indicates where to find the published decision, which court made the ruling, and in what year.

The AWA and AWAR are the foundation of an IACUC's authority and responsibility, so those statutes and regulations clearly apply to all of an IACUC's activities. IACUCs at institutions receiving federal or state funding are also subject to the laws applicable to the funding source, such as the HREA. Additionally, statutes, regulations, and policies governing other activities of an institution may apply to activities involving research animals and thus apply to an IACUC's activities; these could include laws of the Food and Drug Administration, the Environmental Protection Agency, the Department of Veterans Affairs, or the Department of Defense, among others. An IACUC's activities may also be impacted by regulations of the Occupational Safety and Health Administration as they affect the safety and health of persons working with or around animals. The Endangered Species Act or similar laws protecting certain animals may occasionally also factor into decisions by an IACUC.

22:2 What sources of authority other than those noted in 22:1 might apply to an IACUC's activities?

Reg.   The PHS Policy is applicable to activities supported by the PHS. In order to be eligible to receive a grant or contract award from the PHS, an institution must comply with the PHS Policy, as what is sometimes referred to as a *condition of award*. Conversely, without the required Animal Welfare Assurance and verification of IACUC review and approval, an institution is ineligible to receive support or funding from the PHS for any activities involving animals.

Opin.   In addition to the types of legal authority described, other sources of authority may apply, with varying degrees of force, to an IACUC's activities. The *Guide* certainly carries a great deal of authority in any legal setting as the standard for animal care and use, as does the PHS Policy. In addition, a court is likely to consider as presumptively appropriate standards set out by AAALAC or similarly influential professional associations. Furthermore, courts often look for evidence of a "community standard" among similar facilities if there is a dispute as to whether an institution acted appropriately.

Other important sources of authority are any internally adopted policies that apply to the IACUC. These can be policies adopted by the animal facility, the university, the department, or any entity that has authority to set policy for the IACUC, including the IACUC itself. If an IACUC is to be subject to an internal policy that transcends the law's requirements, it is important that the IACUC be able to comply with that policy. Having a policy whose standards are not met can be worse than not having a policy at all.

Finally, there may be contractual obligations concerning the disclosure and treatment of information related to research. Research involving third parties, such as collaborations or sponsored research and research that involves transfers of third-party information or materials, will likely be governed by contracts that restrict the use and disclosure of information. Such contracts may be identified as confidentiality agreements, collaborative research agreements, sponsored research agreements, or material transfer agreements; however, agreement names vary widely, so it is more important to be aware that restrictive agreements may exist than to focus on a particular name. Such agreements usually contain language permitting the use and disclosure of information within an institution and to regulatory bodies in compliance with law; this would permit an IACUC to perform its necessary functions. However, any disclosures beyond those strictly necessary to comply with law may violate such third-party agreements. Violations of contractual confidentiality restrictions can give rise to claims of breach and civil damages.

22:3   What is proprietary or confidential information?

Reg.   Federal law defines certain types of proprietary information, including copyrights, patents, trademarks, and trade secrets. Under federal law, *copyright* protection can attach to "original works of authorship fixed in any tangible medium of expression" (17 U.S.C. §102). A *patent* can be obtained for a "new and useful process, machine, manufacture, or composition of matter, or any new and useful improvement thereof" (35 U.S.C. §101 et seq.). A *trademark* is something that distinguishes the goods or services of the owner from those of others (15 U.S.C. §1052 et seq.). A *trade secret* is information that has economic value by virtue of its not being generally known to or readily ascertainable by the public (18 U.S.C. §1839). *Confidential information* may be defined by federal or state law or by private agreements or policies. Confidential information can include trade secrets, personal information, information protected by a statutorily defined privilege such as the attorney–client or doctor–patient privilege, or other types of information.

Opin.   *Proprietary information* is information that has the legal status of personal property, that is, information that someone exclusively owns, exercises control over, or uses. The term is usually used for trade secrets, confidential business or financial information, or information about an invention that has not yet been disclosed in a filed patent application. IACUCs must be very careful not to disclose such information inappropriately or inadvertently. Proprietary information also includes any intellectual property, including patents, trademarks, copyrights, and proprietary software. Patents, trademarks, and copyrights are publicly known, but they are considered personal property because unauthorized use or copying by someone other than their owner constitutes infringement. Other types of proprietary information are not known by the public and their public disclosure can cause irreparable harm to their owner.

A *trademark* is a word, phrase, symbol, or design that identifies and distinguishes the source of goods or services.

Copyright is an author's exclusive right to copy or modify his or her original expression. Copyright protection can attach to original works of authorship fixed in any tangible medium of expression, including literary works, pictorial and graphic works, and sound recordings. Ideas, procedures, processes, principles, and the like, cannot be protected by copyright.

A *trade secret* may be any secret, commercially valuable information or compilation of information used in a business that gives the owner an advantage over competitors who do not know or use it and is the end product of either innovation or substantial effort. It may include any form or type of financial, business, scientific, technical, economic, or engineering information, including compilations, plans, formulas, patterns, designs, prototypes, devices, methods, techniques, processes, procedures, programs, or codes, whether tangible or intangible, and whether or however stored, compiled, or memorialized. A person who asserts that information qualifies as a trade secret must show that reasonable efforts were made to keep the information confidential, the information has economic value, and the information is not generally known to or readily ascertainable by the public. When properly safeguarded, a trade secret may remain valuable for an indefinite period of time. One of the best-known examples of a well-kept trade secret is the formula for Coca-Cola®.

A U.S. *patent* gives its owner the right to exclude others from making, using, selling, or offering to sell in the U.S., or importing into the U.S., the "process, machine, manufacture, or composition of matter" claimed in the patent. The term of a U.S. patent generally extends from the date of issuance until 20 years from the date of filing of the patent application. A patentable invention must be useful, novel, and nonobvious. Under the novelty requirement, no U.S. patent will be issued if the invention was in public use or described in any printed publication more than one year before the application for the patent was filed; thus, disclosure to the public of information describing an invention can forever preclude the inventor from receiving a patent. In most other countries, this grace period is shorter or does not exist at all. In the early stages of a research project, investigators may not be able to identify reliably all the information that may be important to a future patent application and know which information must be protected to prevent compromising the value of the research, so all potentially patentable information must be kept confidential.

Under the Patent and Trademark Act Amendments of 1980, popularly known as the *Bayh–Dole Act* (35 U.S.C. §200 et seq.), ownership of patentable inventions arising from federally funded research can be vested in the entities performing the research, including universities, other nonprofit organizations, and small businesses. The holder of the patent rights can license those rights to private companies for further development and commercialization and collect royalties if the invention is sold or used. Thus, inappropriate disclosure by an IACUC could result in the foreclosure of an opportunity for a significant income stream to the IACUC's institution. Funding agreements require that if the grantee elects not to file an application for patent, the grant recipient must disclose inventions to the government in time for the government to file a patent application, and that the government be licensed to practice the invention if the grantee obtains a patent. Federal agencies are authorized to maintain the confidentiality of information disclosing an invention the federal government funded for a reasonable period to permit a patent application to be filed. If the grantee patents the invention but does not appropriately utilize it, the Bayh–Dole Act permits the government to step in and grant licenses to other companies to make the products or use the methods it helped finance for the benefit of the public.

Confidential information is any information that is intended to be held in confidence or kept secret pursuant either to state or federal law or by private agreement or policy. Confidential information can include trade secrets and patentable information or educational, health, and personnel records. It can also include research information subject to a confidentiality agreement or nondisclosure agreement. Some information is considered confidential as a result of a statutorily defined privilege such as the attorney–client or doctor–patient privilege; an attorney should be consulted regarding the limits and requirements of these types of privileges because disclosure to another person or to the IACUC can destroy the confidentiality of such privileged information.

Marking something "confidential" is sometimes a condition of protecting information from disclosure under state and federal laws, as well as private agreements. On the other hand, that action alone has no legal significance and does not protect information from disclosure.

22:4   Under what circumstances might proprietary or confidential information disclosed to an IACUC become available to the public?

Reg.   Information disclosed to an IACUC might become available to the public if it is discussed in an open meeting, if the information is included in a document or report submitted by the IACUC to a governmental agency, or if the IACUC itself is considered an agency of the local, state, or federal government. IACUCs of public institutions may be required to open their meetings to the public under state open meetings laws. Agencies of the federal government must disclose records in compliance with the federal Freedom of Information Act (5 U.S.C. §552). Agencies of a state or local government must disclose records in accordance with the public records laws of their particular state or municipality. Documents and records in the possession of a government agency must be made available to the public for inspection and copying, unless some compelling justification exists for keeping them confidential. Whether a document or report was required by law or was submitted or maintained voluntarily is irrelevant to whether it would be disclosed.

Opin.   The confidentiality of proprietary or confidential information must be scrupulously maintained by an IACUC and each individual member of an IACUC. Inappropriate disclosure of confidential or proprietary information can destroy the value of years of research efforts and have devastating effects on the careers of researchers or the profits of a company. Additionally, if investigators lack confidence that their information will be handled appropriately, they may be reluctant to provide information that the IACUC needs to evaluate proposals properly and meet its legal obligations. Moreover, inappropriate disclosures may lead to researchers becoming the object of harassment or terrorism by persons opposed to the use of animals in research, potentially endangering their lives or laboratories. (See 23:39.)

     Trade secrets and other proprietary or confidential information are not entitled to absolute protection. In litigation, confidential or proprietary information may be required to be disclosed to an adverse party when the disclosure is relevant and necessary to the action but may be withheld from public disclosure under a protective order. In some situations, confidential information automatically loses protection after a certain period under an assumption that it is no longer valuable.

     When confidential or proprietary information is disclosed to a government agency, even if that disclosure is required by law, the information may become available to the public under the federal Freedom of Information Act (FOIA) or state or local public records laws. Only information created or obtained by a government agency or government contractor and in the agency's or contractor's possession can be disclosed. Information the government merely has a right to obtain, or may have relied on but did not obtain, is not available to the public. Receiving a grant from a government agency does not make a private organization a government agency subject to FOIA, unless the government extensively supervises the private organization on a day-to-day basis; similarly, receiving a federal grant does not make a state university subject to the federal FOIA.[1] A private organization is not an agency of the government when designated by the government to perform certain functions, provided it has no decision-making authority.[2]

Each state has its own version of a FOIA or Public Records Act that applies to its own government agencies. Additionally, each federal and state government agency operates under a specific set of rules or regulations governing its disclosures of records to the public.[3] Many states also have laws requiring that meetings of government boards, commissions, councils, committees, and the like, be open to the public. The FOIA, public records laws, and open meetings laws are based on the principle that public officials and institutions are accountable to the people. Under this principle, the public is entitled to know how its government makes decisions and to review the documents and data behind those decisions.

Under FOIA and public records laws, every government agency must make available to the public, upon request, any information in its possession or control unless the information falls within the scope of one of several specified exemptions. Courts interpret these exemptions very narrowly. To prevent information from being disclosed under these laws, sufficiently specific arguments must be made regarding the exemption under which the information falls and the substantial harm that disclosure would cause. The institution wanting to withhold the information has the burden of proving that an exemption protects the information from disclosure.

Open meetings laws require that meetings of "public bodies" be open to the public. These laws usually include a provision that the committee or other entity may adjourn into an executive session closed to the public when discussing privileged or confidential information.

The few courts that have considered any of these laws in relation to IACUCs have reached differing conclusions, although it must be understood that they were interpreting different laws. In *Animal Legal Defense Fund, Inc. v. Institutional Animal Care & Use Committee of University of Vermont*,[4] the Supreme Court of Vermont ruled that the IACUC at a state university was a government body subject to the state's open meeting law and a public agency subject to the state's public records act. The Louisiana appellate court in *Dorson v. State of Louisiana*[5] reached the opposite conclusion, holding that the IACUC at a state university was created by the order of the federal government and accountable only to the USDA and NIH, not the state, even though the entity monitored by the IACUC was a state institution. That court concluded that disclosure of IACUC records was governed by the federal FOIA and the state open meetings law had no application to IACUC meetings. In *American Society for the Prevention of Cruelty to Animals v. Board of Trustees of State University of New York*,[6] the highest court in New York also held that meetings of the IACUC of a state university did not need to be open to the public because the IACUC's powers and functions derived solely from federal law. In another decision in the same controversy, the IACUC's research proposal approval applications were required to be disclosed under the state FOI law, with redaction only of the name, department, location, and telephone number of the researcher and the grant number or application number of the funding source—information the court found that if revealed could endanger the person's life or safety and was protected under a different state law.[7]

The court in *S.E.T.A. UNC-CH, Inc. v. Huffines*[8] reached a similar conclusion. That court allowed the nondisclosure for "public policy" reasons of the names and contact information for researchers and all personnel working with the animals, as well as information on their individual training and experience, but required disclosure of the other parts of the applications for approval by the IACUC of the particular experiments at issue in the case. The court found that these application forms revealed no trade secrets or patentable ideas or procedures but maintained the possibility that other applications might contain such information and could be protected from disclosure accordingly. The court also permitted withholding from public disclosure all applications not approved by the IACUC.

In *Washington Research Project, Inc. v. Dep't of Health, Education & Welfare,*[9] the court required disclosure under FOIA of NIH grant applications, reasoning that a scientist's research design was not a trade secret or item of commercial information because a scientist is not engaged in commerce. The court found, however, that summary statements and site visit reports of initial review groups consisting of nongovernmental consultants with no legal decision-making authority were exempt from public disclosure as intraagency memoranda.

The current national standard is to maintain the confidentiality of research proposals until the research has been funded by a governmental agency, but in *Progressive Animal Welfare Society v. University of Washington*[10] the Supreme Court of Washington held that an unfunded grant proposal for the use of animals in research at a state university was subject to public disclosure under the state public records laws. Certain material was found to be exempt from this disclosure, including the names, addresses, and telephone numbers of the researchers; raw data and guiding hypotheses; budget breakdowns; and peer reviewers' comments on "pink sheets." The court noted that the state had an antiharassment law specifically applicable to animal researchers that could help prevent certain portions of records from being released to the public under the public records act. The court rejected the university's arguments that the federal patent and copyright laws, as well as FOIA and the Bayh–Dole Act, preempted application of the state's public records laws to this federal grant proposal. Whether IACUC meetings were subject to the state open meetings laws was not at issue in this case, but the university indicated to the court that the university considered IACUC meetings public.

In *Sangre de Cristo Animal Protection, Inc. v. U.S. Dep't of Energy,*[11] the court required the disclosure under FOIA of all IACUC reports, minutes, and records of a research institution at a military base that was managed and operated under a contract with the government. The institution was permitted to withhold only the names and identities of its employees. In this case, the deciding factor was the research institution's status as a government contractor.

22:5 How specific should IACUC protocols be when proprietary information related to techniques, compounds, or devices is at stake?

Reg. The IACUC must conduct a review of proposed activities related to the care and use of animals and determine that the proposed activities are in accordance with the AWAR (§2.31,d; §2.31,e) and PHS Policy (IV,C,1). Therefore, IACUC protocols must be sufficiently specific to permit this required review.

Opin. Protocols submitted to an IACUC must be sufficiently specific to permit the IACUC to evaluate adequately the proposed procedures involving animals, including understanding any potential physical, physiological, and behavioral discomfort, distress, and pain the animals may experience as a result of the procedures or related husbandry methods. Protocols must include enough detail to assure the IACUC that no alternatives to the use of animals are available and that the procedures will not unnecessarily duplicate previous experiments. IACUC reviewers must be provided sufficient information to understand whether proposed procedures will be performed with appropriate sedatives, analgesics, or anesthetics, or that the withholding of such agents is scientifically justified. Reviewers must also be able to evaluate whether surgery will be performed on animals and, if so, whether appropriate preoperative, operative, and postoperative procedures will be followed and whether any animal will be used in more than one operative procedure. (See 7:2.)

If a proposed activity related to the care or use of animals will involve a proprietary method, compound, or device, proprietary information may necessarily have to be disclosed in the protocol description submitted to the IACUC so that the IACUC can conduct an adequate review of the proposed activity. (See 7:10.)

Protocols must include sufficient specificity to permit the IACUC to understand whether proposed procedures may cause conditions for which animals may need medical care, including the provision of anesthetics, analgesics, or tranquilizers, and whether that medical care will be provided. The method of euthanasia of each animal associated with or involved in the proposed study must also be adequately explained in a protocol submitted to an IACUC.

The IACUC must be provided with sufficient information to be able to evaluate the adequacy of the training and other qualifications of all personnel who will conduct the proposed procedures involving animals.

Including descriptions and explanations of proposed procedures in a protocol submitted to an IACUC that are sufficiently detailed to permit the IACUC to evaluate the protocol adequately in accordance with the IACUC's legal responsibilities as just outlined may require the disclosure of confidential proprietary information. Techniques, compounds, and devices may be patentable or may have significant commercial value if not publicly known, but the IACUC may need information about them to be able to understand properly any potential discomfort, distress, or pain the animals involved may experience as a result of the procedures or related husbandry methods.

Protection of confidential or proprietary information is no excuse for an inadequate review by an IACUC of proposed procedures involving animals. The IACUC must have sufficient facts to be able to perform its legal responsibilities and make informed decisions about the welfare of the research animals. Accordingly, the IACUC, and each individual IACUC member, must follow appropriate procedures to provide assurance to investigators that their secrets will not be revealed in ways that could cause harm to the researchers or loss of the value of their intellectual property.

22:6   Must a sponsor of research be identified on the IACUC protocol forms?

Opin.   There does not appear to be a legal requirement to identify a sponsor on IACUC protocols. However, institutional policies may require disclosure, such as an institutional policy on conflicts of interest in research. Conflict of interest policies require researchers to disclose the identity of sponsors and the nature and extent of any financial relationship between the researcher and sponsor, in order to preserve the integrity of research and prevent potential researcher bias.

Even if the research sponsor's name is not listed on the IACUC documents, sponsor information may be obtainable by a public records request to a public research institution. In addition, research institutions that use a standard grant and contract proposal routing form may require the sponsor to be identified once the research proposal has been approved by the IACUC.

Some agreements concerning the performance of research require that the sponsor's identity be maintained as confidential; this situation arises when both the sponsoring entity and the research institution are private entities and weighs against disclosure of sponsor information on protocol forms to the extent that it is not necessary for regulatory compliance or compliance with institutional policy.

Surv. Does your IACUC protocol form ask for the identification of the financial sponsor of the study?

- Yes   112/156
- No   41/156
- Other   3/156

Other responses indicated that the financial sponsor was identified on another form or that the question was not applicable to the institution.

22:7 Can researchers protect their own identities on IACUC protocol forms?

Reg. Scientists, technicians, and other personnel involved with animal care, treatment, or research must be appropriately qualified and trained in humane practices and research methods (AWA §2143,d; AWAR §2.31,d,1,viii, §2.32 ; PHS Policy IV,C,1,f).

Opin. The IACUC is required to determine that the personnel conducting animal research are properly qualified, including their education and training (see 8:17). The identity of all personnel conducting procedures on animals must be disclosed, in order for the IACUC to make this determination. If a scientist submitting a protocol is not actually performing the work, there is no legal requirement to identify a scientist who merely authors or submits a protocol. Institutional policy would determine whether the identity of the PI were required (consider an example in which a high school student prepares a protocol that will be performed by qualified lab personnel). In the authors' opinion, the identity of the PI overseeing the work of the protocol is relevant to the evaluation of the protocol and should be disclosed to the IACUC. It may also be relevant for purposes of evaluating conflicts of interest in accordance with institutional policies. However, researchers who are not actually performing procedures could request that identifying information be omitted from any reports. The IACUC can accept, require modifications of, or withhold approval of animal activities (AWAR §2.31,c,6; PHS Policy IV,B,6).

If IACUC records are requested by the public under FOIA or a state's public records law, or if IACUC meetings are open to the public, the identities of the researchers may or may not be able to be protected from disclosure, depending on that state's law.

22:8 What procedural mechanisms may an IACUC adopt to flag information considered proprietary by the researchers?

Reg. Under 7 U.S.C. §2157, it is unlawful for any member of an IACUC to release to the public any confidential information of the research facility, including any information concerning or relating to trade secrets or confidential statistical data. It is also unlawful for any member of an IACUC to use or attempt to use to his or her advantage, or to reveal to another person, any information entitled to protection as confidential information of the research facility.

Opin. IACUC members have a duty not to disclose to anyone outside the IACUC any confidential information they learn through service on the IACUC and not to use any such information in their own research without the express permission of the owner of the information. A variety of mechanisms can and should be used to identify and protect from inappropriate disclosure any confidential and proprietary information provided by researchers to an IACUC.

As a general rule, any information provided by researchers to an IACUC should be distributed only to those persons who have a need for the information. No information received by an IACUC member from a researcher should be forwarded, discussed, or otherwise shared with anyone not essential to the review of the procedures or related husbandry, whether or not the information seems confidential or proprietary. To emphasize IACUC members' duty not to disclose information, requiring that they sign confidentiality or nondisclosure agreements is a good practice.

Basic security measures should always be taken to protect information about the use of animals in research or persons conducting such research because of the controversial nature of such activities and the ease with which they can be misinterpreted. These security measures include, at a minimum, building access controls, escorting visitors, and creating restrictions on photography.

Additional security measures should be taken to protect documents, data, and other written or recorded information provided by researchers to an IACUC, as well as IACUC meeting minutes in which such information is discussed. These measures should include storage in a locked cabinet in an area with restricted access and shredding or similar destruction of duplicative or otherwise unnecessary records. IACUC members should be encouraged to destroy their personal copies of IACUC-related documents or return them to the IACUC secretary when they are no longer needed. A clear and concise policy for the retention and destruction of records should be followed, and materials regularly destroyed when the retention period expires.

Electronic materials are readily copied and distributed, so electronic protocol submissions, distribution, and reviews must be tightly controlled. The computers used for electronic storage, transmission, or discussion of researchers' information should be access-controlled and password-protected, including a requirement to enter a password after a period of nonuse, and encryption of IACUC-related documents should be considered. In establishing security policies and procedures for the protection of researchers' electronic confidential or proprietary information, institutions may find useful the specifications for the protection of electronic health information in the security regulations promulgated under the Health Insurance Portability and Accountability Act of 1996 (HIPAA).[12] These regulations describe administrative, physical, and technical safeguards to prevent access to sensitive information by unauthorized persons. Institutions that include or are affiliated with a medical center may have personnel and technical resources experienced in working under the HIPAA regulations available to assist in developing appropriate procedures and policies for the protection of proprietary information.

Although all information submitted to and created by an IACUC should be handled securely, researchers' confidential and proprietary information should be especially protected. Investigators could be requested to flag or mark confidential or proprietary information specially, including a designation on the title page that such information is included. Such information could be included as an attachment to a standard protocol form. In an electronic document, confidential or proprietary information could be included in a separate document that is encrypted or protected with an additional password. Flagging or marking information considered confidential or proprietary, or including such information in an attachment, would facilitate identification and redaction of the information by the IACUC or institution in the event of a request by the public for inspection of IACUC records or the preparation of an IACUC report or submission to a government agency. However, unless information falls within the scope of a defined exemption in the applicable public records law, the information will nevertheless have to be disclosed upon request.

Investigators should be advised that although all information submitted to an IACUC will be treated confidentially to the extent permitted by law, some information may become available to the public through FOIA requests or discussion in open meetings. Only information within the scope of a legal exemption to disclosure can be assured of protection; flagging or labeling portions of a submission as confidential or proprietary cannot guarantee nondisclosure but can assist the IACUC or the institution in reviewing a record in the event of a FOIA request or open meeting. Investigators should be requested to identify and flag selectively only that information that is truly confidential or proprietary.

22:9 Should IACUC members be directed to turn in or destroy copies of protocols reviewed by the committee at the end of a meeting or once a final action has been taken by the IACUC?

Reg. IACUCs are required by law to maintain certain records and reports for at least 3 years, including protocols for proposed activities involving animals (AWAR §2.35; PHS Policy IV,E,2).

Opin. The IACUC is required by law to retain a copy of every approved protocol (see 8:18) and to record and maintain minutes of all committee meetings and deliberations (see 6:21) for at least 3 years (AWAR §2.35,f; PHS Policy IV,E,2). There is no legal requirement for individual members to maintain separate copies of protocols, and maintenance of separate records multiplies the risk of disclosure of confidential information they may contain. IACUCs may have policies and procedures that include collecting copies of protocols and other documents reviewed by committee members, either at the end of the relevant meeting or when the committee's review is completed. These practices would be governed by institutional policy and may vary. The IACUC practices should not conflict with the institution's other record retention requirements, including other federal laws and regulations pertaining to animal-based research with which the IACUC must comply.

Surv. Does your IACUC require that protocols or other sensitive documents be turned in, destroyed, or electronically deleted at the end of an IACUC meeting or once a final action has been taken by the IACUC?

- Yes 44/156
- No 103/156
- Other 9/156

Other responses included that nondisclosure is expected or recommended, but not enforced, or that sensitive materials are collected at the end of meetings or kept securely only in the IACUC office. Some IACUCs have a nondisclosure policy and/or require their members to sign a confidentiality agreement.

22:10   May an IACUC be sued for breach of proprietary confidence?

Reg.   Under AWA §2157, it is unlawful for any member of an IACUC to release to the public any confidential information of the research facility, including any information concerning or relating to trade secrets or confidential statistical data. It is also unlawful for any member of an IACUC to use or attempt to use to his or her advantage, or to reveal to another person, any information entitled to protection as confidential information of the research facility.

Opin.   Yes. IACUCs are created under and obtain their authority from the AWA, which specifically requires that IACUC members maintain the confidentiality of proprietary information they receive in the course of performing their responsibilities. Violation of this law could be the basis for a lawsuit against the IACUC, the research institution, or an individual member of the IACUC.

The last section of the AWA (§2157,a) addresses disclosures or wrongful uses of proprietary information by a member of an IACUC. This statute prohibits the release by an IACUC member of any confidential information that "concerns or relates to—(1) the trade secrets, processes, operations, style of work, or apparatus; or (2) the identity, confidential statistical data, amount or source of any income, profits, losses, or expenditures, of the research facility." This statute (§2157,b) also makes it illegal for any member of an IACUC "(1) to use or attempt to use to his advantage; or (2) to reveal to any other person, any information which is entitled to protection as confidential information" under the statute. Penalties for violation of this law include a fine of up to $10,000 and imprisonment of up to 3 years, as well as removal from the IACUC; the violator also will be required to pay the damages, costs, and attorneys' fees of the injured investigator and/or the research institution (AWA §2157,c; §2157,d). Thus, IACUC members have a duty not to disclose to anyone outside the IACUC any confidential information they learn through service on the IACUC and not to use any such information in their own research without the express permission of the owner of the information.

Additionally, improper disclosure of confidential proprietary information by an IACUC in response to a request under a public records act or submission by the IACUC of unnecessary confidential information in a record to a government agency may be the basis for a legal claim.

Further, inadvertent disclosure of confidential proprietary information may also support a claim for breach of proprietary confidence. Inadvertent disclosure could occur if IACUC records or an IACUC member's notes are not stored securely or are disposed of in a way that permits viewing of confidential information by unauthorized persons.

Although the inappropriate release of confidential proprietary information by an IACUC could have serious consequences, the authors were unaware of any published legal decision on this issue at the time this chapter was written.

22:11   Must an IACUC at a state institution open its meetings to the public?

Opin.   Whether an IACUC at a state institution is required to open its meetings to the public depends on the wording and interpretation of the state's open meetings law. Open meetings laws, also referred to as *open door* or *sunshine laws*, require that meetings of public bodies be open to the public. If either the IACUC or the state institution is considered a public agency or public body as defined under the open meeting law of its state, then the IACUC may be required to make its meetings open to the public. Nevertheless, these laws usually include a provision that the committee or other entity may adjourn into an executive session closed to the public when discussing confidential information or information defined by law as privileged, such as communications between an attorney and client or doctor and patient.

The few state courts that have considered open meeting laws in relation to IACUCs have reached different conclusions, although it must be understood that they were interpreting different state laws (see also 22:4). In *Animal Legal Defense Fund, Inc. v. Institutional Animal Care & Use Committee of University of Vermont,*[13] the Supreme Court of Vermont ruled that the IACUC at a state university was a government body subject to the state's open meeting law.

In *Dorson v. State of Louisiana,*[14] the court held that the IACUC at a state university was created by the order of the federal government and accountable only to the USDA and NIH, not the state, even though the entity monitored by the IACUC was a state institution. That court concluded that the state open meetings law had no application to IACUC meetings. The highest court in New York also held, in *American Society for the Prevention of Cruelty to Animals v. Board of Trustees of State University of New York,*[15] that meetings of the IACUC of a state university did not need to be open to the public because the IACUC's powers and functions derived solely from federal law.

Surv.   Are your IACUC meetings open either in full or in part to the general public?

• Yes	34/154
• We are a state or federal entity	30/34
• We are a private institution	4/34
• No	111/154
• We are a state or federal entity	33/111
• We are a private institution	78/111
• Other	9/154
• We have never been asked	3/9
• We would allow the public if requested	2/9
• Yes, but prior permission is required	1/9
• Yes, unless there was a conflict of interest	1/9
• We occasionally allow a nonmember to attend all or part of a meeting	1/9
• We have no official policy	1/9

22:12   Is an IACUC at a state institution required to make its records available to the public?

Opin.   This issue depends on the wording and interpretation of a state's public records laws, also referred to as *state freedom of information acts*. Public records laws generally require that records in the possession of a public agency be accessible for inspection and duplication by the public. If either an IACUC or the state institution is considered a public agency or public body, as defined in the public records laws, then the IACUC may be required to make its records available to the public. It also depends on whether any exceptions under the public records law apply. Finally, it depends on whether the records have been disclosed to a federal government agency, which may make the records available to the public under the federal Freedom of Information Act (FOIA). When a request for a record is made, the public agency holding the record has the burden of proving why the record should not be released.

The few state courts that have considered public records laws in relation to IACUCs have reached different conclusions, although it must be understood that they were interpreting different state laws (see also 22:4). In *Animal Legal Defense Fund, Inc. v. Institutional Animal Care & Use Committee of University of Vermont*,[16] the Supreme Court of Vermont ruled that the IACUC at a state university was a public agency subject to the state's public records act. Similarly, in *State ex rel. Thomas v. Ohio State University*,[17] the Supreme Court of Ohio ruled that under Ohio law a state university could not withhold the names and work addresses of university researchers in response to a request under the Ohio Public Records Act because they constituted public records to which no disclosure exemption applied.

In *Dorson v. State of Louisiana*,[18] the court held that the IACUC at a state university was created by the order of the federal government and accountable only to the USDA and NIH, not the state, even though the entity monitored by the IACUC was a state institution. That court concluded that disclosure of IACUC records was governed by the federal FOIA. An appellate court in New York also held that records of the IACUC of a state university did not need to be available to the public because the IACUC is not an entity "performing a governmental or proprietary function for the state."[19]

State public records laws typically include a list of exemptions. If one or more of the enumerated exemptions apply, the otherwise presumptively public document (or portions of it) may be withheld from disclosure. For instance, in *In Defense of Animals v. Oregon Health Sciences University*,[20] the names of drug company sponsors and experimental drugs were found to be exempt from disclosure because they constituted sensitive business records that would not ordinarily be disclosed to the companies' competitors. The court also held that staff member names were exempt from disclosure because the public interest in disclosure was not outweighed by the university's interest in nondisclosure. Staff names in that case were held to be exempt under a state law protecting from public disclosure the names, addresses, and locations of persons engaged in research involving animals. States that have such laws provide an additional exemption to disclosure of this type of information under the state public records laws. (See 22:14 for more information on exemptions.)

Although not directly involving IACUC records, the opinion of the Supreme Court of Ohio in *State ex rel. Physicians Committee for Responsible Medicine v. Board of Trustees of Ohio State University*[21] is instructive. An advocacy organization had filed a writ of mandamus to compel a state university to release certain photographic and videotaped records created by the college of medicine in connection with its research on spinal cord injuries using laboratory mice and rats. The applicable public records statute exempted from public disclosure "intellectual property records" created during research, but only if those records had not been "publicly released, published, or patented."[22] In this case, the state university had lent some of the requested records to scientists and research trainees and shown a few of them to scientists at medical conferences. Acknowledging its responsibility to construe the public records act strictly in favor of disclosure, the court nevertheless held that the intellectual property records exception to disclosure was available in this case because of the efforts the state university took to protect the confidentiality of the records. The state university kept the records in secure cabinets within a locked office accessible only to members of the research laboratory, lent them only to persons who signed nondisclosure agreements, and showed them only to small groups of scientists and researchers under controlled circumstances. The court found that this limited sharing, for purposes related to the research, did not constitute prior public release, so the advocacy organization was not entitled to any of the records under the state public records act.

In the event an IACUC or state institution is not considered a public agency under its state public records laws or it is considered a public agency but exemptions apply, the public could still access such records if the records were ever disclosed to the federal government. Once IACUC records are disclosed to the federal government or an agency of the federal government, such as semiannual inspection or annual reports, such records, which were safe under state public records laws, may be considered public records as defined under FOIA and may now be available to the public, unless they are exempt under FOIA. Information the federal government merely has a right to obtain or may have relied on but never actually obtained is usually not available to the public.

22:13   Is an IACUC at a private institution subject to a state's public records or open meeting laws and therefore required to make its records and meetings open to the public?

Opin.   As a general rule, private institutions are not required to make their records or meetings available to the public. However, if a private institution is supported by, affiliated with, or considered a state-aided, state-owned, or public institution, the state open meeting and public records laws may apply. Receiving a grant alone from a government agency does not make a private organization a public agency subject to these laws; nor does a private organization become a public body by performing specific functions for the government.

    When a private institution or IACUC at a private institution discloses any of its records, such as semiannual inspection or annual reports, to the federal government, such records may become available to the public under FOIA, because they now constitute public records. Information the federal government merely has a right to obtain or may have relied on but never actually obtained is usually not available to the public.

22:14   If a state's public records act does apply (see 22:12), what disclosure exemptions might be relevant?

Opin.   State public records acts typically include a list of exemptions. If one or more of the enumerated exemptions apply, the otherwise presumptively public document (or portions of it) may be withheld from disclosure. In addition to the enumerated exceptions, federal law or constitutional principles might preclude disclosure in a particular situation. This section discusses some of the more common enumerated exceptions, as well as the other sources of legal authority that might be relevant. Note that more than one exception or other law might apply and that state laws differ significantly, making it difficult to generalize. For example, in *Progressive Animal Welfare Society v. University of Washington*,[23] the university argued that several different exemptions and other sources of legal authority applied to preclude release of unfunded grant proposals.

Many public records acts include an exception for research-related documents. They differ widely in their wording and interpretation. In *Robinson v. Indiana University*,[24] the Indiana Court of Appeals held that Indiana University records on animal use in research were "information concerning research" and thus exempt under that state's Public Records Act. By contrast, in *Progressive Animal Welfare Society v. University of Washington*,[25] the Washington Supreme Court held that unfunded grant proposals, as a whole, were not "valuable formulae" or "research data" as defined in Washington's act, although much of the material in the grant proposals did fall within that exemption.

Public records acts typically include an exception aimed at protecting the privacy of state employees. Whether an IACUC can delete (redact) names and addresses of researchers or others when responding to a public records request has been the subject of litigation, with differing results primarily stemming from statutory language. In *State ex rel. Thomas v. Ohio State University*, the Ohio Supreme Court required Ohio State University to release the names and work addresses of animal research scientists. The university had, for security reasons, redacted that information from research-related documents provided in response to a public records request. The court acknowledged the likelihood of increased harassment but noted that there did not appear to be the "high potential ... for victimization" required to establish a constitutionally based exemption.[26]

In contrast, in *S.E.T.A. UNC-CH, Inc. v. Huffines*, the North Carolina Court of Appeals ruled that under that state's Public Records Act, the identity of researchers and staff members could be withheld. The court reasoned that "public policy does require that any information contained in the [research] applications relating to the names of the researcher and staff members, their telephone numbers, addresses, their experience, and the department name be redacted from the IACUC applications."[27] That court also ruled that applications that are not approved need not be made public. Many states have exceptions for "preliminary" or "draft" information, and these exceptions might preclude release of unapproved or unfunded proposals. In addition, intellectual property and public policy arguments often have great force when the research is at this early stage.

It is important to keep in mind that other state statutes outside the state public records laws may provide exemptions from disclosure for research records. Statutes protecting trade secrets as well as any antiharassment statutes specifically related to animal researchers may not protect research records in their entirety but may protect various portions of the records in question from disclosure.

Finally, some state courts have determined that the federal FOIA rather than the state public records act applies to the records of an IACUC. The federal FOIA provides nine exemptions to disclosure of records.

One of the nine exemptions from disclosure under the federal FOIA is for "trade secrets and commercial or financial information obtained from a person and privileged or confidential." *Privileged* information is information that would ordinarily be protected from disclosure in civil litigation by a recognized evidentiary privilege, such as the attorney–client privilege. *Confidential* information has been interpreted as information that, if disclosed by the government, (1) may impair the government's ability to obtain necessary information in the future; (2) would substantially harm the competitive position of the person who submitted the information; (3) would impair other government interests, such as program effectiveness and compliance; or (4) would impair other private interests.[28] Another exemption under FOIA is for certain "inter-agency or intra-agency memorandums or letters." Another important exemption is for "personnel and medical files and similar files the disclosure of which would constitute a clearly unwarranted invasion of personal privacy."[29]

The federal FOIA includes no exemption from disclosure for information qualified under the so-called privilege of academic freedom, which proponents trace to the First Amendment of the U.S. Constitution. The few cases that have considered the protection of material under an academic freedom claim in circumstances such as those an IACUC might encounter have rejected claims that this "privilege" prohibits disclosure.[30]

22:15   How does the principle of academic freedom apply to IACUC activities?

Opin.   Researchers are often aware of the principle of academic freedom and are interested in whether it protects from public disclosure information related to their research. Academic freedom is a multifaceted principle, invoked in support of allowing scholars to study, discuss, and publish ideas free of inappropriate restraints. This principle is invoked in a variety of contexts, from debates over tenure to challenges over publication rights.

This principle is explicitly reflected in the policies of numerous universities and academic associations. In addition, it is generally acknowledged to be firmly grounded in the free speech protections of the federal Constitution and state constitutions.[31]

One facet of academic freedom is often referred to as *research scholars' privilege* or *academic privilege*. Some have argued that this privilege is rooted in the First Amendment and that it protects documents that would otherwise be available under public records statutes, because releasing the information would chill research, the unfettered discussion of ideas, and other activities that underlie academic freedom. This is the context in which IACUCs are most likely to face the potential applicability of the principle of academic freedom.

Unfortunately, in this context, the principle of academic privilege has generally not fared well in the courts. In *University of Pennsylvania v. Equal Employment Opportunity Commission*,[32] the U.S. Supreme Court firmly rejected the university's attempts to expand academic freedom to protect confidential peer review materials from disclosure. The university argued that peer review materials from tenure review files deserved protection under First Amendment academic freedom principles, claiming the disclosure of such documents to the EEOC and ultimately the public, pursuant to a subpoena, would have a "chilling effect" on the tenure evaluation process and affect the "quality of instruction and scholarship" at the university. The Court disagreed, finding the university's reliance on past academic freedom cases misplaced since under the facts of the case, there were no attempts by the government to control university speech.

A year later, in *S.E.T.A. UNC-CH, Inc. v. Huffines*,[33] the North Carolina Court of Appeals reviewed the same issue, but this time in the context of IACUC disclosure of records pursuant to public record laws. The court, following *University of Pennsylvania v. EEOC*, rejected the academic privilege argument and ordered the IACUC records to be disclosed. Similar decisions were made by the Supreme Courts of Ohio and Washington,[34] which both refused to expand academic freedom to protect IACUC records from disclosure to the public.

So it appears that the role academic freedom plays in IACUC activities is rather small, at least as far as protecting research records is concerned. Until the Supreme Court rules differently, IACUCs would be better off understanding their state public record laws well in addition to any statutory exceptions that their legislature has seen fit to provide.

22:16   What mechanisms are available to protect against harassment and vandalism?

Reg.   To damage or cause the loss of any property (including animals or records) used by an animal research facility intentionally or to conspire to do so is a federal crime under 18 U.S.C. §43. Punishment can include a fine, imprisonment for a term of years or life, and a requirement for restitution, depending on the extent of the damage or loss caused.

Opin. In recent years, animal research has become controversial and some opponents of animal research have engaged in criminal activity such as vandalism and harassment. Although comparatively infrequent, there have been notable incidents involving bombings, arson, threats of bodily harm, and publication of private information on the Internet.

The IACUC cannot prevent criminal activity by third parties, but it can adopt practices that minimize risk to researchers. For example, researchers' home addresses, telephone numbers, and social security numbers should never be requested or used in IACUC submissions. Files, offices, computers, and animal housing areas and laboratories should be locked and/or password protected. Documents containing confidential information should be shredded before disposal or recycling. Computer records should be backed up in case of theft or destruction of computers, but all computer disks and tapes should be erased or destroyed before disposal or recycling. Academic institutions in particular abhor secrecy in favor of the free exchange of information; however, under current circumstances they might be able to learn from their IACUC counterparts in industry about appropriate security measures. On a personal level, researchers and committee members would be well advised to adopt practices designed to thwart identity theft, such as shredding personal financial information, maintaining unlisted home telephone numbers, and regularly changing passwords.

Another way to protect against crimes such as harassment and vandalism is to educate the public about the nature and purpose of animal research, the care and effort devoted to protocol review, animal care, the prevention or minimization of pain, the fact that animal studies are usually necessary to satisfy Food and Drug Administration (FDA) requirements for safety and efficacy data as a prerequisite to human trials, and the benefit that animal research has given to humankind through the introduction of new drugs, devices, and therapies. (Although FDA regulations and guidance suggest that animal studies should be performed in most cases to demonstrate safety and efficacy prior to human clinical trials of new drugs and medical devices, validated *in vitro* tests can substitute for certain whole animal tests. See, e.g., 21 C.F.R. §§ 312.23(a), 812.27(a). In limited situations, when definitive human efficacy studies cannot ethically be conducted because doing so would expose healthy human volunteers to lethal or permanently disabling toxic substances, studies in animals alone can serve as the basis for FDA approval of a new drug for human use. 21 C.F.R. § 314.610.)

The institution's legal counsel should be educated on the risks to the institution and its employees so that counsel can be prepared if an incident occurs. Counsel should also be encouraged to join the "animal law" committees of the American Bar Association and state and local bar associations, which are deliberating proposed legislation that would impair or eliminate animal research.

These committees are overwhelmingly dominated by opponents of animal research, with little balance to the debate.

If researchers' names and addresses are publicly available or information about a controversial project has been publicly released, facilities can evaluate the applicable state criminal and civil laws regarding trespassing, harassment, and property destruction. If a public facility is involved, the constitutional right to free speech also must be considered, for example, in establishing boundaries for a protest. In addition, many states have laws specifically penalizing vandalism of laboratories and researchers' homes, as well as specifically penalizing certain forms of harassment of researchers. These laws were passed in response to the belief that general state laws against vandalism and harassment provided insufficient protection or compensation. In some states, an individual may obtain an injunction based on the reasonable belief that he or she may be injured by the commission of animal research–related vandalism or harassed because of animal research activities. If an injunction is issued to prevent anticipated vandalism or harassment, the violation of the injunction itself gives rise to legal action. Although this type of statutory provision might be valuable in some circumstances, its practical benefits are limited by the fact that it requires researchers to file court documents identifying themselves, the harassers, and the anticipated unlawful action sufficiently in advance of that action. A number of states are considering antiterrorism legislation that would classify some of the tactics employed by animal rights groups as terrorism. The benefit of these statutes would be that they impose greater criminal penalties and might therefore deter opponents from engaging in illegal behavior. Another benefit is that terroristic threats (a tactic employed by some opposition groups) are actionable. Institutions can protect researchers by lobbying state lawmakers in support of this proposed legislation.

Research associations are often excellent resources for practical steps that can be taken to reduce the level of confrontation over animal research. The National Association for Biomedical Research and its state affiliates, for example, have very helpful resources regarding community outreach, relations with animal rights groups, and response to protests.

References

1. See *Forsham v. Harris*, 445 U.S. 169, 1980.
2. *Public Citizen Health Research Group v. Dept of Health, Educ. & Welfare*, 668 F.2d 537, D.C. Cir., 1981.
3. See, e.g., 7 C.F.R. Part 1 (USDA); 21 C.F.R. Part 20 (FDA); 45 C.F.R. Part 5 (HHS).
4. *Animal Legal Defense Fund, Inc. v. Institutional Animal Care & Use Comm. of Univ. of Vermont*, 616 A.2d 224, Vt., 1992.
5. *Dorson v. State of Louisiana*, 657 So. 2d 755, Ct. App. La., 1995.
6. *Am. Soc'y for the Prevention of Cruelty to Animals v. Bd. of Trs. of State Univ. of N.Y.*, 591 N.E.2d 1169, N.Y., 1992.
7. *Am. Soc'y for the Prevention of Cruelty to Animals v. Bd. of Trs. of State Univ. of N.Y.*, 556 N.Y.S.2d 447, N.Y. Sup. Ct., 1990.
8. *S.E.T.A. UNC-CH, Inc. v. Huffines*, 399 S.E.2d 340, N.C. Ct. App., 1991.
9. *Washington Research Project, Inc. v. Dep't of Health, Educ. & Welfare*, 504 F.2d 238, D.C. Cir., 1974.
10. *Progressive Animal Welfare Soc'y v. Univ. of Washington*, 884 P.2d 592, Wash., 1994.
11. *Sangre de Cristo Animal Protection, Inc. v. U.S. Dep't of Energy*, 1998 U.S. Dist. LEXIS 23505, D.N.M., Mar. 10, 1998.

12. See 45 C.F.R. § 164.302 et seq.
13. *Animal Legal Defense Fund, Inc. v. Institutional Animal Care & Use Comm. of Univ. of Vermont,* 616 A.2d 224, Vt., 1992.
14. *Dorson v. State of Louisiana,* 657 So. 2d 755, Ct. App. La., 1995.
15. *Am. Soc'y for the Prevention of Cruelty to Animals v. Bd. of Trs. of State Univ. of N.Y.,* 591 N.E.2d 1169, N.Y., 1992.
16. *Animal Legal Defense Fund, Inc. v. Institutional Animal Care & Use Comm. of Univ. of Vermont,* 616 A.2d 224, Vt., 1992.
17. *State ex rel. Thomas v. Ohio State Univ.,* 643 N.E.2d 126, Ohio, 1994.
18. *Dorson v. State of Louisiana,* 657 So. 2d 755, Ct. App. La., 1995.
19. *Am. Soc'y for the Prevention of Cruelty to Animals v. Bd. of Trs. of State Univ. of N.Y.,* 184 A.D.2d 508, N.Y. App. Div., 1992.
20. *In Defense of Animals v. Oregon Health Sci. Univ.,* 112 P.3d 336, Or. Ct. App., 2005.
21. *State ex rel. Physicians Comm. for Responsible Med. v. Bd. of Trs. of Ohio State Univ.,* No. 2005-0612, 2006 WL 508325, Ohio, Mar. 15, 2006.
22. Ohio Rev. Code Ann. § 149.43.
23. *Progressive Animal Welfare Soc'y v. Univ. of Washington,* 884 P.2d 592, Wash., 1994.
24. *Robinson v. Indiana Univ.,* 659 N.E.2d 153, Ind. Ct. App., 1995.
25. *Progressive Animal Welfare Society v. Univ. of Washington,* 884 P.2d 592, Wash., 1994.
26. *State ex rel. Thomas v. Ohio State Univ.,* 643 N.E.2d 126, Ohio, 1994.
27. *S.E.T.A. UNC-CH, Inc. v. Huffines,* 399 S.E.2d 340, N.C. Ct. App., 1991.
28. 45 C.F.R. § 5.65(b)(4).
29. 5 U.S.C. § 552(b).
30. See *Univ. of Penn. v. Equal Employment Opportunity Comm'n,* 493 U.S. 182, 1990; *S.E.T.A. UNC-CH, Inc. v. Huffines,* 399 S.E.2d 340, N.C. Ct. App. 1991.
31. See generally Ailsa W. Chang, *Resuscitating the Constitutional "Theory" of Academic Freedom: A Search For a Standard beyond Pickering and Connick,* 53 STAN. L. REV. 915 (2001); J. Peter Byrne, *Academic Freedom: A "Special Concern of the First Amendment,"* 99 YALE L.J. 251 (1989).
32. *Univ. of Penn. v. Equal Employment Opportunity Comm'n,* 493 U.S. 182, 1990.
33. *S.E.T.A. UNC-CH, Inc. v. Huffines,* 399 S.E.2d 340, N.C. Ct. App., 1991.
34. *State ex rel. Thomas v. Ohio State Univ.,* 643 N.E.2d 126 (Ohio 1994); *Progressive Animal Welfare Soc'y v. Univ. of Washington,* 884 P.2d 592, Wash., 1994.

23

General Concepts of the Facility Inspection and Program Review

Stephen K. Curtis and Karen Janes

Introduction

The general aspects of two important responsibilities of the IACUC are addressed in this chapter: program review and facility inspection. As described in greater detail, both the PHS Policy and the AWAR require that the IACUC conduct facility and program evaluations at least once every 6 months.

The material presented in this chapter reflects the procedures used at the authors' institutions. These procedures are commonly used at research facilities regulated under the AWAR and the PHS Policy. State and local laws may add other requirements.

A note is necessary regarding our survey results. In order to reduce the time required to analyze the survey results, we elected to base our survey data on 40 randomly selected surveys from the 152 surveys available. We believe that the information is representative of the whole. However, the reader is cautioned in interpreting the information. In some instances the survey response depends on the scope of the research operation; a facility with only rats and mice, and not subject to the AWAR, would respond differently than a facility with rodents as well as dogs or nonhuman primates. The scope of this text did not allow us to provide information as to the respondent's facility characteristics. Also, we have included information that shows when a respondent did not choose to respond to a particular question. "No response" was a very small percentage of the total responses.

23:1 What is meant by reviewing the program of animal care and use?

Reg. Both the PHS Policy (IV,B,1; IV,B,2) and the AWAR (§2.31,c,1; §2.31,c,2) require the IACUC to review the institution's program of animal care and use and inspect the institution's animal facilities. The PHS Policy (IV,A,1) requires the IACUC to use the *Guide* as a basis for evaluation, whereas the AWAR requires the IACUC to use the standards of the AWAR as a basis for evaluation.

Opin. Institutions and their IACUCs have a great deal of flexibility in determining the method used to conduct the program review. In general, the IACUC should review all aspects of the animal care and use program using the *Guide* as an outline. Program reviews should primarily focus on the administrative policies and procedures for conducting the animal care and use program. Areas to cover include the following:

- Institutional policies, procedures, and responsibilities
- Monitoring of the care and use of animals
- Personnel qualifications and training
- Occupational health and safety
- Animal environment, housing, and management
- Animal housing and physical environment
- Behavioral management and environmental enrichment
- Husbandry practices
- Population and genetic management
- Veterinary medical care
- Animal procurement and transportation
- Preventive medicine
- Surgery and postoperative care
- Management of pain and distress
- Methods of euthanasia
- Physical plant

This last section of the *Guide*, "Physical Plant," is generally reviewed as part of the facility inspection, as discussed elsewhere in this chapter.

23:2 What are the similarities and differences between the program review and the facility inspection?

Reg.   (See 23:1.)

Opin.   Program review and facility inspection are two separate and discrete duties of the IACUC. The program review encompasses all of the policies, plans, standard procedures, and systems under which the institution fulfills its obligation to care for and use animals in a lawful and ethical manner. On the other hand, the facility inspection is a physical and visual assessment of the buildings, equipment, and environment in which the animals are kept. An important aspect of the facility inspection is to look at the physical condition of a representative portion of the animals housed in the facility. (See Table 23.1.)

Of course, there may be some overlap; for instance, if all of the rodent cages are found to be poorly cleaned during a facility inspection, that condition may be due to a poor program element such as an inappropriate SOP or a poor training program. Together, facility inspection and program review provide a good internal audit of the institution's animal care and use program.

23:3 What is the purpose of the facility inspection and the program review?

Opin.   Most animal research facilities (depending on the source of funding and the species of animals used) fall under the jurisdiction of the PHS Policy or the AWAR. Both documents mandate semiannual program and facility evaluations (see 23:1).

TABLE 23.1

Similarities and Differences in the Facility Inspection and Program Review

Feature	Program Review	Facility Inspection
Directed by	As assigned by IACUC Chair	As assigned by IACUC Chair
Focus of review	Policies, plans, and procedures	"Bricks and mortar," equipment, and animals
General reference in the *Guide*	Chapter 1–Institutional Policies and Responsibilities Chapter 2–Animal Environment, Housing, and Management Chapter 3–Veterinary Medical Care	Chapter 4–Physical Plant
General reference in the AWAR	Part 2, Subpart C–Research Facilities	Part 3–Standards
How review is conducted	Talking around a table Asking questions of management; persons who know and supervise the program Reviewing written policies, plans, and procedures Reviewing documentation Receiving reports Reporting on what is read or spoken	Walking around the animal and research facilities Asking questions of the animal care and research staff Visual observation and physical assessment Generally reporting on what you see
Minimal recommended participation	At least two members of the IACUC Generally need more people to conduct the review; representative persons familiar with the policies, plans, and procedures and how they are used: Veterinarians IACUC administrators Maintenance Animal care management Occupational health Research administrators	At least two members of the IACUC Fewer people needed; will talk with persons encountered during the review: Animal care Maintenance personnel Persons doing the work in research laboratories
Inspection of animals	Generally not	Always
Physical review of laboratories	No	Yes
Ratification of findings and reports	Majority of IACUC members' signatures	Majority of IACUC members' signatures

The review helps ensure that animal health and well-being are optimized and pain and suffering are minimized. The inspections give the committee members time to take a firsthand look at the facilities and the animals that they have been discussing during their meetings. The inspections allow investigators to evaluate their own labs continually and ideally maintain or improve the standards within their own area. Monitoring also helps ensure that all personnel are trained in the care and use of animals and protected from occupational hazards associated with animals. In addition, a well-managed and documented animal care and use program helps improve public confidence in the research conducted at the facility and engenders increased support for government-sponsored research in general. Commercial entities may find that clients will seek out laboratories that adhere to the highest standards of animal care, and clients may be willing to pay a premium for high-quality work. In other words, a well-managed animal care and use program is a vital component of good scientific research.

Surv. 1 How useful is the semiannual program review to your IACUC?

- Little value; we do it for compliance purposes   7/40
- Moderate value; a less frequent schedule would be better   20/40
- Very valuable; a vital part of our quality assurance program   12/40
- No response   1/40

Surv. 2 What benefit does your IACUC derive from the semiannual program review? More than one answer is possible.

- Little or none   3/40
- Helps ensure that our program complies with regulatory requirements   33/40
- Helps monitor our training program   17/40
- Helps ensure protection from occupational hazards   18/40
- Improves public confidence in our program   8/40
- Clients have increased confidence in the quality of our research   3/40
- Helps us decide when modification of the protocol form or protocol review criteria is needed   21/40
- No response   1/40

23:4 What laws, guidelines, and policies delineate what to identify when an IACUC inspects an animal facility and reviews the program of animal care and use?

Reg. Two basic documents mandate and govern the administrative aspects of the facility inspection and the program review: PHS Policy (IV,B) and the AWAR (§2.31,c). (See 23:1.)

Opin. Other references also should be used to evaluate an animal care and use program. The primary standard for conducting an animal care and use program is the *Guide*. In fact, the PHS Policy mandates that the *Guide* be used as the basic standard of care (PHS Policy IV,B). It is customary for research facilities to follow the *Guide* even if they have no legal obligation to do so. In addition, if agricultural research is conducted, the *Guide for the Care and Use of Agricultural Animals in Agricultural Research and Teaching (Ag Guide)*[1] should be consulted. Both the *Guide* and *Ag Guide* contain extensive references; a few of the more frequently used references are the following:

- U.S. Government Principles for the Utilization and Care of Vertebrate Animals Used in Testing, Research, and Training (part of the PHS Policy)
- Occupational Health and Safety in the Care and Use of Research Animals[2]
- Biosafety in Microbiological and Biomedical Laboratories[3]
- Guide to the Care and Use of Experimental Animals[4]
- Institutional Animal Care and Use Committee Guidebook[5]

23:5 How frequently should facility inspections and program reviews be conducted?

Reg. PHS Policy (IV,B,1; IV,B,2) and the AWAR (§2:31,c,1; §2.31,c,2) require inspection of the animal facilities and review of the program of animal care and use at least once every 6 months. The *Guide* (p. 9) uses similar language.

Opin.   It is generally accepted that the interval between inspections and review should not exceed 6 months. To minimize the number of evaluations in a given year the evaluations should occur on a regular basis approximately 6 months apart. At least one USDA-registered and AAALAC-accredited facility conducts its inspections and review in April and November, a 7/5-month interval, without criticism to date. Nevertheless, if the inspection and review took place at the beginning of April and the end of November, this would essentially be an 8-month interval and likely lead to a citation on an APHIS/AC inspection report. Inspectors typically allow a 1-month leeway.[6] Another facility conducts their review on an ongoing basis but files a report every 6 months. An individual component of that program may not be reviewed exactly every 6 months. However, conventional wisdom dictates that major departures from the 6-month rule should be prevented.

23:6   Can the twice-yearly inspections occur in November and again during the following January?

Reg.   (See 23:5.)

Opin.   No. (See 23:5.) The regulatory language dictates an inspection every 6 months, not twice a year. The IACUC would have to conduct a third inspection in July and could not wait until the subsequent November. It would be more efficient to conduct a review each May and November or each January and July.

23:7   Must the facility inspection and program review be conducted at the same time?

Opin.   No. (See 23:13.) Small facilities will often conduct both reviews at a single meeting of the full IACUC. At larger facilities it is often necessary, because of time, to conduct the reviews at different times during the reporting period. At very large facilities inspections may be accomplished by inspecting one sixth of the facility each month during a 6-month period. At one of the authors' facility the schedule shown in Table 23.2 is used.

TABLE 23.2

Sample IACUC Inspection and Program Review Schedule

Event/reporting period[a]	November 1–April 30	May 1–October 31
Program review	December	June
Facility inspection	March	September
Semiannual report	April	October

[a]   Do not confuse this reporting period with that for the annual APHIS/AC or AAALAC report. This reporting period is merely a 6-month period for the semiannual report to the IO. Any 6-month period could be used.

23:8   Does the entire IACUC need to participate in the facility inspection and program review?

Reg.   To comply with the AWAR (§2.31,c,3), a subcommittee participating in the facility inspection and program review must be composed of at least two committee members and "no Committee member wishing to participate in any evaluations . . . may be excluded." Under PHS Policy (IV,B,3, footnote 8, 2002 reprint) the IACUC "may, at its discretion, determine the best means of conducting an evaluation of the institution's programs and facilities." The AWAR require that the majority of the members of the IACUC sign the report (§2.31,c,3).

Opin. The entire IACUC need not participate in the facility inspection and program review. It is common practice, especially at large facilities, to appoint a subcommittee to participate in the facility inspections. Generally, the full IACUC will review and ratify the final report before submission to the IO. It should be noted that regardless of the method chosen, the IACUC retains responsibility for the animal care and use program.

23:9 Minimally, who should be included on every inspection and review team?

Reg. (See 23:8.)

Opin. There are wide variations in how institutions conduct their review. It is common practice to have different persons evaluate different parts of the facility and program. Ideally, a typical team consists of several IACUC members with at least one scientist, one veterinarian, and, where possible, one unaffiliated member. Other useful members (not necessarily from the IACUC) are the animal care supervisor, a person with occupational health experience, and a member of the maintenance department. No IACUC member who wishes to participate should be excluded from participation (see 23:8). A useful practice at large facilities is to have as one member of the team with prior knowledge of the facility, someone who participated in the last evaluation, and a new member to provide a fresh or unbiased review. It is important to keep the inspection and review process vibrant so that the process does not become routine and just another walk around the facility or a "rubber stamping" of the program.

Surv. Who participates in the semiannual program review at your institution? More than one response is possible.

• An IACUC subcommittee only	3/40
• The full IACUC	22/40
• Both a subcommittee and the full IACUC	15/40
• Animal care supervisor	18/40
• Occupational health specialist	9/40
• Maintenance department representative	3/40
• Veterinarian	40/40
• IACUC Chairperson	39/40

23:10 Can outside consultants be used by the IACUC during inspection and review?

Reg. PHS Policy (IV,B,3, footnote 8, 2002 reprint) states that the IACUC may invite ad hoc consultants to assist in conducting the evaluation. However, the IACUC remains responsible for the evaluation and report. The AWAR (§2.31,c,3) have similar language.

Opin. Yes, however, two members of the IACUC must participate in the inspection and review to satisfy the AWAR (see 23:8). In practice, outside consultants (other than veterinary consultants employed by some institutions) are not commonly utilized. Most institutions have ample expertise within their own organization.

23:11   Can a recent AAALAC site visit be used in lieu of a semiannual inspection?

Reg.   A site visit by AAALAC could be used in place of the semiannual inspection under defined circumstances when the review is in conformity with the PHS Policy and AWAR.[7] (See 23:8.)

Opin.   Yes, an AAALAC site visit is a comprehensive review of the facility. The PHS Policy specifically allows consultants to assist with the review. To assure that the site visit meets the requirements of the *Guide* and AWAR, several additional procedures must be followed:

- All members of the IACUC must have an opportunity to participate with the AAALAC site visitors.
- At least two members of the IACUC must participate if the facility is subject to the AWAR.
- A properly convened quorum of the IACUC must review the report, determine whether deficiencies are minor or significant, decide on suitable corrective action, and set a schedule for corrections to be completed. Action must be by majority vote and properly recorded in the official minutes of the IACUC.
- Minority opinions must be included if submitted.

It should be noted that the IACUC may add to the list of deficiencies and might decide to exclude other items identified by AAALAC. Action taken by the IACUC to formalize the AAALAC site visit should be completed shortly after AAALAC presents their initial verbal recommendations. Since the AAALAC visit is going to be a form of IACUC subcommittee, it would be best to convene a regular IACUC meeting within several weeks of the AAALAC site visit. The purpose of the meeting would be to discuss the AAALAC findings and endorse the report as a formal IACUC review. If too much time is allowed to elapse, appropriate action may be delayed including the timely correction of deficiencies. Timely committee action will also aid in writing the post–AAALAC visit response, which is usually due to AAALAC within several weeks after the visit.

23:12   What is an effective method for conducting the program review?

Reg.   Footnote 8 of the PHS Policy (August 2002 reprint) states that the IACUC may, at its discretion, determine the best means of conducting an evaluation of the institution's programs and facilities.

Opin.   Lacking specific guidance by the AWAR and PHS Policy, there are wide variations in the methods institutions use to conduct the semiannual program review. However, there are two accepted formats in common use. One involves an ongoing review of the program as part of normal (generally monthly) IACUC meetings, while the other method involves a discrete review at a special meeting or designated regular meeting. In the author's opinion, it is easier to document the program review when using a discrete review conducted over a short period. Regardless of the method used, all areas outlined in 23:1 should be addressed by the committee during the review process. In all instances, the full committee or a designated subcommittee may conduct the review.

Surv.   What methods does your IACUC use to conduct its semiannual program review of humane care and use of animals? More than one response is possible.

- The NIH checklist (http://grants.nih.gov/grants/olaw/ sampledoc/cheklist.htm) 20/40
- An IACUC-prepared checklist 24/40
- Interview of key staff members (e.g., veterinarian, animal care supervisors, members of research teams). 17/40
- Review of selected standard operating procedures, research records, committee reports, and other program-related records 23/40
- Special presentations regarding aspects of the program such as veterinary care and occupational health 8/40
- No response 1/40

23:13 Can the review of animal care and use programs be based on observations made during the inspection of the facilities? (See 23:2; 23:7.)

Opin. In general, more than a visual inspection of the facilities is required to determine whether the animal care and use program is operating in accordance with law and public policy. For example, facility inspection cannot identify a problem with the composition of the IACUC or uncover problems with documenting the occupational health program.

The specific methods for conducting the program review are discussed in 23:12.

A common practice is to conduct a separate program review using an outline or a specific checklist designed to identify areas of review (see 23:42). Checklists help ensure that no important item is overlooked. The *Guide* is generally used as a basic outline for the program review (see 23:1). The NIH/OLAW Web-based training tutorial[8] also lists the following parts of the animal care program:

- Designation of an IO
- Appointment of an IACUC
- Administrative support for the IACUC
- Standard IACUC procedures
- Arrangements for a veterinarian with authority and responsibility for animals
- Adequate veterinary care
- Formal or on-the-job training for personnel who care for or use animals
- An occupational health and safety program for those who have animal contact
- Maintenance of animal facilities
- Provisions for animal care

Surv. Do you conduct your program review at the same time as your facility inspection?

- No, our program review is an entirely different meeting and procedure 12/40
- Yes, it is all part of the same review process 27/40
- No response 1/40

23:14   How much time should the IACUC expect to devote to conducting the program review?

Opin.   An appropriate and complete program review requires the expertise of a large number of people, who are able to compare the actual program as it exists at the institution against the regulatory requirements and other standards. Much of the evaluation of the program should be conducted as these individuals, the area experts, manage their particular aspect of the program. If the appropriate preliminary work has been completed, a formal program review subcommittee meeting will generally be completed within a 2-hour period, even in the most complex institution. Further full committee review will take additional time—up to 2 hours. Time spent beyond this would generally indicate a lack of planning and preparation prior to the formal meeting. Table 23.3 (p. 438) illustrates a typical review process.

Surv.   Every 6 months, approximately how much time does your IACUC devote for the semiannual program review (not inspections)? (See Table 23.4.)

TABLE 23.4

IACUC Time Devoted to Semiannual Program Review

	Subcommittee (if any)	Full Committee
Less than 1 hour	2/18	17/37
1–2 hours	4/18	11/37
3–5 hours	6/18	5/37
More than 5 hours	5/18	0/37
No response	1/18	4/37

23:15   Animal care and use programs often do not change in a given 6-month period. How do you make the IACUC program review interesting and effective under these circumstances?

Opin.   If there were regulations that required that the program review be more interesting and effective, we could count on IACUCs to meet the challenge of making them so. However, in the course of an IACUC's very busy workload, making the process interesting can be of low priority. Needless to say, if a program review is interesting, the participants are more involved in the process and therefore the outcome will be much more effective. If the program of animal care has not improved over the last 6 months, it may be useful to reassess the institutional standards, even if currently meeting all of the requirements; raising the bar at the institution will certainly improve the overall program of animal care.

Simple ideas can be easily implemented to make the program more interesting. The person in charge of leading the discussion should not necessarily be the same person each time. It is best to have some members of the review team remain the same and include new committee members as well, to give the discussion a different perspective. It is helpful to encourage or mandate that new committee members serve on the review team the first year they are on the committee. The IACUC can start by reviewing which items were targeted during the last review and check on progress. If possible, it is useful to demonstrate certain items that are to be evaluated in the program review; for example, a new member may never have seen items that encourage enrichment for a particular species or they may not know what a Microisolator™ looks like. Each member could be responsible for leading a

TABLE 23.3

Typical Steps for the Program Review Process

Time Frame	Activity	Who	What	Time Required
2-4 weeks prior to a meeting	Individual review: remind area experts about pending review; suggest they review appropriate checklist and evaluate their specific part of the program	Area expert: veterinarians, occupational health, facilities maintenance, animal care, IACUC coordinator, training coordinator, other key persons *Note:* Some area experts will also be committee members	Using a checklist or the *Guide*, individual will review his/her aspect of the program along with that of the regulatory requirement	Varies with program; a large dynamic program will take more time Problems identified at this level should be corrected prior to any further action; this part of the review is essentially a self-evaluation
Subcommittee meeting day	Subcommittee review	Should include at least two members of the IACUC as well as area experts as indicated	Collectively reviews the program in terms of the combined and accumulated knowledge of the area experts Makes recommendations to the full committee	Up to 2 hours; longer meeting may be unproductive
Within 30 days of subcommittee review	Full IACUC review	Members of the IACUC and invited area experts as desired	Review findings of the subcommittee; establish corrective action and schedules of completion Significant deficiencies may need more expeditious handling	Up to 2 hours; longer meeting may be unproductive; with good preparation at previous levels meeting may be much shorter
Within 30–45 days of full committee review; some significant deficiencies may need more timely action	Post any administrative paperwork	IACUC Administrator— *Note:* At small facilities a committee member may serve as the Administrator	Complete minutes, send out notice of required corrective action, and complete other required documentation Significant deficiencies may need more expeditious handling	As long as needed; should generally complete process in 30–45 days
See above	Final report	IACUC Chair and IACUC Administrator	Signed by committee and forwarded to the IO	See above

portion of the review. Some IACUCs have found that availability of refreshments keeps the review team energized.

Surv. How do you make the IACUC program review interesting and effective? More than one response is possible.

- Nothing special, we just go through the checklist or other procedure and get it over with   23/40
- We have people make special reports regarding different areas of the program   4/40
- We use a different checklist or procedure each time   1/40
- We rotate the people doing the review so they have a fresh perspective   10/40
- We get outside experts to review our program, much like an AAALAC site visit   1/40
- No response   2/40

23:16 Can alternate members of the IACUC perform semiannual inspections that involve PHS Policy and USDA-covered species?

Reg. The PHS Policy and the AWAR are silent on the use of alternate IACUC members. However, NIH/OLAW and APHIS/AC agree that alternates may be utilized under certain circumstances.[9] Quoting from the OLAW/APHIS policy letter (see 5:23–5:25), conditions that must be met include the following:

- Alternates must have an official appointment, the same as any other member. (See 5:2.)
- There must be a specific one-to-one designation of IACUC members and alternates. For example, an alternate for a non-affiliated IACUC member would need to also meet the non-affiliated member requirements. Use of a pool of alternates would not be consistent with this requirement.
- An IACUC member and his/her alternate may not contribute to a quorum at the same time or act in an official IACUC member capacity at the same time.
- Alternates should receive IACUC training or orientation similar or identical to what is provided regular IACUC members.
- Alternate members would be expected to "vote their conscience" as opposed to representing the position of the regular member for whom they serve.

Opin. It appears that alternates can assist with inspection responsibilities under the PHS Policy and the AWAR. In the absence of the regular voting member, an alternate can step in for any of the tasks the regular member would normally do. It follows that if the regular member could not take part in the inspection, the alternate could go in his or her place. The PHS Policy (see 23:10) provides for great latitude in the way the inspections are performed; therefore, the committee is able to appoint outside members to inspect specific sites on behalf of the committee. This fact provides support for allowing an alternate to be appointed as an inspection team member, regardless of whether the regular voting member is participating or not. In a voting situation, the alternate may not vote if the regular member is present to vote.

Surv. Can an alternate member of your IACUC perform semiannual inspections that involve PHS Policy and USDA-covered species?

- Not applicable as we have no alternate IACUC members   18/40
- Yes   14/40
- No   4/40
- No response   4/40

23:17 What areas should the IACUC include in the semiannual facility inspection?

Reg. Both the PHS Policy (IV,B,1; IV,B,2) and the AWAR (§2.31,c,1; §2.31,c,2) require the IACUC to inspect the institution's research and animal facilities including satellite facilities and animal study areas once every 6 months. The wording is slightly different but the intent is the same. (See 15:2.)

Opin. The areas shown in Table 23.5 are generally included in the inspection process. Note that these areas should have been authorized for animal use in advance of the inspection. There should be no surprises as to what to evaluate.

TABLE 23.5

Typical Areas Reviewed during the IACUC Facility Inspection Process

Centralized animal holding/housing areas	Associated support space used for animal care:
	  feed preparation
	  cage preparation
	  cage washing area
	  storage areas

Research laboratories where live animals are used including areas used for euthanasia only	Surgery areas including central and individual research laboratories regardless of species used
Individual researcher and departmental animal facilities including support areas	Necropsy areas
Areas used for live animal demonstration and teaching such as student science centers	If farm areas are included in the institutions' accreditation application, then farm areas where animal care and housing are provided are included in the IACUC inspection *Note:* It might be impossible to include all production areas such as pastures housing range cattle

Surv. What areas does your IACUC include in your semiannual facility inspection? More than one answer is possible.

- Central animal holding areas   39/40
- Small departmental animal holding areas   24/40
- Laboratories holding animals for any length of time   27/40
- Laboratories holding animals for over 12 hours   4/40
- Necropsy areas and other areas where only dead animals or animal tissues are manipulated   29/40
- Nonanimal areas such as feed storage and cage storage   37/40
- Cage wash and cage preparation areas   36/40
- Survival surgery areas   29/40
- Nonsurvival surgical areas   26/40
- *In vitro* tissue laboratories   3/40
- No response   1/40

23:18   Can an IACUC inspection team inspect areas more often than twice a year?

Opin.   Yes, facilities may be inspected as often as the IACUC deems necessary. The PHS Policy and the AWAR do not preclude IACUC inspections more frequently than once every 6 months. An IACUC may "reset the clock" if it wishes to perform an inspection early and then perform the next inspection 6 months later. More frequent inspections may be required to verify that previously identified concerns have been corrected. Some IACUCs conduct inspections on an ongoing basis. For example, one IACUC sets aside a time each month to visit laboratories where animal research is conducted. Under this system, an individual laboratory may be inspected more frequently. It is also a common practice to evaluate new areas before housing animals.

Members of the animal care staff, such as directors and managers, are often members of the IACUC. In addition, at least one veterinarian should be a member. These individuals are usually in the animal facility on a regular basis between formal IACUC inspections and, therefore, have the opportunity to evaluate and report to the IACUC more frequently than every 6 months.

23:19   Should animal facilities be inspected only if animals are present at the time of inspection?

Reg.   (See *animal facility* definitions in 15:1.)

Opin.   Areas where animals are currently used, even if they are not in that area at the time of an inspection, should be visited and evaluated in relation to their physical structure. If it is likely that animals will never again be housed or used in a particular area, the area can be removed from the list of those locations to be inspected.

It may not be possible at large, decentralized facilities to evaluate every research laboratory where live animal work is conducted. Those areas where animals are housed for extended periods or where surgery is conducted should be inspected semiannually. Laboratories where only minor work is performed, such as terminal tissue collection, could be inspected less frequently. Although this practice is not in keeping with the letter of the law, in the authors' experience, less frequent inspection using these criteria has not created concerns with regulatory agencies. The reader is cautioned, however, that the definition of *minor* work can be open to interpretation. In the example presented, if the animals are euthanized and terminal tissue collection also takes place, it is suggested that this area be inspected every 6 months.[10] NIH/OLAW has suggested that reduced inspections or alternate methods of monitoring might be adopted by the IACUC.[11] It would be wise for the IACUC to establish a specific policy regarding frequency of inspection. Doing so will lead to a more orderly and consistent procedure.

23:20   Should the facility inspection team visit research laboratories where animal research (such as rodent surgery) is conducted?

Reg.   (See 15:1; 24:10; 25:1–25:3.)

Opin.   It is our opinion that all areas where live animals are housed or used (even areas where animals are taken for euthanasia and tissue collection) should be included as part of the semiannual facility inspection. Nevertheless, not all IACUCs strictly follow this practice (see 24:10) and there is leeway within the NIH/OLAW policies. NIH/OLAW[12] states that when considering IACUC responsibilities for semiannual review, it is important to keep in mind that each Assured institution, acting through its IACUC or facility veterinarian, is responsible for all animal-related activities at the institution regardless of where the animals are maintained or the duration of their stay. The PHS Policy allows institutions some discretion regarding specific methods to assure compliance, but the institution is clearly responsible for what happens to animals in investigators' laboratories. The degree, frequency, and method of IACUC oversight often depend on the nature of the activity. For example, satellite holding facilities and areas in which surgical manipulations are performed must always be included in semiannual reviews. Other activities, such as routine dosing, weighing, or immunization of animals in laboratories, may be monitored using other methods such as random site visits and evaluation.

Inclusion of animal research laboratories in the semiannual IACUC review is another way to satisfy the PHS Policy requirements. The IACUC must have access to all investigators' laboratories for the purpose of verifying that activities involving animals are conducted in accordance with the proposal approved by that committee. This implies that IACUCs at institutions regulated by NIH/OLAW have broad discretion in the way they oversee animal work that is conducted in laboratories.

23:21   When doing semiannual inspections, is it a good idea to have some IACUC members "specialize" in certain areas year after year, or is it a better idea to have members rotate among areas?

Opin.   There is benefit to having a combination of people who have inspected the area before and some who have never inspected the area. In this way the IACUC will have someone who has a fresh viewpoint and a person who has had the experience of inspecting a given area before.

Surv.   When performing semiannual inspections or program reviews, is it better to have some IACUC members who specialize in one or more areas, or is it better to have members rotate among areas?

- We try to have the same person(s) look at the same areas at each semiannual inspection   5/40
- We try to rotate members to different areas   17/40
- We make no conscious attempt one way or the other; whatever happens, happens   16/40
- No response   2/40

23:22   Should the IACUC allow a previously unused area to be used for live animal–related work without a formal IACUC inspection?

Reg.   Both the PHS Policy (IV,B,6) and AWAR (2.31,c,6) indicate that the IACUC should review and approve all activities related to the care and use of animals.

Opin.   No research project involving live animals should be conducted prior to a review and approval by the IACUC. The protocol should specify where animals are housed and where animal work will be conducted. As part of the approval process at least two members of the IACUC should evaluate the appropriateness of new animal housing or study areas. Therefore, work should not be conducted until after a formal review by the designated IACUC members. Generally, the AV and one additional member will be designated to review and approve the requested location. (See 25:4.)

Surv.   Does your IACUC allow a previously unused area to be used for live animal–related work without a formal IACUC inspection?

- No   21/40
- Yes   5/40
- We only require that the PI notify the committee prior to use   2/40
- Veterinary review and approval are required by our IACUC   8/40
- One or more IACUC members review and approve the area   2/40
- No response   2/40

23:23   Is there a conflict of interest when the director of the animal facility (an IACUC member) participates on a team that inspects the core animal facility?

Opin.   There might be a perception of a conflict of interest because the director, often a veterinarian, would be in a position of having to criticize his or her own area of responsibility. However, in practice this is seldom a problem, for two reasons. First, animal program directors typically have as their primary goal to have a well-managed and regulation-compliant animal care program and animal facility. They are generally faced with annual unannounced inspections by APHIS/AC and periodic review by AAALAC. It is in their best interest to find and correct any problems that may be present. Second, the director generally seeks to have other eyes view the facility; in this regard the director encourages additional committee members to conduct the review. Directors seldom hide problems; rather, they are the ones to point out problems and potential problems to newer members of the team. The director is often the most avid trainer on the conduct of a facility inspection.

Surv.   Does the director of your animal facility participate on a team that inspects the core animal facility(ies)?

- No   10/40
- Yes, and she or he can inspect the facilities by himself or herself   8/40
- Yes, but along with one or more additional committee members   22/40

23:24   How should the IACUC assure that IACUC members are not in the position of evaluating their own areas of responsibility, such as their own research laboratory?

Opin.   Just as IACUC members should not review their own protocols, they should not evaluate their own research laboratory. If they are on the inspection team, or if they are working in their laboratories when an inspection team arrives, they can provide information to the team. They should be treated the same as any other researcher, since during the inspection of their laboratory, they are in the role of researcher rather than committee member. It would be helpful during the initial orientation session and again during the committee planning process to inform the membership that all committee members are viewed as researchers during the inspection of their own laboratory. Setting this ground rule in a diplomatic manner in the beginning can be very helpful to preventing misunderstandings later.

   The inspections should be "marketed" to the research community as being educational, with the exchange of information going both ways. The committee can learn what is occurring in the laboratories and listen to the concerns of the researchers. The researchers can learn about the regulations and the institution's commitment to proper animal care and use. If done properly, with the right tone, spirit of helpfulness, and a touch of humor, the inspection process can occur in a very collegial manner. This attitude should make it easier for any infraction to be viewed as an educational experience, whether the lab belongs to the researcher or a committee member who happens to be a researcher.

Surv.   Does your IACUC attempt to assure that IACUC members are not in the position of evaluating their own areas of responsibility, such as their own research laboratory?

- They participate fully as desired or assigned   17/40
- They do not participate in the inspection of their own areas
  of responsibility   21/40
- No response   2/40

23:25 Should facility inspections be announced or unannounced?

Opin.   Opinions differ, but most institutions favor announced inspections. There are no requirements in the AWAR or PHS Policy stipulating announced or unannounced inspections. The advantages of both have been briefly discussed.[13]

An unannounced inspection gives the inspection team a clear picture of what is routinely occurring in the facility. On the other hand, announced inspections allow essential persons to be available to answer questions. Announced visits also give investigators the incentive to review their areas and procedures with their lab personnel; they may motivate some PIs to elevate the standards within their labs, especially if they are provided with a listing by the IACUC of what are some common types of infractions. These higher standards will ideally prevail through the entire year. As there are benefits from both announced and unannounced inspections, it may be advisable to conduct one unannounced inspection per year. The choice is that of the individual IACUC. In practice, it is difficult to keep routine evaluation schedules confidential since key members of the animal care staff often serve on the committee. (See 25:6.)

It should be remembered that APHIS/AC inspections are always unannounced while AAALAC evaluations are always scheduled well in advance. PHS visits are generally scheduled with short notice. All three systems have merit.

Surv. 1 Are your facility inspections announced or unannounced?

- Unannounced   7/40
- Announced   18/40
- Both are used   15/40

Surv. 2 If your facility inspections are announced, have you received any benefit from their announcement? More than one answer is possible.

- We see little difference in benefit between announced and
  unannounced inspections   11/33
- No, when inspectors are announced the investigators make
  themselves scarce   2/33
- Yes, the investigators have a better understanding of what is
  expected with announced inspections   14/33

- Yes, there are not as many items on the inspection report and less to correct   3/33
- Yes, the investigators feel more like a part of the process   12/33
- No response   3/33

23:26   Should the IACUC inspectors speak with individuals participating in research using animals?

Opin.   Yes, this is a generally accepted practice and is highly desirable. The PHS Policy and the AWAR do not mandate IACUC discussions with specific individuals nor preclude them. There are two groups of people from whom the committee can obtain information: the research staff and the animal care staff. For example, through informal interviews at the time of the inspection, it is possible to verify that SOPs are being followed, that persons are appropriately trained, and that anesthesia and euthanasia methods are appropriate and in accordance with approved protocols.

Some institutions require that each laboratory have a representative present to answer questions from the inspection team. Other IACUCs schedule inspections at times when persons are normally working and question those persons available on a less formal basis. The inspection process can and should be a learning experience for both the IACUC and the staff. Free and open communication facilitates this learning process.

The IACUC should take time at the end of a visit to give feedback to the persons involved. When warranted, positive as well as negative comments should be made. IACUCs should consider having an "exit interview" with interested persons as a means of exchanging information.

23:27   Should deficiencies and possible corrective actions be discussed at the time of the inspection?

Reg.   Neither the AWAR nor PHS Policy mandates when corrective actions should be discussed. The PHS Policy (IV,B,3) does require that IACUC semiannual program reviews that note program or facility deficiencies contain a reasonable and specific plan and schedule for correcting each deficiency. The AWAR (§2.31,c,3) contain similar language.

Opin.   In general, yes, although opinions vary. If the deficiencies are major (significant), the investigator, facility manager, or veterinarian should be immediately informed of the problem so he or she can correct it as soon as possible. Delaying notification about a significant deficiency until the entire committee is informed only prolongs the problem. The IACUC can always request additional action be taken to correct a deficiency at a later date. By discussing the deficiency with the appropriate persons at the time of the inspection, the inspecting subcommittee can make sure that their recommended course of action and date for correction are workable before presenting them to the entire IACUC for review.

There are times when it might be prudent to withhold discussion at the time of the inspection, pending a full IACUC discussion. If the reviewing subcommittee is unsure of the nature of the deficiency, it is advisable to withhold judgment. On the other hand, if a problem is serious but corrective action is not immediately required, then full committee review is warranted. For example, this could occur if the inspecting subcommittee finds that an investigator is conducting research that was not authorized, but the animals are now only being observed prior to final disposition. Although very serious, in terms of the animals' well-being no immediate action may be required by the reviewers. Nevertheless, the reviewers should immediately report the finding of unauthorized research so that the IACUC can assess the situation and, if warranted, take immediate action. In accordance with NIH/OLAW's guidelines, the IACUC would also determine whether the violation warranted reporting to NIH/OLAW.

Surv. When deficiencies are identified upon facility inspection or program review, how is the required corrective action communicated to the responsible person?

- Verbal (telephone or direct) request for correction is made with the responsible person at the time of the review/inspection — 11/40
- Written (letter or e-mail) request sent from the committee to the responsible person — 32/40
- No response — 1/40

23:28 An IACUC has a policy of requiring investigators to euthanize animals by two methods (e.g., carbon dioxide followed by cervical dislocation) to assure the animals are dead. During a semiannual inspection it is determined that a PI is using only one of the two approved methods. Is this considered a minor or a significant deficiency?

Opin. Minor deficiencies are problems for which immediate resolutions are generally not necessary to protect life or prevent distress. If the IACUC considered it necessary to require two methods of euthanasia to ensure death, it would follow that if only one method were used, there would be a possibility that the animal would be in distress if a quick and painless death did not occur. Thinking in this mode would lead one to list it as a significant deficiency. There would indeed be no question about this if during the inspection animals were found in the freezer still alive. If animals were euthanized using only one method and no distress was found by the inspection team either by direct observation or from speaking to personnel working in the lab, then it could be recorded as a minor deficiency. The infraction would be the investigator's not following the protocol but with no harm occurring to the animals. The IACUC may determine that they were overzealous in asking for both methods and reconsider their policy.

Surv. An approved protocol requires two methods of euthanasia (e.g., pentobarbital followed by cervical dislocation) but only one was used by the PI. Would your IACUC consider this a significant deficiency?

- Not applicable or we have never faced this situation — 10/40
- We would probably consider this a significant deficiency — 15/40
- We would probably not consider this a significant deficiency — 1/40
- We would consider this a significant deficiency only if it resulted in an animal welfare problem — 8/40
- We would not address this type of a problem during an inspection or program review — 2/40
- No response — 4/40

23:29   How are results of inspections and reviews recorded and processed?

Reg.   After the review is completed, a report signed by a majority of the IACUC is forwarded to the IO (AWAR §2.31,c,3; PHS Policy IV,B,3; *Guide*, p. 9). IACUC members have the right to file formal minority opinions that must be included with the final report (PHS Policy IV,E,1,d; AWAR §2.31,c,3). (See 6:1; 23:37–23:39.)

Opin.   Three methods are often used to document the facility inspection and program review:

- Completed checklist
- Written report
- IACUC minutes

The method used and the precise format depend on the complexity of the institution and the preference of the IACUC. Extensive checklists have been written to assist IACUCs in conducting reviews and inspections (see 23:42). The main pitfall with any system is that with repeated use (every 6 months) the process becomes mechanical and members of the IACUC may become complacent. It is best to vary the method of review from time to time to maintain vitality in the review process.

23:30   How should the IACUC decide whether a deficiency item identified during the facility or program review is actually included on the final report?

Opin.   It is common for committee members or others who participate in the semiannual review to identify alleged deficiencies. These items will be of greater or lesser importance and some items may not even be relevant to the animals program. For example, it may be discovered that the sidewalk to an animal area is cracked or separated or that there is a light bulb burned out in an animal room. Should these items be included?

There are a number of methods that could be used to decide whether an identified item is included in the final report. The committee may determine that it will include items

- At the request of least one IACUC member
- At the request of a subcommittee of members
- Only after full committee deliberation and with majority consent

The authors' preferred method is to have one or more IACUC members submit the item for review by the full committee. The committee will decide, on the basis of the alleged deficiency and the regulatory authority, whether the items warrant inclusion in the report. In most cases a formal vote is not taken. However, controversial items may require a vote. Items not included on the report can be corrected by a more informal process. Some items, such as a single burned out light in a room with 40 lights, may not require correction until normal scheduled maintenance.

Surv.   How does your IACUC decide whether a deficiency item identified during the facility or program review is actually included on the final report?

- All items are included provided at least one IACUC member
  requests inclusion   10/40
- A subcommittee of members decides whether an item should be
  included   1/40

- The full committee decides whether an item should be included 4/40
- Anything that was noted as a deficiency is included, whether or not someone requests its inclusion 22/40
- No response 3/40

23:31 Sometimes an item may be insignificant and not warrant inclusion in the final report. An example might be a single crowded cage in a room of 1000 mouse cages. Should the IACUC have a mechanism to track concerns that are not included on the final report?

Opin. Tracking concerns that are not included on the final report can be very useful for staying on top of what is happening in a program and making sure that what may be an insignificant item does not turn into a more significant one. Nevertheless, IACUCs should not exclude items from the report just because they feel the report is too long. It is better to include deficiencies that seem minor so that they can be addressed and do not become significant or more widespread. It is useful to review information that NIH/OLAW provides on their Web site (http://grants.nih.gov/grants/olaw/olaw.htm) regarding what is considered to be a concern and report it accordingly. According to the NIH/OLAW Web site, even a burned out light bulb or paint chips should be included in the report. An overcrowded room would be considered by most IACUCs to be at least a minor problem, and depending on the degree of overcrowding, it could be considered a major problem. If the IACUC does find an item that they feel should not be on the listing, why would they want to keep track of it other than to correct the problem? If the deficiency is something that is important enough to be corrected, then in all likelihood it should be on the official report. IACUCs should avoid the temptation to exclude items because they think they will reflect poorly on the institution. Large complex facilities will have minor items from time to time, and listing them shows that the inspection process is working.

23:32 What criteria are used to categorize deficiencies as minor or major (significant)?

Reg. The AWAR and PHS Policy are in concordance: "A significant deficiency is one which . . . in the judgment of the IACUC and the Institutional Official, is or may be a threat to the health or safety of the animals" (AWAR §2.31,c,3; PHS Policy IV,B,3).

Opin. The IACUC has broad discretion in applying this standard. Deficiencies should be discussed at a regularly convened meeting of the IACUC. At that time, a final determination as to the nature of the deficiency should be made. Formal minority opinions should be attached to the final report if desired by the dissenting committee member (see 23:37). Discrepancies found in the program or the facilities must be identified on the report as either minor or major (significant). The IACUC must devise a "reasonable and specific plan and schedule for correcting each deficiency" (AWAR §2.31,c,3; PHS IV,B,3). (See 23:34.)

23:33 What are some examples of minor and significant deficiencies?

Opin. Any deficiency that would cause injury, death, or severe distress to animals would be considered major. Examples could include failures in heating, ventilating, and air-conditioning systems and the related electrical systems and power failures of sufficient duration to affect important areas such as surgical suites. Not only can program or facility deficiencies be included as major (if they meet the definition of the term) but accidents and natural disasters can result in a reporting of a major deficiency. For example, if a hurricane tore off a roof of an area where animals were kept, and injury, death or severe distress to the animals resulted, it should be reported as a major deficiency.

Examples of minor deficiencies would include expired food, unsealed wood in an animal surgery area, missing ceiling tiles, a dripping faucet, and small cracks in the wall.

Surv.   What are some examples of findings (facility inspection or program review) that your IACUC has considered to be significant deficiencies?

Seven of 40 facilities reported that they had never identified a significant deficiency. Those with significant deficiencies reported items such as human health hazards, unapproved procedures being performed, room temperature problems, avoidable animal injury, unauthorized animals on campus, inadequate housing, and insufficient health records. All facilities had minor deficiencies.

23:34   What regulatory and operational issues must the IACUC address when evaluating minor and significant deficiencies found during the animal facility inspection?

Reg.   In accordance with the PHS Policy (IV,B,3) and AWAR (§2.31,c,3), a reasonable and specific plan and schedule must be given for correction of each deficiency. It does not matter whether the deficiency is designated as minor or significant. Furthermore, "any failure to adhere to the plan and schedule that results in a significant deficiency remaining uncorrected shall be reported in writing within 15 business days by the IACUC, through the IO, to APHIS/AC and any Federal agency funding that activity" (AWAR §2.31,c,3). In accordance with the PHS Policy (IV,F,3), "The IACUC, through the Institutional Official, shall promptly provide OLAW with a full explanation of the circumstances and actions taken with respect to . . . any serious or continuing noncompliance with this Policy [or] any serious deviation from the provisions of the Guide." Furthermore, a recent policy notice from NIH/OLAW states that prompt reporting is required if there is a "failure to correct deficiencies identified during the semiannual evaluation in a timely manner."[14]

Opin.   IACUCs should devise plans that are attainable within the timetable established. Significant deficiencies must be handled immediately and plans formulated to ensure the health and well-being of the animals. The criteria for determining appropriate deadlines for correction are set in concert with a reasonable timetable for the type of correction required. Thus, for significant deficiencies, there may be two timetables—one to ensure the animal's health and well-being immediately and a second to correct the deficiency. The type of deficiency discovered will determine the appropriate timetable for correction. For example, a facility issue (bricks and mortar) may take longer to correct than a programmatic issue in which a correction can be put into motion just by changing a policy. An issue that can be corrected by the IACUC, animal care staff, or investigative staff may be addressed more rapidly than a problem that involves outside contractors and facility issues, which can involve a long procurement process or allocation of funding. Temporary "quick fix" solutions may be used while planning and implementing the final solution.

Technically, the IACUC can revise its plan for correction of deficiencies and thereby avoid the AWAR reporting requirement for uncorrected deficiencies. For example, the original plan to correct an animal holding area floor problem called for resurfacing the floor by August 30. On August 25 it is decided to abandon the area as an animal area. No report would be required. *The reader is cautioned, however, that APHIS/AC disagrees with this interpretation,*[15] noting that the AWAR (§2.31,c,3) require the plan to be "specific" and contain "dates for correcting each deficiency." The section further states that "any failure to adhere to the plan and schedule that results in a significant deficiency remaining uncorrected" must be reported. The main concern in managing deficiencies must be that actions taken by the IACUC are reasonable, are defensible, and protect the health and well-being of the animals in question.

23:35 What action is taken if a significant deficiency is not corrected in accordance with the established schedule?

Reg. (See 23:34.)

Opin. Because a significant deficiency has an adverse effect on animal health and safety it must be corrected quickly. In some instances the initial action need not be the final solution, but the action must protect the animals. For example, if a dog run is found with a broken fence with potential to cause injury, the immediate action would be to remove the dog to a suitable enclosure. The action will correct the significant deficiency. The dog run can then be fixed as time permits.

Because IOs do not like to report items to APHIS/AC or NIH/OLAW, IACUC Chairs should use due diligence to assure that corrective action is taken in a timely manner.

23:36 For how long should facility inspection and program review records be maintained?

Reg. PHS Policy (IV,E,2) states "all records shall be maintained for at least 3 years; records that relate directly to applications, proposals, and proposed significant changes in ongoing activities reviewed and approved by the IACUC shall be maintained for the duration of the activity and for an additional 3 years after completion of the activity." The AWAR (§2.35,f) uses similar wording.

Opin. In practice, there are two opinions regarding record keeping. Many institutions maintain records indefinitely because of the difficulty in segregating records into activities that are either complete or ongoing. At these institutions, records are kept as long as storage is available. However, some institutions have concerns about storing unnecessary records because of freedom of information laws. These institutions generally store minutes and other general records for only 3 years and specific protocols for the duration of the project plus 3 years. It is unusual, in the authors' opinion, for regulatory bodies to request records older than 3 years. (See 24:14.)

23:37 How should the IACUC address minority opinions relative to inspection and review?

Reg. Both PHS Policy (IV,F,4) and AWAR (§2.31,c,3) mandate that minority reports be attached to the semiannual report.

Opin. Minority reports are very infrequent. Most IACUCs are able to devise reasonable compromises to difficult problems. Dissenting members generally do so by voting "no" on an issue and requesting that a notation be made in the regular committee minutes. However, a member may write a full minority opinion if so desired. This opinion then becomes part of the official report.

Surv.   How often does a minority opinion emanate from your semiannual program or facility review?

• It has never happened	19/40
• Very rarely	16/40
• With a low frequency, but every now and then	4/40
• Fairly common	0/40
• No response	1/40

23:38   Who receives the findings from the facility inspection and program review?

Reg.   In accordance with PHS Policy (IV,B,3) and the AWAR (§2.31,c,3), the IACUC must submit a report of the inspection and program review to the IO.

Opin.   It also is useful to submit a copy of the report to those who are responsible for correcting deficiencies, such as the facility management, the AV, and the facilities maintenance department. Members of the IACUC should receive a copy of the report as well.

23:39   Other than the IO and the IACUC, who has legal access to the IACUC's semiannual reports?

Opin.   Legal access depends on whether the facility is private or public and depends on applicable state laws. Legal counsel should be sought to address this issue with respect to local laws and requirements. (See 22:4.)

Surv.   At your institution, who has access to IACUC semiannual reports other than the IACUC and IO? More than one answer is possible.

• Nobody else	3/40
• Some high-level institutional personnel	7/40
• Animal facility administrators	12/40
• Principal investigators	6/40
• The attending veterinarian	22/40
• Anybody within the institution who wants to view them	11/40
• No response	3/40

23:40   Who decides the appropriate resolution of specific deficiencies? How is this done?

Reg.   Both the PHS Policy (IV,B,3) and AWAR (§2.31,c,3) require the IACUC to report on the deficiencies found during the facility and program review and to categorize them as either minor or significant (major). The IACUC is required to develop a reasonable and specific plan and schedule for correcting each deficiency (see 23:28). A deficiency is resolved when the IACUC determines that it has been corrected.

Opin. The IACUC has the ultimate responsibility for determining what actions are needed to correct a deficiency. However, it is best to involve the persons who ultimately will be affected by the committee's mandate. As appropriate, the IACUC should consult the AV, the investigator, and the maintenance department. If major funding is needed to correct an item it will also be necessary to consult the IO or other administrative persons to set a reasonable plan for correction. Such involvement makes for a more collegial atmosphere and ultimately a better solution.

23:41 Is the IACUC responsible for compliance oversight for the proper storage of federally controlled drugs used in animal-based research (e.g., opioids used for analgesia)?

Reg. There is no regulatory requirement that the IACUC serve as the regulatory body for Drug Enforcement Agency–regulated drugs.

Opin. A controlled drug officer should be selected by appropriate management to provide oversight for the ordering, use, and destruction of controlled substances. The IACUC's role would be to make sure that appropriate controlled drugs were available and used in accordance with the approved research protocols and within established expiration dates.

Surv. Does your IACUC take responsibility for the compliance oversight of the proper storage of federally controlled drugs used in animal-based research (e.g., opioids used for analgesia)?

* Not applicable, we do not used controlled substances   1/40
* No   17/40
* Yes   22/40

23:42 What Internet resources are available to help with conducting evaluations?

Opin. Many useful Web sites are available. The following are particularly useful because they provide links to a wide body of information related to the care and use of animals in biomedical research.

* http://grants.nih.gov/grants/olaw/: This page is maintained by NIH/OLAW. NIH/OLAW is the national regulatory agency that oversees the PHS Policy. The NIH/OLAW Web page gives access to a wealth of useful information, including the PHS Policy, the *Guide*, additional policy guidance including articles and NIH/OLAW Reports, IACUC Guidebook, and other related materials, training materials (PHS Policy Tutorial), and sample documents, including forms for conducting the semiannual review.

* http://oacu.od.nih.gov/UsefulResources/index.htm: This site is maintained by NIH, the Office of Animal Care and Use (OACU). OACU is NIH's internal regulatory body; however, the information may be useful for other facilities.

* http://www.aphis.usda.gov/ac/: The USDA/APHIS/AC Animal Care home page. Includes APHIS Animal Policy Manual that clarifies the AWAR.

* http://orsp.rutgers.edu/animal.asp: Rutgers University animal care and use home page. Other major research universities often have laboratory animal care Web sites that may prove useful.

- http://researchtraining.org: Click on the section Animal Research. Public site maintained by the Department of Veterans Affairs. Provides much useful information regarding animal care and use.
- http://www.ccac.ca: Canadian Council on Animal Care.
- http://www.iacuc.org: IACUC.ORG is an information resource for IACUC members and staff, developed with funding from the American Association for Laboratory Animal Science to serve as an organization tool to point to a topic of interest quickly, assisting IACUC members in managing the information available to them on the Web.
- http://www.nal.usda.gov/awic/: The Animal Welfare Information Center is a good starting point, linking to information the IACUC member or administrator needs.
- http://www.ahsc.arizona.edu/uac/: The University of Arizona's University Animal Care page contains links to their IACUC Handbook and to training modules.
- http://ehs.ucdavis.edu/animal/index.cfm: This University of California Davis site includes policy and procedure manual excerpts, several autotutorials, and protocol templates.

Acknowledgment

The assistance of Dawn Myers in the preparation of this document is gratefully acknowledged.

References

1. Committee to Revise the Guide for the Care and Use of Agricultural Animals in Agricultural Research and Teaching, *Guide for the Care and Use of Agricultural Animals in Agricultural Research and Teaching*, 1st rev. ed., Federation of Animal Science Societies, Savoy, IL, 1999.
2. National Research Council, Occupational Health and Safety in the Care and Use of Research Animals, National Academy Press, Washington, D.C., 1997.
3. U.S. Public Health Service, Centers for Disease Control and Prevention, Biosafety in Microbiological and Biomedical Laboratories, 4th ed., U.S. Government Printing Office, Washington, D.C., 1999.
4. Canadian Council on Animal Care, *Guide to the Care and Use of Experimental Animals*, Vols. I and II, Canadian Council on Animal Care, Ottawa, 1993/1984.
5. ARENA/OLAW, Institutional Animal Care and Use Committee Guidebook, 2nd ed., National Institutes of Health, Bethesda, MD, 2002.
6. DeHaven, W.R., personal communication, 1998.
7. Office of Extramural Research Guidance Regarding Reduction of Regulatory Burden in Laboratory Animal Welfare, NIH Guide NOT-OD-00-007, Utilization of AAALAC Activities as Semiannual Program Evaluation, December 21, 1999.
8. NIH/OLAW Tutorial. Available on the World Wide Web at: http://grants.nih.gov/grants/olaw/.
9. Office of Extramural Research Guidance Regarding Administrative IACUC Issues and Efforts to Reduce Regulatory Burden, NIH Guide NOT-OD-01-017, Use of Alternate IACUC Members, February 12, 2001.
10. DeHaven, W.R., personal communication, 1998.

11. Potkay, S., et al., Frequently asked questions about the Public Health Service Policy on Humane Care and Use of Laboratory Animals, *Contemp. Topics Lab. Anim. Sci.*, 36, 47, 1997.
12. Division of Animal Welfare, Office for Protection from Research Risks, National Institutes of Health, The Public Health Service responds to commonly asked questions, *ILAR News*, 33(4), 68, 1991.
13. ARENA/OLAW, Institutional Animal Care and Use Committee Guidebook, 2nd ed., National Institutes of Health, Bethesda, MD, 2002.
14. NIH Guide NOT-OD-05-034, Guidance on Prompt Reporting to OLAW under the PHS Policy on Humane Care and Use of Laboratory Animals, February 24, 2005.
15. DeHaven, W.R., personal communication, 1998.

24

Inspection of Animal Housing Areas

Patricia A. Ward

Introduction

Animal housing facilities must be inspected by the IACUC at least once every 6 months, according to the AWAR, PHS Policy, and the *Guide*. Nevertheless, aside from this simple directive, a few subsequent policy clarifications, and the detailed "standards" of animal housing and care provided in the AWAR and *Guide*, these regulatory authorities give IACUCs little direction in how to carry out this mandate. An IACUC may approach implementation of the inspection requirement in a variety of ways. The information provided in this chapter is intended to assist IACUCs in developing an effective animal housing facility inspection program tailored to the needs of their individual institutions.

Information, ideas, and opinions expressed in this chapter were compiled from the results of three surveys of IACUCs at institutions located throughout the United States. The first survey, conducted in 1997, gathered information from 93 IACUCs on how their inspections were managed and administered. A second survey of 22 IACUCs was conducted in 1998 and focused on how animal facilities were evaluated by IACUCs during inspections. The third survey of 165 IACUCs was conducted in 2005 and collected updated information on IACUC inspection procedures and facility areas inspected. All three surveys included responses from institutions of varying size, type, and culture.

While the information provided in this chapter may represent the way many, or even most, IACUCs conduct animal housing and care inspections, the common practices described should in no way be interpreted as the best practices for every individual institution. The methods that an institution's IACUC employs depend on the size, type, and culture of the institution. The "best" inspection methods are likely to be very different for large versus small facilities, centralized versus decentralized facilities, corporate organizations versus educational institutions, and so forth. IACUCs should be encouraged to exercise their professional judgment and creativity in developing an effective animal housing facility inspection program tailored to the particular needs of their individual institutions.

24:1 What is the definition of an *animal housing facility*?

Reg. (See 15:1.)

24:2   What is the purpose of conducting IACUC inspections of animal housing facilities?

Reg.   The IACUC must fulfill its regulatory responsibility to inspect animal housing areas at least once every 6 months to evaluate compliance with applicable guidelines (AWAR §2.31,c,2; PHS Policy IV,B,2; HREA Section 495,b,3,A). The recommendations of the *Guide* (p. 9) also specify that the IACUC should inspect the animal facilities at least once every 6 months.

Opin. The IACUC must represent its institution both by advancing the institutional mission and by serving as the institutional conscience. By inspecting animal housing areas at least twice a year, the IACUC can ensure that the animal housing and care program is furthering the institutional mission by providing high-quality animals (through proper attention to animal welfare), maintaining a suitable environment for research activities, and safeguarding the health and safety of its personnel. In addition to ensuring the institution's compliance with applicable regulations, the IACUC must be confident that the animal housing and care program is accomplishing these goals in a scientific, humane, and ethical manner.

24:3   Should an institutional guideline, SOP, or policy be developed for IACUC inspection of animal housing facilities?

Opin. There is no regulatory requirement for IACUCs to establish guidelines, SOPs, or policies regarding inspection of animal housing facilities. Nevertheless, most IACUCs have found such documents to be helpful in the administration and management of the inspection process. Clarifying expectations for IACUC members conducting inspections, veterinary and administrative staff supporting inspections, and managers of facilities undergoing inspection can promote consistent and thorough inspections and enhance rapport among all parties. Issues addressed in the guidelines, SOPs, or policies can include the following:

- Who conducts the IACUC inspections
- The orientation or training inspectors receive
- How inspections are scheduled
- Whether inspections are to be announced or unannounced
- How inspectors should prepare for each inspection
- How inspections will be facilitated
- Which sites are to be inspected
- Which general issues are considered during inspections
- Findings of previous inspections as examples
- How findings will be communicated, documented, and distributed
- How the IACUC will follow up deficiencies
- How long inspection documents will be retained

24:4   Who conducts the IACUC inspections of animal housing areas?

Reg.   While the PHS Policy (IV,B,2), the *Guide* (p. 9), and the AWAR (§2.31,c,3) specify animal facility inspection as an IACUC function, the AWAR offer the most explicit "minimal" answer to this question: "No Committee member wishing to participate . . .

Opin.   may be excluded. The IACUC may use subcommittees composed of at least two Committee members and may invite ad hoc consultants to assist."

Opin.   At many institutions, particularly those with smaller or centralized animal facilities, the entire IACUC participates in the animal housing facility inspection. However, in larger or decentralized facilities, the IACUC may divide into subcommittees or appoint a single subcommittee to tackle the job. Such subcommittees must include at least two IACUC members if housing areas for AWA-regulated species will be inspected (AWAR §2.31,c,3). Duly appointed and qualified alternate members of the IACUC may conduct the inspections in place of corresponding regular members.[1] A site visit from the AAALAC may be utilized for the semiannual animal facility inspection, as long as the participation and endorsement of the IACUC meet the requirements of the AWAR and the PHS Policy.[2] Members of the veterinary or IACUC staff also may participate in the inspection to provide expertise, continuity, and support.

24:5 Should IACUC members receive orientation or training in preparation for conducting facility inspections?

Reg.   The *Guide* (p. 9) states that "it is the institution's responsibility to provide suitable orientation, background materials, access to appropriate resources, and, if necessary, specific training to assist IACUC members in understanding and evaluating issues brought before the committee."

Opin.   Most IACUCs arrange for some form of orientation or training for incoming members that includes information about inspection of animal housing facilities. This training is especially beneficial to the nonscientist and nonaffiliated members of the IACUC. Information presented in the inspection training session or packet can include the following:

- An overview of the size and scope of the institution's animal housing facilities, including maps or floor plans
- Institutional guidelines, SOPs, or policies regarding IACUC inspections
- Copies of regulatory, professional, and institutional animal housing and care standards
- Guidelines for evaluating animal well-being
- The institutional inspection checklist, if used
- Sources of additional information (e.g., Web sites or information about contacting key laboratory animal organizations)

Most IACUCs find it worthwhile to spend a significant portion of the training effort reviewing the various established standards that apply to the areas to be inspected. These include the animal housing and care standards provided in the AWAR, the *Guide*, and the *Guide for the Care and Use of Agricultural Animals in Agricultural Research and Teaching* (*Ag Guide*).[3] As applicable, IACUCs may also consider reviewing the various animal management documents produced by, among others, the Institute of Laboratory Animal Resources (ILAR) and the National Research Council (NRC).[4] A working understanding of these standards not only enables IACUC members to evaluate regulatory compliance issues in the facilities inspected, but also helps to develop their professional judgment with regard to the more important issues of animal welfare, research integrity, and personnel health and safety.

24:6 What periods are appropriate for IACUC inspections of animal facilities?

Reg. Both the AWAR (§2.31,c,2) and PHS Policy (IV,B,2) specify that animal facilities be inspected at least once every 6 months. (See 23:5–23:7.)

Opin. For the most part, meeting this requirement is not difficult for institutions where the animal facilities can be inspected in a relatively short time. For larger or decentralized facilities, however, meeting this requirement can be more difficult. If completing the inspection will take several days, the IACUC is faced with management of a burdensome time commitment. Under these circumstances, most IACUCs will establish regular "inspection windows" at 6-month intervals and break the inspection up over several days within each window. Alternatively, an IACUC can opt to distribute inspection trips throughout the entire 6-month period. In this case, or if the inspection window spans more than a few weeks, care must be taken to inspect the various sites in approximately the same order for each 6-month period. Otherwise, the inspection interval for some sites might exceed the 6-month maximum. This could occur if a particular site is inspected early in one 6-month period and late in the next.

24:7 Should inspections be announced or unannounced?

Opin. (See 23:25.)

24:8 How should IACUC inspectors prepare for each inspection?

Opin. Before embarking on each inspection, most IACUC inspectors want to review the sites to be visited with particular respect to entry and exit restrictions, previous inspection findings or deficiency history, and experimental activities being conducted. Entry and exit restrictions may affect the way inspectors dress for the inspection, the order in which various areas are visited, and the extent to which exposure of inspectors to certain other animals is prohibited for a specified period before or after the inspection. Reviewing the previous inspection findings or deficiency history of the sites to be inspected may help IACUC inspectors focus attention on problem areas or remind them to express appreciation for improvements made. Some IACUC inspectors may wish to review experimental protocols approved for use in the areas to be inspected. Doing so can alert them to animal use procedures of particular concern, such as those requiring special consideration during protocol review (*Guide*, p. 10), and allow IACUC inspectors to make pertinent observations or inquiries during the inspection visit. Alternatively, many IACUC inspectors choose to wait until after the inspection to review the experimental protocols of animals noted during the inspection.

24:9 What strategies can be used to facilitate inspections?

Opin. Every IACUC looks for ways to facilitate the inspection process. Including administrative or veterinary staff on the inspection team is one popular strategy used to provide expertise, continuity, and support. Another is the use of the previous inspection report or an inspection checklist, to use as a reference or to record findings during the inspection. NIH/OLAW has made a sample checklist available on its Web site[5] and institutions may download and modify it to suit their own programs. Some IACUC inspection teams carry copies of pertinent regulations and standards, or equipment for recording findings (photo, audio, or video) or validating conditions (temperature or humidity reader, light meter, smoke bottle, etc.). (See 24:11.)

24:10   What sites should be included during the inspection of animal housing facilities?

Reg.   See 12:1 for the AWAR (§1.1, Animal) definition of an *animal*, which does *not* include all vertebrates. The PHS Policy (III,A) applies to any vertebrate animal used in research, research training, experimentation, or biological testing. Most IACUCs resolve this discrepancy simply by including the housing areas of all vertebrate animals, regardless of species or type of use, in their inspections. Some IACUCs elect to inspect the housing areas of invertebrate species as well.

What constitutes an animal housing area is also a matter of regulatory definition (and see 15:1). The AWAR (§2.31,c,2) specify that animal facilities, excluding wild habitats but including study areas, must be inspected. While a definition of an *animal facility* is not provided in the AWAR, a *study area* is defined as any area where animals are housed for more than 12 hours (§1.1).

The PHS Policy (III,B) defines an *animal facility* as "any and all buildings, rooms, areas, enclosures, or vehicles, including satellite facilities, used for animal confinement, transport, maintenance, breeding, or experiments inclusive of surgical manipulation. A satellite facility is any containment outside of a core facility or centrally designated or managed area in which animals are housed for more than 24 hours." The PHS Policy definition is quite comprehensive, including support and experimental surgery areas along with housing areas. (See 23:20; 25:1.)

Opin.   Many IACUCs, particularly those wishing to obtain or maintain accreditation by AAALAC, want to comply with the animal housing inspection requirements of the AWAR, PHS Policy, and the *Guide*. To accomplish this, virtually all IACUCs visit all areas where animals are housed or kept (including study areas, satellite facilities, and laboratories) for more than 12 hours (AWAR-covered species) or 24 hours (all other vertebrate species) during their semiannual inspections. Typically, areas where rats and mice are kept are not inspected if those animals are held there for less than 24 hours. Temporarily unoccupied animal housing areas also are visited by most IACUCs. Corridors and anterooms contiguous with these areas are included in most inspections. Field research locations are not usually inspected by IACUCs unless animals are held in captivity for more than 12 hours, or the IACUC is particularly interested in observing the field research procedures or setting. Cage washing and other sanitation facilities are included in the semiannual inspections, as are diet preparation and food and bedding storage areas. Most IACUCs also inspect areas where caging, equipment, and supplies are stored. Loading docks and transport vehicles, particularly those dedicated to traffic of animals and animal-related supplies and equipment, are usually inspected, as are carcass storage and disposal areas. Animal procedure areas contiguous with the animal housing facilities, including surgical suites, procedure rooms, and euthanasia stations, are also usually included in the semiannual inspection of the animal housing and support areas. (See 23:20.)

24:11   What general issues should be considered when inspecting animal housing facilities?

Reg.   The AWAR (§2.31,c,1) require the IACUC to use the AWAR as a guide for semiannual inspections. PHS Policy (IV,A,1) requires the IACUC to use the *Guide* as a basis for evaluation.

Opin.   Paramount among inspection issues are animal well-being, research integrity, and personnel health and safety. Proper management of animal care and use promotes all three, and IACUC inspectors seek to evaluate the success of the program in this context. To do this during an inspection, most IACUCs try to organize the effort and attention they give to the various aspects of the animal care and use program. Many IACUCs will utilize currently available tabulations, such as the table of contents of the *Guide* or the Facility Inspection Checklist[5] available from NIH/OLAW, to provide such structure. Some IACUCs prefer to focus their attention on issues of particular concern in the type of animal facilities at their institutions. (See also 24:9.)

No matter how the inspection is structured, the many issues that should be considered by IACUC inspectors can be overwhelming once they actually enter an animal housing area to conduct an inspection. One popular approach is for IACUC inspectors to begin with the condition of the animals and expand their focus to the following general categories in any convenient order:

- Animal health and well-being
- Primary enclosures, environmental conditions, and physical plant
- Housekeeping, disinfection, and sanitation
- Water, food, bedding, and other supplies
- Animal euthanasia, carcass storage, and carcass disposal
- Labels, signage, and records
- Personnel, training, and occupational safety

24:12 How should inspection findings be communicated, documented, and distributed?

Reg.   Both the AWAR (§2.31,c,3) and PHS Policy (IV,B,3) require that reports of inspection findings be forwarded to the IO at the close of every 6-month inspection period. This is usually done in conjunction with the IACUC's semiannual evaluation of its institutional program for humane use and care of animals. Inspection findings may be summarized in the semiannual report to the IO, or copies of detailed site-by-site inspection records may simply be attached to the report. In either case, the inspection findings must distinguish significant deficiencies from minor deficiencies (*significant* deficiencies represent a threat to the health or safety of animals; see 23:33) and include a plan and schedule for correction (AWAR §2.31,c,3; PHS Policy IV,B,3).

Opin.   In addition to reporting inspection findings to the institutional official, IACUCs generally communicate their findings to animal facility managers in order to assist them in correcting deficiencies and obtain full compliance. This communication can occur through many means, depending on the culture of the institution. If inspections are arranged with animal facility managers in advance, most IACUC inspectors ask questions and discuss concerns with those managers throughout the inspection visit. Alternatively, and particularly if facility managers are not present during the inspection, IACUC inspectors may arrange for a verbal interview with the managers either at the conclusion of the inspection visit or at a later time. In addition to verbal communication, most IACUCs will present facility managers with a written report of the inspection findings. This may be a simple handwritten list of deficiencies or completed checklist presented at the conclusion of the inspection or a more formal inspection report sent at a later date. In either case, these written inspection reports generally contain the following information:

- Date and site location
- Names of IACUC inspectors
- Deficiencies observed and whether each deficiency was significant or minor
- A directive and date for correction

In addition, many IACUCs also will include one or more of the following in their inspection reports:

- The administrative unit responsible for the facility inspected
- Names of facility representatives present at the inspection
- A list, including description and contents, of all rooms inspected
- Whether deficiencies observed were first-time or previously cited
- The specific standard or regulation violated by each deficiency
- A general summary, comment, or commendation

24:13   How should the IACUC follow up on deficiencies?

Reg.   Both the AWAR (§2.31,c,3) and PHS Policy (IV,B,3) require that deficiencies be included in the semiannual report to the IO, that a plan and date for correction be indicated for each deficiency, and that minor deficiencies be distinguished from significant deficiencies (deficiencies that may threaten the health or safety of animals). Additionally, the AWAR require that failure to adhere to the correction schedule for significant deficiencies be reported within 15 days by the IACUC through the IO to APHIS/AC and any federal agency funding the cited activity (AWAR §2.31,c,3). PHS Policy (IV,F,3,a) requires prompt reporting to NIH/OLAW of serious or continuing noncompliance with the policy. (See 25:12.)

Opin.   When an IACUC encounters a situation that represents a threat to the health or safety of animals, prompt corrective action should be taken and a thorough follow-up conducted in order to ensure that the situation does not recur. However, the vast majority of deficiencies cited by the IACUC on inspection are minor, and the nature of follow-up action depends on the culture of the institution. Many IACUCs reinspect the cited facility at the correction deadline to verify resolution of the deficiency. Most IACUCs specifically assess the correction of previously cited deficiencies at subsequent inspections. Occasionally, an IACUC encounters previously cited minor deficiencies that either have not been corrected or have recurred. These deficiencies usually do not become "elevated" to major deficiencies unless the uncorrected minor deficiency has become a threat to the health or safety of animals. When there are recurrent or uncorrected minor deficiencies, the IACUC can employ a variety of strategies to obtain compliance. Examples include the following:

- Increased frequency of IACUC inspection
- IACUC interview of responsible persons
- Notification of superiors in the chain of command
- IACUC-issued verbal or written warning
- Withdrawal of IACUC approval to house or use animals (which may require notification of sponsoring agencies)
- Discipline of personnel in accordance with institutional policy

24:14   For how long should inspection documents be retained?

Reg.   Both the AWAR (§2.35,f) and PHS Policy (IV,E,2) require that semiannual reports to the IO, which include the IACUC's animal facility inspection findings, be retained by the institution and be available for inspection for at least 3 years.

Opin.   If additional inspection records are created, the length of time they are retained is at the discretion of the IACUC. Many IACUCs retain these records for the same 3 years as the semiannual reports to the IO. Most, however, find both the semiannual reports to the IO and additional inspection records to be a valuable documentation of animal facility management history. These IACUCs are likely to retain such records for an extended period. IACUCs at institutions subject to federal or state Freedom of Information Acts (FOIAs) should consult their institutional FOIA officer for advice on their records retention program. (See 23:36.)

24:15   How should the IACUC assess the well-being of animals during inspection?

Reg.   IACUC inspectors should endeavor to assess the physical and psychological well-being of animals. According to the *Guide* (p. 22), "Animals should be housed with a goal of maximizing species-specific behaviors and minimizing stress-induced behaviors." To accomplish this, the structural environment, social environment, and activity pattern of the animals must be considered (*Guide*, pp. 37–38). The standards of the AWAR (§3.8; §3.81) include specific provisions for the exercise of dogs and environmental enhancement of nonhuman primates.

   Observation of animal activity that "is repetitive, is nongoal-oriented, and excludes other behavior" (*Guide*, p. 38) can indicate to IACUC inspectors that the psychological well-being of the animals demonstrating the behavior, and perhaps the colony in general, has not been adequately addressed.

Opin.   During animal housing facility inspections, the first thing to capture the attention of IACUC inspectors is likely to be the condition of the animals housed. By virtue of training or experience, most IACUC inspectors will readily distinguish normal healthy animals from those exhibiting signs of illness, injury, pain, or distress. Discovery of animals in the latter group should prompt further inquiry into the cause of the condition and the way it is being managed. If the condition is known to be experimentally induced, IACUC inspectors may wish to verify that the condition and its management are consistent with information provided in the IACUC-approved protocol. If the condition is not related to the experimental procedure, inquiry into the colony health status and other recent illnesses or injuries may be in order, and management practices questioned if undesirable patterns are identified. In either instance, IACUC inspectors should be assured that all animals are checked daily, that veterinary care is available around the clock if needed, and that the veterinary staff is promptly notified of animal health problems and closely monitors the effectiveness of the implemented treatment and management plan (*Guide*, pp. 59–60; AWAR §2.33,b).

   If the animal housing and care program is adequately providing for the psychological well-being of animals, IACUC inspectors should expect to see primary enclosures with features that accommodate the animals' species-specific postures and motor activities, management practices that cater to species-specific needs for cognitive stimulation and social interaction, and compatible social groupings where appropriate. (See 27:1.)

24:16   How should the IACUC evaluate animal enclosures during inspection?

Reg.   Both the *Guide* (pp. 23–28) and the AWAR (§3.6; §3.28; §3.53; §3.80; §3.101; §3.104; §3.128) provide specific standards for primary enclosures. To evaluate compliance with these standards, IACUC inspectors generally ask themselves the following questions when examining an animal enclosure:

- Is the enclosure of appropriate design for the comfort, security, and observation of the animals housed?
- Is it constructed of durable and sanitizable materials?
- Is space adequate for the size and number of animals contained?
- Is the enclosure clean and in good repair?
- Are food and water readily accessible?
- Does the enclosure in any way jeopardize the well-being of the animals housed therein?

24:17   What environmental parameters should the IACUC evaluate during inspection of an animal facility?

Reg.   Specific standards for proper animal housing environmental conditions are provided in the *Guide* (pp. 28–36) and AWAR (Part 3). The PHS Policy (IV,A,1) uses the *Guide* as a basis for the development and implementation of an animal care and use program.

Opin.   IACUC inspectors generally find some environmental parameters, such as extremes of temperature and humidity, inadequate ventilation, excessive odor, and chronically loud noise, to be readily assessable the moment they enter an animal housing room. Other parameters require a little more investigation. For example, adequate assessment of the lighting cycle and magnitude and frequency of temperature fluctuation requires more than momentary sensory perception by the IACUC inspectors. Additionally, depending on the type of primary enclosure used, the environmental conditions experienced by the animals within the enclosure may be very different from those perceived by the IACUC inspectors in the larger room environment. Most IACUC inspectors interview animal facility managers about the monitoring methods used for such situations and review the monitoring documentation to ensure that the appropriate environmental conditions are being provided. Included in this discussion and record review should be the results of tests of or actual experience with emergency alarms and response systems.

24:18   How should the IACUC evaluate the animal facility's physical plant during inspection?

Reg.   As indicated in 24:17, IACUC inspectors should consult the *Guide* and AWAR for specific standards regarding the animal housing facility physical plant.

Opin.   In addition to ensuring compliance with these standards, IACUC inspectors generally evaluate the facilities in terms of design, construction, building systems, fixtures and equipment, security, maintenance, and housekeeping. To satisfy most IACUC inspectors, the facility's physical plant should meet the following criteria:

- The design of the structure should be conducive to achieving the goals of animal well-being, research integrity, and personnel health and safety.

- Construction should be structurally sound, surface finishes should be sanitizable, and physical barriers should be in place to prevent entry of vermin.
- Building systems (electrical, plumbing, heating, ventilation, air-conditioning, etc.) should be reliable and sufficient to support facility demands.
- The facility should be outfitted with fixtures and equipment appropriate to the housing, care, and use of the species housed. The surfaces of these items should be sanitizable.
- Equipment requiring periodic service or certification should be properly maintained.
- Appropriate security should include provisions for preventing both escape of animals and entry of unauthorized personnel.
- The facility structure and building systems should be maintained in good repair.
- In general, the facilities should be clean and free of unnecessary clutter.

24:19   How should the IACUC assess the adequacy of housekeeping, disinfection, and sanitation practices during inspection?

Reg.   The *Guide* (pp. 42–44) recommends frequent bedding changes and cleaning and disinfection of primary enclosures and animal housing rooms. The AWAR (§3.11; §3.31; §3.56; §3.84; §3.106; §3.107; §3.131) provide specific minimal requirements for cleaning, sanitization, housekeeping, and pest control in animal housing facilities.

Opin.   During inspections, most IACUC members evaluate these issues through direct observation and examination of records. Animals are examined to ensure that they are clean and dry. Bedding should not be excessively soiled. Primary enclosures should not exhibit an accumulation of soil. Likewise, animal housing rooms should neither exhibit an accumulation of soil or waste, nor be excessively cluttered (most IACUC inspectors discourage storage of equipment and supplies that are not used in the routine care and use of the animals). All surfaces should be sanitizable and agents used for sanitation and disinfection should be appropriate for the purpose and be used according to the manufacturers' directions. IACUC inspectors should see evidence of an effective vermin control program. Waste material should be removed from the animal rooms in a timely fashion and handled in a manner that minimizes the potential to introduce contaminants or attract vermin to the animal facility. Many IACUC inspectors also evaluate the effectiveness of an animal housing facility's housekeeping, disinfection, and sanitation program in terms of colony health. They inquire about recent incidences and spread of disease and identify questionable housekeeping, disinfection, or sanitation practices that may contribute to colony health problems.

Inspection of facilities for washing and sanitizing caging and equipment also are included in the IACUC's evaluation of sanitation practices. Most IACUC inspectors want to see that these facilities do not exhibit an accumulation of soil and that caging and equipment are processed in such a way as to minimize the recontamination of clean items. Chemical agents and automatic washing equipment used for sanitation and disinfection of caging and equipment should be appropriate for the purpose used and be used according to manufacturers' directions. Most IACUC inspectors want to verify the effectiveness of the sanitation process. This can be done by inspecting records of validation test results (e.g., cage washer temperature

indicator strips and cultures of sanitized surfaces). If the process calls for sterilization of caging or equipment, the performance of sterilizers is similarly evaluated
by most IACUC inspectors.

**24:20   How should the IACUC evaluate the adequacy of the water, food, bedding,
medications, and other supplies provided to or used for animals?**

Reg. Prior to inspection, IACUC inspectors are advised to consult the *Guide* (pp. 38–41)
and AWAR (Part 3) for specific standards regarding the quality, storage, and provision of water, food, bedding, medications, and other supplies.

Opin. IACUC inspectors should be assured that all supplies that have contact with animals are appropriate, safe, and effective. Food, bedding, medications, and so on,
can usually be evaluated by direct observation during inspection. Acceptable animal food must be appropriate for the species, within its recommended shelf life,
and stored under conditions that minimize the potential for premature deterioration, spoilage, contamination, or infestation. It should be available to animals *ad
libitum* or provided in an appropriate ration and available to all individuals in a
group unless veterinary or IACUC-approved research protocols dictate otherwise.
Bedding should be absorbent, nonnutritive, nontoxic, and stored in a manner that
minimizes the potential for spoilage, contamination, or infestation. Food and bedding should be free of foreign objects and normal in appearance, texture, and odor.
Medications, treatments, and other veterinary or experimental supplies should be
appropriate for the type of animal and research intended, medical grade, within
designated expiration dates, and stored according to package directions (IACUC
inspectors should be aware that special storage and record keeping requirements
are imposed by the regulations of the Drug Enforcement Agency for controlled
substances).[6] Unless veterinary or IACUC-approved research protocols dictate
otherwise, drinking water should be available to animals *ad libitum*, potable, and
free of contaminants. To assess the latter, most IACUC inspectors interview facility
managers about the source of the water provided (municipal, well, purified, etc.)
and the integrity of the system that delivers the water to the animals (plumbing
systems, manifold flushing practices, backflow prevention, etc.). Inspectors will
also usually examine the results of any water analysis conducted and inquire about
colony health problems in which animal drinking water may have played a role.

**24:21   What aspects of animal euthanasia, carcass storage, and carcass disposal should
concern the IACUC during inspection?**

Opin. Most IACUC inspectors visit areas for animal euthanasia, carcass storage, and carcass disposal in conjunction with their inspection of animal housing facilities.
Euthanasia areas should be clean and the methods employed should be consistent
with the recommendations of the Report of the American Veterinary Medical Association (AVMA) Panel on Euthanasia.[7] (See Chapter 17.)
    Carcasses are generally stored in dedicated refrigerators or freezers, with special
provisions for the labeling and storage of carcasses contaminated with infectious,
radioactive, or chemical hazards, including ether. Whether carcass disposal is handled by the institution internally or contracted to a professional waste management company, IACUC inspectors should be apprised of the disposal methods and
be assured that these methods comply with all applicable regulations, including
local requirements and restrictions. IACUC inspectors also may wish to review any

facility records associated with animal euthanasia and carcass disposal, particularly those required by the AWAR (§2.35,b) for documentation of the final disposition of dogs and cats. Standards on animal euthanasia, carcass storage, and carcass disposal also can be found in the *Guide* (pp. 44–45, 65, 77).

24:22 What labels, signage, and records should be evaluated by the IACUC during inspection?

Opin.   In an animal housing and care facility managed in accordance with the AWAR, PHS Policy, and the recommendations of the *Guide*, most IACUC inspectors can expect to see labels, signage, and records associated with nearly every aspect of the program. Animal enclosures should be labeled with pertinent information (see the *Guide*, p. 46) about the animals and their intended use (e.g., PI- and IACUC-approved protocol number). Animal food should be labeled with the species formulation and the milling or expiration date. Various cleaning, husbandry, veterinary, and experimental substances found in the animal facility should be labeled with the name and opening or expiration date of each item. Usage records should accompany controlled substances as required by the regulations of the Drug Enforcement Administration.[6] Hazardous substances and contaminated animals or materials should be clearly identified and labeled with precautions. Special instructions should be prominently posted to assist personnel to comply with non-routine or critical procedures. Animal housing areas should be posted with emergency contact information. Husbandry activities should be recorded on log sheets. Environmental readings should also be logged or recorded on automatic monitoring system printouts. Animal monitoring records (postprocedural, food and water consumption, enrichment, exercise, etc.) are usually present at the animal housing location and should be reviewed. Animal health, clinical and veterinary treatment records, as well as animal receiving and disposition records, should be maintained in an appropriate location and inspected by the IACUC. Records of sanitation and disinfection, including cage washer and sterilizer performance, also should be available for review by IACUC inspectors.

While not all of these labels, signs, and records are specifically required by the AWAR, PHS Policy, and recommendations of the *Guide*—and certainly not every activity in the animal housing facility need be recorded—many of these records and documents will prove to be useful animal facility management and inspection tools. Some IACUCs choose to require that extensive records be kept of animal housing activities and parameters, particularly to document to outside authorities that there is compliance with standards. Others rely on the state of their facilities and animal care programs to demonstrate compliance and forgo record keeping not specifically required by applicable regulations and standards.

24:23 What personnel issues should concern the IACUC during inspection?

Opin.   One of the highlights of the IACUC animal housing facility inspection process is the opportunity for dialog between IACUC members and personnel working in the animal facility. Such personnel can include members of the research staff, as well as facility managers and animal care and support personnel. In their conversations, most IACUC inspectors try to evaluate the extent to which personnel are qualified, either through experience or training, to perform their duties, and whether or not experimental procedures and work practices are consistent with

IACUC-approved animal use protocols, institutional guidelines, policies, and SOPs.

Some of the most important personnel issues IACUC members consider during their inspections of animal housing facilities concern the occupational health and safety of animal care, use, and support personnel. IACUC inspectors should look for evidence that personnel are informed about and protected from the various health and safety risks associated with exposure to animals and the hazardous experimental agents and procedures used with animals. Examples of such evidence include elements of facility design and management (such as space delineation and air pressure differentials) that separate personnel areas and activities from those of animals, especially where hazardous agents are present. Compliance of personnel with appropriate personal hygiene and protective equipment recommendations also can indicate an effective occupational health and safety program. (See Chapter 20.)

References

1. Office of Extramural Research Guidance Regarding Administrative IACUC Issues and Efforts to Reduce Regulatory Burden, NIH Guide for Grants and Contracts. 2/12/2001, Notice OD-01-017. Available on the World Wide Web at: http://grants.nih.gov/grants/olaw/references/pubartindex.htm#s.
2. Office of Extramural Research Guidance Regarding Reduction of Regulatory Burden in Laboratory Animal Welfare, NIH Guide for Grants and Contracts. 12/21/1999, Notice OD-00-007. Available on the World Wide Web at: http://grants.nih.gov/grants/guide/notice-files/not-od-00-007.html.
3. *Guide for the Care and Use of Agricultural Animals in Agricultural Research and Teaching*, 1st rev. ed., Federation of Animal Science Societies, Savoy, IL, 1999.
4. Institute of Laboratory Animal Resources. Available on the World Wide Web at: http://dels.nas.edu/ilar_n/ilarhome/links.shtml.
5. Office for Laboratory Animal Welfare, National Institutes of Health, Public Health Service, Sample Semiannual Facility Inspection Checklist Animal Housing and Support Areas. Available on the World Wide Web at: http://grants.nih.gov/grants/olaw/sampledoc/cheklist.htm.
6. Office of the Federal Register, Controlled Substances Act, amended May 1, 1987, Title 21, Food and Drugs, Chapter 13, Drug Abuse Prevention and Control, 21 CFR §1301.11–§1308.15, Schedule of Controlled Substances, Washington, D.C., revised April 1, 1988.
7. AVMA Panel on Euthanasia, 2000 Report of the AVMA Panel on Euthanasia, *J. Am. Vet. Med. Assoc.*, 218, 669, 2001.

25

Inspection of Individual Laboratories

Neil S. Lipman and Scott E. Perkins

Introduction

One of the IACUC's principal roles is to ensure that animals used in research, testing, and teaching are used humanely by trained staff. Although IACUC review of proposals describing planned animal activity is an important mechanism for meeting this role, observation of animal use provides the most direct assurance that these goals are attained. At some institutions, animals are never removed from the animal facility as all activities are performed within animal holding or procedure rooms. Nevertheless, it is common for animals to be transported from the animal facility to an investigator's laboratory for experimental use. While animal resource personnel can readily observe the appropriateness of activities conducted within the animal facility, this task becomes significantly more difficult, especially at large institutions, when animal research is conducted in an investigator's laboratory.

Although there is no specific regulatory directive requiring IACUCs to inspect investigators' laboratories and observe ongoing activities (unless animals are held in these areas for more than 12 hours; see 25:1), the IACUC is charged with ensuring that animal use is conducted humanely by appropriately trained personnel and that these activities have been approved, in advance, by the IACUC. It is this chapter authors' opinion that laboratories should be visited by the IACUC periodically in order to meet the spirit of applicable regulations, policies, and the *Guide*.

25:1 Is it necessary to inspect investigators' laboratories where research with animals is conducted?

Reg. The AWAR (§1.1, Study Area; §2.31,c,2) requires inspections if animals are maintained in the laboratory for more than 12 hours, whereas the PHS Policy (III,B; IV,B,2) requires inspections if animals are maintained in the laboratory for more than 24 hours. (See 25:2.)

Opin. Strictly speaking, neither the AWAR nor PHS Policy mandates inspection of investigators' laboratories where research with animals is conducted unless animals are maintained in the laboratory longer than a specified period, as noted. However, the *Guide* does specify in Chapter 1 (Institutional Policies and Responsibilities) that the IACUC inspect animal activity areas every 6 months. Therefore, many institutional IACUCs assume this responsibility in order to meet their regulatory and institutional charge of ensuring that animals are used humanely by trained personnel. Laboratory visitation frequently provides the added benefit of enhancing dialog between the committee and investigative staff, frequently improving the IACUC's understanding of proposals. (See 23:20; 24:10.)

NIH/OPRR (now NIH/OLAW)[1] has addressed the IACUC inspections of laboratories where investigators use animals as follows:

> When considering IACUC responsibilities for semiannual review, it is important to keep in mind that the institution, usually acting through the IACUC and/or the facility veterinarian, is responsible for all animal-related activities of the institution, regardless of where animals are maintained *or the duration of their stay* [emphasis added]. The PHS Policy allows institutions some discretion regarding specific methods to assure compliance, but the institution is clearly responsible for what happens to animals in investigators' laboratories. The degree, method and frequency of IACUC oversight depends a great deal on the nature of the activity. For example, satellite holding facilities or areas where surgical manipulation is conducted should always be included in the semiannual review. Other activities, such as routine dosing, weighing, or immunization of animals in laboratories, may be monitored using other methods, such as random evaluation.
>
> Inclusion of these laboratories in the semiannual IACUC review would be another way to satisfy the PHS Policy requirements. In any case, the IACUC must have access to all investigators' laboratories for the purpose of verifying that activities involving animals are conducted in accordance with the proposal approved by that committee.

APHIS/AC supports the position that although laboratories do not have to be inspected if animals are held there less than 12 hours, IACUC inspection is encouraged to the extent that it helps the IACUC and the institution fulfill its responsibilities under the AWA to oversee all animal care and use activities at the institution.[2]

Surv. Which laboratories that hold or house animals are inspected by your IACUC? More than one response is possible.

- Not applicable as we do not inspect laboratories   16/154
- Inspect all laboratories   73/154
- Only those that hold or house animals for more than 12 hours   32/154
- Only those that hold or house animals for more than 24 hours   16/154
- Inspect laboratories if the laboratory is used for studies on
  USDA-covered species   21/154
- Random selection of laboratories   11/154
- Those where animals are used in studies as APHIS/AC annual
  report category D or E (even if the species is not covered by
  the Animal Welfare Act regulations)   13/154
- Only if the IACUC is unfamiliar with the laboratory where
  research is planned or performed   3/154
- Laboratories where survival surgeries on rodents are
  conducted   37/154
- Laboratories used for postoperative recovery   29/154
- Laboratories conducting studies deemed by the IACUC to be
  of increased concern   28/154
- Other   17/154

25:2 Is it necessary to inspect and evaluate every laboratory where animals are used at the time of each IACUC inspection?

Reg. AWAR (§1.1; §2.31,c,2) require that any *study area* (defined as any building room, area, enclosure, or other containment outside a core facility or centrally designated or managed area in which animals are housed for more than 12 hours) be inspected at least once every 6 months by a subcommittee of at least two IACUC members.

The PHS Policy (III,B) defines an *animal facility* as "any or all buildings, rooms, areas, enclosures, or vehicles, including satellite facilities, used for animal confinement, transport, maintenance, breeding, or experiments inclusive of surgical manipulation. A satellite facility is any containment outside of a core facility or centrally designated or managed area in which animals are housed for more than 24 hours." All facilities, including satellite facilities, must be inspected at least once every 6 months by the IACUC (PHS Policy IV,B,2). (See 15:11; 25:1.)

Opin. As addressed in 25:1, many IACUCs conduct routine laboratory inspections in order to meet their oversight responsibility of the institution's animal care and use program. It should be noted that the AWAR (§2.31,c,2) require that the IACUC inspect all animal facilities where animals are kept for more than 12 hours, including animal study areas, at least once every 6 months. Other animal facilities include surgical rooms, food storage, and cage wash areas.[2] Some of these areas may not house animals for more than 12 hours or may not house animals at all. Nevertheless, as we noted in the introduction to this chapter and in 25:1, it is our opinion that such inspections are appropriate.

25:3 How frequently should each laboratory where animals are used be inspected?

Reg. Research laboratories in which APHIS/AC-covered species are maintained for more than 12 hours must be inspected by a subcommittee of at least two IACUC members every 6 months (AWAR §1.1, Animal; §1.1, Study Area; §2.31,c,2). IACUCs serving institutions required to comply with the PHS Policy (III,B; IV,B,2) must inspect laboratories in which animals are maintained for more than 24 hours at least once every 6 months.

Opin. As there are no other regulatory requirements mandating laboratory inspection, it is at the individual IACUC's discretion whether or not they should inspect animal research laboratories at their institution that do not meet these time requirements. It is the authors' opinion that at least select animal research laboratory inspections should be conducted by the IACUC in order to ensure compliance with both the content and the spirit of applicable regulations and guidelines. (See 25:1.)

Surv. How often does your IACUC inspect laboratories in which animals are used? (*Note:* Several respondents provided multiple responses.)

- Not applicable   20/154
- Every 6 months   112/154
- Periodically, no defined frequency   19/154
- Other   8/154

25:4 Should a laboratory be approved by the IACUC before permitting an investigator to use the laboratory to conduct procedures on animals?

Reg. The AWAR (§2.31,d,1) requires the IACUC, as part of the protocol review process, to ensure that all components of animal care and use are in compliance with the AWAR prior to approval of the protocol. PHS Policy (IV,C,1) requires that the project will be conducted in accordance with the AWA and the *Guide*. (See 25:1.)

Opin.   Other than as noted for the AWAR and the PHS Policy, there are no regulatory requirements mandating the approval of laboratories by the IACUC. Nevertheless, it is the IACUC's responsibility to ensure that the proposed activity is conducted at a suitable location. It is the authors' opinion that there are significant advantages of having the IACUC inspect laboratories before providing approval for work with animals. Inspection of the laboratory provides the IACUC the opportunity to determine whether the facilities and equipment are suitable for the proposed activity. In addition, it provides an opportunity for the investigative staff to gain a better understanding of IACUC functions and provides IACUC members a more thorough understanding of the proposed project. At institutions where IACUC approval of sites is not feasible because of time constraints, the IACUC can delegate this responsibility to a member of the institution's animal resources staff. (See 23:22.)

Surv.   Does your institution require IACUC approval of investigators' laboratories before they are used to conduct procedures on animals in them?

• Approval of laboratory required but not inspected prior to granting approval	43/154
• Must inspect before approving the laboratory	31/154
• Approve only if specific procedures are done, (e.g., survival surgical procedures for nonrodent mammals or areas meeting the AWAR or PHS definition of a study area or satellite facility)	30/154
• Do not approve laboratories	38/154
• Other	12/154

25:5   How does the IACUC determine the location of laboratories used for research with animals?

Opin.   The most direct method for determining which laboratories are used to conduct research on animals is to request that information in the animal use proposal reviewed by the IACUC. This information is frequently used to generate a database. Additionally, laboratories may be identified on the basis of information provided by the animal resources staff or the institution's administration.

Surv.   How does your IACUC determine the location of laboratories used for research with animals? More than one response is possible.

• Not applicable	30/154
• From information on the IACUC application form	101/154
• From a separate database that we update	11/154
• From requests to Principal Investigators to identify their laboratory's location	24/154
• Other	6/154

25:6   Should the IACUC inspection be announced or unannounced? If an IACUC inspection team misses a laboratory during an inspection or if a member of the investigative staff is not present in the laboratory at the time of the inspection, should they go back at an alternative time to complete the inspection?

Opin.   There are advantages and disadvantages of both methods. An announced site visit generally ensures that a knowledgeable staff member is available to provide laboratory access, discuss animal research activities, and answer questions. Staff availability permits dialog, which is advantageous to both the IACUC members and research staff, who frequently gain a better understanding of the IACUC's functions and activities. However, when conducting announced site visits, IACUC members may be exposed to an artificial environment not reflective of the normal state of the laboratory, personnel conduct, or animal use. Further, it is not uncommon for laboratories to cancel animal research activities when expecting a site visit. (See 23:25.)

Unannounced inspections permit the IACUC members to view the laboratory and its activities under normal operating conditions. However, research staff may not be present, potentially limiting laboratory access, or if they are present, ongoing research responsibilities may limit them in interfacing with IACUC members.

Institutions conducting announced inspections may also conduct unannounced inspections if the IACUC has specific concerns that may not be uncovered if inspections are announced.

The determination as to whether a laboratory should be revisited if it is missed is dependent on federal and state regulatory requirements and the regulatory and institutional charge of the IACUC. The AWAR (§2.31,c,2) and the PHS Policy (III,B; IV,B,2) require that study areas be inspected at least once every 6 months by the IACUC to maintain compliance. (See 25:1; 25:2.). Therefore, if the area is mandated for inspection and access cannot be obtained, the authors recommend rescheduling a visit.

Surv.   If there is nobody present at an announced or unannounced laboratory inspection, does your inspection team schedule a return site visit in the immediate time frame or does the IACUC wait until the next inspection cycle? (*Note:* Several respondents provided multiple responses.)

- Not applicable   50/154
- Return for site visit in immediate time frame   57/154
- Wait for next inspection cycle   6/154
- Varies with the inspection team   24/154
- Other   20/154

25:7   Is it necessary for a PI or his or her designee to be present during the inspection?

Opin.   Although not mandated by the AWAR or PHS Policy, there are clear advantages in having a knowledgeable research staff member present during the inspection (see 25:6). If the inspections are announced, the IACUC may stipulate that the PI or designee be present during the inspection. With unannounced visits, the IACUC cannot ensure that access can be gained to the area.

25:8   Prior to the semiannual inspection, should the IACUC members who are visiting the laboratory be briefed on procedures conducted in the laboratory?

Opin.   It is advantageous for IACUC members to be briefed on procedures occurring in the laboratory prior to the inspection, by the AV, a knowledgeable IACUC member, or by consulting the protocol. Alternatively, the PI or his or her designee can provide this information at the time of the inspection.

Surv.   Prior to a semiannual inspection, is your IACUC inspection team briefed on procedures conducted in a laboratory? (*Note:* Several respondents provided multiple responses.)

• Not applicable	24/154
• Yes, and protocols provided prior to inspection	29/154
• Yes, but protocols not provided prior to inspection	43/154
• No briefing	49/154
• Other	9/154

25:9   What areas and issues should the IACUC focus on during inspection of the laboratory?

Opin.   IACUC members should:

- Ensure appropriate facilities, personnel, and equipment are present to provide adequate veterinary care
- Observe the laboratory and equipment used for cleanliness and adequate sanitation
- Assess safety issues such as the following:
  - Food or drink consumption within the laboratory
  - Anesthetic gas scavenging
  - Appropriate use of hazardous substances
  - Use of appropriate animal handling procedures
  - Recapping of needles
- Observe the condition of animals, if they are present
- Confirm compliance with approved IACUC protocols including ascertaining that procedures and staff have been described and included in the approved protocol
- Evaluate the adequacy of training
- If surgery performed, determine adherence to aseptic technique, suitability and understanding of anesthesia monitoring, and postoperative care including pain relief
- Review methods of euthanasia
- Verify that all pharmaceuticals are handled and stored appropriately and are not expired
- Review records, especially when conducting survival surgical procedures (see 25:10)
- If laboratory approved for animal housing, determine that the animals' husbandry needs are met and appropriate records maintained

25:10   What records should the IACUC inspection team review when visiting the laboratory?

Opin.   Record evaluation is influenced by the activities that are conducted in the laboratory. If the laboratory is approved for housing and caring of animals, the PI's staff should maintain and the IACUC members should periodically examine those records indicating the frequency of cage changing, environmental monitoring (e.g., temperature and humidity), laboratory sanitation, and daily animal observations (including weekends). Records should be similar to those maintained by animal resource personnel in the animal facility. The following records, if applicable to the use of the laboratory, should be maintained and made available to the IACUC members during an inspection:

- Numbers of animals used per unit time (period in which the IACUC can compare actual animal usage to the number of animals requested in the protocol)
- Procedures conducted
- Treatment or drug administration
- Anesthetic, surgical, and postoperative care
- Controlled substances log
- Dosimetry for radiation exposure if using radionuclides or exposed to ionizing radiation
- Any institutionally required documentation when using hazardous substances
- Copies of institutionally required policies, handbooks, or guidelines

Surv.   Which records does your IACUC review during a laboratory inspection? (*Note:* Several respondents provided multiple responses.)

- Not applicable   23/154
- Do not routinely review records during an inspection unless observations indicated the need to do so   57/154
- Surgical and postoperative care records   63/154
- Other   29/154

Note: "Other" includes records of drug use, including use of controlled substances; morbidity and mortality logs; applicable standard operating procedures; medical records; guillotine sharpening; vaporizer calibration; and biological safety cabinet and chemical hood certification.

25:11   What criteria can the inspection team use to make sure that acceptable standards are being met in the laboratory?

Opin.   There are many criteria that can and should be utilized to determine that acceptable standards are used in the research laboratory. IACUC members should utilize the *Guide* and the AWAR, as well as IACUC and institutional policies, as the basis for determining the suitability of the laboratory for conducting research with animals.

Depending on the nature of the research conducted and the institution, the Good Laboratory Practices Act,[3] the Occupational Safety and Health Act,[4] the Blood-borne Pathogen Standard,[5] Standards for Protection against Radiation,[6] Lab Safety Standard,[7] Controlled Substances Act,[8] NIH Recombinant DNA Guidelines,[9] and Centers for Disease Control and Prevention's *Biosafety in Microbiologic and Biomedical Laboratories*[10] also may be used as criteria.

25:12   How should the IACUC proceed after identifying an item of noncompliance in a laboratory?

Reg.   The AWAR (§2.31,c,3) require the IACUC to identify deficiency items as significant or minor and develop a timetable for correction. The AWAR (§2.31,c,3) define a *significant deficiency* as "one which, with reference to Subchapter A, and, in the judgment of the IACUC and the Institutional Official, is or may be a threat to the health or safety of the animals." If the established timetable is not adhered to and a significant deficiency remains uncorrected, the deficiency must be reported to APHIS/AC and any federal agency funding the activity within 15 days (AWAR §2.31,c,3). Similarly, the PHS Policy (IV,B,3) requires the IACUC to draft reports that are submitted to the IO of its inspection findings. As with the AWAR, identified deficiencies must be distinguished as significant or minor on the basis of whether they are or may be a threat to the health or safety of the animals. If program or facility deficiencies are noted, the report to the IO must contain a reasonable and specific plan and schedule for correction of each deficiency. (See 24:13; 26:16.)

Opin.   The authors find it most effective for deficiencies to be discussed with the responsible individuals or their designee, if available, at the time of the site visit. The ensuing dialog may clarify misinterpretations. Deficiencies should be reviewed by the IACUC and categorized as described. A timetable for correction should be established. A written notice should be sent from the IACUC to the responsible individual, describing the stated deficiency and timetable provided for correction. The authors advise that the IACUC develop a mechanism by which the responsible individual can contest the deficiency if there is disagreement. Further, the responsible individual should be requested to provide written documentation that the deficiency is corrected.

Alternatively, the authors are aware of institutions that effectively use a multicopy form that is completed by the site visitors at the time of the site visit. The form contains a checklist for the inspectors to use to identify compliance or lack thereof. If deficiencies are noted, a timetable for correction based on preapproved criteria established by the IACUC is provided at the time of the site visit. The form is signed by the responsible individual or designee at the time of the site visit, and a copy is retained by the laboratory. The form stipulates that a formal response is required from the responsible individual for all noncompliant items.

Surv.   How does your IACUC proceed after identifying an item of noncompliance during a laboratory inspection? More than one response is possible.

• Not applicable	23/154
• We contact the responsible individual in writing	99/154
• We contact the investigator by phone	40/154
• We provide the specific time frame in which to rectify the problem	104/154

- We provide a written description of noncompliant items   100/154
- We almost always follow up with a second inspection to
  assure the item is corrected   59/154
- The response varies with the degree and significance of the
  noncompliant item   55/154
- Other   8/154

All communications should convey the date of expected resolution. The authors recommend that the IACUC either reinspect the area to determine whether the item was corrected by the specified date or confirm correction during the next scheduled inspection. Depending on the severity and repetition of the item of noncompliance, the IACUC may impose sanctions on the investigator (see Chapter 29).

References

1. Division of Animal Welfare, Office for Protection from Research Risks, National Institutes of Health, The Public Health Service responds to commonly asked questions, *ILAR News*, 33(4), 68, 1991.
2. DeHaven, W.R., personal communication, 1998.
3. Office of the Federal Register, Code of Federal Regulations, Title 43, Good Laboratory Practice Regulations, Washington, D.C., 1978.
4. Office of the Federal Register, Code of Federal Regulations, Title 29, Part 1910, Occupational Safety and Health Standards, Subpart G, Occupational Health and Environmental Control, Washington, D.C., 1984.
5. Office of the Federal Register, Code of Federal Regulations, Title 29, Part 1910.1030, Bloodborne Pathogens, Occupational Health and Safety Act, Washington, D.C., 1991.
6. Office of the Federal Register, Code of Federal Regulations, Title 10, Chapter 1, Nuclear Regulatory Commission Regulations, Washington, D.C., 1992.
7. Office of the Federal Register, Code of Federal Regulations, Title 29, Part 1910, Department of Labor, Occupational Safety and Health Administration, Occupational exposures to hazardous chemicals in laboratories: final rule, *Fed. Regist.*, January 31, Part 11, Washington, D.C., 1990.
8. U.S. Department of Justice, Drug Enforcement Administration, Office of Diversional Control, Pharmacist's Manual: An Informational Outline of the Controlled Substances Act of 1970, Washington, D.C., 2004. Available on the World Wide Web at: http://www.deadiversion.usdoj.gov/pubs/manuals/pharm2/2pharm_manual.pdf.
9. U.S. Department of Health and Human Services, National Institutes of Health, Guidelines for Research Involving Recombinant DNA Molecules, 2002. Available on the World Wide Web at: http://www4.od.nih.gov/oba/rac/guidelines/guidelines.html.
10. U.S. Department of Health and Human Services, Public Health Service, Biosafety in Microbiological and Biomedical Laboratories, 4th ed., U.S. Government Printing Office, Washington, D.C., 1999.

26

Inspection of Surgery Areas

Scott E. Perkins and Neil S. Lipman

Introduction

The AWAR (§2.31,c,2) require IACUCs to inspect "all of the research facility's animal facilities, including animal study areas" for species covered by the AWAR. The PHS Policy (IV,B,2) requires any and all animal facilities and areas to be inspected at least every 6 months including satellite facilities and surgical areas. The PHS Policy (IV,B,1) also requires adherence to the *Guide* (p. 9), which recommends inspection of the animal facilities and activity areas at least once every 6 months. The *Guide* and AWAR specify requirements for surgical facilities and procedures for conducting survival surgery on rodents and USDA-covered species. The IACUC must approve all surgical procedures involving animals to ensure that appropriately trained individuals conduct the research in a humane manner. This chapter is intended to help the reader to comply with these regulations and recommendations.

26:1 Should all surgical sites be visited and evaluated at the time of each IACUC inspection?

Reg.   There are federal requirements for IACUCs to inspect areas used to conduct survival surgical procedures on USDA-covered species[1] regardless of whether or not the site is located within an animal facility.[2] The AWAR (§2.31,c,2) require that animal facilities, including study areas, be inspected at least once every 6 months.

The PHS Policy (IV,B,2) also requires any and all animal facilities to be inspected at least every 6 months. Specifically, the PHS Policy (III,B) includes "any and all buildings, rooms, areas, enclosures, or vehicles, including satellite facilities, used for animal containment, transport, maintenance, breeding, or experiments inclusive of surgical manipulation." PHS Policy (IV,B,1) also requires adherence to the provisions of the *Guide* (p. 9), which states "The committee should review the animal-care program and inspect the animal facilities and activity areas at least once every 6 months."

Opin.   The AWAR (§2.31,c,2) and PHS Policy (IV,B,2) require semiannual inspection of surgical areas.[1] Therefore, all surgical areas, including surgical facilities outside a core animal facility, should be inspected semiannually by the IACUC to maintain compliance with and meet the spirit of the AWAR, PHS Policy, and *Guide*. Both the AWAR (§2.31,d,1,ix; §2.31,d,1,x) and the *Guide* (pp. 60–64, 78–79) detail facility and procedural requirements for conducting survival surgical procedures on rodents or USDA-covered species.

Surv.   Does your IACUC visit and evaluate all surgical sites at each semiannual inspection? More than one response is possible.

• Not applicable, as we do not do any surgery	15/153
• Inspect all surgical sites used for survival surgery on USDA-covered species	106/153
• Inspect all rodent or nonsurvival surgery sites	113/153
• Do not inspect rodent or nonsurvival surgery sites	3/153
• Other—inspect surgery sites but not every time	2/153

26:2   What is the minimal frequency for evaluating surgical sites?

Reg.   (See 26:1.)
Opin.   (See 26:1.)

26:3   What facilities are required for conducting survival surgery on nonrodent and rodent species?

Reg.   The IACUC must differentiate between major and nonmajor surgical procedures when evaluating facilities for performing survival surgical procedures, as facility requirements depend on the procedure conducted (see 26:5 for definitions).

Both the AWAR (§2.31,d,1,ix; §2.31,d,1,x) and the *Guide* (pp. 78–79) specify facility requirements for conducting major survival surgery on nonrodent species. The AWAR (§2.31,d,1,ix) require that "major operative procedures on non-rodents will be conducted only in facilities intended for that purpose which shall be operated and maintained under aseptic conditions." The AWAR do not require dedicated facilities when conducting major operative procedures at field sites; however, aseptic technique must be followed (§2.31,d,1,ix).

The *Guide* (pp. 78–79) provides significantly more detail pertaining to facility requirements, recommending that nonrodent surgical facilities include a variety of functional components, usually but not always separated by physical barriers. These include areas for surgical support, animal preparation, surgeon's scrub, operating room, and postoperative recovery. The need to provide all of these areas depends on the size and scope of the surgical program. With respect to ventilation, the *Guide* (p. 76) states that "areas of surgery should be kept under relative positive pressure with clean air."

Opin.   It is the chapter authors' opinion that the minimal requirements for most operative suites are

- Area for animal preparation
- Area for surgical scrub
- Operating room

The surgical suite should be designed and operated in a manner to control microbial contamination. This can be accomplished by

- Minimizing traffic in the area
- Adhering to rigorous sanitation and disinfection programs

- Limiting equipment present within the operating room to only essential movable equipment (such as surgical and support tables and anesthesia and monitoring equipment)

- Outfitting the surgical suite with a ventilation system that provides directional airflow maintaining the operating room at positive pressure with respect to the animal preparation and surrounding areas (under most circumstances)

Facility requirements for conducting nonmajor surgery on nonrodents or major survival surgery on rodents are less stringent. The AWAR (§2.31,d,1,ix) do not require dedicated facilities for conducting these procedures but do specify the use of aseptic technique. An area used for such procedures can be a room or portion of a room, which is easily sanitized and is dedicated for that purpose at the time of surgery. The laboratory bench or surgical table should be disinfected prior to and after use, be free of ancillary equipment, be located in an area free of drafts, and have a suitable temperature.

26:4 What facilities are required for conducting *nonsurvival* surgery on animals?

Opin.   There are no facility specific requirements when conducting nonsurvival surgical procedures. The area used for nonsurvival surgery should be a room or portion of a room, which is easily sanitized and at the time of surgery, dedicated for that purpose. The laboratory bench or surgical table should be disinfected prior to and after use, be free of ancillary equipment, and be located in an area free of drafts with a suitable temperature. In general, adherence to aseptic technique is unnecessary. The surgical site should be clipped and the instruments clean, and the surgeon should wear gloves. The IACUC should evaluate the duration of nonsurvival surgery as procedures can occasionally last for sufficient periods for sepsis to develop. In these cases, adherence to aseptic technique may be advisable. Additionally, the IACUC should evaluate the need for aseptic technique or "aseptic" facilities, if certain procedures are being performed (e.g., harvesting an organ for transplantation). (See 18:15.)

26:5 How does one differentiate a major from a minor surgical procedure?

Reg.   The AWAR (§1.1) defines a *major operative procedure* as "any surgical intervention that penetrates and exposes a body cavity or any procedure which produces permanent impairment of physical or physiological functions." (See 18:4.)

        The PHS Policy itself provides no definition of major or minor surgical procedures, but it does require adherence to the provisions of the AWA for covered species (PHS Policy, II) and of the *Guide* (PHS Policy IV,B,1).

        The *Guide* (pp. 61–62) defines a major survival surgical procedure as one that "penetrates and exposes a body cavity and produces substantial impairment of physical or physiological functions." It defines *minor survival surgery* as one in which the procedure "does not expose a body cavity and causes little or no physical impairment."

Opin.   Although both the AWAR and the *Guide* provide a definition of a major operative procedure, the classification of some surgical procedures requires IACUC interpretation, which is based on the procedure's extent, effects, and potential for development of complications. Major operative procedures include laparotomy, thoracotomy, craniotomy, and limb amputation. Minor operative procedures include wound suturing, peripheral vessel cannulations, and certain routine farm animal procedures.

26:6   Assume an animal is covered by the AWAR and the PHS Policy. Under what circumstances can a major survival surgical procedure be performed in an area other than a dedicated surgical suite approved by the IACUC?

Reg.   The AWAR (§2.31,d,1,ix) require that "major operative procedures on non-rodents will be conducted only in facilities intended for that purpose which shall be operated and maintained under aseptic conditions." The AWAR (§2.31,d,1,ix) do not require dedicated facilities when conducting major operative procedures in field studies, although aseptic procedures are required.

PHS Policy (IV,B,1) requires adherence to the *Guide* (p. 62), which states, "In general, unless an exception is specifically justified as an essential component of the research protocol and approved by the IACUC, non-rodent aseptic surgery should be conducted only in facilities intended for that purpose." Nevertheless, the *Guide* (pp. 62–63) does provide exceptions for emergencies and field studies: "Emergency situations sometimes require immediate surgical correction under less than ideal conditions," and "generally, farm animals maintained for biomedical research should undergo surgery with procedures and in facilities compatible with the guidelines set forth in this section. However, some minor and emergency procedures that are commonly performed in clinical veterinary practice and in commercial agricultural settings may be conducted under less-stringent conditions than experimental surgical procedures in a biomedical-research setting."

Opin.   It is the authors' opinion, in accordance with the *Guide*, that major operative procedures can be performed in a nondedicated surgical suite when scientifically justified (e.g., unique instrumentation in the laboratory) and approved by the IACUC. In the event of an emergency when the animal cannot be transported to the dedicated operating room, this should be considered a clinical medical requirement, determined by veterinary medical judgment, and does not require IACUC approval.

Surv.   Under what circumstances does your IACUC allow a major survival surgical procedure to be performed in an area other than a dedicated, IACUC-approved surgical suite? More than one answer is possible.

- Not applicable as we do not perform major survival surgery   33/153
- No policy on this matter   18/153
- Permit this for rodent species without special justification   26/153
- Permit this for rodent species, but only with special justification 39/153
- Allow this for nonrodent species without special justification   0/153
- Allow this for nonrodent species, but only with special justification   12/153
- Allow this for field studies with justification   11/153
- Not permitted   27/153

26:7   Can nonsurvival surgery on nonrodent species, or survival surgery or nonsurvival surgeries on rodents, be performed in survival surgical facilities used for nonrodents? If yes, are there any special procedures that should be followed?

Opin.   IACUCs may permit the conduct of survival rodent surgery or nonsurvival surgical procedures in survival surgical facilities if either of the following conditions is met:

- Nonsurvival procedures that are conducted utilize the same standards that apply to survival procedures in accordance with the standards of the *Guide* (p. 62).
- The operating room is thoroughly sanitized and disinfected prior to conduct of a survival surgical procedure in nonrodents (*Guide*, p. 62).

26:8 What areas and issues should the IACUC focus on during inspection and evaluation of surgical areas?

Reg.   PHS Policy (IV,B,2) requires any and all animal facilities (as defined in PHS Policy III,B) to be inspected at least every 6 months, including "any and all buildings, rooms, areas, enclosures, or vehicles, including satellite facilities, used for animal containment, transport, maintenance, breeding, or experiments inclusive of surgical manipulation." PHS Policy (IV,B,1) requires compliance with the *Guide* (pp. 61–66), which includes IACUC evaluation of aseptic procedures, postoperative evaluation and care, anesthetic monitoring, and other elements. Other sections of the *Guide* provide further information on occupational health and safety (*Guide*, pp. 14–18), design of facilities for aseptic surgery (*Guide*, pp. 78–79), and other considerations. The AWAR (§2.32), PHS Policy (IV,C,1,f), U.S. Government Principle VIII, and the *Guide* (p. 61) all address the need for proper qualifications and experience of personnel working with animals (see 26:1).

Opin.   The AWAR (§2.31,d,1,ix; 2.31,d,1,x) and the *Guide* (pp. 60–64, 78–79) describe facility and procedural requirements for conducting survival surgical procedures in nonrodents.

In the authors' opinion, minimal evaluation of standards for surgical areas should include the following:

- Assessment of personnel training
- Adherence to aseptic technique
- Scavenging of waste anesthetic gases
- Correct use of unexpired pharmaceuticals
- Proper use of equipment and procedures for perioperative monitoring and care
- Maintenance of perioperative records
- Availability of appropriate equipment and pharmaceuticals for emergencies
- Proper disposal of sharps

Surv.   What areas and issues does your IACUC focus on during inspection and evaluation of surgical areas? More than one response is possible.

• Not applicable as we have no surgical areas	22/153
• Physical plant in general	107/153
• Personnel training and surgeons' competence	45/153
• Scavenging of anesthetic waste gases	83/153
• Proper maintenance records for gas anesthesia machines	64/153
• Not applicable, as we do not use gas anesthesia machines	7/153
• Traffic patterns within the surgical suite	35/153

- Availability of appropriate equipment and pharmaceuticals for emergencies 59/153
- Expired pharmaceuticals 103/153
- Drug storage and inventory control 101/153
- Cleanliness and ease of surgical room disinfection 112/153
- Methods and records of disinfection 59/153
- General procedures for sterilizing instruments and surgical supplies 81/153
- Noting of dates when instruments were sterilized 46/153
- Assurance that autoclaves can obtain proper sterilization parameters 55/153
- Adherence to aseptic technique including use of appropriate surgical garb 70/153
- Posting of emergency phone numbers 62/153
- Facilities for postoperative recovery 81/153
- Equipment and procedures for perioperative care 64/153
- Perioperative monitoring and records 60/153

26:9 What reference sources are available to help evaluate the issues in question 26:8?

Opin.   The IACUC should utilize the AWAR, the *Guide,* applicable state regulations, and institutional policies to develop acceptable standards for surgical facilities and their operation. The IACUC can also consult references, as well as expert consultants, to assist in the development of standards for conducting surgery. Additional references include veterinary and human surgery and anesthesia texts, texts and references on experimental surgery, the Good Laboratory Practices Act,[3] the Occupational Safety and Health Act,[4] the Blood Borne Pathogen Standard,[5] Standards for Protection against Radiation,[6] Lab Safety Standard,[7] information on the Controlled Substances Act,[8] Recombinant DNA Guidelines,[9] the Centers for Disease Control and Prevention (CDC) publication *Biosafety in Microbiologic and Biomedical Laboratories,*[10] and references on disinfection and sanitation control.[11]

26:10 What records should the inspection team review when visiting nonrodent and rodent surgical areas?

Opin.   The IACUC should confirm that procedures and staff training, experience, and qualifications have been accurately described and included in the approved animal care and use proposal. The IACUC should review pertinent animal health and experimental records including the following:

- Perioperative records describing the surgical procedure performed
- Experimental agents administered
- Dose and route of anesthetics and analgesics administered
- Frequency and type of intraoperative monitoring procedures performed
- Postoperative monitoring and observations

- Quality assurance records for sterilization of materials (periodically reviewed)
- All records should be made available for IACUC review whether they are maintained with the animal, in the surgical facility, or by the investigator.

Surv. Does the IACUC review records when inspecting surgical areas?

• Not applicable as we have no surgical areas	23/153
• Records are routinely reviewed during inspections	48/153
• Records are not routinely reviewed during inspections	74/153
• No response	8/153

Records for nonrodent APHIS/AC-covered species and for currently active cases were reviewed most frequently.

26:11   How detailed should surgical and postoperative records be? What are the essential elements of an adequate record?

Reg. The *Guide* (p. 46) states, "Clinical records for individual animals can also be valuable, especially for dogs, cats, nonhuman primates, and farm animals. They should include pertinent clinical and diagnostic information, history of surgical procedures and postoperative care, and information on experimental use." The AWAR (§2.33,b,5) require "adequate pre-procedural and post-procedural care in accordance with current established veterinary medical and nursing procedures." APHIS Animal Care Policy 3 states, "Health records are meant to convey necessary information to all people involved in an animal's care. Every facility is expected to have a system of health records sufficiently comprehensive to demonstrate the delivery of adequate health care."[12]

The HREA (Section 495,a,2,B) requires "appropriate pre-surgical and post-surgical veterinary medical and nursing care for animals" used in biomedical and behavioral research. (See 18:31; 18:32.)

Opin. It is the opinion of the authors that the AWAR and HREA regulations could be interpreted as surgical records as a required component of adequate veterinary care. Records should be detailed and contain all relevant information including, but not limited to

- Animal identification
- Project identification
- Detailed description of the surgical procedure
- Experimental agents administered
- Dose and route of anesthetics and analgesics administered
- Frequency (e.g., time interval) and type (e.g., blood pressure, heart rate) of intraoperative monitoring procedures performed
- Postoperative monitoring and observations recorded
- Postoperative treatments administered

Surv. 1 Does your IACUC require detailed surgical and postoperative records?

• Not applicable	29/153
• Require detailed records	58/153

- Do not require detailed records   23/153
- Require detailed records only for USDA-covered species   32/153
- Other—varies with procedure   2/153
- No response   9/153

Surv. 2 Does your IACUC or animal facility provide investigators with specific forms to document the surgical information typically required by your IACUC?

- Not applicable   28/153
- Yes   64/153
- No   53/153
- No response   8/153

26:12 Should surgical records be kept for all species or only certain ones?

Opin. The *Guide* (p. 46) states that records are especially valuable for dogs, cats, nonhuman primates, and farm animals. Although there are no specific record requirements in the AWAR or PHS Policy, APHIS/AC inspectors expect records, in accordance with established veterinary medical practice and as interpreted from the AWAR (§2.33,b,5), to be maintained. PHS Policy also requires adherence to the requirements of the AWAR for covered species (PHS Policy II) and the *Guide* (PHS Policy IV,B,1).

It is the opinion of the authors that surgical records should be maintained for all species to provide documentation that appropriately trained individuals conducted the research in a humane manner. (See 18:30–18:32.)

Surv. For which species does your institution *require* surgical records?

- Not applicable, as we do not perform surgery   19/153
- No policy about required surgical records   32/153
- For USDA-covered species and nonrodents   42/153
- All species, including rodents   60/153

26:13 Should surgical records be kept for each individual animal or can they be maintained by groups?

Opin. There are no regulatory requirements for having individual surgical records for each individual animal. The *Guide* (pp. 46–47) notes that clinical records for individual animals can be useful and states that they should include history of surgical procedures and postoperative care information. It is the opinion of the authors that records can be maintained by groups for rodents, but nonrodents should have surgery records maintained for individual animals. (See 26:10; 27:16.)

26:14 Do records for survival surgery differ from those maintained for animals undergoing nonsurvival surgical procedures?

Opin. Neither the AWAR nor the *Guide* stipulates that record keeping should be less stringent or different for animals undergoing nonsurvival procedures. Records should be maintained in accordance with established veterinary medical standards as noted in 26:10 and 26:12. (See 18:30–18:32.)

Surv.   Are record requirements different for survival versus nonsurvival surgery?

• Not applicable	30/153
• We have no policy for this type of requirement	36/153
• No records kept for nonsurvival surgery	17/153
• Keep records of anesthetics, dosages, and monitoring criteria for survival surgery on some or all species	61/153
• No difference between survival and nonsurvival surgery	6/153
• No response	3/153

26:15   What are the record keeping requirements when performing surgery on rodents?

Reg.   The *Guide* (p. 46) states that clinical records for animals are valuable and describes the kind of information to be maintained. Rodents are not excluded from these statements.

Opin.   At the authors' institutions, records are required for rodents undergoing surgical procedures, although the frequency of monitoring procedures and details for intra-operative monitoring procedures are not required. Additionally, the records may be maintained for groups. Rodent records should include the following:

- A description of the surgical procedure conducted
- Experimental agents administered
- Dose and route of anesthetics and analgesics administered
- Postoperative monitoring and observations
- Treatments administered during the perioperative period

26:16   How should the inspection team proceed after identifying an item of noncompliance in a surgical area?

Reg.   The IACUC should identify the item as significant or minor and develop a timetable for correction as specified in the AWAR (§2.31,c,3) and the PHS Policy (IV,B,3). (See 25:12.)

Opin.   The policy for contacting the appropriate individual varies depending on the magnitude of the item. IACUC options for notification include the following:

- Briefing the responsible individual from the laboratory at the time of inspection, or later by telephone
- Sending a copy of the inspection report to the responsible individual
- Sending a formal letter documenting the noncompliant item

All communications should convey the date of expected resolution. Depending on the severity and repetition of the item of noncompliance, the IACUC may impose sanctions on the investigator such as a letter of reprimand from the IO, cessation of research for a specified time or permanently, mandatory training, or increased inspection frequencies. The IACUC may revisit the area prior to the next site visit to confirm correction. (See 25:12 and Chapter 29.)

Surv.   How does your IACUC proceed after identifying an item of noncompliance in a surgical area? More than one response is possible.

- Not applicable   30/153
- Contact the responsible individual in writing (including e-mail)   94/153
- Contact the investigator by phone   36/153 ·
- Provide a specific time frame in which to rectify the problem   93/153
- Provide a written description of noncompliant items   93/153
- Almost always follow up with a second inspection to assure the item is corrected   50/153
- Response varies, depending on the degree and significance of the noncompliant item   59/153
- No items of noncompliance have been identified in a surgical area   5/153

26:17   Can unrelated major survival surgeries be conducted simultaneously in a single operating room on multiple nonrodent animals of the same or different species? Does the response differ for rodent species?

Opin.   Neither the AWAR nor the *Guide* has restrictions on the performance of more than one surgery, on the same or different species, in the same operating room. There are unique situations (e.g., inter- and intraspecies transplantation) when conducting surgery on more than one animal in the same operating room is justifiable.

     Because of the potential for infectious agent transmission and behavioral incompatibility, it is not advisable to conduct survival surgical procedures on more than one species simultaneously unless the species are of similar microbiological status, for example, specific pathogen-free rodents. Conducting simultaneous multiple survival surgeries on the same species is acceptable as long as the nature of each surgical procedure conducted does not place the other animals undergoing surgery at increased risk (e.g., different microbiological status of animals or an infectious disease study).

26:18   Can a major survival and a nonsurvival operative procedure be performed simultaneously on nonrodent species in the same operating room?

Opin.   When justifiable, as described in 26:17, survival and nonsurvival surgical procedures can be conducted in the same operating room. However, both surgeries should be conducted with aseptic technique as if they were survival procedures because strict aseptic technique should be followed in accordance with the regulations and with the purpose of providing the highest level of sanitation for the animal undergoing the survival procedure.

26:19   What level of training should the IACUC require for individuals performing survival surgical procedures? Should the IACUC verify whether individuals have adequate training and skills to perform major operative procedures?

Reg.   The IACUC has the responsibility for ensuring that individuals conducting surgery, irrespective of species used, have the necessary skills to execute the surgical procedure successfully. Both the AWAR (§2.32) and PHS Policy (IV,c,1,f) require that all personnel involved in animal care, treatment, and use are qualified and trained to perform their duties. U.S. Government Principle VIII stipulates that "investigators and other personnel shall be appropriately qualified and experienced for conducting procedures on living animals." In addition to the conduct of the surgery, individuals associated with the activity must be skilled at administering and monitoring anesthesia and have the capability of providing suitable postoperative care.

Opin.   Although the IACUC's responsibility is clear, neither the PHS Policy nor the AWAR describe implementation methods. There are a variety of mechanisms that the committee can use to meet this responsibility. (See also Chapter 21.) IACUCs frequently garner information pertaining to an investigator's training and skills via required descriptions of them in the initial proposal submitted to the IACUC. Although not all individuals on a specific project may have suitable skills to carry out the project, the IACUC must determine that there is sufficient skilled staff available to meet the project's goals. Clearly, the variety and scope of surgical procedures conducted in a research or training setting vary considerably, requiring the IACUC to make this determination on a case-by-case basis. The IACUC may request additional information or invite the investigative staff to meet with them to review their experience and training.

At many institutions the animal resource program personnel are responsible for the oversight of a centralized surgical facility. The committee frequently relies on the animal resource program's veterinary or paraprofessional staff for information pertaining to an individual's skill or training. Alternatively, the committee may request that one or several of its members observe procedures on a select number of animals before giving the project full approval. Adequacy of training should be ascertained prior to the start of the project; however, it is important for continuing assessment to be conducted. An infrequently used option is that the committee may request the aid of a consultant to make an assessment for projects of a sensitive nature or requiring highly specialized staff.

26:20   What constitutes aseptic technique for major operative procedures in rodents?

Reg.   The AWAR (§2.31,d,1,ix) and the *Guide* (pp. 60–64) require that major survival surgical procedures on all animals be conducted utilizing aseptic technique.

Opin.   Aseptic technique is a body of practices employed to prevent microbial contamination of living tissues or sterile materials by excluding, removing, or killing microbial organisms. Aseptic technique requires strict adherence to practices in several areas. The laboratory bench or surgery table should be clean, disinfected, and free of ancillary equipment. Devices or equipment (e.g., animal restraining devices, monitoring equipment, stereotaxic devices) that will be required in the surgical field should also be disinfected. These practices reduce or eliminate potentially infectious organisms and the substrates on which they grow.[13,14]

All surgical instruments, implantable devices, and materials that will contact the surgical site or are implanted in the animal must be sterilized. The sterilization method selected is dependent on time considerations, specialized equipment available, and the composition of the material to be sterilized. Sterilization monitoring devices should be routinely utilized to validate sterilization techniques. Surgical instruments and supplies should be prepared prior to sterilization so that they can be opened without being touched.[15]

The regulatory requirements for preparation of the surgeon for rodent surgery are less rigorous than those required for nonrodents. Either a clean or disposable laboratory coat or a surgical scrub shirt should be worn by the surgeon and assistants. A surgical mask (which serves to prevent contamination of the wound by droplets of saliva from the surgical team) and a surgical cap, for individuals who have long hair, which can potentially fall into the surgical field, should be worn. The surgeon should wash the hands with a disinfectant and don sterile surgical gloves by a method that maintains sterility.[16]

Preparation of the surgical site involves removal of fur by clipping or using a depilatory, and disinfection of the skin by scrubbing with suitable agents. The surgical site should be covered with a sterile drape. The goal of aseptic technique is to prevent the surgeon, the surgical site, and all instruments and equipment utilized from causing infection and subsequent disease.[15] (See 18:12–18:18.)

26:21 If performing major operative procedures on multiple rodents, can a single set of surgical instruments be used on multiple animals? (See also 18:14.)

Opin. Groups of rodents are frequently surgically manipulated during a single session. Care must be taken to prevent contaminating one animal from another. Sterilized instruments should always be used when initiating a surgical session. New sterile gloves should be donned; preferably, separate sterile surgical instruments should be provided for each animal subject. When a limited number of surgical instruments are available and separate sterile surgical packs are not feasible, the following options may be considered:

- Use two sets of instruments. One set is used for incising and manipulating the skin, which is considered a potentially contaminated site because of microbial flora. Once the initial incision and skin manipulation are complete, these instruments are set aside and not used again without prior sterilization. A second set of instruments is used to manipulate deeper, sterile tissues. The second set of instruments (which is shared) should be rinsed with sterile saline solution between animals. Care must be exercised to maintain the sterility of the surgical gloves by avoiding contacting tissues with fingers when utilizing this option.

- Utilize a glass bead sterilizer to sterilize instrument tips between animals. Recognize that only the instrument tips are sterilized. Care must be taken to prevent touching tissues with instrument handles or other nonsterile instrument parts. Because nonsterile instrument handles are held with gloved fingers, contact of tissues with the fingers is prevented.

- Have two or more sets of sterile instruments available so that the contaminated set can soak in cold sterilant solution for the minimal time required by the solution manufacturer, typically 15 minutes, to kill vegetative bacteria prior to reuse. Instruments must be rinsed thoroughly with sterile saline or water prior to reuse.

References

1. Division of Animal Welfare, Office for Protection from Research Risks, National Institutes of Health, The Public Health Service responds to commonly asked questions, *ILAR News*, 33(4), 68, 1991.
2. DeHaven, W.R., personal communication, 1999.

3. Office of the Federal Register, Code of Federal Regulations, Title 43, Good Laboratory Practice Regulations, Washington, D.C., 1978.
4. Office of the Federal Register, Code of Federal Regulations, Title 29, Part 1910, Occupational Safety and Health Standards; Subpart G, Occupational Health and Environmental Control, Washington, D.C., 1984.
5. Office of the Federal Register, Code of Federal Regulations, Title 29, Part 1910.1030, Bloodborne Pathogens, Occupational Health and Safety Act, Washington, D.C., 1991.
6. Office of the Federal Register, Code of Federal Regulations, Title 10, Chapter 1, Nuclear Regulatory Commission Regulations, Washington, D.C., 1992.
7. U.S. Department of Labor, Occupational Safety and Health Administration, 29 Code of Federal Regulations, Part 1910, Occupational exposures to hazardous chemicals in laboratories; final rule. *Fed. Reg.*, January 31, Part 11, Washington, D.C., 1990.
8. U.S. Department of Justice, Drug Enforcement Agency, Pharmacist's Manual: An Informational Outline of the Controlled Substances Act of 1970, 1986.
9. U.S. Department of Health and Human Services, National Institutes of Health, Guidelines for Research Involving Recombinant DNA Molecules, 2002. Available on the World Wide Web at: http://www4.od.nih.gov/oba/rac/guidelines/guidelines.html.
10. U.S. Department of Health and Human Services, Biosafety in Microbiological and Biomedical Laboratories, 4th ed., U.S. Government Printing Office, Washington, D.C., 1999.
11. Block, S., Ed., *Disinfection, Sterilization, and Preservation,* 5th ed., Lippincott Williams & Wilkins, Philadelphia, 2001.
12. Animal Care Resources Guide, Policies of Veterinary Care, APHIS, Animal Care. January 14, 2000. Available on the World Wide Web at: http://www.aphis.usda.gov/ac/policy/policy3.pdf
13. Cockshutt, J., Principles of surgical asepsis, in *Textbook of Small Animal Surgery,* 3rd ed., Slatter, D., Ed., W.B. Saunders, Philadelphia, 2003, chap. 10.
14. Hobson, H.P., Surgical facilities and equipment, in *Textbook of Small Animal Surgery,* 3rd ed., Slatter, D., Ed., W.B. Saunders, Philadelphia, 2003, chap. 13.
15. Mitchell, S.L., and Berg, J., Sterilization, in *Textbook of Small Animal Surgery,* 3rd ed., Slatter, D., Ed., W.B. Saunders, Philadelphia, 2003, chap. 11.
16. Shmon, C., Assessment and preparation of the surgical patient and operating team, in *Textbook of Small Animal Surgery,* 3rd ed., Slatter, D., Ed., W.B. Saunders, Philadelphia, 2003, chap. 12.

27

Assessment of Veterinary Care

Mark A. Suckow and Bernard J. Doerning

Introduction

Animals used in research or educational programs may experience spontaneous illness or, on occasion, problems related to specific procedures they have undergone. For these reasons, the availability of proper veterinary care is essential to assure the humane treatment of such animals. It must be emphasized that specific treatment of ill animals should always be guided by the professional judgment of the veterinarian.

The overall goal of the program of veterinary care is to assure that the animals remain healthy and are used in a humane and judicious manner. In this regard, the AV has a responsibility to the animals to assure their well-being, to the institution to promote regulatory compliance, and to the investigator to facilitate research. Of these, the well-being of the animals is paramount.

A challenge for the IACUC is to identify means by which to evaluate the scope and adequacy of the program of veterinary care. The IACUC should recognize that a sound program of veterinary care relies on the professional judgment of the AV. Useful parameters for the IACUC include the accuracy and overall scope of records that allow for retrospective evaluation of the program, the number and training of individuals available to assist the AV and investigators in delivery of acceptable care, and the mechanism for reporting problems and the response of the AV and his or her staff to such information.

27:1 What is meant by adequate veterinary care?

Reg.　Adequate veterinary care is an essential part of every animal care program. The AWAR (§2.33,a) state, "Each [registered] research facility shall have an attending veterinarian who should provide adequate veterinary care to its animals." The PHS Policy (IV,C,1,e) requires the IACUC to assure that "medical care for animals will be available and provided as necessary by a qualified veterinarian." Specific descriptions of proper veterinary care are not contained in the PHS Policy; however, the *Guide* (p. 56), which is used as a basis for evaluation by the PHS Policy, lists the following as components of effective veterinary care:

- Preventive medicine
- Surveillance, diagnosis, treatment, and control of disease, including zoonotic disease

- Management of protocol-associated disease, disability, or other sequelae
- Anesthesia and analgesia
- Surgery and postsurgical care
- Assessment of animal well-being
- Euthanasia

The *Guide* (p. 56) also indicates that the program of veterinary care is the responsibility of the AV.

Opin.   *Adequate veterinary care* implies a minimal acceptable standard; however, institutions should strive to exceed that level. General aspects of adequate veterinary care can be considered under three headings: animal husbandry, animal health, and study coordination (see 27:6).

Animal husbandry is much more than cleaning cages and feeding the animals. Prevention of disease and injury is a primary objective of the animal husbandry program. Areas of concern as they relate to veterinary care are listed in the following:

Animal procurement (and see 27:26): Newly acquired animals can introduce disease into established colonies if appropriate effort is not invested in the selection of vendors. A health surveillance program to screen or monitor incoming animals or potential vendors can be used to select appropriate sources of animals. All animals should be observed upon arrival at the facility for signs of illness.

Acclimation and quarantine (see also 27:26): In general, animals should be allowed a period of acclimation and physiologic stabilization after arrival at the facility to allow them to recover from shipping stress and permit them to adapt to their new surroundings. Animals known or suspected to be from contaminated sources should be quarantined in such a way that risk to other animals is minimized. Quarantine often involves special precautions, including restriction of personnel access, maintenance of air pressure differences between facility areas, and strict disinfection procedures.

Separation of species and source: Different species and animals from different sources are best maintained in separate housing units. Separating animals by species is useful to reduce anxiety due to interspecies conflict and to minimize the possibility of interspecies disease transmission. Similarly, animals from different sources should be separated to control possible transmission of infectious disease between groups. Separation can be achieved by housing animals in separate rooms, in cubicles, or in various forms of isolators.

Daily animal care: All animals should be housed in cages of appropriate size and style for the species and cleaned and maintained in accordance with the standards of the AWAR and the *Guide*. In addition, it is imperative that all animals are observed every day to assure that their health and living conditions remain acceptable. An SOP describing methods and equipment to be used in animal care helps specify the standards being followed. It is further useful to utilize a check-off sheet to document when animals or room conditions have been evaluated and when procedures associated with animal care have occurred to provide a permanent record that can be used to evaluate the consistency of animal care over time.

Animal health is intricately related to veterinary care. Ill or injured animals must receive prompt medical attention when needed. Important issues to address include the following:

Observation: All animals must be observed daily and any unexpected deaths and deviations from normal behavior or appearance must be reported to the veterinarian. Sick or injured animals should receive prompt attention.

Reporting of animal health problems (also see 27:16): A mechanism to report and record progress of animals in need of veterinary attention must be defined. Electronic record-keeping systems or notations in notebooks or on cage cards or "animal health forms" can be used (see 27:17–27:18). Documentation by the veterinarian in writing of his or her findings, recommendations, and the progress of the animal is an effective way to assist the IACUC in evaluation of the adequacy of veterinary care.

Anesthesia and analgesia: Animals undergoing procedures that might cause more than momentary pain or distress must receive appropriate anesthetics or analgesics according to the AWAR (§2.31,d,1,iv,A), which state, "Procedures that may cause more than momentary or slight pain or distress to the animals will be performed with appropriate sedatives, analgesics or anesthetics, unless withholding such agents is justified for scientific reasons." The PHS Policy (IV,C,1,b) is essentially the same as the AWAR. Similarly, the *Guide* (p. 64) states, "The proper use of anesthetics and analgesics in research animals is an ethical and scientific imperative." It is the duty of the veterinarian to advise the investigator on proper use of these drugs and to assure that animals are being adequately medicated. Use of such agents should be carefully documented in writing. The IACUC may find evaluation of such records to be useful. (See Chapter 16.)

Survival surgery: A major responsibility of the AV is to assure that survival surgical procedures are conducted in proper facilities by trained individuals using acceptable methods. Again, records of procedures performed and documentation of qualifications of personnel can be used by the IACUC to evaluate this aspect of veterinary care. (See Chapter 18.)

Postoperative and postprocedural care: Additional care and attention are often needed for animals that have had surgical or other invasive procedures that involve anesthesia. For example, care immediately after surgery might include administration of supportive fluids, analgesics, and other drugs as required; monitoring of vital signs such as temperature, pulse, and respiration; provision of supplemental heat if the animal is hypothermic; and observation of the animal until it is awake and ambulatory. Longer-term care might include administration of antibiotics or other drugs, evaluation of the surgical site for signs of dehiscence or infection, and overall continued attention to the general medical needs of the animal. For all postoperative or postprocedural care, it is important that accurate records be maintained for retrospective evaluation if problems are found. This record also allows the IACUC to evaluate this aspect of veterinary care.

Study coordination: *Prestudy veterinary consultation.* Investigators' discussions of research proposals with the veterinarian are a useful component of adequate veterinary care, especially when the research may result in more than momentary pain or distress to the animal (AWAR §2.31,d,1,iv,B). The AWAR (§2.33,b,4) state that the program of veterinary care includes "guidance to principal investigators and other personnel involved in the care and use of animals regarding handling, immobilization, anesthesia, analgesia, tranquilization, and euthanasia."

Training and assessment of skills: Adequate veterinary care is greatly augmented by training and evaluation of skills for anyone planning to conduct animal research. As part of the program of veterinary care, the AV should ensure that a program is in place to accomplish this.

27:2   Must the AV be a laboratory animal specialist?

Reg.   The AWAR (§2.31,b,3,i) and the PHS Policy (IV,3,b,1) indicate that the IACUC must include at least one member who is a doctor of veterinary medicine with training or experience in laboratory animal science or medicine who has direct or delegated responsibility for activities involving animals at the research facility. By definition, this individual is usually the AV, but the AWAR indicate that another veterinarian with delegated program responsibility for activities involving animals can serve as the veterinary representative to the IACUC (§2.33,a,3). The AWAR (§1.1, Attending Veterinarian) also mandate that the AV be a person who has graduated from a veterinary school accredited by the American Veterinary Medical Association's Council on Education, or has a certificate issued by the American Veterinary Medical Association's Education Commission for Foreign Veterinary Graduates, or has received equivalent formal education as determined by the USDA. The PHS Policy (IV,C,1,e) states that the IACUC must determine that "medical care for animals will be available and provided as necessary by a qualified veterinarian," although the Policy does not specifically define *qualified*.

The *Guide* (p. 56) indicates that the AV must be an individual who has a D.V.M. degree and has training or experience in laboratory animal science and medicine. However, the *Guide* does not specifically require the AV to serve as the veterinary member of the IACUC.

Opin.   The answer to this question is yes, if a *specialist* is defined as someone who has specialized training or experience. Individuals may have several areas of specialization, including laboratory animal science and medicine. Thus, the use of a consultant as the AV is acceptable even if he or she devotes major effort to other areas of veterinary medicine. The American Veterinary Medical Association recognizes laboratory animal medicine as a specialty practice area of veterinary medicine. (See 5:27.)

Completion of an internship at a laboratory animal facility as part of the veterinary professional curriculum is one means by which individuals might gain minimal experience in laboratory animal medicine. Sometimes, individuals become specialists as a result of on-the-job experience under the supervision of an experienced AV. Alternatively, some choose to complete residencies in laboratory animal medicine at academic, military, or industrial institutions. Such residencies allow for concentrated study and experience in the specialty. An individual who has completed an approved residency or has approved practical experience, has an approved publication in the field of laboratory animal medicine, and has successfully passed the appropriate written and practical examinations is eligible for certification as a diplomate by the American College of Laboratory Animal Medicine (ACLAM). Diplomate status within ACLAM is widely viewed as evidence of significant training and experience in laboratory animal medicine.

Membership and involvement in relevant professional organizations can be interpreted as evidence of specialized interest in laboratory animal science and medicine. For example, activity in the American Association for Laboratory Animal Science (AALAS) or the American Society of Laboratory Animal Practitioners (ASLAP) would likely improve the professional knowledge of individuals active in the profession.

A veterinary practitioner who lacks specific training in laboratory animal medicine might still ably fill the role of AV as a consultant. For example, if the species maintained at the facility are those with which the practitioner routinely works, his or her experience would be appropriate. In instances when a veterinary practitioner serves as a consulting veterinarian for a research facility, it is important for that individual to supplement his or her knowledge base where needed. This might be done by enrollment in professional associations such as AALAS or ASLAP; attendance at professional meetings that focus on laboratory animal medicine, completion of appropriate online coursework, or provision of appropriate texts or other material for the veterinarian to expand his or her knowledge base.

27:3 Is it necessary for all organizations housing or using laboratory animals to have an AV?

Reg. Any institution that is under the purview of AAALAC, the AWAR, or the PHS Policy is required to have an AV (see 27:1). Other organizations that do not fall under the purview of those standards may still choose to employ an AV on a full-time or consulting basis.

Opin. The AV typically has several functions within the institution[1] (AWAR §2.33,a; §2.33,b). For example, it is the duty of the AV to assure that the husbandry of the animals is appropriate for the species. The AV is responsible for developing and implementing programs for preventive medicine to minimize the likelihood of animals' developing clinical illness. Furthermore, the AV directs the diagnosis and treatment of clinical disease affecting the animals. In addition, the AV typically advises investigators and the IACUC on methods to relieve or alleviate pain and distress, perioperative care, and surgical procedures. The AV must also be familiar with applicable state veterinary acts, inclusive of licensure requirements, particularly in the discharging of certain official duties such as signing interstate health certificates or verifying rabies vaccination or tuberculosis status of animals. In short, the role of the AV is to assure that animals remain healthy and are used in a judicious and humane manner (see 27:6).

Several types of facilities are not required under the existing regulations to have an AV. For example, institutions that use only rats of the genus *Rattus* and mice of the genus *Mus,* bred specifically for use in research or teaching, are not required to have an AV if they lack AAALAC accreditation and do not receive PHS or other federal funds for research or teaching using animals. Such facilities should still develop a written plan for animal health in the event of unexpected illness.[2] In addition, if animals are to be used in surgical or other invasive procedures, it is advisable to identify an individual who can serve as the AV.

27:4 Under what circumstances would a full-time versus a part-time versus a consultant veterinarian be adequate?

Reg. The AWAR (§2.33,a,1) require that institutions using only a consultant veterinarian have a written program of veterinary care and that the AV make regularly scheduled visits to the facility. The PHS Policy (IV,C,1,e) does not specify a requirement for a full-time versus part-time AV. Nevertheless, the policy requires that the IACUC assure that medical care be "available and provided as necessary." PHS Policy (IV,A,3,b,1) also requires that there be a veterinarian with direct or delegated program authority and responsibility who will serve on the IACUC. This suggests that the veterinary staff needs to be adequate to the extent that competent veterinary care can be provided whenever the need arises. In this regard, the IACUC is charged with determining the level of veterinary service that must be procured.

Opin.   PHS Policy (IV,C,1,e) suggests that the veterinary staff must be sufficient to provide the appropriate medical care whenever the need arises. In this regard, the IACUC is charged with determining the level of veterinary service that must be procured.

The need for the services of the AV or other ancillary staff varies extensively with the size and scope of animal research-related operations of the institution. In this sense, it is not possible to outline specific standards for the need of a full-time or other veterinarian.

The American College of Laboratory Animal Medicine (ACLAM) recommends that a formal arrangement for veterinary care exist in all instances, while recognizing that the specific arrangement may vary with the number and species of animals involved and the nature of the experimentation at the institution.[1]

Institutions where research involving survival surgery or other invasive procedures is performed on a routine basis should seek greater veterinary involvement than institutions where this does not occur. Similarly, other conditions that merit increased veterinary involvement are shown in Table 27.1.

TABLE 27.1

Factors That Increase Need for Veterinary Services

Animal-Related Factors	Research-Related Factors
Large number of animals in the colony	Use of biohazards
Use of phylogenetically higher mammals such as dogs, cats, and nonhuman primates; use of immunodeficient animals	Use of chemical hazards
	Use of radiologic hazards
Use of animal models with special medical or biologic needs	Procedures requiring the use of anesthetics or analgesics, particularly survival surgery
	Survival surgical procedures involving extensive perioperative care

The IACUC should consider the factors listed in Table 27.1 and determine whether the available veterinary staff can provide the necessary service and oversight. Each facility has its own unique characteristics and needs. Some small facilities may need a full-time veterinarian, while some larger ones may require only a part-time or consulting veterinarian. Experience of the authors has shown that institutions characterized by three or more of the factors listed can generally justify the need for a full-time veterinarian. Facilities that are characterized by few or none of the listed factors may find a part-time veterinarian (who is a regular employee of the institution) or even a consultant veterinarian (one who provides veterinary care on a fee-for-service basis to the institution) to be sufficient.

Veterinary support in addition to the AV may be needed as the number of animals in the facility and the number of relevant factors increase. Such support can be in the form of additional veterinarians or, if appropriate, ancillary veterinary staff such as veterinary technicians or trained animal care technicians (see 27:8). Veterinary activities performed by technicians should be conducted under the direction of a qualified veterinarian. Communication between the veterinarian and technicians should occur regularly to ensure that the veterinarian is aware of any animal health or care issues and can advise the technical staff accordingly

(AWAR §2.33,b,3; *Guide*, p. 56). Under any arrangement, veterinary services must be available at all times, including evenings, weekends, and holidays (AWAR §2.33,b,2; and see 27:5). It is important that the information for contacting the veterinarian (telephone or pager numbers) is communicated to all research and animal care personnel.

27:5   If a consultant is used, how frequent should visits be and how can emergency care be provided if the consultant is not available?

Reg.   The frequency of visits to a facility by a consultant can vary significantly with the size and scope of animal use activities (*Guide*, p. 12). The AWAR state that "regularly scheduled" visits should be made to the facility (§2.33,a,1).

Opin.   Visits should be frequent enough to provide the same degree of oversight to each study as would be provided to studies in a larger facility employing full-time veterinary staff. The experience of the authors is that on-site visits that occur less frequently than once a month diminish the effectiveness of the AV if animals are housed continuously at the facility. As the scope and complexity of the facility increase, so should the frequency of on-site visits by the consultant. Visits by the consulting veterinarian should be regularly scheduled and supplemented with additional visits as needed.

   The availability of consultants to respond to emergency situations may be limited by time or distance constraints. Nonetheless, emergency veterinary care must be provided (AWAR §2.33,b,2; *Guide*, p. 46). Trained veterinary or animal care technicians can provide such care under the direction of the veterinarian, through instructions communicated by such means as telephone, fax, e-mail, or prior written instructions and guidelines. In contrast, some facilities make prior arrangements with a local practicing veterinarian to provide occasional emergency care on-site. Any veterinarian employed in this capacity should be provided with initial training regarding the species of animals and the type of research at the facility. Both the veterinarian and the facility should be comfortable with each other's expectations before entering this relationship.

27:6   What are the AV's responsibilities?

Reg.   The principal responsibility of the AV is to develop and implement an effective program of veterinary care (see 5:28; 7:1; 27:26). Such a program has multiple components, including preventive medicine; proper use of anesthetics, analgesics, and tranquilizers; perioperative care; euthanasia; diagnostic procedures; and provision of veterinary care, emergency care, and aspects of animal husbandry[1] (AWAR §2.33,b,1–§2.33,b,5; *Guide*, p. 56). These responsibilities and associated aspects are summarized in Table 27.2.

Opin.   Although the AV has a principal responsibility to assure the well-being of the animals being used at the institution, he or she also has responsibilities to the institution and to the investigators. Ultimately, the AV should facilitate the ability of the investigator to perform animal research in a humane manner and consistent with regulatory principles. Typically, the IACUC and the AV have a close working relationship. The IACUC relies on the professional judgment of the AV in animal health issues, and the AV turns to the IACUC for support if resistance to proper animal care and use is encountered.

TABLE 27.2

Responsibilities of the Attending Veterinarian

Type of Responsibility	Specific Aspects of Responsibility
Diagnostic procedures	Direct or perform procedures to identify disease etiologies
Provision of medical care	Prescribe treatments to treat ill animals
Preventive medicine	Develop programs for quarantine and isolation, monitoring of vendors, monitoring of colony health, and routine vaccinations and parasite control
Use of anesthetics, analgesics, sedatives, and tranquilizers	Advise investigators on and assure proper use of methods to relieve or reduce pain and distress
Perioperative care	Oversee presurgical preparation, surgical procedures, and postsurgical care of animals
Euthanasia	Assure that animals are euthanized when appropriate and in a humane manner
Animal husbandry	Assure that programs for disinfection, housing, nutrition, breeding, and environmental enrichment are appropriate
Use of hazards	Work with hazards oversight professionals to assure safe use of biologic, radiologic, and chemical hazards
Use of animals	Advise investigators and the IACUC on the appropriateness of specific techniques and methods in animals, and the availability of alternative animal and nonanimal models
Occupational health	Advise IACUC and health professionals on aspects of occupational health program

27:7   What is an appropriate chain of command for decision making with respect to veterinary care, and what is an appropriate chain of communication with respect to animal health issues?

Reg.   The AWAR (§2.33,a,2) state with respect to the AV and adequate veterinary care that the facility should have an AV who provides adequate veterinary care and that the AV have appropriate authority to ensure the provision of adequate veterinary care and to oversee the adequacy of other aspects of animal care and use. Further, both this section of the AWAR (§2.33,b,3) and the *Guide* (p. 56) state that a mechanism of direct and frequent communication is required to provide timely and accurate information concerning animal health to the AV.

Opin.   Clearly, the AWAR indicate that the AV has the appropriate authority to ensure the provision of adequate veterinary care. Most facilities interpret this to mean that in an emergency or life-threatening situation, or one in which an animal is clearly suffering and study personnel cannot be contacted, the designated veterinary staff has the authority to initiate treatment that may or may not include euthanasia (also see 27:23–27:24). It should be stressed, however, that every effort should be made to contact study personnel prior to initiating treatment or euthanasia. There may be a simple explanation for clinical signs being observed–such as a lethargic animal's recovery from anesthesia. Further, communication with the study personnel might be critical if vital blood or tissue samples need to be collected from the animal.

     A good system of communication between the facilities' veterinary staff and research personnel is critical. The veterinarian needs to be aware of study details to diagnose and develop a treatment plan effectively. The study director must be aware of veterinary care diagnostics and treatment plans in order to consider potential study interactions.

     The ideal situation would develop as follows: problem identified, veterinary observation, study director contact and discussion, diagnostic and treatment plan discussed, mutually acceptable concurrence, implementation and completion of plan, detailed notes to the medical records. This works well for most nonemergency situations, but when something must be done immediately and either the veterinarian or the research personnel cannot be reached it is best to have a policy to delegate authority and responsibilities.

Occasionally Good Laboratory Practice (GLP) regulations that state that a study director has overall responsibility for the technical conduct of the study and therefore veterinary involvement is prohibited are cited. This misconception has been addressed by authorities from the USDA and FDA on numerous occasions. GLP regulations in no way restrict or prohibit veterinarians from fulfilling their responsibilities as defined in the AWAR. It is important to communicate and work with the user groups to develop a mutually acceptable pathway for providing animal health services.

27:8 What staff, other than veterinarians, are useful in providing adequate veterinary care? What level of support staff is considered adequate?

Opin. Ancillary personnel such as veterinary technicians, research technicians, and trained animal care technicians can provide valuable support to the program of veterinary care (see 27:4). Such individuals often interact closely with the animals on a daily basis and can serve as a primary point for observation of animal health.

The size of the staff needed to provide adequate veterinary care is completely dependent upon the size and complexity of the animal facility. Any animal housed in a research facility must be checked at a minimum of once per day, every day (AWAR §2.33,b,3; *Guide*, p. 46). This need can increase with the invasiveness or potential for untoward outcomes of the study. Unreported or untreated illness, insufficient feed or water, or signs of insufficient husbandry can all indicate inadequate observation of (or response to) the animals and may signal a need for additional staff.

27:9 What type of training should be provided to animal care and research staff with respect to veterinary care issues?

Opin. A variety of personnel, as described in 27:8, might potentially examine animals exhibiting signs of illness. It is therefore critical for the IACUC to assure that these individuals are familiar with both normal and abnormal appearances and behaviors of animals (Table 27.3). Training provided by the veterinary staff should ensure that personnel are able to recognize that an animal may require veterinary attention. While useful, it is not critical that ancillary personnel be able to diagnose illness;

TABLE 27.3

Some Indications of Illness in Animals

Body System	Abnormal Clinical Signs
Respiratory system	Nasal discharge, labored breathing, blue tint to mucosal membranes
Gastrointestinal system	Vomiting, diarrhea, thin appearance
Integumentary system	Bleeding or seeping wounds, alopecia, observed persistent scratching, closed wounds
Nervous system	Inability to move normally, overly aggressive behavior, lethargy
Circulatory system	Edema in extremities, ascites, labored breathing

instead, they should understand the procedures for notifying the veterinary staff if an animal is even suspected of being ill. Though a clear mechanism to report sick animals to the AV must exist, the rigor of reporting may vary with the nature of the specific study; thus, a dermal abrasion in an otherwise normal animal should be called to the attention of the veterinary staff but does not likely constitute an emergency. For certain animal models, it is important that personnel also learn to recognize medical nuances specific to that model. For example, in an animal model of hemophilia it would be important for the staff to understand the urgent nature of even a seemingly minor wound. Similarly, in an animal model of Parkinson's disease, it would be important to instruct personnel that tremor of limbs is an expected outcome.

Some staff may also benefit from training related to provision of veterinary care. In this regard, it may be useful to train personnel with respect to administration of drugs by various routes, bandaging of surgical wounds, or other common veterinary care procedures. The IACUC should assure that proficiency has been documented for ancillary personnel expected to provide veterinary care.

27:10   To what extent and under what circumstances can the AV delegate clinical responsibilities to other veterinarians or nonveterinarians?

Opin.   The AV may delegate the responsibility for observation of animals and "hands-on" care to other veterinarians or technicians provided that it can be documented that such individuals are adequately trained in the care of the species and in the procedures involved. All animals should be observed every day by someone trained to evaluate the health and well-being of that species (AWAR §2.33,b,3; *Guide*, p. 46; also see 27:1). The ultimate responsibility for assuring proper clinical care remains with the AV, and that person should establish a mechanism for receiving animal health information that will ensure fulfillment of this responsibility (AWAR §2.33,b,3; *Guide*, p. 56).

27:11   Are PIs responsible for any aspect of adequate veterinary care?

Opin.   Investigators play an indirect but significant role in veterinary care. It is critical to recognize that veterinary care is not exclusively the treatment of sick animals. It includes the proper care and use of animals in such a way that animals remain healthy and recover normally from experimental manipulations. Beginning with protocol initiation, an appropriate animal model must be chosen to ensure a project is successful. Potential adverse reactions or outcomes should be identified by the investigator and plans made to address them. Investigators and their technicians may observe animals frequently during the procedure and are, therefore, in an excellent position to report potential problems to the veterinary staff. For example, weight loss in a cage of group-housed animals may not be apparent to a caretaker whereas review of the investigator's study notes with individual body weights of animals could detect the problem.

Depending on the staffing of the facility, PIs or their technicians may be responsible for administering treatments prescribed by the veterinarian. As part of this, they should be instructed on procedures to contact the veterinary staff when animals are found in poor clinical condition.

For these reasons, it is critical to view the role of the PI and his or her staff, along with the AV, as providers or facilitators of veterinary care. Although the AV is ultimately responsible for provision of proper veterinary care, the IACUC should recognize that there are others who have an ethical, if not a regulatory, responsibility to impact the health and welfare of research animals positively.

27:12   What aspects of a veterinary care program can and should be addressed under SOPs?

Reg. Federal regulations (AWAR §2.33,a,1) require that each research facility maintain the services of an AV. When the AV is employed part-time or under consultation arrangements a written program of veterinary care must be maintained at the research facility. The PHS Policy (IV,A,1) requires that the Assurance statement fully describe the institution's program for animal care and use, using the *Guide* as a basis for the development of that program. Though the *Guide* is silent on the need for a written program of veterinary care, one can assume that a full description of the institution's program would include a description of the program of veterinary care.

Opin. It is worthwhile for institutions to develop a written program of veterinary care; such a document should serve as the basis for the facility animal care program and can be reviewed by the IACUC during the semiannual program review. While the contents of this document should be customized for the facility, it should address areas such as daily assessment of animal health; prevention, control, diagnosis, and treatment of animal disease and injury; guidance to researchers on proper handling, restraint, anesthesia, analgesia, and euthanasia methods; training of research personnel in appropriate surgical techniques and procedures; emergency preparedness; and monitoring of surgical procedures and postsurgical care. Additionally a facility may utilize or augment their program of veterinary care with documents typically referred to as *standard operating procedures* (SOPs). In general terms, the function of an SOP is to provide written guidance in sufficient detail so that a person with the appropriate education, training, and experience can perform the procedure as described in the SOP. Research facilities that wish to conduct animal studies under good laboratory practices (GLP) conditions are required by the Food and Drug Administration to maintain SOPs (21 CFR Part 58). Specifically, Sect. 58.81 requires the following:

(a) A testing facility shall have standard operating procedures in writing setting forth nonclinical laboratory study methods that management is satisfied are adequate to ensure the quality and integrity of the data generated in the course of a study.

(b) Standard operating procedures shall be established for, but not limited to, the following:

1. Animal room preparation
2. Animal care
3. Receipt, identification, storage, handling, mixing, and method of sampling of the test and control articles
4. Test system observations
5. Laboratory tests
6. Handling of animals found moribund or dead during study
7. Necropsy of animals or postmortem examination of animals
8. Collection and identification of specimens

9. Histopathology
10. Data handling, storage, and retrieval
11. Maintenance and calibration of equipment
12. Transfer, proper placement, and identification of animals
13. Emergency preparedness

Non-GLP facilities often operate under SOPs as well. Alternatively, and sometime additionally, both GLP and non-GLP facilities may develop guidelines or policies. These are documents that help define the conduct of procedures or situations that may have been a source of controversy or earlier deliberation for which the IACUC has reached a consensus. For example, the IACUC might approve a policy to allow with certain restrictions the controversial practice of chronic housing of rodents in wire bottom cages.

27:13 What equipment and pharmaceuticals might be considered essential items for an adequate program of veterinary care?

Reg. The AWAR (§2.33,b,1–§2.33,b,2) state that, among other considerations, appropriate facilities and equipment must be available to prevent, control, diagnose, and treat diseases and injuries. The *Guide* (p. 56) describes components of an effective program of veterinary care to include preventive medicine; surveillance, diagnosis, treatment, and control of disease; management of protocol-associated disease or disability; anesthesia and analgesia; surgery and postsurgical care; assessment of well-being; and euthanasia. Thus, these documents do not specify the essential items required for adequate veterinary care and allow the AV and the institution to decide which items are essential to achieve the programmatic objectives described.

Opin. The specific items needed to provide adequate veterinary care will vary widely with the species housed within the facility and the type of research being conducted. For example, a program in which specific pathogen-free rodents are used and euthanized for collection of tissues within several days of arrival will have a much different need for specialized equipment and supplies than a program in which nonhuman primates undergo survival surgery and are maintained for several years afterward. In general, the facility should rely upon the advice of the AV in deciding which equipment and supplies are necessary to the needs of the animals and the research in a way that meets the objectives stated in the AWAR and the *Guide*. Most veterinary care programs find that routine supplies include items such as topical or injectable antibiotics, bandage materials, materials for euthanasia and necropsy, material and equipment for surgical repair of wounds in an aseptic manner, supplies for microscopic examination of samples, sterile fluids, antiseptics, syringes and needles, and means to provide supplemental warmth, such as a circulating water heating pad. These items are only examples of the kinds of supplies that would be useful to address the typical clinical needs of animals in a research facility. The list is by no means complete and each veterinary care program should have on hand supplies to meet its specific needs. For example, radiographic and other imaging equipment is essential to some programs, but not for others. Though availability of needed supplies is essential and suggests that the veterinary care program need not necessarily have all equipment and supplies on hand but rather only have ready access to such resources, adequate veterinary care is facilitated to a far greater extent when the veterinary care program has direct ownership of these resources.

27:14   Are there any circumstances under which veterinary care can be provided at a facility not affiliated with the institution, such as a private veterinary clinic? If so, what sort of criteria might the IACUC use to determine the acceptability of such sites and what sort of oversight should the IACUC provide?

Reg.   The *Guide* (p. 60) states that methods of disease prevention, diagnosis, and therapy should be those currently accepted in veterinary practice. The AWAR (§2.33,b,2; §2.33,b,3) state that a program of veterinary care includes the availability of appropriate facilities, personnel, and equipment to prevent, control, diagnose, and treat diseases and injuries. Though no specific mention is made of utilizing off-site resources, these passages imply that the importance to animal health offered by certain equipment, methods, and expertise is substantial. One might infer, then, that if the necessary resources cannot be practically located at the institution, the primacy of animal health would lend merit to the use of off-site resources.

Opin.   At some institutions, particularly small ones, the expertise or equipment needed to make accurate diagnoses or to provide proper treatment may not exist. For example, a small institution that maintains rabbits producing a valuable polyclonal antibody may not have radiographic equipment or orthopedic expertise to diagnose and treat a rabbit suspected of fracturing its leg. Though euthanasia is an option in such cases, that approach may not represent judicious use of animals if the condition is one that could be reasonably treated so that the animal regains health as well as research usefulness. In such cases, diagnostic and treatment options at other facilities, such as nearby academic institutions, zoos, or private veterinary clinics, are an acceptable means for providing veterinary care if the animal can be easily and safely transported to such a site. Further, it could be useful for veterinary professionals from such sites to visit the location of the ill animal if needed equipment is mobile.

The IACUC should focus on several oversight aspects when it is determined that the medical needs of an animal are best served by transport to an off-site location. First, it should be clear that there is benefit to the animal in instances when off-site locations may be utilized. For example, the need for specialized imaging equipment needed for the diagnostic effort of a nonhuman primate may be evident, but if the animal is moribund and and has a poor prognosis, there may be limited advantage to such an effort. Second, the IACUC should be certain that transport of the animals does not place personnel at risk. Transport of animals infected with a biohazardous agent, for example, would be unwise. Third, the specific methods for transport, the location and business entity of the off-site facilities, and the overall infrastructure, including expertise, being made available should be understood by the IACUC. Though specific instances of need for use of off-site facilities cannot always be anticipated, the IACUC should expect the veterinary staff to generate a written plan if the possibility exists for use of such facilities. While it may not be necessary for the IACUC to conduct an inspection of seldom used off-site facilities, it is advisable, and an inspection is certainly warranted, if use of such sites is likely to occur. If off-site facilities are to be used as part of a contingency plan for veterinary care, the IACUC could, instead, evaluate such sites by written description from the AV after his or her visit to the site.

Surv.   If your animal facility uses the clinical facilities of an off-site private veterinary clinic, does your IACUC oversee this and what criteria are used to evaluate the acceptability of this practice? More than one response possible.

- Not applicable; we do not use off-site private practices   140/153
- We use off-site practices on occasion, but our IACUC has
  not evaluated the practice   10/153

- We use off-site practices and the IACUC has inspected them
  and approved their general veterinary care and facilities   4/153
- We use off-site practices and the IACUC expects them to
  abide by the general provisions of the *Guide*   9/153
- We use off-site practices and the IACUC requires them
  to be accredited by the American Animal Hospital
  Association (AAHA)   4/153
- Other (one response each):
  - Used primarily for animals involved in field study emergencies
  - All animals are rodents; on site-care only
  - We have purchased blood from an emergency clinic for cardiac surgery
    requiring CPB [cardio-pulmonary bypass]
  - Our vet is off-site and on call
  - Off-site facilities used for emergencies with primates only

27:15   Is it acceptable for private pets to receive veterinary care at the veterinary facility of a research institution? Is the IACUC responsible for oversight of such activity?

Opin. There is nothing in the AWAR, PHS Policy, or the *Guide* that specifically prohibits an animal facility from providing care to private pets. Because these animals are not research animals we can only assume that IACUC involvement should center on the program of adequate veterinary care under which the facility would be operating. Even though pet animals may not be covered under regulatory requirements set forth by APHIS/AC, there are a number of other issues that must be addressed, including the following:

- The appropriate state and federal licenses of those conducting the procedures
- The liability to the individual and institution in case of problems
- The potential for cross-contamination of the in-house research animals
- A fair and equitable program describing whose pet animals will be treated
- The time required for the veterinary staff to provide such care and the possibility that other duties of the veterinary staff might not be completed as a result

The IACUC may be asked to monitor issues such as these in order to provide assurances and safeguards for the research facility.

Surv. Does the veterinary care unit at your institution ever provide care to private pets of employees or others?

- We do not provide care to private pets   144/153
- We do provide care and the IACUC oversees this activity   3/153
- We do provide care and the IACUC does not oversee
  this activity   3/153
- Other   3/153

27:16   Which animal health and veterinary care records should be kept? How detailed should these records be? (See also 18:31; 18;32; 26:10–26:15; 27:21; 27:29.)

Reg. APHIS/AC Policy 3[2] specifies that records must be legible and include the following:

1. Identity of the animal

2. Descriptions of any illness, injury, distress, and/or behavioral abnormalities, and the resolution of any noted problem

3. Dates, details, and results (if appropriate) of all medically related observations, examinations, tests, and other such procedures

4. Dates and other details of all treatments, including the name, dose, route, frequency, and duration of treatment with drugs or other medications

5. Treatment plans that include a diagnosis and prognosis, when appropriate; they must also detail the type, frequency, and duration of any treatment and the criteria and/or schedule for revaluation by the AV; in addition, it must include the AV's recommendation concerning activity level or restriction of the animal

Opin.   The purpose of any record keeping system is to document the occurrence of a procedure, result, or event. All records must be detailed enough to explain adequately to a person knowledgeable in the subject what occurred to the animal. Records associated with adequate veterinary care serve two main purposes. They assist in facility management and document an adequate veterinary care program to regulatory officials. All records should be maintained as long as they may be useful to the facility. It should be noted that veterinary medical records only document animal health; many research animals can be maintained in a state of good health without maintenance of a medical record.[3]

Some regulatory documents have sections that pertain to the retention of records. The majority of such requirements relate to the product submission and approval process. In many instances, particularly for records related to studies conducted under Good Laboratory Practice (GLP) standards, permanent archiving of records is done (21 CFR Part 58.195). In the case of the Federal Insecticide, Fungicide, and Rodenticide Act (FIFRA), records need to be retained for the life of the product (40 CFR Part 160.195). It stands to reason that these retained records should include those related to animal health. By law, certain records, such as acquisition records of dogs and cats, are required to be retained for a minimum of 3 years after completion of the study (AWAR §2.35,f).

Records having import with respect to veterinary care include the following:

Standard procedures and animal health: Activities such as observation of animals, provision of feed and water, exercise, provision of environmental enrichment, and evaluation of environmental parameters are usually checked daily and should be documented. Many facilities have found it useful to develop an SOP detailing what is to be included in routine animal husbandry. A daily check-off form can then be used to document that a procedure has occurred. Many facilities employ a monthly room care sheet to document activities occurring less frequently such as cage changing or disinfection.

Veterinary care records (see 27:1; 27:21; 27:22): Any animal exhibiting abnormal behavior or signs suggestive of illness should be reported to the veterinary staff. Written documentation should describe the veterinarian's findings, diagnostic procedures, treatments, diagnosis and prognosis, follow-up care, and progress of the animal. It is also important for the records to note the provision of preventive medical care to the animal;[3] this would include items such as physical examination, vaccination, dosing with anthelmintics, hoof or nail trimming, grooming, and other procedures typical of health maintenance for the species. The records should also indicate the study protocol number and include notation for experimental procedures performed on the animal. Euthanasia and the euthanasia method should be noted and the results of any necropsy included.

Colony animals (see 27:21; 27:26): Maintenance of long-term animal colonies requires an additional level of record keeping. For example, production records in a breeding colony may offer clues on disease transmission or genetic abnormalities. It is typical to maintain individual animal records on dogs, cats, nonhuman primates, and occasionally farm animals. These individual records serve as a reference for growth, development, health-related observations, and preventive medical care.

Surgery and postsurgical and postprocedural care: Records should detail preoperative preparation of the animal; use of anesthetics, analgesics, and other medications; and monitoring of the animal during surgery and during recovery. Records should be sufficiently detailed and organized for the IACUC to determine readily that procedures are being conducted in a manner that minimizes the risk of pain or distress to the animals. In addition, it should be clear that the veterinary staff has responded expeditiously to any significant problems.

Surv.   What are the constituent parts of a complete animal health record at your institution? Check all that apply.

• We do not have any animal health records	14/153
• This is not an IACUC concern at our institution	15/153
• We have no policy or the IACUC is unaware of any policy	17/153
• Medical treatments	90/153
• Veterinarians' medical and surgical orders	83/153
• Clinical observations	93/153
• Surgical records	81/153
• Anesthesia/analgesia records	83/153
• Exercise times	26/153
• Enrichment plan/activities	38/153
• Grooming of animals	19/153
• Other:	
• Sentinel health screens only	1/153
• Health records from vendor	2/153
• Daily monitoring of condition	1/153
• Health checks, weights, nail clipping	1/153
• No records maintained for rodents	2/153
• Diagnostic reports and morbidity reports	1/153
• Euthanasia records	1/153
• Content left up to PI	1/153

27:17   Are veterinary medical records best maintained in hard copy or an electronic format?

Reg.   APHIS/AC Policy 3[2] requires that information contained in the veterinary record must be readily available, but it can be maintained in whatever format is convenient for the licensee or registrant. Similarly, the *Guide* (p. 46) states that clinical history information should be readily available to investigators, veterinary staff, and animal care staff. The *Guide* does not prescribe a specific format for such records but recognizes that information can be appropriately recorded in formats ranging from notation on cage or identification cards to detailed computerized records.

Opin. Several factors influence the format in which veterinary medical records are maintained.

For large, complex animal care programs with a broad scope involving a multitude of species, invasive procedures, chronic studies, and procedures likely to result in substantial pain or distress to the animals, an electronic record system would have clear advantage in that such a format can allow rapid and efficient data synthesis and access. Such systems can be used, for example, to determine quickly how many animals undergoing a certain procedure have had postoperative infections within a given time frame. For programs of smaller scope, hard copy records are usually sufficient as long as the information that has been entered can be easily accessed and allows retrospective evaluation of veterinary care. Hard copy records can be maintained effectively as detailed written records in individual files for each animal, as written records in laboratory notebooks, or even as notations made on an identification or cage card. The specific degree to which hard copy records are maintained will depend upon the scope of the program and the typical severity of clinical problems that might be expected.

Surv. In what format does your veterinary care unit maintain medical records?

- No records maintained   23/153
- Paper format   76/153
- Electronic format   1/153
- Combination of paper and electronic   52/153
- Other (have no animals)   1/153

27:18 Are cage cards an acceptable location for recording veterinary medical information?

Reg. There is no specific regulatory passage in either the AWAR or the PHS Policy that specifically forbids notation of veterinary medical information on cage cards. The *Guide* (p. 46) states only that "identification cards should include the source of the animal, the strain or stock, names and locations of the responsible investigators, pertinent dates, and protocol number, where applicable."

Opin. Though use of cage cards for recording of veterinary medical information is not specifically mentioned in the AWAR, PHS Policy, or the *Guide*, the amount and scope of information that can be noted on a cage card are substantially limited by the size of the card. As described in 27:16, the expectations of APHIS/AC in terms of information included in a medical record is broad enough to rule out use of a cage card generally for animals under AWAR purview and maintained for extended periods; still, notes made on a cage card in those cases might serve as a useful summary of relevant clinical events that are recorded in greater detail in a more formal clinical record. In contrast, for animals such as rodents or others that will not be maintained for long periods and are generally free of illness, notations made directly on the cage card may be an efficient way to record veterinary medical information; with large colonies of such animals, it can be efficient to have the limited clinical information available with the animal.

27:19 Who should be allowed access to veterinary care records and where should the records be maintained?

Reg.   APHIS/AC Policy 3[2] states that animal health records must be "readily available," and the presumption is that this policy refers to availability of the records to representatives of APHIS/AC acting under the auspices of the AWAR.

Opin.   Veterinary care records exist to guide the clinical care of an individual or group of animals. For this reason, such records should be readily accessible to members of the veterinary staff, including veterinary technicians. When animal health records are maintained by the PI, in a research notebook, for example, some mechanism should be established so that the veterinary staff can easily access the clinical information. It is the difficulty with providing for this sort of seamless flow of information that argues against maintaining health records in research notebooks or research labs. For obvious reasons, it is critical that both the IACUC and representatives from regulatory or accrediting agencies be allowed easy and complete access to veterinary care records; without such access it is impossible to determine accurately whether veterinary care is being provided at an acceptable level. Further, it is important for animal care and research personnel to be provided access to veterinary care records relevant to animals under their charge. Such personnel often are the initial point at which ill animals are identified, and these same personnel typically both are curious about and have some degree of responsibility to assure that sick or injured animals are receiving appropriate care and veterinary attention. For all of these reasons, it is logical for most veterinary records to be physically maintained within the veterinary services unit.

Surv.   At your institution, who has access to animal health records? Check all appropriate boxes.

• We do not have any animal health records	14/153
• The attending veterinarian	132/153
• The principal investigator	112/153
• The IACUC (committee members)	104/153
• IACUC administrators	64/153
• Veterinary technicians	79/153
• Animal care technicians	87/153
• The Institutional Official	84/153
• Other institutional administrators	30/153
• Other:	6/153
• All have access through the attending veterinarian's office	1/153
• IACUC has access if animals are in the central facility	1/153
• Facility managers	1/153
• Everyone—available on public Web site	1/153
• Anyone with a need to know may gain access	1/153
• Anyone who asks (publicly funded institution)	1/153

27:20 For how long should veterinary records be maintained?

Reg.   APHIS/AC Policy 3[2] states that health records must be held for a minimum of 1 year after disposition or death of an animal. In contrast, other records for individual animals are required to be maintained for a minimum of 3 years after completion of the activity (§2.35,f). The *Guide* does not suggest a specific period for which records should be maintained. (See 16:31.)

Opin. The purpose of veterinary records is to document the health of animals on a specific study and of animals under the purview of the animal care program in general. In the former instance, it stands to reason that veterinary records should be maintained for a period that will allow the scientific community to review the results of the research. In the case of records for animals being used to generate data to meet the needs of a regulatory directive (e.g., Food and Drug Administration), the records should be permanently archived. Beyond the regulatory aspects, an archiving system for veterinary records is useful in that it allows retrospective review and analysis of animal health trends. Such information can help guide and direct veterinary approaches to health care for the broader population of animals within the program. From this latter perspective, it may not be necessary to archive the entire veterinary record for every animal, but a database of clinical information distilled from the records could still prove very useful.

Surv. For what length of time does your institution maintain animal health records? More than one response is possible.

• We have no policy or the IACUC is unaware of any policy	46/153
• We do not have any animal health records	9/153
• This is not an IACUC concern at our institution	11/153
• One year after the last entry into the record	2/153
• One year after the study is completed	3/153
• Three years after the last entry into the record	47/153
• It varies by species	10/153
• Indefinitely	8/153
• Other:	
• Six years beyond disposal of animal (Washington state law)	1/153
• Current year plus previous year	1/153
• Five years	3/153
• Varies by contract with sponsor	1/153
• Ten years after completion of study	1/153

27:21 Should every animal in the facility have a health record or should every colony have one? (See also 27:10; 27:11.)

Reg. APHIS/AC Policy 3[2] essentially states that health records sufficiently comprehensive to demonstrate the delivery of adequate health care are required for species regulated under the AWAR. Though the policy stops short of requiring entirely separate records for each animal, it is clearly expected that notations for individual treatment of an animal must be made as entries specific to that animal.

Opin. In general, all observed illness or abnormalities should be entered into a clinical record; however, an alternate view holds that many animals, such as rodents, can be successfully maintained without formal medical records.[3] Records are frequently more detailed for some species than for others. For example, individual health records for dogs, cats, nonhuman primates, and farm animals are useful (*Guide*, p. 46).

A useful rule of thumb is that individual records should be maintained for those animals that receive routine periodic individual health evaluations. For example, individual records should be maintained for dogs, detailing vaccinations, parasite evaluations, any treatments administered and periodic physical examinations. Separate records are not needed for animals that may have periodic evaluation by examination of several representative individuals from the colony. Rather, the health information of such animals can be recorded as a colony health record. An alternative rule of thumb is that individual records should be maintained for long-lived animals (*Guide*, p. 47). In either case, the interpretation is that individual records are useful for species such as dogs, cats, nonhuman primates, rabbits, and farm animals, and of less value for rodents or common nonmammalian species. Alternatively, individual surgical records can be very useful for animals that have had invasive procedures, regardless of species.

Surv. For which species are individual health records maintained at your institution? Check all appropriate responses.

- We do not maintain individual animal health records   24/153
- This is not an IACUC concern at our institution   15/153
- All species covered by the AWAR   40/153
- All species covered by the AWAR, except rodents   26/153
- All mammalian species that we house   26/153
- All vertebrate species, mammals or nonmammals   19/153
- Common laboratory rats and mice   13/153
- Other:
  - Primates, dogs, and swine   1/153
  - All vertebrates on clinical or surgical rounds   1/153
  - Any animal with a clinical problem   1/153
  - Left up to PIs   2/153
  - Animals on water and food restriction, behavioral studies   1/153
  - Cotton rats   1/153
  - For mice and rats if ill or if surgical procedure has been done   1/153
  - All AWAR-regulated species except rabbits, guinea pigs, and hamsters   2/153

27:22 How sick or abnormal should an animal be to merit veterinary attention?

Opin. Every animal that is found to be ill, is behaving abnormally, or is suspected of being ill should be reported to the AV or to the veterinary staff as directed by the AV. In some cases, the veterinary staff may determine that no specific treatment is merited, while in other cases treatment may be needed. It is the veterinary staff, under the direction of the AV, that should make a judgment regarding the severity of illness and subsequent course of action.

27:23   Are there any circumstances under which an investigator may rightfully refuse veterinary care for his or her research animals?

Reg.   The AWAR state that the AV has appropriate authority to ensure adequate veterinary care (§2.33,a,2). Similarly, the *Guide* (p. 12) states that adequate veterinary care must be provided to animals and that the veterinary care program is the responsibility of the AV (*Guide*, p. 56). Because the PHS Policy (IV,A,1) requires adherence to the AWAR and principles in the *Guide*, clearly the expectations with respect to veterinary care are the same.

Opin.   It is unequivocal that the AV has been granted authority by APHIS/AC and PHS to make decisions regarding the veterinary care provided to animals. The *Guide* (p. 60) indicates that treatment should be done in consultation with the investigator and, when possible, in such a way that no undesirable experimental variable results. Essentially, then, the AV has the final say in matters related to veterinary care, but it is preferable that agreement between the veterinarian and the investigator be reached. This principle must be understood not to endow the AV with authority to make veterinary decisions that contradict veterinary parameters outlined in an approved protocol. For example, if the protocol has approved a procedure involving unrelieved pain and distress (support with scientific justification), the AV would not have the authority to administer treatments to mitigate that pain and distress. In instances in which the animal is in immediate peril, the AV may need to make clinical decisions without consulting the investigator.

It is conceivable that an investigator may choose to euthanize animals rather than pursue the option of treatment recommended by the veterinarian. Depending upon the clinical circumstances and the nature of the research, such a decision may be practical and reasonable. Similarly, it is within the authority of the veterinarian to euthanize an animal for clinical reasons

Surv.   Who is in charge of making decisions regarding the veterinary care provided to animals at your institution? More than one response is possible.

- The AV or designee for routine and emergency needs   77/153
- The AV or designee in consultation with investigator   80/153
- The investigator alone for routine and emergency needs   10/153
- The AV in emergencies when investigator cannot be reached   64/153
- Other:
  - Use of a form with instructions, personnel allowed to make decisions on veterinary care, phone numbers, and so on   1/153

27:24   If an animal is clearly suffering, can the IACUC allow a veterinarian to euthanize the animal if the research laboratory personnel cannot be contacted?

Reg.   The AWAR state that the AV has authority to ensure adequate veterinary care (§2.33,a,2). The *Guide* states that adequate veterinary care must be provided to animals (p. 12) and that the veterinary care program is the responsibility of the AV (p. 56). Because the PHS Policy (IV,A,1) requires adherence to the AWAR for AWAR-regulated species and the principles in the *Guide*, it is clear that the expectations with respect to veterinary care are the same.

Opin. *Veterinary care* is defined in the AWAR (§2.33,b,4) as including provision of guidance by the AV to PIs regarding euthanasia. Though this could be taken to mean that the AV only provides recommendations to the investigator, it is more commonly interpreted to mean that the AV provides guidance only in terms of methods for euthanasia rather than the decision to euthanize a suffering animal. This interpretation is supported by an earlier section in the AWAR (§2.31,d,v) that states that animals experiencing severe or chronic pain that cannot be relieved will be euthanized. Likewise, the *Guide* (p. 56) considers euthanasia to be a component of the veterinary care program for which the AV has responsibility; since the *Guide* (p. 12) also states that adequate veterinary care must be provided to animals, it is apparent that the intent of this document is to assign to the AV authority to euthanize animals that are suffering and for which the IACUC has not approved procedures that involve suffering or death as an end point.

Surv. If an animal is clearly suffering, does your institution allow a veterinarian to euthanize the animal if the research laboratory personnel cannot be contacted?

- We have no policy on this matter   16/153
- Yes, under all circumstances   121/153
- Yes, if prior understandings have been made with
  the investigator   13/153
- Yes, but only with IACUC approval (Chair, subcommittee,
  and so on)   1/153
- No   0/153
- Other:
  - Yes, unless death as an end point has been approved   1/153
  - Facility manager would euthanize   1/153

27:25 How often should notations be made in the animal health care records for an animal reported to be ill?

Opin. Clinically ill or abnormal animals should be followed until resolution, either through return to clinical normalcy or euthanasia or determination that the relevant condition is clinically minor with little likelihood of increased severity. In this regard, records should document that follow-up care and evaluation have occurred.

The exact frequency at which follow-up evaluations should be made varies considerably with the clinical condition and depends upon the sound professional judgment of the veterinarian. Severe conditions might necessitate evaluation several times daily, while other conditions might be appropriately evaluated once every several weeks. In general, follow-up evaluations should occur and be documented in the records at a frequency such that significant increases in the severity of the condition could reasonably be expected to be detected. For severe conditions, this usually involves at least daily evaluation with more frequent evaluations if necessary. For less threatening conditions, weekly or biweekly evaluations are often sufficient.

Repeated instances of ill animals' progressing to severe disease states in the absence of corresponding notations in the record may indicate that animals are not being examined frequently enough. Conversely, if notations exist for such animals, they could suggest that the decline in clinical condition is being duly noted but ignored by the veterinary staff. In either circumstance, the possibility that records are being maintained or utilized inappropriately exists, and the IACUC must further investigate the situation.

27:26   How can the IACUC evaluate programs for preventive medicine and health monitoring? What type of programs should be expected?

Opin.   Preventive medicine and health monitoring programs are designed to assure that healthy animals are acquired initially and that these animals remain healthy. To accomplish this, regularly scheduled evaluations of animals and vendors should be performed. The IACUC should inspect the veterinary care records (see 27:29) for accordance with a written description of the preventive medicine program, as well as visually examine animals for general appearance as an indicator of health status. Further, the IACUC should ascertain that animal health evaluations indicate the animals remain healthy and a timely and appropriate response is made when evaluations indicate a problem (see 27:1).

The program of preventive medicine should begin with evaluation of the vendor or source of the animals. Regular reports of the health status of the vendor's animals should be reviewed by the AV or his or her designate to determine that healthy animals are being acquired by the institution. If this review determines that health problems exist at the vendor's facilities, then the AV should coordinate arrangements to identify alternate vendors or other means that ensure that healthy animals are being used.

At the time of arrival at the facility, animals should be inspected by the veterinary staff and then separated from other more established groups of animals (see 27:1) until, in the professional judgment of the veterinary staff, the newly arrived animals do not pose a threat to the established animals. For example, animals such as dogs, cats, and nonhuman primates often receive a physical examination upon arrival as part of the preventive medicine program and are then separated from established animals.

The program of preventive medicine should also include procedures such as appropriate vaccinations, parasite evaluations, and testing for tuberculosis for nonhuman primates. Livestock, such as sheep and cattle, should have routine deworming. Other procedures are often included depending on the facility and the species involved. These procedures should be repeated periodically, per veterinary medical standards. Periodic, often annual, physical examinations should be performed for animals such as cats, dogs, and nonhuman primates. Routine physical examinations on other animals such as rabbits, guinea pigs, and other small animals can be useful for animals maintained for extended periods. Often, such examinations are brief and may be conducted by trained personnel other than the AV.

Rodent colonies are often evaluated for microbial contamination (*Guide*, p. 60). In general, a small number of animals (often sentinel animals placed in the room specifically for the purpose of health evaluation) are sacrificed, blood is collected for serologic detection of microbial contaminants, tissues are evaluated histologically, and the gastrointestinal tract and pelt are evaluated for parasites. Recommendations for the exact procedures and number of animals evaluated in this way are presented in detail elsewhere.[4]

Examples of activities that can be associated with a preventive medicine program are shown in Table 27.4.

The exact procedures and scope of activities vary significantly with the species and type of animals and the type of research. Activities listed are examples only.

TABLE 27.4

Typical Components of a Preventive Medicine Program

Aspect of Preventive Medicine	Typical Activity
Animal vendor surveillance	Review of vendor health status reports, on-site visits to vendor when possible
Animal transportation	Evaluate method of transportation for minimization of animal stress and risk of disease
Processing of newly received animals	Inspection and examination of animals, initial vaccinations, medications, parasite evaluations
Periodic reevaluation (usually every 6–12 months)	Physical examination, vaccination, tuberculosis testing (nonhuman primates), dental prophylaxis
Rodent health surveillance	Serologic and histopathologic evaluation for microbial contaminants, pelt examination for ectoparasites, examination of gastrointestinal tract for endoparasites

27:27   Should the IACUC attempt to evaluate animal health? If so, how?

Opin. Yes, the IACUC should periodically evaluate the health of the animals, since the IACUC is responsible for evaluating the program of veterinary care (AWAR §2.31,c,1 and §2.31,d,1,C,vii; §2.33,a; PHS Policy IV,B,1 and IV,C,1,e; *Guide*, pp. 9, 12). This can be troublesome since the IACUC typically relies on the AV for advice with respect to animal health concerns, thereby creating a conceivable conflict of interest if the AV is asked to evaluate his or her own program of veterinary care.

The most basic way for the IACUC to evaluate animal health is to observe the animals during the semiannual inspections. Table 27.3 lists some clinical observations that might be evidence of an animal health problem.

The number of animals with indications of illness is a valuable measure of animal health the IACUC can use. Of course, animals found moribund or dead can also signal an animal health problem.

The IACUC should also examine animal health, anesthesia and analgesia, and surgical records for possible evidence of health-related problems (see 27:16). Would it be appropriate for the IACUC to check anesthesia and analgesia records at the same time, as part of adequate veterinary care? In this regard, it is useful to spot-check records detailing animal health over the preceding 6 months, including records of animal health monitoring such as routine fecal examinations for endoparasites or routine serologic monitoring of rodent colonies. It is worthwhile to query personnel involved with animal care to determine whether they are aware of any animal health problems.

If the preventive care aspect of the program of veterinary care (including careful vendor selection) is effective, the IACUC should find few problems. However, it is not unusual for the IACUC to identify occasional problems, since spontaneous disease and consequences of experimental procedures can sometimes lead to clinical illness even in well-managed animal facilities. The key for the IACUC is to evaluate the *adequacy of the mechanism for reporting problems and the response to any problems* (see 27:29). For example, efforts to diagnose a problem, medications administered, supportive care provided, advice to investigators on technique refinement, and efforts to isolate sick animals to prevent disease spread could all be considered evidence of an appropriate response to a clinical problem. The precise response will vary with the specific problem.

27:28   How can the IACUC determine that the veterinary response to health monitoring findings is reasonable and adequate?

Opin.   Having determined that the program for preventive medicine and health monitoring is properly designed and implemented, the IACUC must further ensure that results of these efforts are interpreted correctly and that appropriate action is taken when potential problems are identified. To accomplish this, the IACUC must evaluate the response of the veterinary staff, in particular the AV.

In general, the IACUC should expect evidence of a plan to respond expeditiously to identified perturbations in health monitoring and surveillance. The scientific objectives of the study, the potential consequences of infection with a particular pathogen, and the possible adverse effects of the infectious agent on other studies within the facility should all be considered when determining the appropriate response (*Guide*, p. 60). The precise response can vary and can include continued monitoring of animals, isolation of suspect animals, specific treatment of animals, or euthanasia of all affected or exposed animals. Again, the appropriateness of the response depends on the type and species of animals, the nature of the research, and the professional judgment of the AV and the veterinary staff.

Additionally, the IACUC should further examine health records to determine whether the response of the veterinary staff has been successful. If not, the IACUC should determine that the AV is developing other strategies to resolve the problem. The IACUC must be assured that problems are not being ignored, and that the veterinary staff does not abandon the care of an animal.

27:29   How can the IACUC assure that veterinary care records are adequate?

Opin.   Records for veterinary care should be evaluated for completeness and for evidence of an adequate response to reported problems (see 27:16; 27:21). For example, records should clearly identify the specific animals involved, the clinical history, the source of the animal, the investigator, the age and gender of the animal, the date of animal receipt, and the types and dates of experimental or therapeutic procedures that have been performed on the animals (*Guide*, p. 46).

Appropriate diagnostic efforts and prescribed treatments should be documented. In addition, the record should indicate each time the animal is medicated or otherwise treated. The clinical progress of the animal should be noted until the clinical problem has resolved or the animal has been euthanized. Evidence that an animal languished until it died or was finally euthanized without reasonable attempts to diagnose and treat the problem can be considered as indicative of inadequate veterinary care (see 27:27). In addition, proper records should reflect the substance of any discussions with investigators or other staff on ways to minimize problems in the future. Individuals examining animals or providing treatment or other actions should sign or initial their notations.

Records can be considered to be inadequate if the IACUC cannot readily understand the steps taken to identify, report, diagnose, discuss, treat, and continue to evaluate an animal until the humane resolution of the case.

27:30   For what type of studies does the IACUC require direct veterinary involvement (e.g., requiring a veterinarian to check Category E animals or new surgical procedures)? For these studies, what is an appropriate period for this oversight?

Reg.   The IACUC is charged with ensuring that "personnel conducting procedures on the species being maintained or studied will be appropriately qualified and trained in those procedures" (AWAR §2.31,d,1,viii; PHS Policy IV,C,1,f).

Opin.   A commonly used system, employed by many IACUCs, is to require direct veterinary involvement in evaluating pain and in training and documenting the surgical skills of the researchers. It is important to remember that this requirement for training as specified in the AWAR does not only apply to APHIS/AC annual report Category E protocols (i.e., those that involve significant pain and distress that are not relieved by anesthetics, analgesics, or euthanasia) and new surgical procedures, but to all procedures being conducted on animals. Because of expected pain, potential pain, and the skill required for surgical procedures, an IACUC may impose additional means of oversight for Category E protocols and those requiring new surgical procedures and in this way guarantee that those conducting the procedures are adequately trained and competent. Although direct veterinary involvement is effective, other methods of performing this task should not be overlooked. Experienced in-house investigators or external resources may be more familiar with the specific procedures and better suited for training or qualifying the competency of the person performing the technique. They may also be more attuned to the subtle signs of pain that may occur. Duration of this oversight should be directly related to the skill and competence of the individuals performing the procedures. Some leeway must be given during the training period to those learning the procedures, but as a rule, individuals learning techniques should have a documented experienced trainer nearby, preferably in the same room, who can provide guidance, suggestions, and assistance as necessary. The role of the veterinarian or his or her designee should conclude the process by assessing and documenting the appropriate skill levels and reporting this to the IACUC.

Surv.   Does your IACUC require direct veterinary involvement for certain types of research?

• We have no formal policy	41/153
• Yes (new surgical models, USDA Category E, complaints)	63/153
• Occasionally if IACUC believes it to be appropriate	41/153
• No	8/153

27:31 What is an appropriate organizational relationship among the IACUC office, the veterinary care staff, and the IO?

Reg.   The AWAR (§2.31,a) state that the CEO of the research facility is the official who appoints the IACUC, qualified through the experience and expertise of its members to assess the research facility's animal program, facilities, and procedures. The IACUC reports its evaluations to the IO as required by §2.31,c,3. The *Guide* (p. 9) likewise states that the IACUC reports to "responsible institutional officials."

Opin.   Often it is difficult to distinguish among the roles of the IACUC, veterinary staff, facility operations, and even the IO. Some smaller facilities with limited staffing may require one individual to fill multiple roles. To guard against possible conflict of interest—where those inspecting and assessing the operations are also those in charge of them—the trend in larger facilities has been to create an obvious separation of responsibilities. This system provides for the checks and balances similar to those we have seen in government. The executive branch appoints (CEO) and supports (IO) the IACUC and facility. The judicial branch (IACUC) interprets and enforces regulations, and the legislative branch (veterinary care and facilities operations) oversees the daily operational needs. The trick then becomes one of effectively maintaining good communications and close working relations. A key point to feedback to each group and to the research community is that the overall goal of the process is to provide for the optimal research conditions.

27:32   How might a veterinary services unit work for the IACUC, with the IACUC, or against the IACUC?

Opin.   Nearly all the respondents to this question indicated that the veterinary services unit worked with the IACUC; although this scenario may seem ideal in some ways, it raises the issue of how separate the two groups are. Do the groups work well together because they are made up of the same players? If the groups tend to be the same, then they are really working with themselves. It could be argued that an efficient IACUC and veterinary services unit will at times find that they are working for and against each other.

  Expanding on the checks and balances analogy from the previous question, each group (IACUC office, the veterinary care staff, facility operations, and the IO) has the responsibility of to ensure that a single component of the program does not gain power to the point at which they can control all functions. There are times an IACUC may call on the veterinary staff to work for them, utilizing the expertise and training of the laboratory animal veterinarian to help develop policies or detail programs of care and use. Likewise, the groups could work against each other if either is becoming too complacent or authoritarian in their activities. A diverse IACUC through their interactions and resulting discussions will perform a valuable service to the institution.

Surv.   In your personal opinion, do your veterinary service personnel work for the IACUC, with the IACUC, or against the IACUC?

- Works with the IACUC to try to reach common goals   149/153
- Works for the IACUC, in that veterinary services goals appear to be subordinate to the IACUC's goals   3/153
- Works against the IACUC, appearing to have very different goals from those of the IACUC   1/153

27:33   What types of issues might the IACUC commonly find during the semiannual programmatic review and facility inspection?

Opin.   The IACUC is required to review the program of veterinary care to assure that it is designed and implemented in such a way as to provide an adequate level of veterinary care. This means that programs, personnel, and other resources are available for the diagnosis, treatment, control, and prevention of animal disease and injury. Veterinary records detail delivery of this program and should therefore be reviewed to assess adequacy of implementation by the veterinary and research staffs. While planned programs are sometimes lacking in scope and breadth, underfunding of the veterinary effort often results in lack of needed equipment. Records, particularly those maintained by research staff, are frequent sources of concern as insufficient detail or data entry indicates a lack of attention to provision of animal health support and/or a lack of attention to proper record keeping.

Surv. When your IACUC reviews the program of veterinary care during the semiannual program review, what types of programmatic problems have you identified? Check all appropriate responses.

- We do not review the program of veterinary care   12/153
- We have not identified any problems   105/153
- Inadequate veterinary care by veterinary staff   2/153
- Inadequate veterinary care by research staff   3/153
- Inadequate pre- or postprocedural care by veterinary staff   0/153
- Inadequate pre- or postprocedural care by research staff   19/153
- Inadequate medical records by veterinary staff   5/153
- Inadequate records by research staff   26/153
- Other:
  - Institution needs veterinarian dedicated to animal welfare   1/153
  - Insufficient veterinary staff   1/153
  - Failure of research staff to wean rodents on time   1/153
  - Expired pharmaceuticals   1/153
  - Lack of sentinel program   1/153
  - Use of unapproved animal vendor   1/153

References

1. Report of the American College of Laboratory Animal Medicine on Adequate Veterinary Care in Research, Testing and Teaching, American College of Laboratory Animal Medicine, Cary, NC, 1996.
2. Animal and Plant Health Inspection Service, United States Department of Agriculture, Animal Care Resource Guide, Policy 3, Veterinary Care, January 14, 2000.
3. American College of Laboratory Animal Medicine Medical Records Committee, Public Statement: Medical Records for Animals Used in Research, Teaching, and Testing, 2004. Available on the World Wide Web at: http://www.aclam.org/PDF/pub_med_records_2.pdf.
4. Health Surveillance Programs, in Infectious Diseases of Mice and Rats, National Research Council, A Report of the Institute of Laboratory Animal Resources Committee on Infectious Diseases of Mice and Rats, National Academy Press, Washington, D.C., 1991, p. 21.

28

Laboratory Animal Enrichment

Kathryn A. L. Bayne

Introduction

In 1985, Congress passed amendments to the AWA (PL 99-198) that included a provision mandating an environment "adequate to promote the psychological well-being of nonhuman primates." Six years later, the USDA published implementing regulations pertaining to this amendment.[1] In December 1996, APHIS/AC published the results of an internal survey, "USDA Employee Opinions on the Effectiveness of Performance-Based Standards for Animal Care Facilities,"[2] which included questions pertaining to environmental enrichment programs to improve the psychological well-being of nonhuman primates. Approximately 45% of respondents to the survey indicated that the criteria for primate enrichment are not sufficiently clear, and about 50% said the criteria were not useful. APHIS/AC inspectors responding to the survey indicated that there should be a clearer definition of the requirements and enhanced documentation to ensure that enrichment plans were being followed. Some suggested stricter guidelines or policies pertaining to group housing, cage space, and enrichment. The USDA team assigned to review the survey results made several recommendations. Those relevant to environmental enrichment programs for nonhuman primates included the following: "For primates, the enrichment options should be grouped into elements and a set of examples given for each category. Facilities should be required to provide something in each category. They should document consideration of different species' needs, efforts on behalf of individuals with behavior problems, periodic review of the results of plan implementation, and improvements made after review."[2]

The 1996 edition of the *Guide* includes the section Behavioral Management. It describes potential enhancements to the animal's cage environment, social environment, and activity level. The National Research Council's publication *The Psychological Well-Being of Nonhuman Primates*[3] highlights the increasing need for IACUC involvement in, and oversight of, laboratory animal enrichment programs.

The field of environmental enrichment is still evolving. Empirical measures of assessing animal well-being are not completely defined and the roles of genetic, developmental, and environmental influences on well-being require further study. The *Guide* (pp. 1, 36–38) recommends expanding the enrichment program to all laboratory animals rather than restricting enrichment provisions to nonhuman primates as required by law. The institution should not take a minimalist approach to implementing the enrichment program(s) and the IACUC should take a proactive role in its oversight.

For the following questions, 150 completed survey responses were tallied. Some questions allowed the respondent to choose more than one answer. Accordingly, percentages of the total number of responses are provided as often as possible to give a better sense of weight for each answer.

28:1 What is meant by laboratory animal enrichment?

Opin. Environmental enrichment is an approach to animal housing in which complex stimuli are provided to stimulate the expression of species-typical behaviors and reduce the occurrence of abnormal behaviors.[4] It is a component of animal husbandry that is designed to enhance the quality of captive animal care by identifying and providing the environmental stimuli necessary for optimal psychological and physiological well-being. In practice, this covers a multitude of innovative and imaginative techniques, devices, and practices aimed at keeping captive animals occupied, increasing the range and diversity of behavioral opportunities, and providing more stimulating and responsive environments.[5]

Care should be taken to enrich the physical environment in the primary enclosure by providing means of expressing only noninjurious species-typical activities. Species differences should be considered when determining the types or methods of enrichment, and thus a basic understanding of the normal behavior of the species in question is requisite to a successful enrichment program. Examples of environmental enrichments include perches, swings, mirrors, nesting material, hiding places, and other cage complexities; provisions of objects to manipulate; varied food items; social opportunities; use of foraging or task-oriented feeding methods; provision of control over some aspect of the environment, such as visual access to other animals, music, or video; and interaction with the caregiver or other familiar and knowledgeable person consistent with personnel safety precautions. Of course, provision of the presumed enrichment must be well thought out and with a rather detailed knowledge of the species's normal behavior, constraints imposed by the research, safety considerations, and the caregiver's level of experience.[6]

Surv. What is meant by laboratory animal enrichment?

The most common response to this question was a combination of the definitions provided as possible responses: the enrichment program promotes physical and psychological well-being, the program promotes species-typical behavior, the program provides cage complexity, the program provides toys or similar diversions to animals, and the program decreases aberrant behaviors. The second most common response was that the institution did not have a formal definition of laboratory animal enrichment. A large proportion of these respondents did not use AWA-regulated species, thereby indicating that most rat-only or mouse-only enrichment programs are less formal than enrichment programs for AWA-regulated species. Slightly more than one fifth of respondents checked off all the definitions of enrichment.

- Our institution has no formal definition of enrichment   40/150 (26.7%)
- Checked off all possible answers   33/150 (22.0%)
- Checked off a variety of combinations of the responses   73/150 (48.7%)
- Did not respond   4/150 (2.6%)

28:2 Besides the federal mandate for canine exercise (AWAR §3.8) and improving the psychological well-being of nonhuman primates (AWAR §3.81), what else is required by an institution for laboratory animal enrichment?

Reg. The *Guide* (p. 36) states, "Depending on the animal species and use, the structural environment should include resting boards, shelves or perches, toys, foraging devices, nesting materials, tunnels, swings, or other objects that increase opportunities for the expression of species-typical postures and activities and enhance the animals' well-being." It further states, "It is desirable that social animals be housed in groups; however, when they must be housed alone, other forms of enrichment should be provided to compensate for the absence of other animals, such as safe and positive interaction with the care staff and enrichment of the structural environment. ... Animals should have opportunities to exhibit species-typical activity patterns."

Opin. It also should be noted that institutions that are accredited by AAALAC are required to conform with the *Guide*, as are institutions that receive PHS funds (PHS Policy IV,A,1). Although the provisioning of environmental enrichment to laboratory animals other than nonhuman primates is a relatively new endeavor, there are a number of good references to use when developing such a program.[7-9]

Surv. Besides the federal mandate for providing the opportunity for canine exercise and for addressing the psychological well-being of nonhuman primates, what else is required by your institution for laboratory animal enrichment?

The majority of respondents to this survey question stated that they had no further institutional requirements. Those respondents that had additional institutional requirements for enrichment noted a variety of reasons: enrichment was provided to all species on the basis of corporate policy; there was a written institutional policy to provide enrichment to all species but fish; all species were enriched without a formal institutional policy; and enrichment was provided to rodents, rabbits, and cats (in addition to that provided in response to federal regulations).

- No further requirements   78/150 (52%)*
- Institution has further requirements   26/150 (17.3%)**
- No response   46/150 (30.7%)

Note: * When only tallying those individuals who responded, this figure becomes 75% of cases. ** When only tallying those individuals who responded, this figure becomes 25% of cases.

28:3 What should be the goals of a laboratory animal enrichment program?

Opin. Mench[10] has suggested that enrichment programs should provide opportunities for exploration for species that are generalists or "are adapted to environments that are highly variable in terms of resource availability." Enrichment also is needed for species that exhibit "complex antipredator behaviors" and species that have a complex social order. Mench encourages viewing enrichment from the perspective of the information-gathering needs of the animal. She points to successful enrichments that allow animals to carry out appetite-related components of behavior (e.g., foraging boards). Goals of an enrichment program might include the following:

- Promoting the well-being of the research animal and providing a more refined animal model for the research
- Increasing knowledge of the environment's impact on behavior and stress reduction for personnel
- Providing enrichments that are meaningful to the animals and do not compromise personnel safety, animal health and safety, or the research aims

Goals should not include provision of enrichments that are aesthetically pleasing to staff but have no species relevance (e.g., a brightly colored ball may look attractive in the cage but may not prove to be behaviorally stimulating to some species of animals).

In *The Psychological Well-Being of Nonhuman Primates*[3] a sample of goals and aims of the enrichment program (for nonhuman primates) is offered. In addition to those objectives described, this report suggests providing cognitive stimulation and opportunities for animals to alter their environment as well as decreasing self-injurious behavior. The report emphasizes training of personnel in the natural history, behavior, and husbandry of the species as an essential means of achieving enrichment program goals.

Surv.   What should be the goals of a laboratory animal enrichment program? Multiple answers were acceptable for this question.

Although a surprising number of respondents had not given thought to the goals of their enrichment program, by far the most common response was that the goal of the program was to promote species-typical behavior. The next most common response was that the program should decrease the expression of abnormal behavior. Other responses stated that the goal was to maximize animal well-being or to provide an environment as similar to a natural environment as possible. Some respondents commented that since they did not have nonhuman primates or dogs, having a goal for the program was not applicable. In many cases, meeting regulatory requirements was the sole response given.

- Meet regulatory requirements   37/254 (14.6%)
- Promote species-typical behavior   80/254 (31.5%)
- Decrease abnormal behavior   56/254 (22.1%)
- Goals change over time, depending on the needs of the program   17/254 (6.7%)
- The goals are defined by the veterinarian   9/254 (3.5%)
- The respondent had not thought about goals for the program   9/254(3.5%)
- No response   46/254 (18.1%)

28:4   What species does your institution include in the laboratory animal enrichment plan?

Reg.   The regulatory requirement for enrichment is currently restricted to promoting the psychological well-being of nonhuman primates (AWAR §3.81) and providing canines the opportunity to exercise (AWAR §3.8). The *Guide* (pp. 36–38) strongly recommends addressing the structural environment, social environment, and activity of laboratory animals as they pertain to an overarching behavioral management program. PHS Policy (II; IV,A,1) requires institutions to comply with the AWAR and to follow the *Guide*.

Opin.   Reference documents[7,11,12] that are available contain numerous suggestions and literature citations pertaining to environmental enrichment of nonhuman primates, birds, cats, dogs, farm animals, ferrets, rabbits, and rodents. The large number of references contained in these resources, as well as more recent publications,[13–21] suggest that whether or not laboratory animals other than primates are covered by an enrichment plan, enrichment strategies should be developed for them. (See 28:5.)

Surv. What species does your institution include in a laboratory animal enrichment plan? Multiple answers were acceptable for this question.

A variety of species are provided environmental enrichment, though not all may be covered by a formal enrichment plan. Respondents indicated that the most common species that receive enrichment were rodents. However, this response may simply reflect the greater use of rodents than of the other species of animals listed.

- We do not provide enrichment programs for any species   24/404 (5.9%)
- We have no formal enrichment plan but provide enrichments for one or more species   39/404 (9.7%)
- We have a formal enrichment plan that covers one or more species   26/404 (6.4%)
- Nonhuman primates   37/404 (9.2%)
- Dogs   45/404 (11.1%)
- Cats   31/404 (7.7%)
- Rabbits   54/404 (13.4%)
- Guinea pigs   33/404 (8.2 %)
- Rodents in general   90/404 (22.3%)
- Farm animals   20/404 (4.9%)
- No answer   5/404 (1.2%)

28:5 Where can help be found to develop a laboratory animal enrichment plan?

Opin. Assistance in developing an enrichment plan may be found in appendix A of *The Psychological Well-Being of Nonhuman Primates*.[3] The report includes two sample plans that provide general information on creating an enrichment plan for a large, complex nonhuman primate program (e.g., many animals, diverse species, diverse research uses) and a smaller or less diverse primate program (e.g., few animals or one species). A checklist of items the plan should address is also presented in the report (pp. 21–24). (See 28:4.)

The AWAR (§3.81) identify several key points that must be considered when developing an enrichment plan for nonhuman primates.[1] At a minimum, the plan must contain "specific provisions" to address the social needs of nonhuman primates in accordance with currently accepted professional standards. The plan also must address the physical environment of the primates in terms of environmental enrichment. Consideration must be given to the special needs of some categories of primates such as infants and young juveniles, those that show signs of being in distress, those that experience protocol-related restricted activity, those housed in a manner that precludes seeing or hearing primates of their own species, and the needs of great apes that weigh more than 50 kg. The use of restraint devices for medical treatment or IACUC-approved research activities and a contingency for at least one continuous hour of unrestrained activity when primates are held in restraint devices for more than 12 hours also should be indicated in the plan. Finally, identification and monitoring of any animals that are exempted from the plan as a result of health or well-being or research constraints are considered. In addition, the National Institutes of Health *Nonhuman Primate Intramural Management Plan*[22] may be obtained from the National Agriculture Library (*Federal Register*, Vol. 56, No. 32, Friday, February 15, 1991).

28:6   What are the types of enrichment animals might receive?

Opin.   Regardless of the species for which the enrichment plan is being considered, strategies can be categorized as social or nonsocial. Social enrichment includes both contact and noncontact elements and can be applied to interactions between animals and personnel (including training of the animals where applicable), animals of the same species, and occasionally animals of different species. Nonsocial enrichment techniques include cage complexities, bedding, feeding techniques, toys, and other environmental conditions (e.g., lighting, sound). The enrichment technique(s) selected should be safe for both the animal and personnel, should be species-relevant, and should not conflict with the goals of the research study.

Surv.   What are the types of enrichment animals might receive?

Both social and nonsocial enrichment strategies were described. Enrichment provided, other than conspecific contact, is itemized in Table 28.1. Excellent references for enrichment techniques are available.[3,7–9,11,12,21]

TABLE 28.1

Examples of Enrichment Provided to Laboratory Animals

Nonhuman primates	Perches, wooden gnawing blocks, varied food items, chew toys, mirrors, foraging boards, puzzle feeders, videos, positive human interaction, nesting box (if appropriate for species), music, ladders/swings, ice cubes, perfume samples, sun lamps (marmosets)
Dogs	Toys, exercise outside cage, positive human interaction, occasional food treat, resting shelf, wood shavings or shredded paper bedding
Cats	Multilevel resting surfaces, climbing posts, toys, food treats, positive human interaction, music, scratching posts, cat condos, group exercise
Rabbits	Toys (cat bell ball, syringe cases, plastic soft drink bottle), food treats, positive human interaction, exercise areas, music, chewing/gnawing toys, chain in cage
Guinea pigs	Toys (see rabbit examples), food treats, positive human interaction.
Rats, mice, hamsters, chinchillas	Polyvinylchloride (PVC) tubes, Nylabones®, carrot bones for chewing, Nestlets®, chew blocks (chinchillas), hiding spaces, paper towels or tissue if housed on wire-bottom cages
Chickens, pigeons	Perches, pecking objects, complex diet for seed searching, hanging toys
Frogs and turtles	Large cage, dry and wet surfaces, blocks or rocks to climb on
Pigs	Toys (e.g., balls, chains), back-scratch device, positive human interaction, chew toys

28:7a   How should the institution assess an investigator's request for his or her animals to be exempted from all or part of the enrichment program?

Reg.   The AWAR (§3.81,e,1) state that "the attending veterinarian may exempt an individual nonhuman primate from participation in the environment enhancement plan because of its health or condition, or in consideration of its well-being. The basis of the exemption must be recorded by the attending veterinarian for each exempted nonhuman primate. Unless the basis for the exemption is a permanent condition, the exemption must be reviewed at least every 30 days by the attending veterinarian." The IACUC also may exempt an individual nonhuman primate for scientific reasons (AWAR §3.81,e,2).

PHS Policy (II; IV,A,1) requires institutions to comply with the AWAR and the *Guide*. The *Guide* (p. 37) also addresses circumstances in which exceptions might be necessary.

Opin. Exemptions from an enrichment program should not be considered as an "all or nothing" participation. An animal may be exempted from only a portion of the program, while still enjoying the benefits of other elements of the program. The ILAR report[3] affirms that an animal's role in research involving infectious diseases, atypical rearing conditions, physical restraint, surgery, pain, substance abuse, or aggression should not be excluded a priori from the enrichment program. Indeed, the authors of the report note that protocols that may pose restrictive environments for animals should be reevaluated periodically to determine whether new technologies that could reduce the restrictions placed on the animal may be available.

The report recommends four criteria for assessing animal well-being in addition to general physical health:

- The animal's ability to cope effectively with environmental changes
- The animal's ability to engage in beneficial species-typical behavior
- The absence of maladaptive or pathological behavior
- The presence of a balanced temperament

When evaluating an animal for an exemption based on health or behavior, it is important to understand fully what is "normal" for an animal in that condition (e.g., an aged animal frequently has reduced perceptual and locomotor capabilities[23] and, thus, the enrichment program may need to be modified to accommodate these limitations).

Surv. How does your institution assess an investigator's request for his or her animals to be exempted from all or part of the enrichment program?

Many respondents indicated that either they did not have enrichments or that there was no formal process for exempting an animal from the enrichment program. When there was a formal system, exemptions were typically reviewed and authorized by either the IACUC or the veterinarian. Exemption from the enrichment program based on a scientific justification was typically handled on the IACUC protocol form.

- Not applicable, as we have no specific enrichment activities   52/150 (34.7%)
- The request is made on the protocol form and reviewed and approved by the IACUC   62/150 (41.3%)
- Review and approval of the request are done by the veterinarian   11/150 (7.3%)
- Review and approval of the request are done by the facility manager   4/150 (2.7%)
- There is no formal process; the investigator may provide or remove enrichments   11/150 (7.3%)
- No response   10/150 (6.7%)

28:7b What criteria are used for animal exemptions?

Opin. The management of an enrichment program includes, of necessity, provisions for allowing exemptions for select animals from participating in aspects of the program. Exemptions may be based on concerns related to the health of the animal (either related to the experiment or due to natural causes) or the scientific goals of the study. Often, an animal does not need to be excluded from the entire enrichment program but rather may be offered elements of the program (e.g., the animal may be precluded from social housing but offered a variety of nonsocial enrichments, such as food treats or toys). Thus, the animal should initially be evaluated for exemption from certain parts of the program, rather than in terms of participation or nonparticipation.

Surv. What criteria are used for animal exemptions? Multiple answers were acceptable for this question.

The most common justification given for exempting an animal from enrichment was for scientific or medical reasons.

- We have no enrichment program   41/254 (16.1%)
- Scientific justification based on research needs   87/254 (34.3%)
- Medical exemption based on veterinary assessment   60/254 (23.6%)
- Behavioral exemption, based on incompatibility for
  social housing   52/254 (20.5%)
- No exemptions have been requested   3/254 (1.2%)
- No response   11/254 (4.3%)

28:7c   Who grants the exemption?

Opin. According to the AWAR (3.81,e) the AV is responsible for granting an exemption for nonhuman primates participating in the enrichment program based on reasons of health, condition, or well-being. It is logical to extend this responsibility of the AV to other species at the institution that may have health considerations that would be impacted by the provision of enrichments. Because exemptions based on a scientific justification would likely be contained in the protocol form submitted by the investigator, the IACUC is the appropriate agent of the institution to be charged with granting exemptions from the enrichment program when such participation would confound research data or otherwise compromise the integrity of the study or the welfare of the animal.

Surv. Who grants the exemption? Multiple answers were acceptable for this question.

Most frequently, the IACUC and the AV granted the exemption from the enrichment program. As expected, the AV typically granted the medical exemptions, and the IACUC granted exemptions based on a scientific justification. In addition to the following responses, other individuals involved in granting exemptions were the principal investigator, the facility director, the animal care staff, the facility supervisor, a consensus of the AV/PI/IACUC, and the IACUC Chair.

- Not applicable   47/177 (26.5%)
- The IACUC, with or without the veterinarian's input,
  depending on the situation   69/177 (39.0%)
- A subcommittee of the IACUC   2/177 (1.1%)
- The AV   50/177 (28.3%)
- No response   9/177 (5.1%)

28:8   Does your institution have an enrichment committee?

Opin.   For the enrichment program to be implemented in a coordinated and efficient manner, there must be a responsible party. If an individual is made responsible for the
program, that person should seek input from the scientific, veterinary, and animal
care staffs. A committee that comprises these categories of personnel also will benefit from the diversity of input requisite for the program to be successful. A committee may gain the added benefit of greater involvement and vesting in the
enrichment program by committee members if they not only assist in developing
the enrichment program but participate in keeping it current and vital. Stewart and
Raje[24] have recently provided a case for using an enrichment committee to develop
an effective enrichment program.

Surv.   Does your institution have an enrichment committee?
   Most respondents' institutions did not have an enrichment committee.

- We do not have an enrichment committee   125/150 (83.3%)
- We have an enrichment committee   12/150 (8.0%)
- No response   13/150 (8.7%)

28:9a   How is the efficacy of the enrichment program assessed?

Reg.   The AWAR (§2.31,c,1) and PHS Policy (IV,B,1) require that the IACUC review at
least once every 6 months the program for humane care and use of animals. The
Guide (p. 9) states that the IACUC has oversight and evaluation responsibility for the
animal care and use program and the components of the program described in the
Guide. The *Guide* (p. 9) echoes the regulatory requirement for IACUC review of the
program and facilities at least once every 6 months. (See Chapter 23; Chapter 27.)

Opin.   Morgan and associates[25] reviewed various methods of assessing enrichment programs. Methods included reliance on others to inform the IACUC, common sense,
and empiricism. They noted that a significant disadvantage of reliance on others is that
acceptance is often based on the authority of the other person (the tendency to accept
that a point of information is true simply because it is that of an authority figure). They
suggested that common sense is a good starting point for evaluating enrichment,
although its main disadvantage is that common sense is grounded on individual interpretations and biases. However, they strongly argued that empirical testing of enrichment strategies is necessary to validate their use both from the standpoint of the value
to the animal and for judicious expenditure of limited resources.
   An assessment strategy specific to the psychological well-being of nonhuman
primates is available.[3] Four factors are proposed for consideration when assessing
psychological well-being:

1. The animal's ability to cope with day-to-day changes in its environment
2. The animal's ability to exhibit beneficial species-typical activities
3. The absence of maladaptive or pathological behaviors that could result
   in self-injury or other "undesirable consequences"
4. The presence of a balanced temperament

These assessment variables may be sufficiently broad to be applicable to other
laboratory animal species.

The IACUC has a responsibility to review the animal care and use program semiannually and to ensure its completeness, appropriateness, and currency.

Surv. How is the efficacy of the enrichment program assessed? Multiple answers were acceptable for this question.

Respondents most commonly reported that the AV or animal care staff performed the assessment of the success of the enrichment program. However, a substantial proportion of respondents either found this question not applicable to their situation or did not assess the enrichment program (a total of 42.9% of the responses). Assessments by AAALAC were relied on by several institutions as the means of ensuring an effective program. Other responses included evaluations by compliance officers, investigator input, and an improvement in observed behaviors expressed by the animal.

- Not applicable; we have no enrichment program   47/189 (24.9%)
- We do not specifically assess the efficacy of our enrichment program   34/189 (18.0%)
- The veterinarian or animal care staff do the assessment   57/189 (30.2%)
- The IACUC (or a subcommittee of the IACUC) assess the program as part of the semiannual review process   28/189 (14.8%)
- We consider the program successful if AAALAC is pleased with it during the site visit   15/189 (7.9%)
- No response   8/189 (4.2%)

28:9b How should the IACUC monitor the enrichment program?

Opin. Monitoring of the enrichment program should be conducted by different constituents of the animal care and use program. For example, implementation of the enrichment program can be assessed by the IACUC during the semiannual facility inspections and the scope of the program can be reviewed by the IACUC during semiannual program reviews. The animal care staff and AV should be assessing the success of the program periodically. This assessment should include whether animals use the enrichments, whether the enrichments are safe for the animals, and whether the enrichments result in improved well-being of the animals. Outside consultants are often retained to assist in either establishing this system of monitoring or periodically conducting the assessment.

Surv. How do you monitor the enrichment program? Multiple answers were acceptable for this question.

As noted elsewhere in this chapter, one of the more striking features of the responses to the survey questions is the important role that the AV plays in providing oversight of the enrichment program and related training activities. In this case, if the enrichment program was monitored, the veterinarian was most often reported to be the individual or entity doing the monitoring, supplemented by the animal care staff.

- Not applicable as we either do not have enrichment activities or we do not monitor them   56/216 (25.9%)
- This is a component of the IACUC's semiannual facility inspection and program review   47/216 (21.8%)

- The animal care staff informally report behavioral
  abnormalities to the veterinarian   53/216 (24.5%)
- The animal care staff record their observations on animal
  behavior and enrichment use in the animal's record   29/216 (13.4%)
- A staff member conducts routine behavioral
  observations as part of his or her duties and follows
  up on any observed abnormalities   21/216 (9.7%)
- No response   10/216 (4.6%)

28:10   Should the institution's enrichment plan and/or associated records be reviewed by the IACUC semiannually as part of the program review?

Reg.   The AWAR (§2.31,c,1) state that the IACUC shall "review at least once every six months the program for humane care and use of animals using title 9, chapter I, subchapter A-Animal Welfare, as the basis for evaluation." This reference includes the sections pertaining to dog exercise and environmental enhancement to promote psychological well-being. The *Guide* (p. 9) ascribes to the IACUC the responsibility for "oversight and evaluation of the animal care and use program and its components as described in the *Guide*," which includes the behavioral management section. The PHS Policy states (IV,B,1) that the IACUC must "review at least once every six months the institution's program for humane care and use of animals, using the *Guide* as a basis for the evaluation."

Opin.   Institutions that have an enrichment plan for nonhuman primates and/or a canine exercise program should review those aspects of the animal care and use program semiannually. Similarly, if the institution holds a PHS Assurance or is accredited by AAALAC, the behavioral management program described in the *Guide* should be reviewed as part of the semiannual program review. Some elements of these programs may also be assessed during the facility inspections, assessing the degree of implementation of the plan and gauging its effectiveness in terms of the animals' behavior or other measures of well-being.

Surv.   Are your institution's enrichment plan and/or associated records reviewed by the IACUC semiannually as part of the program review?
          Of those respondents who have an enrichment plan and/or related enrichment records, the majority review these documents during the semiannual review by the IACUC.

- Not applicable as we have no enrichment plan   50/150 (33.3%)
- We have no enrichment plan records   28/150 (18.7%)
- We evaluate our plan and/or records as part of the
  semiannual program review process   51/150 (34.0%)
- We do not evaluate our plan and/or records as part
  of the semiannual review process   13/150 (8.7%)
- No response   8/150 (5.3%)

28:11   How is the IACUC trained so that the enrichment program may be appropriately evaluated?

Reg.   The *Guide* (p. 9) states that the IACUC "is responsible for oversight and evaluation of the animal care and use program and its components described in this Guide."

The *Guide* (p. 9) further states that "it is the institution's responsibility to provide suitable orientation, background materials, access to appropriate resources, and, if necessary, specific training to assist IACUC members in understanding and evaluating issues brought before the committee."

Opin. The *Guide* clearly places responsibility for all elements of the animal care and use program with the IACUC (as designated by the IO). The laboratory animal enrichment program is one component of the entire program and, thus, should be included in the semiannual program review performed by the IACUC. To accomplish this task, the IACUC should be provided with the tools to evaluate the behavior of the animals they see during the facility inspection, to understand the rationale for the enrichment program's design, and to assess the program's efficacy.

Surv. How is the IACUC trained so that the enrichment program may be appropriately evaluated? Multiple answers were acceptable for this question.

Of those respondents who provide training to the IACUC, of note is the important role the AV plays in providing this training. Other ways in which the IACUC was familiarized with the enrichment program was through review of SOPs, through discussions during IACUC meetings, and by attendance at national meetings.

- Not applicable; we have no enrichment program 45/170 (26.5%)
- We have no formal training to evaluate the enrichment program 57/170 (33.5%)
- We have an environmental enrichment committee that helps train the IACUC on how to evaluate the quality of the program 7/170 (4.1%)
- We depend on the veterinary staff to inform us on what a high-quality program should encompass 46/170 (27.1%)
- We depend on outside experts to inform us on what a quality program should encompass 7/170 (4.1%)
- No response 8/170 (4.7%)

28:12a What should enrichment records reflect?

Reg. The AWAR (§3.81) require that dealers, exhibitors, and research facilities develop, document, and follow an environment enhancement plan that is adequate to promote the psychological well-being of nonhuman primates. They further require that records of exemptions of animals from the plan be maintained (§3.81,e,3) and be made available to USDA officials or representatives of federal funding agencies upon request.

PHS Policy (II; IV,A,1) requires institutions to comply with the AWAR as applicable and to follow the *Guide*. Institutions must keep records of departures from the *Guide* and IACUC approval of any departures (PHS Policy IV,B,3).

Opin. Documentation not only serves as a verifiable record of enrichment activities, but also validates the effectiveness of ongoing enrichments or provides evidence of the need to modify the enrichment program. It has been suggested[3] that nonhuman primate enrichment programs should include a mechanism whereby there are "protocols for diagnosing the cause of physical impairments and abnormal behavior, determining when remediation is necessary, developing remediation

Surv. plans, assessing the effectiveness of remediation, and maintaining appropriate records."

Surv. What do your enrichment records reflect? Multiple answers were acceptable for this question.

The majority of respondents (principally rodent users) indicated they did not maintain enrichment records. When records were kept, they described the provision of toys and food treats, exercise opportunities for canines, exemptions from the program, and effects of the enrichment on the animals.

- No enrichment records   100/225 (44.4%)
- Rotation schedule of manipulative objects used for
  enrichment   32/225 (14.2%)
- Food treat use   33/225 (14.7%)
- Exemptions from the enrichment program   28/225 (12.4%)
- Observed effects of the enrichment program   19/225 (8.4%)
- No response   13/225 (5.8%)

28:12b For what length of time should enrichment records be kept?

Reg. The AWAR (§3.81,e) require that records be maintained by the AV for nonhuman primates that are exempted from the enrichment program that describe the basis for the exemption. The AWAR do not address record keeping pertinent to enrichment for other species of animals.

Opin. The length of time for which records that pertain to an animal's enrichment should be maintained will likely vary with the species, the type of use of the animal, potentially the types of enrichment that an animal received (e.g., rotating social partners may be more important to track than the types of inanimate enrichment an animal received), the type and length of study in which the animal is used, and whether the research is being conducted under Good Laboratory Practices (GLP) standards, to name a few. Therefore, such decisions should be discussed and determined by the IACUC, as this committee is in the best position to consider all the potential variables and make a well-considered judgment.

Surv. For what length of time do you keep enrichment records?

Most respondents indicated that if they maintained enrichment records, those records were archived for 3 years or longer. Approximately equal numbers of respondents archived the enrichment records for the length of the study or indefinitely. Those that did not maintain enrichment records were principally rodent users.

- No enrichment records   100/150 (66.7%)
- One year   2/150 (1.3%)
- Three years   8/150 (5.3%)
- More than 3 years   8/150 (5.3%)
- Duration of the study   9/150 (6.0%)
- Indefinitely   10/150 (6.7%)
- No response   13/150 (8.7%)

28:13a   How are financial resources allocated to support the environmental enrichment program?

Opin.   Since the provision of an enrichment program is mandated by federal law (see 28:4) for some species, it is critical that adequate resources be identified to ensure compliance with this institutional responsibility. An ongoing budgetary commitment to the enrichment program will allow for the program to evolve as new information becomes available and to stay dynamic. Inadequate funding can cause a program to be ineffective and not meet regulatory requirements or recommendations of the *Guide*. The *Institutional Administrator's Manual for Laboratory Animal Care and Use*[24] states that "sustained and visible support from institutional officials is absolutely essential to establishing and maintaining a high quality animal care and use program. … They can assure sufficient monetary and personnel resources are allocated to the institution's program." The enrichment program should not be viewed as separate or an "extra" program but rather as an integral component of the entire animal care and use program.

Surv.   How are financial resources allocated to support the environmental enrichment program? Multiple answers were acceptable for this question.

   Fiscal support for enrichment programs was principally derived from the animal facility budget; the next most common source of funding was through per diem charges. Other responses included costing the enrichment out for each study, deriving fiscal support from a core fund for nonhuman primates, obtaining it through the investigator's laboratory budget, or using other investigator-managed funds.

- Not applicable; no enrichment program   47/156 (30.1%)
- From per diem charges   33/156 (21.2%)
- From the IACUC's budget   3/156 (1.9%)
- From the animal facility's budget   63/156 (40.4%)
- No response   10/156 (6.4%)

28:13b   How are human resources allocated to support the environmental enrichment program?

Opin.   The way in which institutions allocate human resources for the implementation of the enrichment program often varies with the size of the institution, the species used at the institution, and whether it is an academic institution. For example, large programs may have the funds to employ a staff member dedicated to the behavior and enrichment program (e.g., primate research centers), whereas smaller programs may exclusively rely on the animal care staff to create, implement, monitor and update the enrichment program. Institutions that use only rodents are more likely to rely on current staff to manage the enrichment program. Academic institutions have the opportunity to involve students in the support of aspects of the program. Regardless of the type of individual who manages the program, the IACUC should ensure that he or she is adequately qualified through training or experience.

Surv.   How are human resources allocated to support the environmental enrichment program? Multiple answers were acceptable for this question.

   The majority of respondents indicated that providing enrichment was encompassed in the animal care staff's job description. Students and staff hired specifically to manage the enrichment program were far less utilized. Only two respondents indicated that they used outside volunteers.

- Not applicable; we do not have an enrichment program 46/159 (28.9%)
- We have one or more people hired specifically for the enrichment program   5/159 (3.1%)
- Implementing the enrichment program is part of the job description of our animal care staff   89/159 (56.0%)
- We use students to implement the enrichment program   5/159 (3.1%)
- We use outside volunteers to implement the enrichment program   2/159 (1.3%)
- No response   12/159 (7.6%)

28:14   How are staff trained to implement the enrichment program and to identify animals requiring special consideration?

Reg.   The AWAR (§2.32,b; §2.32,c) require training and instruction of personnel to include "humane methods of animal maintenance." The PHS Policy (IV,A,1,g) stipulates inclusion of a "synopsis of training or instruction in the humane practice of animal care and use" in the NIH/OLAW Animal Welfare Assurance document. Personnel qualifications and training also are emphasized in the *Guide* (p. 13), which states that "personnel caring for animals should be appropriately trained … and the institution should provide for formal or on-the-job training to facilitate effective implementation of the program and humane care and use of animals." The *Guide* lists behavioral management as one area in which personnel with expertise may be required.

Opin.   Regarding nonhuman primate enrichment, the ILAR report[3] notes that the success of an enrichment program is dependent on personnel having knowledge of and experience with nonhuman primate behavior. The report also states that "periodic training of staff to acquaint them with advances in the field is essential." Safety training also may be considered an essential topic in a behavioral management training program. Seminars at local, regional, or national meetings; scientific publications, autotutorial slide sets and videos; consultants; and on-the-job training are good means for providing staff training. The value of providing training on enrichment and recognition of signs that animals need additional attention may vary with the species used at the institution because of the inherent complexities of providing enrichment to nonhuman primates versus rats and mice.

Surv.   How are staff trained to implement the enrichment program and to identify animals requiring special consideration? Multiple answers were acceptable for this question.

Given the regulatory requirements for training and strong emphasis on training received in the *Guide* and other reports, it is somewhat surprising that less than 50% of those who responded provide either formal or informal training to facilitate implementation of the enrichment program. More than one quarter of respondents relied on SOPs to ensure appropriate implementation of the program.

- Not applicable   46/178 (25.8%)
- We provide formal training sessions   18/178 (10.1%)
- We send people to various meetings   18/178 (10.1%)

- We expect supervisors to train employees   33/178 (18.5%)
- We use outside experts to provide training   6/178 (3.4%)
- We have no formal training; it is part of our standard operating procedures   48/178 (27.0%)
- No response   9/178 (5.1%)

28:15   How are investigators notified of the necessity to consider enrichment for their animal subjects?

Opin.   Keeping investigators informed about the enrichment program through the protocol form, newsletters, Web pages, training programs, and dialog fosters their involvement in the program and will likely result in improvements to the program due to their diverse scientific expertise. As the successful enrichment program depends on a strong team approach among various institutional sectors (administrators, investigators, and animal care personnel), the better informed these parties are about the program, the more effective the enrichments will be for the animals.

Surv.   How are investigators notified of the necessity to consider enrichment for their animal subjects? Multiple answers were acceptable for this question.

Most respondents to this question indicated that investigators were informed of the enrichment program in an informal manner; however, IACUC-related documents—either the protocol form or direct correspondence—were an important method of informing investigators about the program. The AV again plays an important role in oversight of the enrichment program by serving as a key source of information about the program to investigators. Other responses included notification of investigators by the facility supervisor or manager and through SOPs.

- Not applicable   43/187 (23.0%)
- It happens, but not in a formal manner   41/187 (21.9%)
- It is noted on the protocol form   30/187 (16.0%)
- Direct correspondence from the IACUC   21/187 (11.2%)
- Direct information from the veterinarian   30/187 (16.0%)
- Via an intranet site   4/187 (2.1%)
- Via a required or voluntary training program   8/187 (4.3%)
- No response   10/187 (5.4%)

28:16a   What objections are typically expressed by PIs concerned about using animal enrichment?

Surv.   If your PIs are concerned about using animal enrichment, what is their major objection? Multiple answers were acceptable for this question.

The most common answer provided by respondents who had investigators who had concerns about the enrichment program was related to the possible effect on research results. Related to this was the concern that a change in husbandry would be implemented without objective data to support such a change. The cost of the enrichment program, possible concerns about maintaining a sanitary environment that has enrichment devices, and any potential complications for accessing the animals were not considered important by most respondents.

- Not applicable   78/187 (41.7%)
- Possible effects on research results   50/187 (26.7%)
- Cost   11/187 (5.9%)
- Ease of access or handling of the animal(s)   10/187 (5.4%)
- Hygiene   8/187 (4.3%)
- They are concerned about change without an associated
  scientifically documented outcome   18/187 (9.6%)
- No response   12/187 (6.4%)

28:16b   Are investigators concerned that enrichment may introduce a scientific variable into their work?

Opin.   The potential for environmental enrichment techniques to cause unintended consequences in research has recently been reviewed.[26] Such unintended consequences can be categorized as physical harm or change, physiological changes, immunological changes, and altered response to disease and other conditions. A careful review of the literature should be undertaken when introducing enrichment into the environment of research animals, and a determination made whether some or all aspects of the program are unsuitable for the particular study animals.

Surv.   Are your investigators concerned that enrichment may introduce a scientific variable into their work?

Of those who responded, 53.9% indicated that either the provision of enrichment had not generally been a concern for the investigators, or the issue had simply not arisen. However, approximately a fifth of those who responded (21%) had investigators who expressed concern about the impact of enrichment on their research.

- Not applicable; we have no enrichment activities   36/150 (24.0%)
- The issue has not arisen   36/150 (24.0%)
- Yes, in general   29/150 (19.3%)
- No, in general   40/150 (26.7%)
- No response   9/150 (6.0%)

28:17a   What circumstances might merit single housing of nonhuman primates?

Reg.   The AWAR state (3.81,a), "The environment enhancement plan must include specific provisions to address the social needs of nonhuman primates of species known to exist as social groups in nature." The AWAR allow exemption from social housing if the animal has an infectious disease, shows aggressive behavior, or is incompatible with other animals. The AWAR (§3.81,a,2) state that individually housed nonhuman primates must be able to see and hear nonhuman primates of their own or compatible species, unless the veterinarian determines that this contact would compromise the well-being of the animal.

Opin.   The provision of social housing of compatible animals is perhaps the most effective means of environmental enrichment for nonhuman primates. A compatible social partner is an ever-changing stimulus that can also provide comfort through mutual grooming, and in the case of younger animals, huddling. Because social relationships among primates can change, ongoing monitoring of the animals' compatibility is essential to prevent harm to the animals.

Surv.   Which answer best describes the reason for single housing your nonhuman primates? Multiple answers were acceptable for this question.

Concerns about the effect of social housing on the study, due either to introduction of a confounding variable by social housing or the potential for or actual aggression that can occur between animals was the most common concern expressed by the respondents. This concern speaks to the importance of ensuring ongoing compatibility between and among socially housed animals.

- Not applicable (do not single house or have no nonhuman primates)   106/205 (51.7%)
- Study concerns (possible effect of group housing on study results)   27/205 (13.2%)
- Concerns about aggression/injury before study begins   24/205 (11.7%)
- Concerns about aggression or injury that occurs during study   25/205 (12.2%)
- Current cage space limitations   6/205 (2.9%)
- Ease of access to individual animals   8/205 (3.9%)
- Cost of social cages   3/205 (1.5%)
- No response   6/205 (2.9%)

28:17b   What circumstances might merit single housing of rodents?

Reg.   The *Guide* (p. 37) states, "Consideration should be given to an animal's social needs. … When it is appropriate and compatible with the protocol, social animals should be housed in physical contact with conspecifics." The authors of the *Guide* recognized that social housing is not always possible for experimental, health, and behavioral reasons.

Opin.   Some strains of rodents engage in barbering behavior and/or aggression that results in wounding of cage mates. In addition, as noted previously, there may be unintended consequences of social housing.[26] Therefore, personnel should be familiar with the research objectives and the possible consequences of social housing on the research objectives, as well as the proclivity of the particular strain of rodent for aggression before automatically social housing rodents.

Surv.   Which answer best describes the reason for single housing your rodents? Multiple answers were acceptable for this question.

Similarly to the responses for single housing of nonhuman primates, the most significant concern about social housing of rodents was the potential impact on the research. Concerns about the potential for aggression and injury either before the study was initiated or during the course of the study were also common among respondents. Cage space limitations and ease of access to individual animals were of only minor concern.

- Not applicable (do not single house or have no rodents) 31/288 (10.8%)
- Study concerns (possible effect of group housing on study results)   105/288 (36.5%)
- Concerns about aggression or injury before study begins   50/288 (17.4%)

- Concerns about aggression or injury during study   74/288 (25.7%)
- Current cage space limitations   12/288 (4.29%)
- Ease of access to individual animals   11/288 (3.8%)
- No response   5/288 (1.7%)

28:17c What circumstances might merit single housing other research animal species?

Surv.   Which answer best describes the reason for single housing any of your other research animals? Multiple answers were acceptable for this question.

     The preceding pattern of concern (28:17b), which was principally directed to the impact of social housing on the research goals, also carried through to other laboratory animal species.

- Not applicable (do not single house other research animals)   45/244 (18.4%)
- Study concerns (possible effect of group housing on study results)   64/244 (26.2%)
- Concerns about aggression or injury before study begins   43/244 (17.6%)
- Concerns about aggression or injury during study   53/244 (21.7%)
- Current cage space limitations   15/244 (6.2%)
- Ease of access to individual animals   12/244 (4.9%)
- No response   12/244 (4.9%)

28:18a Are wire-bottom cages an acceptable form of housing for rats and mice?

Reg.   The *Guide* (p. 24) states that because some evidence suggests that rodents prefer bedded, solid-bottom caging, this type of caging is recommended. The *Guide* further states that the IACUC should review this aspect of rodent housing and ensure that cage systems enhance animal well-being "consistent with good sanitation and the requirements of the research project."

Opin.   Since the recommendation for using bedded solid-bottom cages appeared in the 1996 edition of the *Guide*, several additional studies have evaluated this husbandry parameter. While these studies have contributed to the information base regarding the different effects of wire-bottom and solid-bottom caging on rodent activity patterns and physiological processes, a clear recommendation of one type of caging over the other cannot be made. For example, while the *Guide* cites references[27–29] that indicated that ulcerative pododermatitis was a possible health consequence of the use of wire-bottom caging, Peace and colleagues[30] determined from a retrospective analysis of data from a chronic study that gross lesions were not identified in rats until they had been housed for more than a year in wire-bottom cages. Histological changes, such as those described in the *Guide* references, were not available from this retrospective study.

     Rock and colleagues[31] identified less motor activity in rats housed in solid-bottom cages, which they report has been cited as an indication of a relaxed emotional state. They also found that rats preferred to rest and groom on solid floors, while climbing, eating, and locomotion occurred preferentially in wire-bottom cages. Manser and associates[32,33] similarly observed that rats preferred to rest on solid floors, even if they had to work (i.e., lift a barrier) to access them. However, tests have shown that there was no statistical difference in the frequency or duration of resting behavior between rats housed in solid floor cages and rats housed in wire-bottom cages, and in fact, rats played more when housed in wire-bottom cages.[34] However, telemetry studies have documented physiological differences with housing condition, specifically higher blood pressure and heart rate in rats housed in wire-bottom cages,[35,36] although corticosterone levels were not significantly different in rats housed in the two cage systems.[37] Because of these mixed results, investigators, IACUCs, and veterinarians should closely evaluate documented effects of different types of cage systems and choose the system that has minimal impact on the animals' health and the scientific objectives.

Surv. Are wire-bottom cages the default type of caging for housing rats and mice at your institution?

When a respondent indicated that wire-bottom caging was the default system for housing rodents, the same respondent tended also to indicate that a scientific justification for this type of housing was not required by the IACUC. However, an overwhelming majority of respondents indicated that wire-bottom caging was *not* the default system for rodent housing.

- Not applicable (do not house rats or mice)   9/150 (6.0%)
- Yes   15/150 (10%)
- No   121/150 (80.7%)
- No response   5/150 (3.3%)

28:18b If an institution uses wire-bottom caging for rodents, is a scientific justification required in the protocol for the IACUC to approve this type of housing?

Opin. Although there may not be health consequences for animals housed in wire-bottom cages for short periods, the range of behaviors that these animals can express is limited by the cage environment. It is not known whether this more limited environment results in a decrement in animal well-being; however, what has been demonstrated is that bedding, nesting material, and hiding places are routinely used by rodents when available in the cage. Again, however, it has not been documented whether such use is equivalent to improved welfare. The expression of additional species-typical behaviors is judged to be beneficial for the animal. The IACUC should evaluate the circumstances surrounding rodent housing at the institution (e.g., length of the studies, weight of the animals, any evidence of behavioral abnormalities or other health effects) and then determine whether the committee will require a scientific justification or not.

Surv. Is scientific justification required in the protocol for your IACUC to approve the use of wire-bottom housing?

Of those respondents who responded to the question and found it to be applicable to their situation, the IACUCs of a significant majority of respondents (86.1%) required a scientific justification for the use of wire-bottom caging, in accordance with the *Guide*'s recommendation.

- Not applicable   69/150 (46.0%)
- Yes   62/150 (41.3%)
- No   10/150 (6.7%)
- No response   9/150 (6.0%)

References

1. Office of the Federal Register, 9 Code of Federal Regulations, Part 3, Animal Welfare; Standards; Final Rule, *Fed. Regist.*, 56(32), February 15, 1991, 6369.
2. U.S. Department of Agriculture, USDA Employee Opinions on the Effectiveness of Performance-Based Standards for Animal Care Facilities, U.S. Department of Agriculture, Animal and Plant Health Inspection Service, Animal Care, Riverdale, MD, December 1996.
3. The Psychological Well-Being of Nonhuman Primates, National Research Council, Washington, D.C., 1998.
4. Bayne, K., et al., The use of artificial turf as a foraging substrate for individually housed rhesus monkeys (*Macaca mulatta*), *Anim. Welfare*, 1(1), 39, 1992.
5. Shepherdson, D.J., Tracing the path of environmental enrichment in zoos, in Second Nature: Environmental Enrichment for Captive Animals, Shepherdson, D.J., Mellen, J.D., and Hutchins, M., Eds., Smithsonian Institution Press, Washington, D.C., 1998, p. 1.
6. Wolfle, T.L., Psychological well-being: the billion dollar solution, in *Through the Looking Glass: Issues of Psychological Well-Being in Captive Nonhuman Primates*, Novak, M.A., and Petto, A.J., Eds., American Psychological Association, Washington, D.C., 1991, p. 119.
7. Environmental Enrichment Information Resources for Laboratory Animals: 1965–1995, AWIC resource series, No. 2, Animal Welfare Information Center, National Agriculture Library, USDA, Washington, D.C., 1995.
8. Canadian Council on Animal Care, *Guide to the Care and Use of Experimental Animals*, Vol. 1, 2nd ed., Olfert, E.D., Cross, B.M., and McWilliam, A.A., Eds., Canadian Council on Animal Care, Ottawa, 1993.
9. BVAAWF/FRAME/RSPCA/UFAW Joint Working Group on Refinement, Refinements in rabbit husbandry, second report of the BVAAWF/FRAME/RSPCA/UFAW Joint Working Group on Refinement, *Lab. Anim.*, 27, 301, 1993.
10. Mench, J.A., Environmental enrichment and the importance of exploratory behavior, in Second Nature: Environmental Enrichment for Captive Animals, Shepherdson, D.J., Mellen, J.D., and Hutchins, M., Eds., Smithsonian Institution Press, Washington, D.C., 1998, p. 30.
11. Environmental Enrichment Information Resources for Nonhuman Primates: 1987–1992, Animal Welfare Information Center, National Agriculture Library, USDA, Washington, D.C., 1992.
12. Reinhardt, V., Reinhardt, A., and Selig, D., *Environmental Enrichment for Nonhuman Primates: An Annotated Bibliography for Animal Care Personnel*, Animal Welfare Institute, Washington, D.C., 1998.
13. Armstrong, K.R., Clark, T.R., and Peterson, M.R., Use of corn-husk nesting material to reduce aggression in caged mice, *Contemp. Topics Lab. Anim. Sci.*, 37(4), 64, 1998.
14. Arnold, C., and Westbrook, R.D., Enrichment in group-housed laboratory golden hamsters, *Anim. Welfare Inform. Cent. Newslett.*, 8(3–4), 22, 1997/1998.
15. Task Force Report No. 130, *The Well-Being of Agricultural Animals*, Council for Agricultural Science and Technology, Ames, IA, 1997.
16. Coviello-McLaughlin, G.M., and Starr, S.J., Rodent enrichment devices—evaluation of preference and efficacy, *Contemp. Topics Lab. Anim. Sci.*, 36(6), 66, 1997.
17. Lidfors, L., Behavioral effects of environmental enrichment for individually caged rabbits, *Appl. Anim. Behav. Sci.*, 52, 157, 1996.
18. Turner, R.J., et al., An immunological assessment of group-housed rabbits, *Lab. Anim.*, 31, 362, 1997.
19. van de Weerd, H.A., et al., Nesting material as enrichment in two mouse strains, *Scand. J. Lab. Anim. Sci.*, 23(1), 119, 1996.
20. van Loo, P.L.P., van de Weerd, H.A., and Baumans, V., Short- and long-term influence of an easy applicable enrichment device on the behavior of the laboratory mouse, *Scand. J. Lab. Anim. Sci.*, 23(1), 113, 1996.
21. Stewart, K.L., and Bayne, K., Environmental enrichment for laboratory animals, in *Laboratory Animal Medicine and Management*, Reuter, J.D., and Suckow, M.A., Eds., International Veterinary Information Service (www.ivis.org), Ithaca, NY, 2004, B2520.0404.

22. Office of Animal Care and Use, National Institutes of Health, Nonhuman Primate Intramural Management Plan, Bethesda, MD, 1991.

23. Bayne, K., Qualitative observations of idiosyncratic behavior in old monkeys, in *Behavior and Pathology of Aging in Rhesus Monkeys*, Davis, R.T., and Leathers, C.W., Eds., Alan R. Liss, New York, 1985.

24. Stewart, K.L., and Raje, S.S., Environmental enrichment committee: its role in program development. *Lab Anim.* (NY), 30(8), 50, 2001.

25. Institutional Administrator's Manual for Laboratory Animal Care and Use, Office for Protection from Research Risks, National Institutes of Health, Bethesda, MD, NIH Publication No. 88-2959, 1988.

26. Morgan, K.N., Line, S.W., and Markowitz, H., Zoos, enrichment, and the skeptical observer: the practical value of assessment, in Second Nature: Environmental Enrichment for Captive Animals, Shepherdson, D.J., Mellen, J.D., and Hutchins, M., Eds., Smithsonian Institution Press, Washington, D.C., 1998, p. 153.

27. Bayne, K., Potential for unintended consequences of environmental enrichment for laboratory animals and research results, *ILAR J.*, 46(2), 129, 2005.

28. Fullerton, F.R., and Gilliatt, R.W., Pressure neuropathy in the hind foot of the guinea pig, *J. Neurol. Neurosurg. Psychiatry*, 30, 18, 1967.

29. Grover-Johnson, N., and Spencer, P.S., Peripheral nerve abnormalities in aging rats, *J. Neuropathol. Exp. Neurol.*, 40(2), 155, 1981.

30. Ortman, J.A., Sahenk, Z., and Mendell, J.R., The experimental production of Renault bodies, *J. Neurol. Sci.*, 62,233, 1983.

31. Peace, T.A., et al., Effects of caging type and animal source on the development of foot lesions in Sprague Dawley rats (*Rattus norvegicus*), *Contemp. Topics Lab. Anim. Sci.*, 40(5), 17, 2001.

32. Rock, F.M., et al., Effects of caging type and group size on selected physiologic variables in rats, *Contemp. Topics Lab. Anim. Sci.*, 36(2), 69, 1997.

33. Manser, C.E., Morris, T.H., and Broom, D.M., An investigation into the effects of cage flooring on the welfare of laboratory rats, *Lab. Anim.*, 29, 353, 1995.

34. Manser, C.E., Morris, T.H., and Broom, D.M., The use of a novel operant test to determine the strength of preference for flooring in laboratory rats, *Lab. Anim.*, 30, 1, 1996.

35. Stauffacher, M., Comparative studies on housing conditions, in *Proceedings of Sixth FELASA Symposium: Harmonization of Laboratory Animal Husbandry*, O'Donoghue, P.N., Ed., Royal Society of Medicine Press, London, 1997, p. 204.

36. Krohn, T.C., Hansen, A.K., and Dragsted, N., Telemetry as a method for measuring impacts of housing condition on rats, *Anim. Welfare*, 12, 53, 2003.

37. Krohn, T.C., and Hansen, A.K., The application of traditional behavioural and physiological methods for monitoring of the welfare impact of different flooring conditions in rodents, *Scand. J. Lab. Anim. Sci.*, 29(2), 79, 2002.

29

Animal Mistreatment and Protocol Noncompliance

Jerald Silverman

Introduction

One of the most contentious roles of the IACUC is its mandate to review allegations of animal mistreatment and protocol noncompliance. This must be done carefully and with the utmost concern for due process and confidentiality as careers may be severely hurt by unfounded accusations and unsubstantiated rumors. Nevertheless, the welfare of animals cannot be compromised. The result is the thin line on which the IACUC walks. The intent of this chapter is to provide the reader with guidance to many of the problems that IACUCs face when confronted with allegations of animal mistreatment and protocol noncompliance. It is important to remember that allegations remain nothing more than allegations until proven otherwise.

Because of the seriousness and sensitivity of this IACUC function, it is very important that the procedures it will use to review and investigate complaints be formalized ahead of time. The time to hunt for the most appropriate operating procedure is not when a complaint reaches the IACUC. Know what you will do ahead of time. Know what authority you are given under the AWAR and PHS Policy, and through your own institution. The AWAR and PHS Policy do not provide all the answers to the myriad problems the IACUC faces when complaints are presented to it. Although IACUC policies can never account for all contingencies, it is wise to have a reasonable number available.

No IACUC works in a vacuum. Rules, regulations, policies, and the like, require a conscientious IACUC and strong institutional support. Without the latter, the most conscientious IACUC is severely handicapped when reviewing and investigating allegations of protocol noncompliance and animal mistreatment.

29:1 What is meant by animal mistreatment and protocol noncompliance?

Reg.   Neither the PHS Policy nor the AWAR directly define *animal mistreatment*; however, U.S. Government Principle IV states that the proper use of animals includes the avoidance or minimization of discomfort, distress, and pain when consistent with sound scientific practices. The AWA (Sect. 1,b,1) states that a purpose of the Act is "to insure that animals intended for use in research facilities or for exhibition purposes or for use as pets are provided humane care and treatment." AWA Sect. 13,a,3,A states that animals used in experimental procedures will be covered by practices causing minimal pain and distress. The *Guide* (p. 1) notes that its basic goal is to promote the humane care of animals used in biomedical and behavioral research, teaching, and testing.

Likewise, *protocol noncompliance* is not directly defined; however, since both the PHS Policy (IV,B,6) and the AWAR (§2.31,c,6) require IACUC approval for animal use activities, it follows that any animal use activity not approved by the IACUC constitutes noncompliance.

Opin. Animal mistreatment is physical or psychological wrongful or abusive treatment of an animal.[1] Examples include hitting animals, taunting animals, or not providing food for punitive reasons.

Protocol noncompliance indicates that procedures or policies approved by the IACUC are not being followed.[1] Examples include performing of unauthorized surgery, participation of unauthorized persons in a research project, or injection of drugs that the IACUC has not approved. It is not unusual to have some overlap between mistreatment and protocol noncompliance. For example, the unauthorized restraint of a nonhuman primate for an unusually long time can potentially entail both animal mistreatment and protocol noncompliance.

When faced with protocol noncompliance the IACUC's first step, if possible, should be to find a way to make the protocol compliant. On the other hand, when faced with animal mistreatment, the IACUC must take immediate action to stop the mistreatment and provide for the needs of the animal.

There may be instances when issues not directly affecting animals become part of an IACUC approved protocol and can lead to compliance concerns. For example, as part of the protocol review process questions may arise about aspects of occupational health and safety (OHS). The PHS Policy (IV,A,1), which uses the *Guide* as the basis for implementing an animal care and use program, requires the implementation of an OHS program (*Guide*, p. 14). The establishment and functioning of this program are the responsibility of the research institution and may encompass multiple departments; nevertheless, the IACUC often acts as a gatekeeper for animal-related issues, not approving an animal activity until the proper OHS or other assurances are established. Discussions in this chapter concerning protocol noncompliance are limited to issues directly affecting animals.

29:2 What role does an institution's administration have in the overall concept of animal mistreatment or protocol noncompliance?

Reg. Under the PHS Policy, the IO, by signing an Animal Welfare Assurance with NIH/OLAW, commits the institution to compliance with the PHS Policy. Noncompliance (such as animal mistreatment or protocol noncompliance) is ultimately the responsibility of the IO who represents the institution's administration (PHS Policy III,G; IV,A). The IO has a similar responsibility for assuring compliance with the AWAR (AWAR §1.1, Institutional Official).

Opin. The IACUC is a regulatory committee that is an agent of the institution (PHS Policy IV,B; AWAR §2.31,c). The institution must make it known that the IACUC not only helps to assure animal welfare, but is an indispensable link in helping protect the researcher against unwarranted accusations and assuring the integrity of the institution.

Because regulatory committees are often disparaged by investigators, it must be understood that an IACUC cannot properly function without general institutional ethical and administrative support. The institution, by actions and words, must stand behind the IACUC. The IO can be a powerful ally and should have a clear understanding of his or her responsibilities and authority under the AWAR and PHS Policy and communicate this support to the IACUC and investigators.

29:3   What is the initial responsibility of the IACUC relative to allegations of animal mistreatment or protocol noncompliance according to the AWAR and the PHS Policy?

Reg.   Under AWAR (§2.31,c,4), the IACUC must "review and, if warranted, investigate concerns involving the care and use of animals at the research facility." This includes complaints from the public and from laboratory or research facility personnel or employees. The PHS Policy (IV,B,4) is slightly broader and states that the IACUC must "review concerns involving the care and use of animals at the institution."

Opin.   A key phrase in the AWAR is *if warranted*, as not all complaints need to be fully investigated. The IACUC Chair, the entire IACUC, or appropriate designees should make such a decision on the basis of the nature of the complaint. For example, the complaint "The dogs prefer Brand A dog food to Brand B" requires far less action than "Dr. X is performing unapproved splenectomies on dogs." Nevertheless, all concerns and complaints must be at least reviewed. (See 29:11.) In both the AWAR and PHS Policy, it is prudent for the IACUC to interpret *institution* and *research facility* as any site where the IACUC has jurisdiction.

29:4   Who can take allegations of animal mistreatment or protocol noncompliance to the IACUC?

Reg.   The AWAR (§2.31,c,4) state that the general public and institutional employees may make complaints. The PHS Policy is silent on this issue; however, it does state that "all institutions are required to comply, as applicable, with the Animal Welfare Act and other Federal statutes and regulations relating to animals" (PHS Policy, II). Further, there is nothing in the PHS Policy that precludes the IACUC from acting upon concerns raised from any source. In addition, it is not unusual for APHIS/AC veterinary medical officers to identify such problems to the IACUC Chair or the AV.

Opin.   Free and open communication is the first step in building the trust that is necessary for people to voice their concerns. Any limitations on who can present allegations to the IACUC sends a message to employees and the public that the IACUC or the institution has something to hide or does not have a full commitment to animal welfare.

Surv.   At your institution, who can bring allegations of animal mistreatment or protocol noncompliance to the IACUC (more than one response is possible)?

- No policy in place   8 /142
- Any employee   141/142
- The public in general   71/142
- Students   75 /142
- Other   2 /142

29:5   What communication pathways should be followed in order to bring allegations of animal mistreatment or protocol noncompliance to the attention of the IACUC?

Reg.   The AWAR (§2.32,c,4) state that research facility personnel must receive training on how to report alleged deficiencies in animal care and treatment, but there are no specific statements in the AWAR as to how to comply with this requirement. PHS Policy is silent on this question.

Opin. Institutions have developed various procedures of their own and most are focused on providing open and easy communication. At least one institution has an ombudsperson who can accept concerns about animal mistreatment or protocol noncompliance. Any allegation must eventually reach the Chair of the IACUC. If the allegation is against the Chair, a written policy should state who will help adjudicate the complaint. As an example, this might be the Vice-Chair of the IACUC, the AV, or the IO.

Typical communication methods include direct verbal conversations and written letters to the AV, IACUC Chair, IO, IACUC members, college deans, and others. This information must be efficiently passed on to the committee chair. It is helpful to have a written IACUC-initiated policy on how to do this.

Surv. In your institution, by which means can people bring allegations of animal mistreatment or protocol noncompliance to the attention of the IACUC? Check all appropriate answers.

• No mechanism in place	16/142
• Through the AV	127/142
• Through any veterinarian or animal facility director	116/142
• Through the IACUC Chairperson	136/142
• Through an IACUC member	129/142
• Through the IO	123/142
• Through an institutional administrator (e.g., dean, IACUC administrator)	112/142
• Through a telephone hotline	35/142
• Through a computer hotline	10/142
• Other	7/142

29:6 How are employees, students, researchers, and others trained and informed that they have the right to bring complaints to the IACUC about animal care and use?

Reg. The AWAR (§2.32,c,4) state that research facility personnel (i.e., all persons, not just those in the animal housing areas) must receive training on how to report alleged deficiencies in animal care and treatment. The means of providing this training is left up to the institution. PHS Policy is silent on this question.

Opin. It is suggested that training be provided for all institutional personnel even if the research facility is only under the auspices of PHS Policy.

Surv. At your institution, how are people notified of the means to bring allegations of animal mistreatment or protocol noncompliance to the attention of the IACUC? More than one response is possible.

• No mechanism or formal policy	27/142
• During a formal training session	76/142
• From handouts given to investigators	32/142
• Through a newsletter	15/142
• Through word of mouth	30/142

- Through posted signs in the animal facility   78/142
- Through written SOPs   52/142
- Through a computer Web site   53/142
- Other   5/142

29:7   Must a complainant be identified to the IACUC or the IACUC Chair?

Opin.   In general, no. There are no federal animal welfare laws, regulations, or policies requiring the identification of a complainant. Required identification might potentially deter certain people from making complaints. Some academic or other institutions have policies that require the identification of a complainant before action can be taken. Although this requirement can lead to a conflict between the IACUC and the institution, in practice it need not. The IACUC itself can act as the complainant in order to maintain the confidentiality of the true complainant (whether known or not). It is prudent for the IACUC and the institution to develop policies for handling such procedural conflicts well before a problem arises. It is suggested that confidentiality be maintained when requested by the complainant.

Surv.   At your institution, must a complainant identify himself or herself before the IACUC is willing to consider a complaint?

- No policy in place   33/146
- Must openly identify himself or herself   1/146
- Need only be identified to the IACUC Chairperson   6/146
- Need only be identified to the AV or IACUC Chairperson 6/146
- Need not be identified   97/146
- Other   3/146

29:8   Is there protection against repercussions if a person makes a complaint to the IACUC; that is, are "whistle-blowers" protected?

Reg.   The AWAR (§2.32,c,4) provide specific protection for employees, IACUC members, or laboratory personnel against discrimination or other reprisals for reporting violations of the AWAR or the AWA itself. Although the PHS Policy makes no definitive statements about reprisals, it does state that all institutions are required to comply, as applicable, with the AWA and other federal statutes and regulations (PHS Policy II; U.S. Government Principle I).

Opin.   It is advisable for institutions to develop policies against repercussions.

29:9   How common are investigations of allegations of animal mistreatment or protocol noncompliance?

Opin.   In this author's experience, allegations are not common. Although it is appropriate to handle problems as expeditiously as possible, there are times when problems should be called to the attention of the IACUC (e.g., unauthorized surgery). To do otherwise makes the IACUC somewhat of a paper tiger and defeats one of the intents of the AWAR and PHS Policy. Thus, the fact that some institutions in the

survey that follows resolve problems without IACUC input should not be construed as being fully appropriate.

Surv.   At your institution, has the IACUC ever been involved with investigating allegations of animal mistreatment or protocol noncompliance?

• No, the issue has never arisen	53/148
• No, the issue has arisen but the problem was resolved without its reaching the IACUC	7/148
• Yes, but very rarely	63/148
• Yes, occasionally	22/148
• Yes, fairly often	2/148
• Other	1/148

29:10   To what extent should allegations of animal mistreatment or protocol noncompliance be documented?

Reg.   The PHS Policy (IV,E,1,b) requires the institution to maintain minutes of IACUC
meetings, activities of the IACUC, and IACUC deliberations. To the extent that
allegations are considered by the IACUC at meetings or acted upon by the IACUC,
records of such actions must be maintained. The AWAR (§2.35,a) have a similar
requirement.

Opin.   It is unlikely that all complaints taken to the IACUC will be fully documented in
writing and signed. Thus, the person who receives the complaint (typically the AV
or IACUC Chair) must get as much information as possible. If the complaint is verbal, questions such as, Did you see this yourself? When did this happen? and
Where did this happen? are appropriate. If possible, one should obtain the name of
a contact person (such as the complainant) and a means of contacting that person.[2]
(See 29:23.)

Surv.   At your institution, how much documentation from the complainant does the
IACUC require before starting an investigation?

• Not applicable because it has never happened	44/140
• No policy in place	14/140
• We will act on any hearsay or other skimpy documentation	26/140
• We will act on hearsay or other skimpy documentation only if it suggests a significant problem	32/140
• We require sufficient documentation to support an allegation before starting an investigation	22/140
• We require heavy documentation	0/140
• Other	2/140

29:11   When should the IACUC take investigative action on allegations of animal mistreatment or protocol noncompliance?

Reg.   The IACUC is responsible for determining whether an activity is in accordance with the AWA, the *Guide*, the institution's Assurance, the PHS Policy

(PHS Policy IV,C,6), or the AWAR (AWAR §2.31,d,6). Thus, the IACUC must consider all allegations of noncompliance and determine whether there is sufficient reason to investigate further.

Opin. Although the IACUC must review all allegations, it need not investigate all of them (see 29:3). There are at least four circumstances when IACUCs should consider becoming actively involved.[2]

- When an allegation should be, but is not, satisfactorily resolved at the local level.

- When any reasonable person would consider that gross mistreatment or protocol noncompliance has occurred. This is based on the allegation itself, although it may subsequently be found to be untrue.

- When there are repeated minor instances of noncompliance or mistreatment-involving the same person or research group. These complaints are often from the AV. (See 29:43.)

- When the IACUC cannot clearly decide whether an allegation is worth investigating, it is probably better to investigate.

The NIH/OLAW provides examples of situations that often result in an IACUC investigation.[3, 4] Any investigation of an allegation should be done in a timely manner. Memories can quickly fade and the desire to move forward can ebb as time passes. (See 29:43; 29:60.)

29:12 What types of documentation might an IACUC need to investigate a complaint fully?

Opin. Necessary documentation varies with the specific situation. Assuming the IACUC as a whole or the designated person (e.g., the Chair) determines that further investigation is warranted, the investigating persons can determine what information they must collect to investigate an allegation properly. It may be necessary for the IACUC to interview people, examine animals, or obtain surgical records, housing records, and even an investigator's research records (in the last example, confidentiality must be assured). (See 29:23.) The IACUC's own communications or policies may have to be examined. Any records obtained should be directly related to the allegation made. Because the IACUC does not have any authority to subpoena people or records, generalized institutional support is crucial to the investigative process. (See 29:2; 29:10.)

29:13 Should the IACUC let all concerned persons know the basis of a complaint and the procedures to be followed? Who might be "concerned persons"?

Opin. Assuming the IACUC has determined that a complaint should be fully investigated, it is prudent to inform all involved persons about the nature of the complaint and procedures to be followed. This method will help assure due process (the protection of the legal and ethical rights of all concerned individuals).[2] Concerned persons can include the alleged violator and that person's immediate superior (if the alleged violator is not the PI). The IO is often informed of the problem at or before this time. Clearly, individual circumstances will dictate final decisions on whom to notify.

Surv.   At your institution, when do you notify an investigator of an allegation of animal mistreatment or protocol noncompliance against him or his staff?

• No policy in place	42/142
• Notify only when an initial inquiry suggests reason to proceed; no details or very limited details provided at this time	28/142
• Notify only when an initial inquiry suggests reason to proceed; full details provided at this time	37/142
• Notify immediately, irrespective of the quality of the allegation; no details or very limited details provided at this time	17/142
• Notify immediately, irrespective of the quality of the allegation; full details provided at this time	18/142

29:14   Once a complaint has reached the IACUC, what is the responsibility of the IACUC Chair?

Opin.   The Chair oversees the efforts described in 29:3. The Chair should assure that the investigation moves forward and due process is followed.

29:15   Is it necessary to investigate fully all complaints taken to the IACUC, even those that seem frivolous?

Reg.   (See 29:11.)
Opin.   No. (See 29:3.)

29:16   Should the IACUC approach each complaint on an individual basis or is a standard operating procedure for handling complaints more appropriate?

Reg.   Although each complaint is likely to have its own unique needs, the AWAR and PHS Policy require the IACUC to review all complaints (see 29:3; 29:11).

Opin.   It is suggested that the IACUC have an SOP to handle complaints. This helps prevent careless errors in due process or confidentiality. It is important that the persons formulating the policy for handling complaints and investigations be representative of the institution's research community as a whole in order to achieve broad support for its recommendations.

29:17   Can the IACUC Chair appoint an investigative subcommittee?

Opin.   Yes. There is nothing in the AWAR or PHS Policy that details how an IACUC should proceed with an investigation if the IACUC determines one is needed. The choice of procedures is entirely at the discretion of the IACUC. It is strongly recommended that an IACUC policy on this topic be established well before a potential problem arises.

Surv.   Once a complaint has reached the IACUC, what procedures does your IACUC Chairperson follow? Check all appropriate answers.

• Not applicable because it has never happened	39/143
• No policy in place	11/143
• Chairperson does his or her own preliminary investigation	48/143
• Chairperson establishes a subcommittee to investigate	50/143
• Chairperson informs IO, who then establishes a subcommittee	13/143
• Chairperson assigns AV to investigate	33/143
• Varies with the problem	55/143
• Entire IACUC investigates	26/143
• IACUC administrative staff investigates	16/143
• Other	4/143

29:18   What might be the composition of an appropriate investigative subcommittee?

Opin.   There are no federal regulations or policies defining the composition of an investigative subcommittee of an IACUC. Investigative subcommittees can be composed of one or more persons, including non-IACUC members. Some IACUCs assign one person (often a veterinarian) to do all (or at least preliminary) investigations, while others establish a larger subcommittee (see 29:17). If the IACUC is small, the entire IACUC may act to investigate an allegation. Some IACUCs attempt to assure that a scientist be on an investigative subcommittee if the allegation is against a scientist. Most investigative subcommittees include a veterinarian. It is suggested that in a large IACUC there be broad enough representation on a subcommittee to help assure due process and acceptance of its recommendations.

Surv.   If an investigative subcommittee (one or more persons) is established, what is a typical composition of this subcommittee? Check all appropriate answers?

• Not applicable at our institution	61/141
• We have a single investigative person who initially handles all initial allegation inquiries	9/141
• Our subcommittee always includes a veterinarian	48/141
• Our subcommittee always includes a representative from the accused's department	7/141
• Our subcommittee tries to have very broad representation	28/141
• Our subcommittee always has a scientist	35/141
• Our subcommittee always has a nonscientist	16/141
• Our subcommittee can have persons not on the IACUC	12/141
• Rarely includes a veterinarian	0/141
• Always includes nonaffiliated member	8/141
• Other	10/141

29:19   Can the IACUC use non-IACUC members as part of an investigative subcommittee?

Opin.   Yes (see 29:18). There is no federal regulation or policy prohibiting this. However, since the IACUC is charged with reviewing all claims of animal mistreatment or protocol noncompliance (see 29:11), it is suggested that only the IACUC appoint such member(s) and that they report to the IACUC directly or through the IACUC subcommittee of which they are a part. The IACUC retains the responsibility for investigative subcommittees under its auspices and any subsequent IACUC actions or determinations.

29:20   Should the IACUC keep a complainant informed of the progress of an IACUC investigation?

Opin.   Assuming the complainant is known, this becomes an IACUC and institutional choice. It is this author's opinion that limited feedback should be provided to a complainant during the course of an investigation. That feedback should be little more than updates (if requested) on where the IACUC is in the investigation and a possible date for the conclusion of the investigation. Full feedback, in this author's opinion, can potentially lead to unwarranted pressure on the investigative committee and intentional or unintentional leaks of confidential information.

Surv.   At your institution, is a complainant kept informed of the progress of an ongoing IACUC investigation? Check all appropriate answers.

• Not applicable because it has never happened	44/149
• No policy in place	32/149
• No feedback is provided	3/149
• We provide limited feedback	21/149
• We provide full feedback	20/149
• Depends on the situation	33/149
• Other	1/149

29:21   Should the IACUC keep a complainant informed of the results of an IACUC investigation?

Opin.   Assuming the complainant is known, this decision becomes an IACUC and institutional choice. In this author's opinion, providing limited results to a complainant once an investigation is completed is appropriate. It reinforces the credibility of the IACUC to both the institution and the public. There is a downside to full disclosure; the complainant may not be satisfied with the IACUC's determination, and he or she may initiate actions against the institution or an individual. Likewise, the alleged violator may do the same and also claim defamation of character. It is recommended by this chapter's author that, in most instances, the results of the investigation should be made known to the full committee and placed in the IACUC minutes (e.g., at the next scheduled meeting), although the specific details to be included become a matter of institutional policy. Institutional policy must be established well in advance. See 27:47–27:53 for federal reporting requirements.

Surv. At your institution, is a complainant informed of the findings of an IACUC investigation?

- Not applicable because it has never happened   44/161
- No policy in place   21/161
- We provide limited results to complainant   11/161
- We provide full results to complainant   42/161
- Depends on the situation   42/161
- No results are given to complainant   0/161
- Other   1/161

29:22   How might the AV report activities not included in an approved IACUC protocol?

Opin. The nature of the activity often dictates actions. An extra toy placed in a cage by a researcher's technician might be welcomed rather than vilified and likely will lead to no report. If a protocol does not include walking a dog up and down a hallway in the animal facility, but it is being done for the benefit of the animal, the AV might suggest to the PI that a protocol addendum (amendment) be made. For significant activities not included in an IACUC protocol (e.g., a major surgical procedure) the most direct route is personal contact with the IACUC Chair followed by a written description of the alleged violation. The AV also may contact the IO or request another IACUC member to contact the Chair. If the AV is the IACUC Chair, he or she should call (if necessary) an emergency meeting of the committee. It is not necessary for the AV to be identified as the source of the complaint. (See 29:5.)

In order to correct the problem rapidly, immediate communication between the AV and PI is strongly advised, if at all possible.

29:23   What can the IACUC do to ensure confidentiality of research or other records during the course of an investigation?

Opin. Various procedures are in use. Some IACUCs keep no records of investigative proceedings, some keep limited records, and many instruct people involved with the proceedings to maintain confidentiality. Two IACUCs, shown on the following survey in the Other category, have formal confidentiality agreements. It must be remembered that allegations remain allegations until proven otherwise, and unintentional leaks of information can adversely affect people's lives. It is suggested that records be kept as they may be crucial for further reference, particularly when APHIS/AC or NIH/OLAW needs to conduct their own investigation. The availability of records should be strictly restricted to those persons having a legitimate need to have access to them. Use of locked cabinets or secured computers may be necessary. (See 29:10.)

Surv. How does your IACUC ensure the confidentiality of research or other records during the course of an investigation? Check all appropriate answers.

- Not applicable at our institution because we have never had an investigation   45/159
- No policy in place   34/159
- No recordings or other records are kept by the IACUC   1/159

- All records and notes are collected after a meeting   23/159
- Only a limited number of people have access to records   52/159
- Instructions are given not to discuss the proceedings   47/159
- Other   5/159

29:24 Is legal or other representation on behalf of the accused person or the institution allowed during the course of an IACUC investigation?

Opin. PHS Policy and the AWAR are silent on this question. As shown by the following survey, most IACUCs have not addressed this issue and there is no AWAR or PHS Policy guidance on the issue. One suggestion is to allow legal or other representation as advisory only to the accused or the IACUC, not as an active participant in an investigation or hearing.

Surv. Is legal or other representation on behalf of the accused person or on behalf of your institution allowed during the course of an IACUC investigation?

- No policy in place   109/142
- We allow legal representation for the accused only   0/142
- We allow legal representation for the institution only   2/142
- We allow legal representation for all parties   23/142
- We do not allow any legal representation   4/142
- Other   4/142

29:25 What is meant by *sanctions*?

Reg. There is no definition of the word *sanction* in the AWAR, and the word does not appear in the PHS Policy. The PHS Policy (IV,C,6; IV,C,7) addresses *suspension of an activity* by the IACUC. NIH/OLAW has defined a *suspension* as any IACUC intervention that results in the temporary or permanent interruption of an animal activity.[3] If the IACUC places an ongoing project on hold or requires a temporary cessation, these actions are synonymous with suspension. Furthermore, PHS Policy, the AWAR, and the *Guide* presume that all ongoing animal activities have received prospective review and approval. Accordingly, the IACUC's authority to suspend unauthorized activities is always implied, if not explicit.[5]

Opin. For the purposes of the questions that follow, *sanctions* refer to any penalty or coercive action taken by the IACUC to help ensure compliance with the AWAR or PHS Policy. Sanctions are only one of the many tools the IACUC can use to help assure the welfare of laboratory animals and rarely should have to be the first used. The IACUC's suspension of a previously approved activity can be considered a type of sanction. (See 29:26.) In the discussions that follow, we often segregate suspensions from other forms of sanctions.

The IACUC always has the authority to make recommendations to the IO regarding any aspect of the institution's animal care and use program (PHS Policy IV,B,5; AWAR §2.31,d,5). If the IACUC believes that the institution should impose institutional sanctions (e.g., a reprimand for repeated violations or revocation of a PI's privilege to conduct research with animals), the IACUC should make those recommendations to the IO.

29:26   What sanctions may the IACUC impose if allegations of animal mistreatment or protocol noncompliance are verified?

Reg.   AWAR (§2.31,c,8) and PHS Policy (IV,B,8) only allow for suspension of an activity previously approved by the IACUC. Thus, any other sanctions imposed by the IACUC, the IO, or the institution are imposed through institutional policy. (See 29:25; 29:33; 29:48; 29:49).

Surv.   Please provide examples of sanctions you have used, irrespective of the verified allegation:

- Not applicable because we have never used sanctions   66/145
- Suspension of all of the investigator's protocols until problem(s) is corrected   36/145
- Suspension of all parts of a specific protocol   34/145
- Suspension of part of a specific protocol, but not the entire protocol   22/145
- Denied access to animal facility   29/145
- Retraining of personnel   61/145
- Permanent suspension of animal use privileges   7/145
- Technical ability required to be approved by a veterinarian   39/145
- Written reprimand   55/145
- Written warning   51/145
- Other research staff required to take over work temporarily   11/145
- Other research staff required to take over work permanently   4/145
- Must attend IACUC meetings   9/145
- Financial charge to investigator for extra work done by veterinary or animal care staff   10/145
- Significant salary reduction for a period of time   1/145
- Monetary fine   2/145
- Dismissal of personnel   8/145
- Loss of research funds   2/145
- Other   2/145

29:27   If an allegation of animal mistreatment or protocol noncompliance is verified, must the IACUC apply sanctions?

Reg.   PHS Policy (IV,C,6) states that the IACUC may suspend an activity if it determines that the activity is not being conducted in accordance with applicable provisions of the AWA, the *Guide*, the institution's Assurance, or IV,C,1,a–IV,C,1,g of the PHS Policy. The AWAR (§2.31,d,7) and the PHS Policy (IV,B,8) state that the IO, in consultation with the IACUC, must take appropriate corrective action in the event of a suspension of a previously approved activity.

Opin.   If, in the opinion of the IACUC, sanctions are not appropriate, they need not be applied. A clearly minor and unintentional misinterpretation of an IACUC policy that has created no problem for an animal is an example of a verified allegation of protocol noncompliance that might lead to an explanation, not a sanction. Interestingly, a comparison between the data reported in the 29:26 and 29:48 surveys revealed that 11 of the approximately 145 responding institutions informed either APHIS/AC or NIH/OLAW of a significant incident but did not impose any sanctions of any sort (comparative data not shown).

For sanctions other than a suspension of a previously approved activity, institutional and IACUC policy should be established and followed. It is strongly suggested that any authority the IACUC, IO, or any other institutional representative has to impose sanctions (other than suspensions) be unambiguous.

Surv. If an allegation of animal mistreatment or protocol noncompliance is verified, does your IACUC apply sanctions?

- Not applicable because problem has never occurred   62/142
- We have never applied sanctions   11/142
- We have occasionally applied sanctions of some sort   36/142
- We always apply sanctions of some sort   30/142
- Other   3/142

29:28 Does the IACUC, the IO, or both decide on the sanctions to be taken?

Reg. The AWAR (§2.31,d,7) and the PHS Policy (IV,C,7) state that the IO, in consultation with the IACUC, must take appropriate corrective action after a suspension of a previously approved activity. (See 29:25.)

Opin. The wording of the AWAR and PHS Policy suggests that the IO has somewhat more authority than the IACUC in the development of corrective actions after suspensions. In practice, most IACUCs decide what sanctions are to be taken and only the IACUC can suspend a previously approved activity (see 29:26). A smaller number of IACUCs recommend a sanction to the IO, who then imposes the same.

Surv. At your institution, who typically decides what sanctions are to be taken? Check all appropriate responses.

- Not applicable because problem has never arisen   53/140
- No policy in place   10/140
- The investigative subcommittee of the IACUC only   1/140
- The investigative subcommittee with full IACUC approval 14/140
- The full IACUC   50/140
- The IACUC Chairperson with IACUC approval   10/140
- The IO only   1/140
- The IO and the IACUC   34/140
- Other   4/140

29:29 If sanctions are imposed by the IACUC, do they occur via a formal full committee vote or can the IACUC Chair take action on behalf of the committee?

Reg. A formal full committee vote is required for the suspension of a previously approved activity (AWAR §2.31,d,6; PHS Policy IV,C,6).

Opin. Neither the AWAR nor PHS Policy addresses the question of sanctions (as defined in 29:25) other than suspension of a previously approved activity. The method for approval of the imposition of sanctions (other than a suspension) to help correct a problem should become a policy of the IACUC in consultation with the IO. It is suggested that any sanctions taken (whether a suspension of a previously approved activity or any other form of sanction) require an affirmative vote by a majority of a quorum of the full committee.

29:30   Can an appeal be made to the IACUC or other institutional authority if sanctions are imposed?

Reg.   The AWAR and PHS Policy are silent on this issue. Nevertheless, officials of an institution cannot approve an activity that has not been approved by the IACUC (AWAR §2.31,d,8; PHS Policy IV,C,8).

Opin.   An institution, as part of its own policy, may allow appeals to be made to the IACUC, the IO, or other persons. This may be appropriate as the IACUC is not infallible or (for example) may not have followed the correct procedures for suspending a protocol. Another pair of eyes or the presentation of a different perspective on a problem might suggest alternative solutions to the IACUC. If appeals are permitted through institutional policy for sanctions other than the suspension of an animal activity, then that same policy can permit the IO or other persons to override that sanction. Nevertheless, this is a dangerous policy, and not one that can be recommended. (See 9:56–9:57.)

For the IACUC to run smoothly and accomplish its task, it is strongly recommended that all policies concerning sanctions be clearly understood by the IACUC, IO, and all animal users.

Surv.   Does your IACUC allow appeals of any sanctions imposed?

- Not applicable because the problem has never arisen   69/143
- No policy in place   32/143
- Appeals allowed to IACUC only   12/143
- Appeals allowed to the IO only   3/143
- Appeals allowed to IACUC and IO   17/143
- Appeals allowed to IO and higher administrators   5/143
- No appeals allowed   4/143
- Other   1/143

29:31   Can an appeal of sanctions imposed by the IACUC be made to an APHIS/AC administrator or to the NIH/OLAW?

Opin.   No. There are no provisions for such appeals. The PHS Policy (V,A,5) allows the NIH/OLAW to waive provisions of its policy, but that section refers to general programmatic decisions, not decisions on a given sanction.

29:32   Can an institution or IO override the IACUC and impose lesser sanctions on a person?

Reg.   An IO may not override an IACUC suspension (IACUC withdrawal of approval). In the event of an IACUC suspension, the IO is required to take appropriate corrective action and report that action with a full explanation to NIH/OLAW (PHS Policy IV,C,7) or APHIS/AC (AWAR §2.31,d,7).

Opin. Although the IO may not override an IACUC suspension, other sanctions do potentially open the door to overriding. As noted in 20:30, this is not a recommended procedure for the IO.

29:33 Can an institution or IO impose sanctions that are more severe than those imposed by the IACUC?

Reg. The PHS Policy and the AWAR do not preclude any institutional or IO sanctions. Therefore, these may be even more severe than those imposed by the IACUC.

Opin. Yes. There is nothing to prohibit the institution or its authorized representative from imposing a sanction more severe than that recommended by the IACUC. The AWAR (§2.31,d,7; §2.31,d,8) imply that the research institution may do this as does PHS Policy (IV,C,7; IV,C,8).

29:34 Can an IACUC impose sanctions if the vertebrate animals used are not currently covered by the AWAR (e.g., common laboratory rats) and funding for the study is from nonfederal sources?

Opin. The AWAR (§1.1, Animal) and PHS Policy (III,A) clearly state which nonhuman animals are covered by their respective agencies. (See 12:1.) Most IACUCs, as an institutional policy, oversee the care and use of all vertebrate animals in biomedical research, teaching, and product testing, whether or not those animals are under the auspices of pertinent federal regulations or policies. Indeed, many institutional Animal Welfare Assurances to the NIH/OLAW are worded in a manner that includes all vertebrate animals, not just those of studies funded by the PHS. If, as part of an institutional policy, an IACUC is empowered to apply sanctions (in addition to the suspension of a previously approved activity), then in many instances those sanctions can be applied to all vertebrate animals.

29:35 Can the IACUC Chair apply sanctions (other than suspension of a previously approved protocol) without full IACUC approval if the IO is in agreement with the sanctions?

Opin. The only sanction the IACUC is empowered to employ under the AWAR and PHS Policy is the suspension of a previously approved activity, and a vote for suspension must occur at a meeting of the full committee (see 29:26). Thus, the application of other sanctions by the Chair, with or without full IACUC approval, is a matter of institutional policy. (See 29:30.) There may be unique circumstances under which a Chair might unilaterally impose a sanction, pending rapid review by the IACUC (see 29:38; 29:55).

29:36 Can the AV suspend an ongoing IACUC-approved research project or related activity?

Reg. There is nothing in the AWAR or PHS Policy that specifically authorizes the AV to suspend a research project. However, the AWAR (§1.1 Attending Veterinarian; §2.33,a,2) state that the AV must have direct or delegated authority for activities of animal care and use and must be able to provide adequate veterinary care. If this means stopping an activity to provide adequate veterinary care, then the AV has that authority.

The PHS Policy has a similar statement (IV,A,3,b,1). The IACUC clearly has authority to suspend a research protocol (AWAR §2.31,d,6; PHS Policy IV,B,8).

Opin.   These regulations have been interpreted by some IACUCs as giving the AV authority to suspend an ongoing project, subject to IACUC review. Such an in-house policy can become part of an adequate veterinary care statement for APHIS/AC (for consulting AVs) or the PHS Assurance statement. It should be disseminated to investigators. The IACUC should be immediately notified if a suspension occurs. In this author's opinion, the AV should have the authority to suspend an ongoing study, subject to rapid review by the IACUC. Any such suspension, even if not sustained by the IACUC, must be reported to the federal government.[4,6] (See 29:55.)

Surv.   At your institution, does the AV have the authority to suspend a research or teaching project?

• No policy	16/142
• No	28/142
• Yes, and the IACUC will quickly meet and review	78/142
• Yes, and review at the next IACUC meeting	18/142
• Other	2/142

29:37   Can a veterinarian other than the AV suspend a previously approved research or teaching project?

Opin.   There is nothing in the AWAR or PHS Policy that specifically authorizes the AV or any other veterinarian to suspend a previously approved activity. Any such authorization is a matter of institutional policy. (See 29:36.) In this author's opinion, the veterinarian with primary responsibility for laboratory animal science issues should have the authority to suspend an ongoing project, subject to subsequent rapid review by the IACUC. Any such suspension, even if not sustained by the IACUC, must be reported to the federal government.[4,6] (See 29:55.)

Surv.   At your institution, does a veterinarian other than the AV have the authority to suspend a research or teaching project?

• Not applicable as we only have one veterinarian	55/142
• No policy	10/142
• No	45/142
• Yes, and the IACUC will quickly meet and review	26/142
• Yes, and IACUC will review at the next scheduled meeting	5/142
• Other	1/142

29:38   Can the IACUC Chair suspend an ongoing IACUC-approved research project prior to a meeting of the full IACUC? (See 29:35.)

Opin.   There is nothing in the AWAR or PHS Policy that gives such authority to the IACUC Chair. Nevertheless, an institution may delegate such authority to the IACUC Chair pending review by the full committee. Such an in-house policy should become part of the PHS Assurance statement and should be disseminated to investigators. It also should become part of an adequate veterinary care statement for the APHIS/AC (if a consulting veterinarian is used). In the opinion of this author, the IACUC Chair should have the authority to suspend an ongoing study, subject to subsequent rapid review by the IACUC. Any such suspension, even if not sustained by the IACUC, must be reported to the federal government.[4,6] (See 29:55.)

Surv. At your institution, does the IACUC Chairperson have the authority to suspend a research or teaching project?

- No policy   17/137
- No   48/137
- Yes, and the IACUC will quickly meet and review   61/137
- Yes, and the IACUC will review this at the next
  scheduled meeting   11/137

29:39 Can the IACUC suspend a previously approved research project?

Reg. Yes. The AWAR (§2.31,d,6) and PHS Policy (IV,B,8; IV,C,6) specifically allow the IACUC to suspend a previously approved activity after review of the matter at a convened quorum of the IACUC, and with a suspension vote of a majority of the quorum present.

29:40 Must the IACUC suspend an entire previously approved activity or can it suspend parts of a previously approved activity?

Opin. The AWAR and PHS Policy are silent as to whether the IACUC has the authority to suspend only portions of a previously approved activity. It, therefore, may be interpreted to allow the IACUC the authority to suspend either a full activity or a portion of such an activity. As an example, an IACUC can determine that only a surgical component of an ongoing research protocol need be suspended. (See 29:49.)

29:41 An investigator performs an animal activity that does not have IACUC approval. Must the IACUC take a formal vote in order to suspend an activity that it never approved?

Reg. The purpose of the IACUC is "to assess the research facility's animal program, facilities and procedures." (AWAR §2.31,a; PHS Policy IV,A,3,a, uses similar language.) The AWAR (§2.31,c,6) and the PHS Policy (IV,B,6) authorize the IACUC to approve activities involving the care and use of animals. Suspension of an ongoing animal activity requires a formal vote of a quorum of the full committee. (See 29:26; 29:29.)

Opin. NIH/OLAW and APHIS/AC have written, "While it is debatable whether the IACUC can technically suspend an activity that it has not previously approved, PHS Policy, USDA Regulation, and the *Guide* language presume that all ongoing animal activities have received prospective review and approval. Accordingly, the IACUC's authority to suspend unauthorized activities is always implied, if not explicit."[3] On the basis of this interpretation, and the IACUC's responsibility to oversee animal-based research, the IACUC can suspend an activity it never formally approved. Further, it would require a vote of a quorum of the full committee to suspend that activity. This opinion is based on the premise, noted previously, that IACUC approval is presumed to be necessary to begin an animal activity. It therefore follows that the suspension of that activity must have a properly taken suspension vote from the IACUC. Another option would be for the institution itself to give an individual, such as the IACUC Chair or IO, the authority to suspend an animal activity. (See 29:38; 29:58.) Such authority should have strict limitations as it can override an important function of the IACUC. It would still require notification of the suspension to the appropriate federal agencies.[4,6]

29:42   An IACUC voted to suspend a researcher's animal use activity if any instance of protocol noncompliance were to be confirmed during a certain period. Another instance of noncompliance was confirmed during that period. Is it necessary for the IACUC to take another vote to suspend the protocol?

Reg.   A formal vote of a majority of the quorum present at a meeting of the IACUC is required for the suspension of a previously approved activity (AWAR §2.31,d,6; PHS Policy IV,C,6).

Opin.   This scenario approaches an unclear area of compliance, as the IACUC had already voted to suspend the researcher's animal use activity if a confirmed instance of protocol noncompliance occurred. Nevertheless, it would be appropriate to take a revote since noncompliance can range from a very minor incident to a significant incident. Further, the circumstances surrounding a particular incident can influence the committee's decision making. Because the IO, in consultation with the IACUC, has to review the reasons for the suspension and take appropriate corrective action (AWAR §2.31,d,7; PHS Policy IV,C,7) it might be inappropriate to put the IO in this position if the incident was quite minor. There is nothing in the AWAR or PHS Policy that prohibits the IACUC from reversing a decision if it is correctly done and documented. Indeed, the AWAR (§2.31,d,4) and PHS Policy (IV,C,4), referring specifically to withheld approval of an animal activity, allow for the IACUC to reconsider its decision. One can extrapolate from this that reconsidering a previous action is a reasonable and proper function of the IACUC; however, if there is to be a reversal or other change to a previous action, the IACUC must do so by voting.

29:43   For helping to determine when instances of continuing protocol noncompliance or animal abuse require direct IACUC involvement and possible sanctions, a point system can be considered. What is meant by a *point system* and how might it be implemented?

Reg.   PHS Policy (IV,F,3,a) requires prompt reporting to NIH/OLAW of any continuing noncompliance with the PHS Policy. The PHS Policy is silent on the types of continuing noncompliance that might prompt such a report. The AWAR (§2.31,d,7) requires notification of APHIS/AC and any federal funding agency of a suspension of a previously approved animal activity and the reason for the suspension. There is no mention of "continuing noncompliance."

Opin.   Neither the PHS Policy, the *Guide*, nor the AWAR provide specific examples of what constitutes serious enough incidents of continued protocol noncompliance to report to APHIS/AC or NIH/OLAW. The *Guide* (p. 3) notes that the general evaluation of the animal care and use program is left to the judgment of the IACUC and others involved in the program. NIH/OLAW has provided, as an example, that failure to correct situations identified in previous semiannual evaluations as significant deficiencies would constitute continued noncompliance.[3] Other examples of reportable situations have been identified by NIH/OLAW.[4] (See 29:48.) Whereas certain actions, either individually or repeated, are clearly significant enough to consider evoking sanctions (e.g., performing major survival surgery without IACUC approval), others are not quite as clear (e.g., forgetting to replace a watering device promptly after an animal has fully recovered from anesthesia, not observing postoperative animals as often as stated in the approved protocol). Repeated problems of the latter sort in the same research group, even if they do not cause an animal well-being problem, can present a problem for an IACUC. Assigning a point value for certain types of noncompliance actions, with a predetermined total "score" that triggers the IACUC to consider sanctions or a report to a federal agency, can help an IACUC decide when enough is enough. Thus, repetitive problems are summed up over a predetermined period.

For example, an IACUC might use a total score of 100 points within 6 months to trigger committee deliberations on a possible sanction. This can be the suspension of an animal activity or a lesser sanction if the committee has its organization's authority to impose lesser sanctions. Not replacing a water bottle immediately after full recovery from anesthesia might be assigned 5 points, not observing an animal as often as stated in the protocol might be assigned 20 points, and missing one scheduled analgesic administration might be assigned 35 points. It is suggested that if a point system is to be considered, general categories of noncompliance activities be defined ahead of time to prevent the IACUC from becoming bogged down with determining the relative importance of every noncompliance scenario that reaches the committee. Further, a point system, if used, should not be considered so inflexible as to preclude the IACUC from taking an alternate course of action.

29:44 Is there any regulatory requirement to have a "face-to-face" meeting of a quorum of the IACUC in order to vote on the suspension of an animal activity?

Reg.   PHS Policy (IV,C,6) and the AWAR (§2.31,d,6) state that a vote to suspend an animal activity can occur "only after review of the matter at a convened meeting of a quorum of the IACUC and with the suspension vote of a majority of the quorum present."

Opin.   The regulatory requirement noted previously suggests, but does not explicitly state, that a face-to-face committee meeting is required. NIH/OLAW has written that a *convened quorum* generally means the traditional gathering of people in a meeting room at the same time; however, in exceptional circumstances "Innovative mechanisms [such as videoconferencing] that meet the description of convened quorum could be used to perform this IACUC function."[7] APHIS/AC has also implied that a videoconference might be able to satisfy the requirement.[8] Given the seriousness of suspending an animal activity, it is this author's opinion that in all but the most exceptional of circumstances there should be a face-to-face meeting of the IACUC when a vote to suspend an animal activity is needed.

There may be unique circumstances in which there is an immediate need to suspend an animal activity without a meeting of the full IACUC because of a significant danger to animal welfare. This is discussed in 29:36–29:39.

29:45 How long should a suspension of a previously approved activity last?

Opin.   The duration is at the discretion of the IACUC. It is unusual for an activity to be permanently suspended. Most activities are suspended for the length of time that the IACUC believes will prevent recurrence of the problem and allows any remedial actions to occur. Any suspension must include the enactment of corrective actions to help prevent future problems (AWAR §2.31,d,7; PHS Policy IV,C,8).

29:46   If a protocol is suspended, how is it reactivated?

Reg.   The AWAR and the PHS Policy are silent on this issue.

Opin.   The IACUC should lift a suspension only after it has determined that the activity is, or will be, in full compliance with the PHS Policy, the NIH/OLAW Animal Welfare Assurance, AWAR, and the *Guide*. The IACUC should establish a policy to provide for reactivation. If the suspension is for a finite period, the committee might vote to have it automatically lifted when that period ends. If the suspension is open ended until a particular action occurs, the IACUC can vote (at the time of the suspension) for the IACUC Chair to be empowered to reactivate the protocol at his or her discretion, or it can vote that the committee itself must approve reactivation.

29:47   If sanctions are imposed, which persons or agencies must be informed and who does the informing?

Reg.   PHS Policy (IV,C,7; IV,F,3) and the AWAR (§2.31,d,7) refer only to the suspension of previously approved activities. Under that circumstance, the IO and the IACUC must promptly report the suspension along with a full explanation to NIH/OLAW. PHS Policy (IV,F,3) also requires the IACUC, through the IO, to report to NIH/OLAW any serious or continuing noncompliance with the PHS Policy or serious deviations from the provisions of the *Guide*. If the species used is under the auspices of the AWAR, the suspension and a full explanation must be sent to APHIS/AC and any federal agency funding the suspended activity.

Opin.   Sanctions taken by the IACUC (other than a suspension of a previously approved activity) need not be reported to federal agencies. However, some institutions routinely report any sanctions, even minor ones, to the IO and in certain cases other officials of the institution.

29:48   What types of problems should be reported to the NIH/OLAW and APHIS/AC?

Reg.   PHS Policy (IV,F,3) requires the IACUC, through the IO, to report to NIH/OLAW any serious or continuing noncompliance with the PHS Policy or serious deviations from the provisions of the *Guide*. The requirement to report suspensions of previously approved activities (PHS Policy IV,C,7; IV,F,3; AWAR §2.31,d,7) has been previously noted (see 29:47; 29:49).

Opin.   NIH/OLAW offers examples of actions that require reporting.[3,4] They include (but are not limited to) the following (and see 30:16):

- IACUC suspension or other institutional intervention that results in the temporary or permanent interruption of an activity due to noncompliance with the PHS Policy, AWA, the *Guide*, or the institution's Animal Welfare Assurance
- Failure to adhere to IACUC-approved protocols
- Serious or continuing noncompliance with the PHS Policy

- Serious deviations from the provisions of the *Guide*, such as shortcomings in programs of veterinary care, occupational health, or training
- Failure to maintain appropriate animal-related records
- Failure to ensure the death of an animal after euthanasia
- Failure of animal care and use personnel to carry out veterinary orders (e.g., treatments).

The AWAR (§2.31,c,3) also require reporting of any uncorrected significant deficiency (found on semiannual review) to be reported to APHIS/AC and any federal funding agency.

Surv.   Has your IACUC ever informed NIH/OLAW or APHIS/AC (because of regulatory requirements) about an activity involving animals (or an animal facility) that led to a suspension, investigation, or other reportable action?

• This has never happened	74/142
• This has happened, but very infrequently	56/142
• This happens with moderate frequency	9/142
• This happens fairly often	0/142
• Other	3/142

29:49   If only parts of a previously approved activity are suspended by the IACUC (e.g., the surgical portion of a large project), must the IACUC notify the NIH/OLAW or APHIS/AC?

Reg.   PHS Policy (IV,F,3) requires the IACUC, through the IO, to report promptly any serious or continuing noncompliance with PHS Policy or serious deviation from the provisions of the *Guide* or any suspension of an activity by the IACUC, along with a full explanation of the circumstances and actions taken by the IACUC.

The AWAR (§2.31,d,6) allow the IACUC to suspend an activity if it is not being conducted in accordance with the description that was provided by the PI and approved by the IACUC. The AWAR (§2.31,d,7) require a full explanation as part of the reporting.

Opin.   The wording in the AWAR and PHS Policy refers to an *animal activity*, which need not be the entire IACUC protocol. Many suspensions of a part of an ongoing approved activity would meet one or more of the concerns mentioned in the regulatory section of this question and, therefore, would be reportable even if the entire protocol were not suspended. In some instances (e.g., excessive mortality or morbidity rate), the IACUC might suspend part of an ongoing activity until more information could be gathered. If all or part of an animal activity is suspended, even if temporarily, NIH/OLAW, or APHIS/AC and the appropriate federal funding agency, should be notified.[4,6] (See 29:40; 29:48; 29:55.)

On the basis of the results of the surveys in 29:26 and 29:48, it is apparent that not all organizations are cognizant of the reporting requirement described herein. Although not shown, a comparative analysis of the data from those surveys revealed that ten of the approximately 145 responding institutions had suspended all or part of an IACUC protocol without reporting this to either NIH/OLAW or APHIS/AC (nearly 100% of survey respondents were required to comply with the PHS Policy and AWAR). The comparison also showed that eight additional institutions imposed significant sanctions (other than suspensions of previously approved activities) yet did not report any associated incident to either NIH/ OLAW or APHIS/AC. These sanctions included denied access to animal facilities, a significant monetary fine, and dismissal from employment.

29:50 Must the IACUC inform a federal funding agency for any sanction that is imposed or only if certain classes of sanctions are imposed?

Reg.   (See 29:47; 29:49.)

29:51 Must a federal funding agency be informed if allegations are verified, but sanctions are not imposed?

Reg.   No. A federal funding agency must be informed only if a suspension of a previously approved activity occurs and the species involved are under the auspices of the AWAR (§2.31,d,7). The NIH/OLAW must be informed of certain circumstances that may not necessarily evoke sanctions. (See 29:47–29:49.)

Opin.   The requirement to inform federal agencies about the suspension of an animal activity should not be confused with the need to notify appropriate federal funding agencies and the NIH/OLAW if major deficiencies were noted during the semiannual program review and inspection of animal care and use areas that were not corrected in the appropriate period (AWAR §2.31,c,3; PHS Policy IV,F,3). (See 29:48.)

29:52 Must nonfederal funding agencies be informed of full or partial suspensions of a previously approved activity?

Opin.   Nonfederal funding agencies may establish whatever criteria are appropriate to their needs. These criteria may be binding on an institution. The IACUC may be requested by the nonfederal agency to provide information to it about suspensions or other sanctions.

29:53 If an IACUC suspension of a previously approved activity occurs, what information should be provided to NIH/OLAW, APHIS/AC, and federal funding agencies?

Reg.   (See 29:47; 29:49.)
Opin.   NIH/OLAW has provided guidance on this issue.[4] It is appropriate to include the federal grant or contract number, the Animal Welfare Assurance number (when applicable), and the category of people involved (e.g., research technicians or veterinarians). It is also suggested that a full and clear statement of the problem be provided along with the planned (or completed) corrective action. Details of the suspension (e.g., length of time) should be included.

29:54 An institution has an IACUC but it does not house or use animals that fall under the auspices of the AWAR or PHS Policy. Must the institution inform APHIS/AC and NIH/OLAW of any suspensions of a previously approved activity?

Reg.   No. If no activity involving PHS-supported animals is conducted at the institution, the institution does not have reporting obligations to NIH/OLAW. Likewise, if animal use does not include species covered by the AWAR, there is no reporting obligation to APHIS/AC.

Opin. Under the circumstances described, the applicable federal regulations and policies do not pertain to the institution. The words *institutional animal care and use committee* do not guarantee that a facility has indirect federal oversight of its animal care and use activities. Nevertheless, the NIH/OLAW Assurance may include vertebrate animals in studies that are not PHS funded. Under those circumstances, NIH/OLAW must be notified. (See 29:34.)

29:55 Should NIH/OLAW or the APHIS/AC be informed that a project has been suspended pending the completion of an IACUC investigation?

Reg. (See 29:35–29:38.)

Opin. There is no AWAR or PHS Policy that specifically requires APHIS/AC or NIH/OLAW notification about an activity that is under investigation. However, a gray area arises if the IACUC has properly discussed and voted to suspend an activity until an investigation is completed. Technically, that suspension must be reported to APHIS/AC or NIH/OLAW, but because full details are not yet known or verified, the IACUC cannot properly comply with the AWAR (§2.31,d,7) or PHS Policy (IV,C,7) on reporting the suspension along with a full explanation of the problem and corrective actions. Therefore, it is suggested that an institutional policy be formulated to permit the IACUC to suspend an activity pending the results of a full investigation. The NIH/OLAW and APHIS/AC have suggested that they be informally advised of any suspension that occurs under these unique circumstances, and formally notified under the previously noted regulation and policy if the IACUC verifies the complaint leading to the suspension.[4,9] (See 29:60.)

29:56 A PI begins a study without IACUC approval. The IACUC Chair becomes aware of this and asks the PI to halt all animal work. The PI responds that if he complies, all the data collected to date will be meaningless and the lives of the animals will have been wasted. What is the appropriate IACUC response?

Reg. The AWAR (§2.31,c,6) and the PHS Policy (IV,B,6) authorize the IACUC to approve activities involving the care and use of animals. Beginning an animal activity without IACUC approval constitutes serious noncompliance with the AWAR and PHS Policy. Suspension of an ongoing animal activity requires a formal vote of a quorum of the full committee (see 29:26; 29:29).

Opin. The lives of the animals should not be wasted, yet the IACUC should not accept the PI's explanation without investigating the validity of the claim in a timely manner. In the interim, the committee should institute detailed IACUC oversight of the research, define any limitations on further animal use within the study, and require the immediate submission of a protocol for IACUC review. If it is determined that the PI's claim is not valid, the project must be suspended until (and if) it is approved by the IACUC. The only sanction allowable under the PHS Policy and AWAR is the suspension of the ongoing animal activity (see 29:26). However, the committee can impose any sanctions it deems appropriate via authority it may have received from the research institution.

The IACUC Chair has no authority under the PHS Policy or AWAR unilaterally to suspend an animal activity. Any such authority would emanate from internal institutional policies. Whether or not all or part of the project is formally or informally suspended, or whether the IACUC uses institutional authority to impose other sanctions on the PI, the fact remains that this action constitutes serious noncompliance with the PHS Policy and AWAR and should be reported to those agencies (PHS Policy IV,F,3; AWAR §2.31,d,7).[4,6] (See 29:38; 29:49; 29:57.)

29:57   A PI gently takes a small amount of blood from a dog's cephalic vein but does so without having an IACUC-approved protocol. Is this considered serious noncompliance with PHS Policy and the AWAR and should it be reported to those agencies and the appropriate funding agencies?

Reg.   The AWAR (§2.31,c,6) and the PHS Policy (IV,B,6) authorize the IACUC to approve activities involving the care and use of animals. The PHS Policy (IV,F,3,a) requires the prompt reporting of any serious noncompliance with the PHS Policy.

Opin.   Previous guidance from NIH/OLAW[3,4,10] makes it clear that performing research without IACUC approval is serious noncompliance and subject to reporting of the action to NIH/OLAW. The same rationale can be extrapolated for the purposes of reporting such an activity to APHIS/AC and any sponsoring federal agencies, although the AWAR does not specifically mandate this unless the IACUC also suspends all or part of the approved protocol. Nevertheless, as noted in 29:1, this is considered protocol noncompliance under the PHS Policy and AWAR. Further, the AWA (Section 13,a,7,B,iii) requires a research facility to provide an explanation for any deviation from the standards of the provisions of the AWA during an APHIS/AC inspection and provide a similar explanation on the APHIS/AC annual report. The latter requirement is reiterated in the AWAR (§2.36,b,3). Also, Section 13,b,4,A,ii of the AWA requires that during the IACUC's semiannual inspections of research facilities the committee is to note any violation of the standards or deviations of research practices from the originally approved proposals that might adversely affect animal welfare. The fact that the procedure was a common one and was performed gently does not negate the fact that there was no IACUC oversight of the research.

29:58   An IO suspends an animal activity even though the IACUC voted not to suspend the same animal activity. Is this suspension reportable to the NIH/OLAW and/or APHIS/AC since it did not arise from an IACUC vote?

Reg.   The PHS Policy and the AWAR do not preclude any IO sanctions. (See 29:33.) A formal full committee vote is required for the suspension of a previously approved activity (AWAR §2.31,d,6; PHS Policy IV,C,6).

Opin.   Although the IO cannot approve an animal activity that has not been approved by the IACUC, there is nothing in the PHS Policy or AWAR that prohibits the IO from suspending an animal activity if the IO has the institutional authority to do so (see 29:33). In a scenario similar to that described in this question, APHIS/AC responded that although a suspension did not occur strictly according to regulatory requirements, it was done with the institution's decision to allow the IO to suspend the activity and therefore would have to be promptly reported to the USDA and any federal agency funding that activity.[6] The same rationale is maintained by NIH/OLAW.[4] This is further supported by NIH/OLAW position that judgments of the degree of seriousness associated with noncompliance with the PHS Policy or deviation from the provisions of the *Guide* are not the exclusive prerogative of the IACUC.[10]

29:59   The PHS Policy (IV,F,3) speaks of providing NIH/OLAW with a full explanation of the circumstances and actions taken with respect to any serious or continuing noncompliance with PHS policy or any serious deviation from the provisions of the *Guide.* What are some examples of serious IACUC protocol noncompliance?

Opin.   *Noncompliance* in the context of PHS Policy IV,F,3 refers to the entirety of the PHS Policy. Protocol noncompliance is a subcategory. NIH/OLAW has provided many examples of noncompliance with the overall PHS Policy and some specific examples of protocol noncompliance.[4] They include (but are not limited to) beginning of projects without IACUC approval, inadequate postprocedural care, failure to maintain appropriate animal-related records, and failure to ensure death of animals after euthanasia procedures.

      Although *serious protocol noncompliance* is *not* synonymous with *significant deficiencies* found during semiannual IACUC inspections, the concept that a *significant deficiency* is any action that "is or may be a threat to the health or safety of animals" can be applied to certain instances of *serious noncompliance* with an IACUC protocol. Therefore, additional examples of serious protocol noncompliance are not providing anesthetics or analgesics as required in the protocol, performing unapproved multiple survival surgical procedures, using an unapproved species, and using an unapproved euthanasia method. Further examples that constitute serious protocol noncompliance include having improperly trained persons working with animals, changing the PI without IACUC approval, and using far more animals than approved.

29:60   How soon after a previously approved activity is suspended must that suspension be reported to APHIS/AC or NIH/OLAW?

Reg.   The PHS Policy (IV,F,3) requires institutions to report "promptly" any serious or continuing noncompliance with the PHS Policy or serious deviation from the provisions of the *Guide.* This includes the temporary or permanent interruption of an activity involving animals by the IACUC.[4,11] The AWAR (§2.31,d,7) does not provide any general or specific time frame for notifying APHIS/AC of a suspension of an activity involving animals.

Opin.   Although the AWAR do not specify any time frame for reporting a suspension to APHIS/AC, this should be done promptly, along with the reasons for the suspension and the specific activities or plans for corrective action (see 29:53). NIH/OLAW requests similar information[11] and has emphasized that *promptly* means without delay.[4] NIH/OLAW understands that it may take some time to develop a full report, as required by PHS Policy (IV,F,3), and therefore recommends that a preliminary report be transmitted to NIH/OLAW as soon as possible, and followed up by a thorough report once action has been taken.[4] (See 29:55.)

29:61   Is continued or serious noncompliance with the PHS Policy to be construed as research misconduct under the auspices of the NIH Office of Research Integrity?

Opin.   No. The Office of Research Integrity (ORI), working under the regulations of 42 CFR Part 50, Subpart A, directs the PHS research integrity activities on behalf of the Secretary of Health and Human Services (with the exception of the regulatory research integrity activities of the Food and Drug Administration). The ORI oversees *research misconduct* defined as fabrication, falsification, plagiarism, or other practices that seriously deviate from those that are commonly accepted within the scientific community for proposing, conducting, or reporting research. Its charge does not include overseeing compliance with regulations or policies governing the proper use of animals in biomedical research.

References

1. U.S. Department of Health and Human Services, Public Health Service, National Institutes of Health, Institutional Animal Care and Use Committee Guidebook, NIH Publication No. 92-3415, 1992, Chapter D-1.
2. Silverman, J., IACUC handling of mistreatment or noncompliance, *Lab Anim.* (NY), 23(8), 30, 1994.
3. OPRR Reports, 94-02, January 12, 1994. Available on the World Wide Web at: http://grants2.nih.gov/grants/olaw/references/dc94-2.htm.
4. Wolff, A., Guidance on Prompt Reporting to OLAW under the PHS Policy on Humane Care and Use of Laboratory Animals, NIH Notice NOT-OD-05-034. Available on the World Wide Web at: http://grants.nih.gov/guide/notice-files/NOT-OD-O5-034.html.
5. Garnett, N., and DeHaven, W.R., The view from USDA and OPRR, *Lab Anim.* (NY), 27(9), 17, 1998.
6. Garnett, N.L., and Gipson, C.A., A word from OLAW and USDA, *Lab Anim.* (NY), 32(9), 19, 2003.
7. Garnett, N., and Potkay, S., Use of electronic communications for IACUC functions, *ILAR J.*, 37(4), 190, 1995.
8. Garnett, N., and Gipson, C., Suggestions to bring electronic protocol system into compliance, *Contemp. Topics Lab. Anim. Sci.* 40(6), 8, 2001.
9. DeHaven, W.R., and Garnett, N., personal communication, 1998.
10. Garnett, N.L., A word from OLAW, *Lab Anim.* (NY), 31(5), 19, 2002.
11. Potkay, S., et al., Frequently asked questions about the Public Health Service Policy on Humane Care and Use of Laboratory Animals, *Lab Anim.* (NY), 24(9), 24, 1995.

30

Postapproval Monitoring

Ron E. Banks

Introduction

Regulatory agencies have provided the framework for the operation of animal care use programs; the institution has a clear and unmitigated obligation to oversee and manage animal research with a sense of integrity and focus. The IACUC, charged by the institution with the practical performance of these tasks, has the overarching responsibility to assure animal welfare and well-being in the research environment, whereas the PI is focused on moving the boundaries of science in a humane and ethical manner. However, sometimes these varied requirements and orientations do not connect sufficiently and may result in a dysfunctional or disconnected program of oversight and management. For example, the IACUC may approve activities for a maximal period of 3 years. Nevertheless, the PI often follows experimental outcomes that he or she may not be able to forecast. Sometimes, with the enthusiasm to perform additional studies subsequent to encouraging preliminary research results, an investigator may fail to submit a protocol amendment. It is this sort of scenario, multiplied by many instances, that can lead to program confusion, program dysfunction, and program distress.

This chapter begins with the premise that the AWAR and amendments, the PHS Policy, and the various funding agency requirements for assurance of animal use oversight remain valid in the central expectation that the institution will provide sufficient integrity of process to ensure that animals are being used as approved by the IACUC and as anticipated by the funding and regulatory agencies. While the time-tested activity of the IACUC semiannual review continues to be a necessary function of the animal care and use program, many animal research programs are larger and more diverse and complex than in the past. The "old ways" are simply no longer sufficient to assure granting agencies, the public, regulatory agencies, and legislative bodies of the appropriate quality of care and use for animals used in research, testing, or teaching. The postapproval monitoring (PAM) process provides a simple and effective way for assuring the institution's good name, extending the oversight of the IACUC in a practical and reasonable manner, enhancing the effective utilization of the AV and the clinical care staff, and providing collegial facilitation for the researcher.

30:1 What are the goals of a PAM program?

Reg. According to the AWAR (§2.31,c,1; §2.31,c,2), the IACUC should review, at least once every 6 months, the research facility's program for humane care and use of animals and inspect, at least once every 6 months, all of the research facility's animal facilities, including animal study areas. The PHS Policy (IV,B,1) similarly notes that the IACUC should, with respect to PHS-conducted or -supported activities, review at least once every 6 months the institution's program for humane care and use of animals, using the *Guide* as a basis for evaluation.

The PHS Policy (IV,B,1) also indicates that the IACUC should review proposed projects or significant changes in ongoing research projects for conformity with the Policy and the *Guide*. It also states that the IACUC shall conduct continuing review of activities covered by the Policy at appropriate intervals as determined by the IACUC, but not less than once every 3 years (PHS Policy IV,C,5). Further, the IACUC may suspend an activity that it previously approved if it determines that the activity is not being conducted in accordance with applicable provisions of the AWAR, the *Guide*, the institution's Assurance, or the PHS Policy (PHS Policy IV,C,6).

The U.S. Government Principles for the Utilization and Care of Vertebrate Animals Used in Testing, Research, and Training state in Principle VIII, "Investigators and other personnel shall be appropriately qualified and experienced for conducting procedures on living animals. Adequate arrangements shall be made for their in service training, including the proper and humane care and use of laboratory animals." Without having a mechanism to confirm statements by investigators or research staff as to their qualification, the institution cannot appropriately fulfill this obligation.

Opin. While there is no specific regulatory requirement for a PAM program, against the backdrop of increasing complexity and depth of regulations and research programs, it is clear that the ability of many IACUCs to assure all aspects of the regulatory requirements is being stretched. As such, institutions must look to adjunctive measures to extend the oversight capabilities of the IACUC. One such measure is the development of a PAM program. The goals of such a program focus on

- Ensuring animal well-being
- Keeping the IACUC informed about program status and process while communicating IACUC positions on matters of animal care, education, or training needs
- Protecting the institution's good name with clear investment of IACUC practices in all laboratories and facilities
- Serving as a resource to the research community to assist with collegial facilitation of reports, rewrites, and new submissions
- Supporting the advancement of strong science through a coordinated and sound basis for animal use
- Providing regulatory compliance by becoming the eyes and ears of the IACUC

The attention and focus of the PAM should be in the order as outlined. Programs that focus on compliance as the first goal often become adversarial in nature and not well supported by the research community. Programs that focus on assuring high-quality research, proper animal care, and communication between researchers and the IACUC tend to achieve regulatory compliance as a by-product in a more collegial environment. Programs of PAM must resist the natural development of characterizations such as "animal police" or "compliance watchdogs." A concerted effort of the institution, the IACUC, and the PAM participants should be to facilitate good behavior rather than trying to find and punish poor behavior.

30:2   Is PAM the same as the IACUC semiannual review?

Reg.   According to the AWAR (§2.31,c), with respect to activities involving animals, the IACUC, as an agent of the research facility, shall

1. Review, at least once every 6 months, the research facility's program for humane care and use of animals
2. Inspect, at least once every 6 months, all of the research facility's animal facilities, including animal study areas
3. Prepare reports of its evaluations and submit the reports to the IO
4. Review, and, if warranted, investigate concerns involving the care and use of animals at the research facility resulting from public complaints received and from reports of noncompliance received from laboratory or research facility personnel or employees
5. Make recommendations to the IO regarding any aspect of the research facility's animal program, facilities, or personnel training

Opin.   The goals (and the practical processes) of the IACUC semiannual review and of PAM are parallel and mutually supportive, yet they remain entirely separate activities. Because of the nature of the IACUC semiannual process, most IACUC semiannual reviews tend to focus assessment on facilities, environment, storage, refrigeration, controlled drugs, and record keeping, especially if the semiannual review involves laboratory visits by appointment. There is usually far too little time to address the process and procedure of the approved protocol, and even less opportunity to view ongoing research. Only rarely do IACUCs happen upon a rodent surgical procedure in progress, and they are even less likely to observe the preparatory steps, the intraoperative measures taken, and the postsurgical recovery procedure. PAM allows the IACUC to review those same laboratories, but to focus on procedures and activities, comparing the approved protocol or SOP with the observed practice. The construction and use of an automobile provide an analogy to the similarity between the semiannual review and PAM: The IACUC semiannual review is the quality assurance of the finished product sitting on the factory floor; the PAM is the road test, the real test for the viability of the product. The most important outcome of an effective PAM program is that cultural change results in improved animal care and more consistent practices.

30:3   Now that I know what PAM is (see 30:2), what is PAM not?

Reg.   There is no regulation that directly addresses the need for PAM (see 30:5). However, according to the AWAR (§2.31,c) the IACUC should (1) be empowered to review and approve, require modifications in (to secure approval), or withhold approval of proposed significant changes regarding the care and use of animals in ongoing activities; and (2) be authorized to suspend activities involving animals. In addition, the PHS Policy (IV,C,5; IV,C,6) states that the IACUC is responsible for (1) the review of care and treatment of animals in all animal study areas and facilities of the institution at least semiannually and (2) maintenance of appropriate records of reviews conducted. These regulatory passages suggest the need for ongoing, active oversight of animal research activities.

Opin. It is sometimes easier to define an activity in terms of what it is *not*, and this may be the case with PAM:

- PAM staff is not the "animal cops" of the institution. It is not the job of the PAM staff to be the "heavy" for the IACUC. The IACUC is charged with overseeing, and in some cases managing, the program for animal care and use.

- PAM staff is not the hatchet arm for the IO or the AV. Attitude to the contrary will severely disrupt and even destroy the ability of PAM to serve its most valuable function, that of extending the oversight and educational functions of the IACUC.

- PAM is not required by the APHIS/AC, NIH/OLAW, or AAALAC. Even so, all three agencies have recognized the value and benefit of a well-engaged PAM process.

- PAM is also not a mandate by any funding agency. As with the oversight bodies, most if not all funding agencies recognize the value of using PAM but leave the ultimate expression of program oversight to the institution and the IACUC.

- PAM is not any easier at large institutions than it is at small institutions. While the nature and expression of PAM may differ among institutions, all have the same obligations to assure competent and appropriate animal use oversight, and all have essentially the same tools to accomplish this task. At a smaller institution, the PAM staff may be part-time and also engaged in other aspects of program function other than IACUC-related efforts. In this case, perceptions of conflict of interest may be greater than if the PAM staff resided entirely within the "IACUC office." At larger institutions, multiple PAM staff may be required to oversee the animal use enterprise on campus effectively, but the cost of this activity will be greater.

- PAM is not new ground or new regulation. The federal regulations have remained essentially unchanged since 1985 (the *Guide*, since its most recent edition in 1996). In this time, most programs have grown in scope and complexity, and as the clothes of a child no longer fit a mature adult (while the need for protection from the elements remains the same), so the institution must consider new methods to meet the same need for covering the institution's vulnerable nature. While the process in the operational sense may be new to some, the requirement for a program with sufficient integrity and assurance to confirm that funds provided for research involving animals are being used in the manner in which those funds were intended, and a program assuring the level of care and use that was approved by the IACUC is actually being performed, are long-standing expectations. What is new is that the public now expects an increased level of confidence that animal care and use activities are being properly conducted. The development of a PAM process is a logical extension of the IACUC's obligation with respect to these concerns, while supporting the continued growth and maturation of the research enterprise.

30:4 Are quality assurance (Q/A) and PAM the same thing?

Opin. Maybe. The question is really one more of process than nomenclature. If the process is strictly regulatory in nature, then the research community may regard the

PAM as an adversarial process or as an inspection (for the purpose of this discussion, such a process is referred to as *Q/A*). If the process is more collegial, providing clear commendation of positive aspects of research, offering clear advice toward resolution of concerns, providing (or scheduling) education as necessary, and assisting the research community with working effectively and appropriately under IACUC approval, then the program will assume a greater degree of respect, consistency and integrity (for the purpose of this discussion, this is referred to as *PAM*).

30:5 Is a PAM program a requirement?

Reg.   The AWAR (§2.31,c,8) state that IACUCs must be authorized to suspend activities involving animals in accordance with certain specifications. The PHS Policy (IV,B,1–IV,B,3) states that the IACUC must evaluate and prepare reports on all of the institution's programs and facilities (including satellite facilities) for activities involving animals at least twice each year. Similarly, the *Guide* (p. 9) states that the animal care program should be reviewed and the animal facilities inspected once every 6 months. (See 30:3.)

Opin. There is a clear expectation that the IACUC will employ a mechanism to assure that animal care and use are in accordance with regulations and guidelines. All institutions using animals under the purview of the AWAR, the PHS Policy, or the *Guide* are required by law and policy to assure that animals are being used in a manner that is approved by the IACUC. Funding agencies require that animals be used in a manner in which the grant funds were provided. The degree, depth, or complexity of the institution's performance of PAM will be in response to the size of the program, the numbers of species, the programs using animals, the degree of invasiveness of approved activities, the level of care required, and many other variables. While none of the regulatory guidance requires a PAM program, all of them require oversight of the process at an appropriately detailed level to ensure compliance and humane care of animals. In certain instances, the IACUC may be able to oversee the needs of the institution's program fully; in others, it may not have enough members or time to accomplish the task adequately. In these cases, a PAM process can allow the IACUC fully to assure the legal, and more importantly, the implied ethical aspects of animal care and use. While PAM programs will look slightly different, all institutions must address the same issues. Recognizing this common requirement, the question then becomes, By what method are you assuring that what was approved by your IACUC is being performed as approved? Is your methodology working? What is an appropriate tool for assessing the effectiveness of your training program? For most programs, the use of the semiannual review alone is not sufficient. For those institutions that have taken a serious look at all aspects of their program, the discovery has routinely been that a PAM would significantly improve institutional assurance of the requirements in the law and precepts of the policy.

30:6 What methods are useful for implementation of PAM?

Reg.   The AWAR (§2.31,c,3) make three statements related to the methods used for PAM:

- The IACUC may invite consultants to assist in the review of complex issues arising out of its review of proposed activities. Consultants may not approve or withhold approval of an activity and may not vote with the IACUC unless they are also members of the IACUC. The PHS Policy (IV,C,3) also indicates that the IACUC may use consultants to assist in the review of complex issues.

- The IACUC may determine the best means of conducting evaluations of the research facility's programs and facilities; no committee member wishing to participate in any evaluation conducted under this subpart may be excluded. The IACUC may use subcommittees composed of at least two committee members and may invite ad hoc consultants to assist in conducting the evaluations; however, the IACUC remains responsible for the evaluations and reports as required by the act and regulations.

- The IACUC must review, and, if warranted, investigate concerns involving the care and use of animals at the research facility resulting from public complaints received and from reports of noncompliance received from laboratory or research facility personnel or employees.

None of these references prohibit the use of a PAM, and all provide sufficient grounds for employment of an "extra-IACUC" process for assuring the integrity of program oversight. The AWAR do not, however, prescribe a specific method for the IACUC to undertake PAM should it choose to do so.

Opin.   There are three general methods for conducting PAM:

- IACUC members conduct monitoring
- Veterinary staff conducts monitoring
- The institution employs a dedicated staff to perform PAM

Sometimes, institutions may choose a hybrid of these basic models, although a simple approach is usually best. The key is not so much the selection of a specific method, but choice of the method that best assures and supports the institutional objectives. As a general rule, the more complex or intensive research enterprises benefit by employing a dedicated staff for PAM. Smaller, more limited programs might be able to accomplish the task successfully by having the IACUC members conduct PAM. In contrast, the option of having the veterinary staff conduct PAM is the least desirable choice regardless of program size, as making the veterinary staff the central monitor will unnecessarily create roadblocks to successful engagement of veterinary care and collegial interaction between the veterinary staff and the research staff. The use of the veterinary staff for PAM should be avoided.

30:7 What are the advantages and disadvantages of having IACUC members perform PAM?

Opin.   It really boils down to a few simple issues: time availability, consistency, and conflict of interest. In most situations, the disadvantages outweigh the advantages of IACUC members' conducting the PAM. For example, most IACUCs consist of researchers who give their time in support of the institutional objectives of animal care and use. This usually includes review of animal use proposals, review of amendments or modifications, participation in semiannual program reviews and facility inspections, and many other special assignments that occur during the year. While critically important activities, committee membership and participation are not income-generating activities for the researcher or the institution; nor do they typically count toward tenure. Most institutions are best served by utilizing the researchers' strengths in the laboratory rather than their spending hours observing the research techniques of others. Second, while IACUC members are usually well intended, they generally do not have the depth of familiarity with regulatory requirements and generally do not possess sufficiently broad background in animal research and care to provide the institution with consistent process or procedure in PAM. Having a large number of individuals (e.g., IACUC membership) performing PAM will introduce a degree of inconsistency of process that is usually discouraging to the researcher and marginally productive for the institution. Additionally, since IACUC members are often specialized researchers, well versed in specific research areas, the most productive use of their skills and limited time should result in the focused review of areas of expertise. For example, assigning a behaviorist to observe surgical activities would not produce a fair or consistent evaluation in contrast to assigning the review of the activity to a research surgeon or technician well versed in such procedures. Finally, for many IACUCs, maintaining reviewer anonymity is an important protection for the IACUC membership, one that discourages institutional territorial battles; therefore, sending a colleague into another researcher's laboratory may present conflicts that interfere with consistent and accurate assessment, effective communication, and teamwork.

30:8   Is it advisable for the veterinary staff to serve as the eyes and ears of the IACUC?

Reg.   The AWAR (§2.31,d,1,vi; §2.31,d,1,vii) note that the AV or other scientist trained and experienced in the proper care, handling, and use of the species being maintained or studied must direct the housing, feeding, and nonmedical care of the animals, and that medical care will be provided by a qualified veterinarian. The *Guide* (p. 56) states that the AV is responsible for the overall program of veterinary care. The PHS Policy indicates that medical care for animals will be provided by a qualified veterinarian. None of these documents prescribe a role for the veterinary staff in performing PAM for the IACUC.

Opin. Though the veterinary staff often has the most extensive background in animal health and welfare, the demands on them are often such that there is little time during the course of the day to trade hats, stop providing veterinary and husbandry guidance and direction, and start overseeing the process of animal use. While the veterinary staff is often in a position to provide assessment of the use of animals, a significant conflict may arise as a result. Using the veterinary staff as the sole (or primary) mechanism for assuring adherence to an IACUC-approved protocol can become self-defeating. The institutional goal should be for the veterinary staff to assist researchers while encouraging, improving, and assuring high-quality animal care. Placing the staff in a position of reporting deviations of process discourages open and engaged communication between the research and veterinary staffs and may foster an adversarial relationship with the research staff. Additionally, certain research activities do not occur at times when the veterinary staff is present, a significant complication for sole reliance upon the veterinary staff to perform PAM. Further, the veterinary staff usually has specified obligations to assure and oversee certain parts of the program, such as managing the program for veterinary care and operating the core vivarium. Because PAM should include the entire animal care program (including care practices within the core vivarium) and because the IACUC has an obligation for all animal care and all animal use, assigning the veterinary staff to perform PAM poses a conflict of interest. It is preferable to prevent any conflicts of interest and maintain PAM and veterinary care as separate and distinct operations.

30:9   What is meant by a *dedicated PAM staff*? The institution is not large enough to employ someone full time, nor does it have the resources to employ someone full time.

Opin.   The term *dedicated* does not refer to a personnel allotment but rather an assignment of function. The importance of having dedicated staff is focused more on the assignment of responsibility than on the length of engagement. A dedicated staff member may spend only a percentage of his or her time performing the activities of PAM. Other appropriate duties of the PAM employee may include coordinating IACUC business, providing training for new faculty or staff, serving as IACUC recorder, or assisting with report generation to the regulatory agencies. It is generally considered more appropriate *not* to have the PAM staff as voting members of the IACUC. The role of the PAM staff is to serve as the eyes and ears of the committee. They should be held at arm's length from the decision making activities of the IACUC in order to protect the relationship between PAM staff and the research staff, so that a distinction is clear and that corrective actions required of the research staff are understood to be from the IACUC and not the PAM staff.

30:10   What are the advantages of using the dedicated staff model of PAM?

Opin.   Having a dedicated staff to perform PAM allows other groups, such as the IACUC and veterinary staff, to focus on the aspects of animal care and use for which they are responsible. For example, a dedicated PAM staff allows the veterinary staff to focus on animal health and facility management and to ensure a stable and effective platform for progressive and valuable research. Additionally, having a dedicated PAM staff allows the veterinary staff to serve more directly as facilitators to enhance animal care and assist the PIs with proper animal use.

A dedicated PAM staff can also serve as a focused continuing education and training service, addressing specific issues of noncompliance as opportunities for training. For example, when problems are identified in a laboratory, the PAM staff might provide information on required, improved, or enhanced alternate techniques or processes. In this way, the PAM staff can become a user-friendly resource and at the same time help to assure that activities are performed in conformance with the requirements and expectations of the IACUC. In most circumstances, PAM staffs interact with the research associates and technicians more than PIs. This establishes communications networks that are not routinely employed by the IACUC for sharing information and assuring productive communication. Over time, the PAM staff gains an institutional memory and can provide a sense of stability to the IACUC.

The most valuable role of PAM staff is providing the IACUC with firsthand reports of ongoing activities in the campus laboratories on a regular and recurring basis. The amount of PAM time should be variable and determined by the needs of the program. For example, in institutions where rodent breeding represents the majority of the animal activity, the time needed to perform a reasonable level of PAM will likely be less than at an institution where the principal activity involves survival surgery.

30:11 What types of individuals are best suited to serve as PAM staff?

Opin.   The selection of individuals who will perform PAM is less concerned with credentials than with personality. The IACUC requires clear and accurate information;
therefore, individuals who have excellent observation and good communication
skills are ideal. The best PAM staff observe small aspects of laboratory activity that
impact on the care or use of animals—such matters as whether animals are checked
with sufficient frequency while in the laboratory, whether each instance of drug
administration is recorded or such use is just bunch-recorded at the end of the
study, the presence and reading of thermometers located near the animals (versus
in the room but distant to the animals), and whether proper aseptic technique is
being used for all survival surgeries.

   The PAM staff also requires good communication skills. The researchers may at
times be less than supportive of the perceived intrusion into their laboratories, and
PAM staff members who have a firm but nonadversarial style of communication
are worth their weight in institutional gold.

   Beyond these skills, individual programs might have specific issues that would
benefit from individuals with certain skills and backgrounds. For example, an
institution having a surgical program would be well served to find someone with
a good understanding of surgical technique and related skills (e.g., veterinarian,
technician with surgery experience, or M.D. surgeon). An institution conducting a
significant amount of antibody production would be best served by having a PAM
staff member who has relevant experience and knowledge regarding expected
standards and procedures in this area.

   Generally speaking, "nice people" who have good observation skills can be
trained to assess the accuracy of IACUC approvals and expectations with the performance of the procedures in the laboratory or animal care facility. It is far more
difficult to teach people to speak well and present a pleasing style, no matter what
their degree of technical expertise.

30:12 What might be the job description for PAM staff?

Opin.   While specific job tasks may vary from institution to institution (and be based on
specific needs of the institution's program for animal care and use), a general template for performing protocol assurance and other IACUC-related duties might
include general categories of responsibility that are highly parallel to the obligations of the IACUC, such as the following:

- Audits for consistency with approved activities
  - Communicate effectively with PIs, lab managers, veterinary staff, and
    IACUC and animal care program leadership concerning observations
    made
  - Conduct observations and audits of approved animal use activities by
    monitoring and auditing approved animal use activities of the types noted
    (e.g., compliance review, procedure observation, surgical facility assess-
    ment, laboratory assessment) (also see 30:17)
  - Report activities in a clear, concise, and accurate manner to the IACUC,
    laboratory managers, PIs, and veterinary staff
- Communication and program enhancement ideas for the IACUC

- Develop (or assist in development) of systems and relationships to support communication and information exchange between the IACUC and the research community
- Assist PIs in developing and implementing corrective action for relevant findings in the laboratory assessment reports
- As instructed by the IACUC, prepare and assist in accomplishing corrective actions of IACUC-defined noncompliant items
- Identify and report other program concerns that may arise through the work process
- Draft revised SOPs
- Recommend development of other activities or training events that enhance the institution's program for animal care and use
- Benchmark procedures against other institutions and determine best practices for the local campus construct, making recommendations as appropriate
- Training of research staff
  - Identify training and educational needs regarding species-specific and procedural practices based upon auditing activities
  - Draft enhancements to the institution's program of training as observations indicate necessary
  - Implement and enhance programs of training for animal care and use personnel
  - Coordinate the implementation of other global institutional training programs (e.g., orientation to the institution, regulations, rules, protocol preparation) for the animal care and use community
  - Coordinate with the animal care unit to assure an integrated and mutually supportive training program for all animal handlers and users on campus
- Serve as a reference resource
  - Review technical publications, articles, and abstracts to stay abreast of current regulations, trends, and "official guidance"
  - Draft program enhancements in response to the reviewed literature
  - Monitor incorporation of actions into workplace environment
  - Advise the IACUC on trends within the field
  - Collaborate (as necessary) with the Office of Compliance or legal office in matters that require their attention
  - Serve as liaison to other institutional monitoring and audit groups

30:13 Is it advisable for PAM staff to serve as voting members of the IACUC?

Reg.   Neither the PHS Policy nor the AWAR address the specific role of IACUC members in monitoring activities after approval. The AWAR (§2.31,d,1,xi,3) state that consultants may be used to review proposed activities but that these consultants may not vote with the IACUC unless they are also members of the IACUC.

Opin.   While it is appropriate for PAM staff to attend IACUC meetings, it is not necessarily in the best interest of the IACUC to have PAM staff serve as voting members, though there is a wide array of opinion on this matter. The PAM staff should be considered an extension of the IACUC and not a replacement for the IACUC. If the PAM is regarded as serving in a consultant role to the IACUC, then it is not appropriate (according to the AWAR) for PAM staff to be voting members of the IACUC.

Of those institutions having some form of process oversight, this author is aware of only a few that have PAM staff as voting IACUC members; most simply have PAM staff attend the IACUC meetings and provide insight and counsel to the IACUC without voting. Some institutions reported a concern with the PAM staff having a vote, since PAM staffs often have knowledge of studies and investigators that might introduce bias to a vote. Institutions should be very cautious of engendering more authority in the PAM staff than in the IACUC.

Serving as IACUC members may complicate the job of the PAM staff. It is far more productive for them to relate the decisions of the IACUC to the researcher than to defend the decisions of the IACUC as members who participated in the development of that decision. The purpose of the PAM staff is to serve as the eyes and ears of the IACUC: to be consultants to the IACUC, not to supersede the IACUC or become the committee's gatekeeper.

30:14   Is it advisable for the PAM staff to work directly for the AV?

Opin.   While possible, it is probably not the preferred approach. The AV oversees some programs (e.g., animal husbandry, veterinary care) that should be under review by the PAM staff. For purposes of process review, the AV (and the husbandry units) should be considered as another PI and therefore should be viewed as another activity under the oversight of the IACUC.

If the PAM staff reported to the AV, they would be in the difficult position of reporting process deviations to the individual who was both culpable for the deviation as well as responsible for PAM staff and activity. This would represent a significant conflict of interest. In other words, the PAM staff would be reporting concerns directly to an individual who could produce adverse administrative conditions on the PAM staff or discourage accurate reporting to the IACUC. This would lead to a biased and unproductive process. For a program to argue internal institutional integrity, process must be stable, consistent, applicable to all animal handlers and users, and without regard to credentials. The PI community will become more supportive of the process of PAM if there is a clear expectation that all animal user groups, including the veterinary staff, are being held accountable to the expectations of the PAM program.

30:15   Should the PAM staff work directly for the IACUC?

Opin.   In regulatory principle, this is the only appropriate option as it is the IACUC that has authority for oversight, but practically this may not be the best design. The IACUC has the obligation for oversight of approved activities, not the PAM staff. Further, the IACUC is not a person, but rather a committee of the institution. Its membership has other duties that consume most of their time. While they are members every day of the month, they likely only perform IACUC functions as a small part of their normal work week. In some instances, the institution may have a central program office to assist with program management, and when this occurs the PAM would logically be located there, supervised by the director of that office. Such an arrangement provides the level of supervision and oversight necessary to assure a proper level of accountability, direction, and focus.

Sometimes, the institution may choose to have the PAM staff report directly to the IACUC chairperson. Nevertheless, this can be a problem if the IACUC Chairperson is an animal user or serves only part-time and is not always available. Alternatively, the institution may choose to have the PAM staff report to another member of the animal program (essentially a resource, a mentor, or a guide for the PAM to assist in development of clear, accurate, and comprehensive reports to the IACUC). There often is a clear understanding that all matters of concern are to be reported to the IACUC. In all cases, however, the PAM staff functions in support of the IACUC.

30:16   Which type of issues would typically be reported by the PAM staff to the IACUC?

Reg.   Notice NOT-OD-05-034 (*Guidance on Prompt Reporting to OLAW under the PHS Policy on Humane Care and Use of Laboratory Animals*; February, 24, 2005) provided guidance to assist IACUCs and IOs in determining what, when, and how situations should be reported to NIH/OLAW under the PHS Policy (IV,F,3), and to promote greater uniformity in reporting. PHS Policy (IV,F,3) requires that "the IACUC, through the Institutional Official, shall promptly provide OLAW with a full explanation of the circumstances and actions taken with respect to:

a) any serious or continuing noncompliance with this Policy;

b) any serious deviation from the provisions of the *Guide*; or

c) any suspension of an activity by the IACUC."

Opin.   A comprehensive list of definitive examples of reportable situations is impractical. Therefore, the following examples do not cover all instances but demonstrate the threshold at which NIH/OLAW expects to receive a report; thus, it is reasonable that these same types of situations are those that the PAM staff should report to the IACUC. The NIH/OLAW insists that institutions use rational judgment in determining what situations meet the provisions of PHS Policy IV,F,3 and fall within the scope of the examples that follow. NIH/OLAW has also determined that situations that meet the provisions of IV,F,3 and are identified by external entities such as the USDA or AAALAC or by individuals outside the IACUC or outside the institution are not exempt from reporting under IV,F,3. This list, while not exhaustive, should be used as a starting point for what should be reported to the IACUC for review and consideration (and see 29:48):

• Conditions that jeopardize the health or well-being of animals, including natural disasters, accidents, and mechanical failures, resulting in actual or potential harm or death to animals

• Conduct of animal-related activities without appropriate IACUC review and approval

• Failure to adhere to IACUC-approved protocols

• Implementation of any significant change to IACUC-approved protocols without prior IACUC approval

• Conduct of animal-related activities beyond the protocol expiration date established by the IACUC

• Conduct of official IACUC business requiring a quorum in the absence of a quorum

- Failure to correct deficiencies identified during the semiannual evaluation in a timely manner
- Chronic failure to provide space for animals in accordance with recommendations of the *Guide* unless the IACUC has approved a protocol-specific deviation from the *Guide* based on written scientific justification
- Participation in animal-related activities by individuals who have not been determined by the IACUC to be appropriately qualified and trained
- Failure to monitor animals post-procedurally as necessary to ensure well-being (e.g., during recovery from anesthesia or during recuperation from invasive or debilitating procedures)
- Failure to maintain appropriate animal-related records (e.g., identification, medical, husbandry)
- Failure to ensure death of animals after euthanasia procedures (e.g., failed euthanasia with carbon dioxide)
- Failure of animal care and use personnel to carry out veterinary orders (e.g., treatments).
- Performance of animal work after IACUC suspension or other institutional intervention that results in the temporary or permanent interruption of an activity due to noncompliance with the PHS Policy, AWAR, the *Guide*, or the institution's Animal Welfare Assurance

Examples of PAM observations that might be reported to the IACUC, but would not result in reporting to NIH/OLAW, APHIS/AC, or AAALAC (but might be tracked on the PAM data matrix) include the following:

- Death of animals that have reached the end of their natural life spans
- Death or failures of neonates to thrive when husbandry and veterinary medical oversight of dams and litters was appropriate
- Animal death or illness from spontaneous disease when appropriate quarantine, preventive medical, surveillance, diagnostic, and therapeutic procedures were in place and followed
- Animal death or injuries related to manipulations that fall within parameters described in the IACUC-approved protocol
- Infrequent incidents of drowning or near-drowning of rodents in cages when it is determined that the cause was jamming of water valves with bedding (frequent problems of this nature, however, must be reported promptly along with corrective plans and schedules)

30:17 What process should be used for monitoring and assessing activities assessed as part of PAM?

Reg.   There are no specific regulatory or other guidelines to direct an institution with respect to the process used for PAM.

Opin.   Over the last few decades, a clear and observable shift has occurred in the philosophy of animal use oversight, from one referred to in the *Guide* (p. 3) as the engineering standard to a performance standard. In an operational sense, this means that the oversight agencies recognize that one size does not fit everybody, and that programs must be built and managed to achieve an appropriate outcome, rather than have duplicative or nonfunctional institutional offices. Recognizing that the desire is for an appropriate outcome, a single methodology for PAM is likely to be insufficient for the breadth of activities that occur in most programs for care and use. PAM staff typically benefit from using a variety of methods to complete the task of accurate assessment for the IACUC. These include the following:

- Comparative review: A relatively formal process, comparative review is intended to be nonthreatening and collegial in nature. Of the various types of PAM review, comparative review is the closest to an IACUC semiannual review, though it is usually performed with greater intensity than an IACUC semiannual review. The various activities that typically occur as part of comparative review include the following:
  - Review of the approved protocol(s) in advance, to understand fully the focus of the laboratory and activity, and to allow a fair and accurate review of the procedures
  - Comparison of animal usage (purchase or breeding) numbers to date, with the approved numbers in the protocol (to assess laboratory activity from a program perspective)
  - Provision of an entrance interview with researchers (or senior lab managers) to explain the PAM process and the potential outcomes of the process
  - Inquiry about any observed, unexpected deaths or other adverse events, and the outcome for such observed animals
  - Review of procedure records, looking for consistency of procedures with the approved protocol
  - Review of animal use and study areas to assess proper animal care
  - Review of laboratory training records to determine what training may be needed for the procedures being performed (and subsequent scheduling of that training)
  - Confirming of medical surveillance participation for all members of the laboratory having risk of animal-related injury or illness (note: PAM staff usually do not have access to personal health care information; instead, the question is whether the laboratory individuals are enrolled in an appropriate medical surveillance program or not)
- Procedure observation: The procedure observation is a critical aspect of the process of effective PAM. It may be integrated into, or used as a direct follow-on to the comparative review or in some instances as a substitute for the comparative review. The procedure observation provides unique and special information for the IACUC and should be used to the greatest extent possible. It is important to observe activities being performed on animals and assure that they are consistent with those described in the approved protocol. The process of procedure observation may be randomly conducted and exercised on all protocols over a given period; or the process may be selective, on a risk assessment of institutional concerns. The procedure observation must be transparent to laboratory routine and research procedures. Injecting too many questions into a procedure or asking for a demonstration of an activity injects artificial variables into the observed procedure that alter the "normal practice" and may create a noncompliant activity. A procedure observation can be applied to both IACUC-approved experimental protocols and animal care

SOPs (such as those describing routine husbandry and medical care). Procedure observation might review the following aspects of animal care and use:

- Presence and condition of equipment and supplies required to assure care of the research animals
- Husbandry and care practices for animals before, during, and after the procedure
- Animal manipulations: observing specific procedures and confirming they are being performed as approved in the protocol
- Transport of animals to and from housing to the experimental areas to determine whether the transport is being conducted consistently with institutional practices
- Personnel qualifications with respect to the procedures being performed to establish that the individuals working with the animals exhibit the behaviors and expertise expected and consistent with those stated on their personnel qualifications form
- Assessment of findings, potentially sharing any concerns with the IACUC, the AV, and the compliance officer for the institution

- Surgical facility (and use) assessment: The PAM staff should also monitor the use of animal surgical facilities. While the IACUC performs a similar function as part of the IACUC semiannual assessment, rarely will the IACUC see a surgery in progress. The PAM should schedule a time to observe ongoing activities and provide the institution with the confidence that the space, equipment use, and the activities that occur in that space and with that equipment are appropriate and are being conducted as approved by the IACUC. While significant differences will occur, and professional opinion may be required (this could be a situation in which someone on the PAM staff who has credentials relevant to a specific activity might be of value), this type of review may be conducted in conjunction with, or separately from, the other forms of assessment. The general concerns that would typically be considered as part of this type of review include the following:
  - Physical plant appropriateness, to assure that the operative area has the surfaces and equipment necessary to perform sterile surgical procedures
  - Housekeeping and sanitation procedures, to assure that the operative areas have been prepared appropriately prior to utilization
  - Drug inventory and utilization
  - Sterilization method and packs
  - Personnel skills and training
  - Emergency protocols for adverse procedural events
  - Anesthetic equipment and scavenging
  - Anesthetic and analgesic regimen
  - Postprocedural monitoring, nursing, and supportive care
  - Pre-, intra-, and postprocedural methods appropriate for the species in use
  - Overall conduct of major operative surgeries
- Laboratory assessment: The PAM staff may also visit laboratories to assure the approved practices are being performed as anticipated by the IACUC. These visits may be random (provided to all protocols over a given period)

or selective (based upon a risk assessment of institutional concerns) and may include both procedure and study areas. The general items that could be considered include the following:

- Physical plant appropriateness (e.g., for rodent surgery performed in the laboratory, verification that the operative area has the surfaces and equipment necessary to perform sterile surgical procedures)
- Housekeeping and sanitation procedures, determine that animal use areas are orderly and prepared appropriately prior to utilization
- Drug inventory and utilization
- Sterilization method and surgical packs
- Personnel skills and training
- Emergency protocols for adverse procedural events
- Anesthetic equipment and scavenging
- Anesthetic and analgesic regime
- Assurance that postprocedural monitoring, nursing, and supportive care are provided
- Assurance that pre-, intra-, and postprocedural methods are appropriate for the species in use

30:18 Should PAM visits be scheduled or unscheduled? (See 23:25.)

Reg.   There are no regulatory directives or other guidelines that suggest that visits by either PAM staff or the IACUC should be scheduled versus unscheduled.

Opin.   Scheduling visits encourages a sense of collegiality and institutional support for the research enterprise by meeting the research team on their grounds at a convenient time. On the other hand, scheduled visits may not provide an unvarnished assessment of the animal care and use procedures. Anonymous concerns reported to the IACUC may require unscheduled visits. These must be handled graciously, as no one is guilty of any noncompliance until the IACUC has made that decision. The role of the PAM staff is not to make any decision, but to gather information and present that information to the IACUC in a clear and impartial manner. Sudden and frequent appearance by the PAM staff in any laboratory may be viewed as harassment, so the unscheduled option should be as a focused surgical event, with a defined purpose.

30:19 Once observations have been made by the PAM staff, how should that information be handled?

Opin.   All PAM activities in research or vivarium spaces should conclude in a standard manner:

- An initial assessment of the observations: In some instances, the PAM staff may desire to provide a quick exit briefing to the laboratory manager or PI prior to departing the laboratory. In other instances, there may be need for discussion of the observations with other members of the program staff prior to making any assessment.
- Consultation with a veterinary member of the institution for clarification on issues related to animal welfare and well-being: If there are observations that

are not clear or require background clarification, it is advisable to discuss those observations with members of the institution's veterinary staff.

- Consultation with the PI (or laboratory manager) for accuracy of observations: This should be a written (or e-mail) exchange with the laboratory representative to assure the observations are accurate. This is not an opportunity to negotiate the corrective actions—that is the responsibility of the IACUC.

- Consultation with the IACUC Chairperson (if regulatory noncompliance issues have been observed): It is always advisable to keep the IACUC's leadership apprised of any questionable situations. This will provide the IACUC leadership with the opportunity to review relevant background or regulatory information prior to IACUC discussion of the observations.

- Reporting to the IACUC: The PAM staff report should occur as part of the IACUC business session. The IACUC can then determine whether specified corrective measures are required, the duration or degree of corrective measures needed, or whether the observations confirm the need for focused training or other types of program enhancement tools. The PAM staff (or the veterinary staff) may provide background for the IACUC decision.

- Report to the PI and laboratory manager.

Once the steps have been completed, the PAM staff should assist with preparation of the findings, IACUC decision, and the required corrective (if applicable) measures. The report along with needed corrective actions should not carry the signature of the PAM staff, but rather that of the IACUC or the IACUC Chairperson, since the final decisions on matters of noncompliance are made by the IACUC. However, the PAM staff can play an integral role in facilitating the laboratories' full completion of the corrective actions.

30:20 Should IACUCs develop a method for the timely self-reporting by PIs of adverse events (e.g., a large number of postoperative complications, excessive mortality rate)?

Opin. Yes. Institutions that have developed a collegial and mutually supportive PAM program have reported that PIs often self-report as a natural extension of the process, especially if there has been a strong culture of compliance. As with most issues, simple is best. Some institutions have fostered a culture of compliance by producing a "program business card" that can be laminated and distributed to all laboratories at the time of the PAM visit. The research staff knows how to report concerns and to whom concerns should be reported in a timely manner. Other institutions provide a special Web address for these self-reports. Still others have developed a special hotline for self-reporting as well as anonymous allegations of misconduct. Regardless of the mechanism developed, all IACUCs should work diligently to encourage communication from the PI to the committee. Engendering active participation in process oversight is the goal and the evidence of success of program integrity. (See 29:5.)

30:21 Should the PAM staff audit and evaluate the IACUC?

Reg. The AWAR (§2.31,c,1) require that the IACUC review, at least once every 6 months, the research facility's program for humane care and use of animals.

Opin.   Maybe. While all aspects of an institution's program for care and use must be evaluated, the IACUC usually performs this task as part of the semiannual program
review. The IACUC might consider itself a poor reviewer of its own processes and
so could utilize the PAM staff to perform a review of the IACUC process. The committee may also use a subcommittee (e.g., PAM staff) of at least two people to perform the program review. An internal audit by the PAM staff can serve as an
effective tool in assuring the committee's process of oversight is consistent with
regulatory expectations.

30:22 Should the animal care facilities be included in PAM?

Reg.   The AWAR (§2.31,c,2) state that it is a responsibility of the IACUC to inspect, at
least once every 6 months, all of the research facility's animal facilities, including
animal study areas. The PHS Policy (IV,C,1,d) also notes that the review should
include an assessment to assure that the living conditions of animals will be appropriate for their species and contribute to their health and comfort.

Opin.   The better question is, Why should the animal care facilities not be included in
PAM? The animal care program is an institutionwide program; there are no parts
of the institution's program for animal care and use that should be exempted from
a clear and concise review. Yes, all animal care spaces and processes, including animal care facilities, should be included in PAM. What is also critical (and sometimes
sensitive) is that the review focuses on processes and procedures and not on professional veterinary judgment of clinical care decisions. The PAM review of animal
care facilities must be consistent with the review of experimental procedures,
focusing on

- Spaces being used and the appropriateness of those spaces
- Adherence to protocols (or SOPs) for care
- Training or qualifications of the activities being performed by care staff

30:23 "Animal police" versus "research liaisons"—what is the difference?

Opin.   On one hand, not much; on the other hand—the world! Perception is often reality,
and if the research community believes that they are not trusted, that they are being
accused of poor animal care practices by having the "animal police" go to the laboratory, then the ability to develop a strong and mutually supportive program is
marginal at best. On the other hand, if the research community views the PAM staff
as members of the institution's support team, which supports and assists the
researchers with their research efforts, assuring that the research stays compliant
with IACUC approvals and current on recent enhancements of process, then the
chances of success are much greater, and the outcome of a collegial and supportive
environment more likely. A critical aspect early in the phase of PAM development
is the selection of the proper name. The preference for terms such as *liaison* rather
than *officer* engenders a more collegial and supportive atmosphere. No one likes to
live fearful of receiving a ticket for speeding; almost everyone will be nervous
when being watched by the animal cops. Most are willing to accept suggestions
and assistance for keeping the research on track with help from a colleague located
within the animal program.

30:24   How can institutional, faculty, or IACUC resistance to PAM be handled?

Opin.   Regardless of the real or perceived value of any process, change is not easy for people. Appealing to the scientific mind and using a clear discussion of the value of the process is usually the most productive method. It is helpful to focus on the advantages of PAM, and how this will protect and defend the rights and privileges of continued animal use at the institution while extending the ability of the IACUC to achieve its regulatory obligations.

It is further helpful to discourage the tendency to look at PAM as "another regulation" or institutional invasion of the laboratory. The truth is that this is not a new regulation; nor is it an invasion. Rather, it is a desire of the institution to guarantee the continued use of animals for research; it is the hope of the IACUC that the PAM staff will report evidence of good performance; and it is the desire of the public that animals receive good care and that the care is confirmed.

There are three common questions that arise when introducing PAM into a new institution:

What can PAM do? The PAM process can provide the IO and the IACUC with a level of assurance of overall program process that cannot be obtained in any other manner. It can identify deficiencies in the laboratory and encourage collegial correction. It is a method by which the PAM staff can help the researcher stay fully compliant with federal, state, and local requirements for animal use. PAM can repair the deficiency while it is a small issue.

The PAM process can facilitate the IACUC program assessment. Most IACUCs struggle with the federal requirement to perform a "program review" on a semiannual basis. The use of the PAM reports over the previous 6 months provides a markedly enhanced level of program assessment. In other words, PAM provides the IACUC with information not available from other sources, and the opportunity to direct training toward trended deficiencies in the program.

Why is it important? The PAM process is critical to the long-term protection and integrity of the animal care and use program. In a world where we hear on occasion of poor judgment or faulty process in animal care or use, PAM can be the least expensive manner to keep the program on track.

Having a PAM program makes a strong statement that the institution is dedicated not just to staying out of the news, or keeping the AAALAC accreditation plaque, but to assuring that the care and use of animals on a daily basis are real and valued parts of the research program. Some institutions have blindly signed letters of Assurance to NIH/OLAW as a statement of program integrity, without any clear and definable evidence for that assurance. Many institutions have claimed that the absence of a program disaster is evidence of a strong and reliant program, or that the accreditation plaque on the wall is evidence of a stellar research enterprise. While these may be true in part, the absence of a disaster may be pure and simple luck, and the highly valued accreditation plaque an indication that the institution is meeting minimal standards and performed well at the last triennial site visit. Neither of these can provide the daily, on-the-ground assurance that processes and procedures as approved by the IACUC and intended by the veterinary staff are a reality.

What may happen if we do not have a PAM program? Maybe nothing. After all, many institutions have never had a program disaster thus far. But the more critical question is, What would be the effect upon the research enterprise and the institution's good name if a preventable issue erupted? What if the institution could have prevented the highly emotive issue of inappropriate animal care or use had the problem been identified at an early stage? While the cost of such a situation is not easily documented, it is clear that the outcome could be catastrophic. Potentially, the institution's animal research program could be suspended, interrupting the research efforts of not only the guilty but also the innocent and well-intended researchers. Possibly, there would be a disruption in core fund support for the institution's research program. Potentially, the institution's alumni would be discouraged and benefactor support would suffer. Without question, the prestige of the institution would suffer from a program disaster.

Finally, most importantly, and most critically, Are there instances of animal misuse, neglect, or abuse that the institution failed to identify? Such a situation indicates overall program failure. Failures to assure that principles of humane care and compassionate use are practiced are ethical failures as well. The ultimate concern is that institutions that do not look for evidence of good performance do not fully ensure a necessary level of engagement in the program. Institutions should establish and monitor benchmarks, without which there are no reliable methods for improving compliance.

30:25 How does PAM encourage compliance with regulatory requirements? How can PAM assist in preventing noncompliance?

Opin. Training of the institution's research staff is critical to a compliant and humane program for animal care and use. In many instances, the IACUC requires the PI to confirm on the protocol application that the PI and the research associates are properly trained. However, unless there is a recognized laboratory failure, the training is never questioned and focus is almost always on technical training rather than procedural training. PAM does not replace any of the existing requirements for training or oversight, but PAM can assist with assuring the integrity of the institutional program as required by the existing regulations by serving as an ongoing training activity, sharing new learning opportunities, and updating the laboratory members on modified or enhanced review or reporting requirements. PAM provides a special and focused tool to identify concerns and problems at an early level and provides the institution an opportunity to correct the dysfunction before it becomes a larger noncompliance issue. PAM provides clear and documented evidence that the institution is not simply striving to meet the minimal requirements for animal care and use, but is rather working toward a reasonable and practical level of care and oversight that also assures integrity of research data. If the assumption is true that research staff do not willfully intend to violate accepted standards for animal care and use, then PAM provides the communication bridge that is often missing, providing the IACUC with focused information and providing the research community with focused training.

30:26 What internal controls for PAM should an institution implement?

Reg. The PHS Policy (IV,B) indicates that the IACUC has overarching authority for the institution's program for humane care and use of animals. The AWAR (§2.31,a) note that "the Chief Executive Officer [CEO] of the research facility shall appoint an Institutional Animal Care and Use Committee (IACUC), qualified through the experience and expertise of its members to assess the research facility's animal program, facilities, and procedures."

Opin. There is one absolute: the authority lies not with the PAM, but with the IACUC and the CEO (or IO). For successful performance, the PAM must not be placed in a situation of bias —either perceived or real. This can best be effected by the PAM staff's

- Having sufficient authorization to enter and observe all animal care practices at the institution, regardless of location or persons involved
- Performing their task unobtrusively
- Protecting any and all observations with the greatest confidentiality
- Not working directly for an animal user (especially if the animal user is the IACUC Chairperson or the AV)
- Not having any authority to determine outcomes and corrective actions, since such determinations are IACUC functions, but rather making observations and providing reports of those observations to the IACUC

30:27　How is PAM of animal health different from the AV's routine "well-animal" monitoring?

Reg. Both the PHS Policy (IV,C,1,d; IV,c,1,e) and the AWAR (§2.31,d,1,vi; §2,31,d,1,vii) state that "the living conditions of animals will be appropriate for their species and contribute to their health and comfort. The housing, feeding, and non-medical care of the animals will be directed by a veterinarian or other scientist trained and experienced in the proper care, handling, and use of the species being maintained or studied. Medical care for animals will be available and provided as necessary by a qualified veterinarian."

Opin. There is no clear requirement for the AV to perform any monitoring that is separate from the IACUC. The regulations prescribe for the AV the task of implementing a program for animal care and health but do not indicate a requirement to monitor animal care or use. The PAM staff typically does not directly participate in clinical animal care and can therefore provide objective oversight of the animal care program. The orientation of PAM assessment is different from that of the clinical veterinary staff or the AV. The clinical veterinary staff and AV are focused on animal disease prevention, husbandry support, and care facility management. In contrast, PAM focuses on adherence to institutional policy or process as defined by the IACUC and institutional SOPs. Confidence and support for PAM activity are critically dependent upon community acceptance of the process and are best accomplished if the AV is considered as another PI whose activities are those of animal care program coordination. The clinical decisions of the AV are not the issue; rather, the PAM process provides a consistent method for addressing concerns with the animal care process. For example, PAM might determine whether animal physical examinations are performed as scheduled or in the manner prescribed by veterinary SOPs or whether aseptic technique is performed in the same rigorous manner required of PIs when the veterinary staff performs surgery. With respect to routine animal care, a SOP can be considered reasonably similar to a research protocol for purposes of PAM.

30:28　How is the PAM process different from IACUC semiannual inspections and program review?

Opin. The goals of the IACUC semiannual review process are markedly different from those of PAM. The IACUC attends to facilities and documented evidence of performance. In contrast, PAM attends to ongoing care and use practices and experimental congruence with IACUC approvals. The IACUC semiannual inspection and review and PAM are not redundant activities, but complementary activities. Together, they provide an enhanced foundation for research integrity by identifying concerns at a very low risk level and correcting these prior to significant program impact. The IACUC semiannual process is required, focused, routine, and performed by very busy volunteer members of the committee. The IACUC inspection process does not routinely capture animal use activities in progress; nor does the semiannual inspection evaluate the preparation, performance, or postprocedure care practices. The IACUC process often rotates the duty of inspection among its members and therefore injects a level of inconsistency into a review system that is uncomfortable for the researchers.

PAM, on the other hand, provides a resource that can dedicate the time to observe all phases of an experiment, giving a top-to-bottom assessment of protocol adherence. PAM can provide a desired level of consistency, allowing the research staff to become comfortable with PAM observations, to understand the expectations of the PAM process, and to be assured that the application of oversight is institutionwide. PAM facilitates communication with the research technical staff (below the level of the PI) that will encourage enhanced communication since the research staff will know exactly whom to call with questions.

30:29 How do smaller institutions get the funding for PAM staff positions?

Opin. The truth is that larger institutions have larger bills, not easier solutions. Funding of any program at the institution is always an issue. Some would consider that this funding is easier for larger institutions than smaller institutions. Every institution has the same issue—assuring animal care and experimentation are conducted as approved by the IACUC. All institutions have the same obligation to appropriate care and use of animals used in research, testing, or teaching. All institutions have limited money and resources, and those never appear to be enough for all of the unfunded mandates or program needs. Some institutions may believe that "they do not have a problem," but what is categorically clear is that most institutions that have taken an honest look at their program have found issues that require attention, in some cases critical and immediate attention. The assertion "We do not have a problem" is a euphemism for "We have not looked or are unwilling to do so." There is no such thing as a perfect program. For those institutions that perform federally funded work, the indirect cost recovery is money partially contributed to the institution to provide for animal program oversight. In some cases, institutions should consider reassessing the allocation of those resources toward infrastructure for the overall program needs. For institutions that do not use federal funds, a reassignment of specified PAM responsibilities to existing staff may be appropriate. Perhaps, nearby institutions might consider sharing a common resource (and therefore distributing costs) of a PAM staff person who would spend a specified number of hours per month in each.

Appendix A

Animal Welfare Act

Public Law 89-544 Act of August 24, 1966 as amended (selected portions and sections of the Act are presented here).

To authorize the Secretary of Agriculture to regulate the transportation, sale, and handling of dogs, cats, and certain other animals intended to be used for purposes of research or experimentation, and for other purposes.

Be it enacted by the Senate and House of Representatives of the United States of America in Congress assembled. That, in order to protect the owners of dogs and cats from theft of such pets, to prevent the sale or use of dogs and cats which have been stolen, and to insure that certain animals intended for use in research facilities are provided humane care and treatment, it is essential to regulate the transportation, purchase, sale, housing, care, handling, and treatment of such animals by persons or organizations engaged in using them for research or experimental purposes in transporting, buying, or selling them for such use.

Sect. 13. The Secretary shall promulgate standards to govern the humane handling, care, treatment, and transportation of animals by dealers, research facilities, and exhibitors. Except as otherwise provided in this Act, nothing in this Act shall be construed as authorizing the Secretary to promulgate rules, regulations, or orders with regard to the performance of actual research or experimentation by a research facility as determined by such research facility.

Nothing in this Act shall be construed as authorizing the Secretary to promulgate rules, regulations, or orders for the handling, care, treatment, or inspection of animals during actual research or experimentation by a research facility as determined by such research facility.

Sect. 16. The Secretary shall make such investigations or inspections as he deems necessary to determine whether any dealer or research facility has violated or is violating any provision of this Act or any regulation issued thereunder. The Secretary shall promulgate such rules and regulations as he deems necessary to permit inspectors to confiscate or destroy in a humane manner any animals found to be suffering as a result of a failure to comply with any provision of the Act or any regulation issued thereunder if (1) such animals are held by a dealer, or (2) such animals are held by a research facility and are no longer required by such research facility to carry out the research, test, or experiment for which such animals have been utilized.

Sect. 19. (a) If the Secretary has reason to believe that any research facility has violated or is violating any provision of this Act or any of the rules or regulations promulgated by the Secretary hereunder and if, after notice and opportunity for hearing, he finds a violation, he may make an order that such research facility shall cease and desist from continuing such violation. Such cease and desist order shall become effective fifteen days after issuance of the order.

Any research facility which knowingly fails to obey a cease-and-desist order made by the Secretary under this section shall be subject to a civil penalty of $1,500 for each offense, and each day during which such failure continues shall be deemed a separate offense.

Sect. 21. The Secretary is authorized to promulgate such rules, regulations, and orders as he may deem necessary in order to effectuate the purposes of this Act.

Appendix B

Animal Welfare Act Regulations

Selected portions of the AWAR (Title 9 of the Code of Federal Regulations ([9 CFR]), Chapter 1, Subchapter A, Parts 1, 2, and 3 are presented here. For ease of reading, the appropriate section from the AWAR is indicated; however, information within such sections may be in abridged form. The original document should be consulted for the full text.

§1.1 Definitions

Animal means any live or dead dog, cat, nonhuman primate, guinea pig, hamster, rabbit, or any other warm-blooded animal, which is being used, or is intended for use for research, teaching, testing, experimentation, or exhibition purposes, or as a pet. This term excludes: birds, rats of the genus *Rattus*, and mice of the genus *Mus*, bred for use in research, and horses not used for research purposes and other farm animals, such as, but not limited to livestock or poultry, used or intended for use as food or fiber, or livestock or poultry used or intended for use for improving animal nutrition, breeding, management, or production efficiency, or for improving the quality of food or fiber. With respect to a dog, the term means all dogs, including those used for hunting, security, or breeding purposes.

Attending veterinarian means a person who has graduated from a veterinary school accredited by the American Veterinary Medical Association's Council on Education, or has a certificate issued by the American Veterinary Medical Association's Education Commission for Foreign Veterinary Graduates, or has received equivalent formal education as determined by the Administrator; has received training and/or experience in the care and management of the species being attended; and who has direct or delegated authority for activities involving animals at a facility subject to the jurisdiction of the Secretary.

Euthanasia means the humane destruction of an animal accomplished by a method that produces rapid unconsciousness and subsequent death without evidence of pain or distress, or a method that utilizes anesthesia produced by an agent that causes painless loss of consciousness and subsequent death.

Farm animal means any domestic species of cattle, sheep, swine, goats, llamas, or horses which are normally and have historically, been kept and raised on farms in the United States, and used or intended for use as food or fiber, or for improving animal nutrition, breeding, management, or production efficiency, or for improving the quality of food or fiber. This term also includes animals such as rabbits, mink, and chinchilla, when they are used solely for purposes of meat or fur, and animals such as horses and llamas when used solely as work and pack animals.

Field study means a study conducted on free-living wild animals in their natural habitat. However, this term excludes any study that involves an invasive procedure, harms, or materially alters the behavior of an animal under study.

Housing facility means any land, premises, shed, barn, building, trailer, or other structure or area housing or intended to house animals.

Institutional official means the individual at a research facility who is authorized to legally commit on behalf of the research facility that the requirements of 9 CFR parts 1, 2, and 3 will be met.

Major operative procedure means any surgical intervention that penetrates and exposes a body cavity or any procedure which produces permanent impairment of physical or physiological functions.

Painful procedure as applied to any animal means any procedure that would reasonably be expected to cause more than slight or momentary pain or distress in a human being to which that procedure was applied, that is, pain in excess of that caused by injections or other minor procedures.

Positive physical contact means petting, stroking, or other touching, which is beneficial to the well-being of the animal.

Principal investigator means an employee of a research facility, or other person associated with a research facility, responsible for a proposal to conduct research and for the design and implementation of research involving animals.

Quorum means a majority of the Committee members.

Research facility means any school (except an elementary or secondary school), institution, organization, or person that uses or intends to use live animals in research, tests, or experiments, and that (1) purchases or transports live animals in commerce, or (2) receives funds under a grant, award, loan, or contract from a department, agency, or instrumentality of the United States for the purpose of carrying out research, tests, or experiments: *Provided*, That the Administrator may exempt, by regulation, any such school, institution, organization, or person that does not use or intend to use live dogs or cats, except those schools, institutions, organizations, or persons which use substantial numbers (as determined by the Administrator) of live animals the principal function of which schools, institutions, organizations, or persons, is biomedical research or testing, when in the judgment of the Administrator, any such exemption does not vitiate the purpose of the Act.

§2.31 Institutional Animal Care and Use Committee (IACUC)

(a) The Chief Executive Officer of the research facility shall appoint an Institutional Animal Care and Use Committee (IACUC), qualified through the experience and expertise of its members to assess the research facility's animal program, facilities, and procedures. Except as specifically authorized by law or these regulations, nothing in this part shall be deemed to permit the Committee or IACUC to prescribe methods or set standards for the design, performance, or conduct of actual research or experimentation by a research facility.

(b) IACUC membership.

(1) The members of each Committee shall be appointed by the Chief Executive Officer of the research facility;

(2) The Committee shall be composed of a Chairman and at least two additional members;

(3) Of the members of the Committee:

(i) At least one shall be a Doctor of Veterinary Medicine with training or experience in laboratory animal science and medicine, who has direct or delegated responsibility for activities involving animals at the research facility;

(ii) At least one shall not be affiliated in any way other than as a member of the Committee, and shall not be a member of the immediate family of a person who is affiliated with the facility. The Secretary intends that such person will provide representation for general community interests in the proper care and treatment of animals;

(4) If the Committee consists of more than three members, not more than three members shall be from the same administrative unit of the facility.

(c) IACUC functions. With respect to activities involving animals, the IACUC as an agent of the research facility shall:

(1) Review, at least once every six months, the research facility's program for humane care and use of animals, using title 9, chapter I, subchapter A—Animal Welfare, as a basis for evaluation;

(2) Inspect, at least once every six months, all of the research facility's animal facilities, including animal study areas, using title 9, chapter I, subchapter A—Animal Welfare, as a basis for evaluation; *Provided, however,* That animal areas containing free-living wild animals in their natural habitat need not be included in such inspection;

(3) Prepare reports of its evaluations conducted as required by paragraphs (c) (1) and (2) of this section, and submit the reports to the Institutional Official of the research facility; *Provided, however,* That the IACUC may determine the best means of conducting evaluations of the research facility's programs and facilities; and *Provided, further,* That no Committee member wishing to participate in any evaluation conducted under this subpart may be excluded. The IACUC may use subcommittees composed of at least two Committee members and may invite *ad hoc* consultants to assist in conducting the evaluations, however, the IACUC remains responsible for the evaluations and reports as required by the Act and regulations. The reports shall be reviewed and signed by a majority of the IACUC members and must include any minority views. The reports shall be updated at least once every six months upon completion of the required semiannual evaluations and shall be maintained by the research facility and made available to APHIS and to officials of funding Federal agencies for inspection and copying upon request. The reports must contain a description of the nature and extent of the research facility's adherence to this subchapter, must identify specifically any departures from the provisions of title 9, chapter I, subchapter A—Animal Welfare, and must state the reasons for each departure. The reports must distinguish significant deficiencies from minor deficiencies. A significant deficiency is one which, with reference to Subchapter A, and, in the judgment of the IACUC and the Institutional Official, is or may be a threat to the health or safety of the animals. If program or facility deficiencies are noted, the reports must contain a reasonable and specific plan and schedule with dates for correcting each deficiency. Any failure to adhere to the plan and schedule that results in a significant deficiency remaining uncorrected shall be reported in writing within 15 business days by the IACUC, through the Institutional Official, to APHIS and any Federal agency funding that activity;

(4) Review, and, if warranted, investigate concerns involving the care and use of animals at the research facility resulting from public complaints received and from reports of noncompliance received from laboratory or research facility personnel or employees;

(5) Make recommendations to the Institutional Official regarding any aspect of the research facility's animal program, facilities, or personnel training;

(6) Review and approve, require modifications in (to secure approval), or withhold approval of those components of proposed activities related to the care and use of animals, as specified in paragraph (d) of this section;

(7) Review and approve, require modifications in (to secure approval) or withhold approval of proposed significant changes regarding the care and use of animals in ongoing activities; and

(8) Be authorized to suspend an activity involving animals in accordance with the specifications set forth in paragraph (d)(6) of this section.

(d) IACUC review of activities involving animals.

(1) In order to approve proposed activities or proposed significant changes in ongoing activities, the IACUC shall conduct a review of those components of the activities related to the care and use of animals and determine that the proposed activities are in accordance with this subchapter unless acceptable justification for a departure is presented in writing; *Provided, however,* That field studies as defined in part 1 of this subchapter are exempt from this requirement. Further, the IACUC shall determine that the proposed activities or significant changes in ongoing activities meet the following requirements:

(i) Procedures involving animals will avoid or minimize discomfort, distress, and pain to the animals;

(ii) The principal investigator has considered alternatives to procedures that may cause more than momentary or slight pain or distress to the animals, and has provided a written narrative description of the methods and sources, *e.g.,* the Animal Welfare Information Center, used to determine that alternatives were not available;

(iii) The principal investigator has provided written assurance that the activities do not unnecessarily duplicate previous experiments;

(iv) Procedures that may cause more than momentary or slight pain or distress to the animals will:

(A) Be performed with appropriate sedatives, analgesics, or anesthetics, unless withholding such agents is justified for scientific reasons, in writing, by the principal investigator and will continue for only the necessary period of time;

(B) Involve, in their planning, consultation with the attending veterinarian or his or her designee;

(C) Not include the use of paralytics without anesthesia;

(v) Animals that would otherwise experience severe or chronic pain or distress that cannot be relieved will be painlessly euthanized at the end of the procedure or, if appropriate, during the procedure;

(vi) The animals' living conditions will be appropriate for their species in accordance with part 3 of this subchapter, and contribute to their health and comfort. The housing, feeding, and nonmedical care of the animals will be directed by the attending veterinarian or other scientist trained and experienced in the proper care, handling, and use of the species being maintained or studied;

(vii) Medical care for animals will be available and provided as necessary by a qualified veterinarian;

(viii) Personnel conducting procedures on the species being maintained or studied will be appropriately qualified and trained in those procedures;

(ix) Activities that involve surgery include appropriate provision for pre-operative and post-operative care of the animals in accordance with established veterinary medical and nursing practices. All survival surgery will be performed using aseptic procedures, including surgical gloves, masks, sterile instruments, and aseptic techniques. Major operative procedures on non-rodents will be conducted only in facilities intended for that purpose which shall be operated and maintained under aseptic conditions. Non-major operative procedures and all surgery on rodents do not require a dedicated facility, but must be performed using aseptic procedures. Operative procedures conducted at field sites need not be performed in dedicated facilities, but must be performed using aseptic procedures;

(x) No animal will be used in more than one major operative procedure from which it is allowed to recover, unless:

(A) Justified for scientific reasons by the principal investigator, in writing;

(B) Required as routine veterinary procedure or to protect the health or well-being of the animal as determined by the attending veterinarian; or

(C) In other special circumstances as determined by the Administrator on an individual basis. Written requests and supporting data should be sent to the Animal and Plant Health Inspection Service, Regulatory Enforcement and Animal Care, Animal Care, 4700 River Road, Unit 84, Riverdale, Maryland 20737-1234.

(xi) Methods of euthanasia used must be in accordance with the definition of the term set forth in 9 CFR part 1, §1.1 of this subchapter, unless a deviation is justified for scientific reasons, in writing, by the investigator.

(2) Prior to IACUC review, each member of the Committee shall be provided with a list of proposed activities to be reviewed. Written descriptions of all proposed activities that involve the care and use of animals shall be available to all IACUC members, and any member of the IACUC may obtain, upon request, full Committee review of those activities. If full Committee review is not requested, at least one member of the IACUC, designated by the chairman and qualified to conduct the review, shall review those activities, and shall have the authority to approve, require modifications in (to secure approval), or request full Committee review of any of those activities. If full Committee review is requested for a proposed activity, approval of that activity may be granted only after review, at a convened meeting of a quorum of the IACUC, and with the approval vote of a majority of the quorum present. No member may participate in the IACUC review or approval of an activity in which that member has a conflicting interest (e.g., is personally involved in the activity), except to provide information requested by the IACUC, nor may a member who has a conflicting interest contribute to the constitution of a quorum;

(3) The IACUC may invite consultants to assist in the review of complex issues arising out of its review of proposed activities. Consultants may not approve or withhold approval of an activity, and may not vote with the IACUC unless they are also members of the IACUC;

(4) The IACUC shall notify principal investigators and the research facility in writing of its decision to approve or withhold approval of those activities related

to the care and use of animals, or of modifications required to secure IACUC approval. If the IACUC decides to withhold approval of an activity, it shall include in its written notification a statement of the reasons for its decision and give the principal investigator an opportunity to respond in person or in writing. The IACUC may reconsider its decision, with documentation in Committee minutes, in light of the information provided by the principal investigator;

(5) The IACUC shall conduct continuing reviews of activities covered by this subchapter at appropriate intervals as determined by the IACUC, but not less than annually;

(6) The IACUC may suspend an activity that it previously approved if it determines that the activity is not being conducted in accordance with the description of that activity provided by the principal investigator and approved by the Committee. The IACUC may suspend an activity only after review of the matter at a convened meeting of a quorum of the IACUC and with the suspension vote of a majority of the quorum present;

(7) If the IACUC suspends an activity involving animals, the Institutional Official, in consultation with the IACUC, shall review the reasons for suspension, take appropriate corrective action, and report that action with a full explanation to APHIS and any Federal agency funding that activity; and

(8) Proposed activities and proposed significant changes in ongoing activities that have been approved by the IACUC may be subject to further appropriate review and approval by officials of the research facility. However, those officials may not approve an activity involving the care and use of animals if it has not been approved by the IACUC.

(e) A proposal to conduct an activity involving animals, or to make a significant change in an ongoing activity involving animals, must contain the following:

(1) Identification of the species and the approximate number of animals to be used;

(2) A rationale for involving animals, and for the appropriateness of the species and numbers of animals to be used;

(3) A complete description of the proposed use of the animals;

(4) A description of procedures designed to assure that discomfort and pain to animals will be limited to that which is unavoidable for the conduct of scientifically valuable research, including provision for the use of analgesic, anesthetic, and tranquilizing drugs where indicated and appropriate to minimize discomfort and pain to animals; and

(5) A description of any euthanasia method to be used.

§2.32 Personnel Qualifications

(a) It shall be the responsibility of the research facility to ensure that all scientists, research technicians, animal technicians, and other personnel involved in animal care, treatment, and use are qualified to perform their duties. This responsibility shall be fulfilled in part through the provision of training and instruction to those personnel.

(b) Training and instruction shall be made available, and the qualifications of personnel reviewed, with sufficient frequency to fulfill the research facility's responsibilities under this section and §2.31.

(c) Training and instruction of personnel must include guidance in at least the following areas:

    (1) Humane methods of animal maintenance and experimentation, including:

        (i) The basic needs of each species of animal;

        (ii) Proper handling and care for the various species of animals used by the facility;

        (iii) Proper pre-procedural and post-procedural care of animals; and

        (iv) Aseptic surgical methods and procedures;

    (2) The concept, availability, and use of research or testing methods that limit the use of animals or minimize animal distress;

    (3) Proper use of anesthetics, analgesics, and tranquilizers for any species of animals used by the facility;

    (4) Methods whereby deficiencies in animal care and treatment are reported, including deficiencies in animal care and treatment reported by any employee of the facility. No facility employee, Committee member, or laboratory personnel shall be discriminated against or be subject to any reprisal for reporting violations of any regulation or standards under the Act;

    (5) Utilization of services (e.g., National Agricultural Library, National Library of Medicine) available to provide information:

        (i) On appropriate methods of animal care and use;

        (ii) On alternatives to the use of live animals in research;

        (iii) That could prevent unintended and unnecessary duplication of research involving animals; and

        (iv) Regarding the intent and requirements of the Act.

§2.35 Recordkeeping Requirements

(a) The research facility shall maintain the following IACUC records:

    (1) Minutes of IACUC meetings, including records of attendance, activities of the Committee, and Committee deliberations;

    (2) Records of proposed activities involving animals and proposed significant changes in activities involving animals, and whether IACUC approval was given or withheld; and

    (3) Records of semiannual IACUC reports and recommendations (including minority views), prepared in accordance with the requirements of §2.31(c)(3) of this subpart, and forwarded to the Institutional Official.

(f) All records and reports shall be maintained for at least three years. Records that relate directly to proposed activities and proposed significant changes in ongoing activities reviewed and approved by the IACUC shall be maintained for the duration of the activity and for an additional three years after completion of the activity. All records shall be available for inspection and copying by authorized APHIS or funding Federal agency representatives at reasonable times. APHIS inspectors will maintain the confidentiality of the information and will not remove the materials from the research facilities' premises unless there has been an alleged violation, they are needed to investigate a possible violation, or for other enforcement purposes. Release of such materials, including reports, summaries, and photographs that contain trade secrets or commercial or financial information that is privileged or confidential will be governed by applicable sections of the Freedom of Information Act. Whenever the Administrator notifies a research facility in writing that specified records shall be retained pending completion of an investigation or proceeding under the Act, the research facility shall hold those records until their disposition is authorized in writing by the Administrator.

Appendix C

Health Research Extension Act of 1985 (P.L. 99-158, November 20, 1985, "Animals in Research")

Sec. 495.

(a) The Secretary, acting through the Director of NIH, shall establish guidelines for the following:

   (1) The proper care of animals to be used in biomedical and behavioral research.

   (2) The proper treatment of animals while being used in such research. Guidelines under this paragraph shall require—

      (A) the appropriate use of tranquilizers, analgesics, anesthetics, paralytics, and euthanasia for animals in such research; and

      (B) appropriate pre-surgical and post-surgical veterinary medical and nursing care for animals in such research.

      Such guidelines shall not be construed to prescribe methods of research.

   (3) The organization and operation of animal care committees in accordance with subsection (b).

(b)

   (1) Guidelines of the Secretary under subsection (a)(3) shall require animal care committees at each entity which conducts biomedical and behavioral research with funds provided under this Act (including the National Institutes of Health and the national research institutes) to assure compliance with the guidelines established under subsection (a).

   (2) Each animal care committee shall be appointed by the chief executive officer of the entity for which the committee is established, shall be composed of not fewer than three members, and shall include at least one individual who has no association with such entity and at least one doctor of veterinary medicine.

   (3) Each animal care committee of a research entity shall—

      (A) review the care and treatment of all animal study areas and facilities of the research entity at least semiannually to evaluate compliance with applicable guidelines established under subsection (a) for appropriate animal care and treatment.

      (B) keep appropriate records of reviews conducted under sub-paragraph (A); and

      (C) for each review conducted under sub-paragraph (A), file with the Director of NIH at least annually (i) a certification that the review has been conducted, and (ii) reports of any violations of guidelines established under subsection (a) or assurances required under paragraph (1) which were observed in such review and which have continued after notice by the committee to the research entity involved of the violations.

Reports filed under subparagraph (C) shall include any minority views filed by members of the committee.

(c) The Director of NIH shall require each applicant for a grant, contract, or cooperative agreement involving research on animals which is administered by the National Institutes of Health or any national research institute to include in its application or contract proposal, submitted after the expiration of the twelve-month period beginning on the date of enactment of this section—

(1) assurances satisfactory to the Director of NIH that—

(A) the applicant meets the requirements of the guidelines established under paragraphs (1) and (2) of subsection (a) and has an animal care committee which meets the requirements of subsection (b); and

(B) scientists, animal technicians, and other personnel involved with animal care, treatment, and use by the applicant have available to them instruction or training in the humane practice of animal maintenance and experimentation, and the concept, availability, and use of research or testing methods that limit the use of animals or limit animal distress; and

(2) a statement of the reasons for the use of animals in the research to be conducted with funds provided under such grant or contract. Notwithstanding subsection (a)(2) of section 553 of title 5, United States Code, regulations under this subsection shall be promulgated in accordance with the notice and comment requirements of such section.

(d) If the Director of NIH determines that—

(1) the conditions of animal care, treatment, or use in an entity which is receiving a grant, contract, or cooperative agreement involving research on animals under this title do not meet applicable guidelines established under subsection (a);

(2) the entity has been notified by the Director of NIH of such determination and has been given a reasonable opportunity to take corrective action; and

(3) no action has been taken by the entity to correct such conditions; the Director of NIH shall suspend or revoke such grant or contract under such conditions as the Director determines appropriate.

(e) No guideline or regulation promulgated under subsection (a) or (c) may require a research entity to disclose publicly trade secrets or commercial or financial information which is privileged or confidential.

Appendix D

Public Health Service Policy on Humane Care and Use of Laboratory Animals

I. Introduction

It is the Policy of the Public Health Service (PHS) to require institutions to establish and maintain proper measures to ensure the appropriate care and use of all animals involved in research, research training and biological testing activities (hereinafter referred to as activities) conducted or supported by the PHS. The PHS endorses the "U.S. Government Principles for the Utilization and Care of Vertebrate Animals Used in Testing, Research, and Training" developed by the Interagency Research Animal Committee. This Policy is intended to implement and supplement those Principles.

II. Applicability

This Policy is applicable to all PHS-conducted or supported activities involving animals, whether the activities are performed at a PHS agency, an awardee institution, or any other institution and conducted in the United States, the Commonwealth of Puerto Rico, or any territory or possession of the United States. Institutions in foreign countries receiving PHS support for activities involving animals shall comply with this Policy, or provide evidence to the PHS that acceptable standards for the humane care and use of the animals in PHS-conducted or supported activities will be met. No PHS support for an activity involving animals will be provided to an individual unless that individual is affiliated with or sponsored by an institution which can and does assume responsibility for compliance with this Policy, unless the individual makes other arrangements with the PHS. This Policy does not affect applicable state or local laws or regulations which impose more stringent standards for the care and use of laboratory animals. All institutions are required to comply, as applicable, with the Animal Welfare Act, and other Federal statutes and regulations relating to animals.

III. Definitions

A. *Animal*—Any live, vertebrate animal used or intended for use in research, research training, experimentation, or biological testing or for related purposes.

B. *Animal Facility*—Any and all buildings, rooms, areas, enclosures, or vehicles, including satellite facilities, used for animal confinement, transport, maintenance, breeding, or experiments inclusive of surgical manipulation. A satellite facility is any containment outside of a core facility or centrally designated or managed area in which animals are housed for more than 24 hours.

D. *Animal Welfare Assurance or Assurance*—The documentation from an institution assuring institutional compliance with this Policy.

F. *Institution*—Any public or private organization, business, or agency (including components of Federal, state, and local governments).

G. *Institutional Official*—An individual who signs, and has the authority to sign the institution's Assurance, making a commitment on behalf of the institution that the requirements of this Policy will be met.

IV. Implementation by Institutions

A. Animal Welfare Assurance—No activity involving animals may be conducted or supported by the PHS until the institution conducting the activity has provided a written Assurance acceptable to the PHS, setting forth compliance with this Policy. Assurances shall be submitted to the Office of Laboratory Animal Welfare (OLAW), Office of the Director, National Institutes of Health. The Assurance shall be signed by the Institutional Official. OLAW will provide the institution with necessary instructions and an example of an acceptable Assurance. All Assurances submitted to the PHS in accordance with this Policy will be evaluated by OLAW to determine the adequacy of the institution's proposed program for the care and use of animals in PHS-conducted or supported activities. On the basis of this evaluation OLAW may approve or disapprove the Assurance, or negotiate an approvable Assurance with the institution. Approval of an Assurance will be for a specified period of time (no longer than five years) after which time the institution must submit a new Assurance to OLAW. OLAW may limit the period during which any particular approved Assurance shall remain effective or otherwise condition, restrict, or withdraw approval. Without an applicable PHS-approved Assurance no PHS-conducted or supported activity involving animals at the institution will be permitted to continue.

1. Institutional Program for Animal Care and Use—The Assurance shall fully describe the institution's program for the care and use of animals in PHS-conducted or supported activities. The PHS requires institutions to use the *Guide for the Care and Use of Laboratory Animals (Guide)* as a basis for developing and implementing an institutional program for activities involving animals. The program description must include the following:

   a. a list of every branch and major component of the institution, as well as a list of every branch and major component of any other institution, which is to be included under the Assurance;

   b. the lines of authority and responsibility for administering the program and ensuring compliance with this Policy;

   c. the qualifications, authority, and responsibility of the veterinarian(s) who will participate in the program and the percent of time each will contribute to the program;

   d. the membership list of the Institutional Animal Care and Use Committee(s) (IACUC) established in accordance with the requirements set forth in IV.A.3 of this Policy.

   e. the procedures which the IACUC will follow to fulfill the requirements set forth in this Policy;

   f. the health program for personnel who work in laboratory animal facilities or have frequent contact with animals;

g. a synopsis of training or instruction in the humane practice of animal care and use, as well as training or instruction in research or testing methods that minimize the number of animals required to obtain valid results and minimize animal distress, offered to scientists, animal technicians, and other personnel involved in animal care, treatment, or use;

h. the gross square footage of each animal facility (including satellite facilities), the species housed therein and the average daily inventory, by species, of animals in each facility; and

i. any other pertinent information requested by OLAW.

2. Institutional Status—Each institution must assure that its program and facilities are in one of the following categories:

Category 1—Accredited by the Association for Assessment and Accreditation of Laboratory Animal Care International (AAALAC). All of the institution's programs and facilities (including satellite facilities) for activities involving animals have been evaluated and accredited by AAALAC, or another accrediting body recognized by PHS. All of the institution's programs and facilities (including satellite facilities) for activities involving animals have also been evaluated by the IACUC and will be reevaluated by the IACUC at least once every six months, in accordance with IV.B.1. and 2. of this Policy, and reports prepared in accordance with IV.B.3. of this Policy.

Category 2—Evaluated by the Institution. All of the institution's programs and facilities (including satellite facilities) for activities involving animals have been evaluated by the IACUC and will be reevaluated by the IACUC at least once every six months, in accordance with IV.B.1. and 2. of this Policy. The most recent semi-annual report of the IACUC evaluation shall be submitted to OLAW with the Assurance.

3. Institutional Animal Care and Use Committee (IACUC)

a. The Chief Executive Officer shall appoint an Institutional Animal Care and Use Committee (IACUC), qualified through the experience and expertise of its members to oversee the institution's animal program, facilities, and procedures.

b. The Assurance must include the names, position titles, and credentials of the IACUC chairperson and the members. The committee shall consist of not less than five members, and shall include at least:

(1) one Doctor of Veterinary Medicine, with training or experience in laboratory animal science and medicine, who has direct or delegated program responsibility for activities involving animals at the institution (see IV.A.1.c);

(2) one practicing scientist experienced in research involving animals;

(3) one member whose primary concerns are in a nonscientific area (for example, ethicist, lawyer, member of the clergy); and

(4) one individual who is not affiliated with the institution in any way other than as a member of the IACUC, and is not a member of the immediate family of a person who is affiliated with the institution.

c. An individual who meets the requirements of more than one of the categories detailed in IV.A.3.b.(1)–(4) of this Policy may fulfill more than one requirement. However, no committee may consist of less than five members.

B. Functions of the Institutional Animal Care and Use Committee (IACUC)—As an agent of the institution, the IACUC shall with respect to PHS-conducted or supported activities:

1. review at least once every six months the institution's program for humane care and use of animals, using the *Guide* as a basis for evaluation.

2. inspect at least once every six months all of the institution's animal facilities (including satellite facilities) using the *Guide* as a basis for evaluation;

3. prepare reports of the IACUC evaluations conducted as required by IV.B.1. and 2. of this Policy, and submit the reports to the Institutional Official. (Note: the reports shall be updated at least once every six months upon completion of the required semiannual evaluations and shall be maintained by the institution and be made available to OLAW upon request. The reports must contain a description of the nature and extent of the institution's adherence to the *Guide* and this Policy and must identify specifically any departures from the provisions of the *Guide* and this Policy and must state the reasons for each departure. The reports must distinguish significant deficiencies from minor deficiencies. A significant deficiency is one which, consistent with this Policy, and, in the judgment of the IACUC and the Institutional Official, is or may be a threat to the health or safety of the animals. If program or facility deficiencies are noted, the reports must contain a reasonable and specific plan and schedule for correcting each deficiency. If some or all of the institution's facilities are accredited by AAALAC International or another accrediting body recognized by PHS, the report should identify those facilities as such.);

4. review concerns involving the care and use of animals at the institution;

5. make recommendations to the Institutional Official regarding any aspect of the institution's animal program, facilities, or personnel training;

6. review and approve, require modifications in (to secure approval) or withhold approval of those components of PHS-conducted or supported activities related to the care and use of animals as specified in IV.C. of this Policy;

7. review and approve, require modifications in (to secure approval) or withhold approval of proposed significant changes regarding the use of animals in ongoing activities; and

8. be authorized to suspend an activity involving animals in accordance with the specifications set forth in IV.C.6 of this Policy.

C. Review of PHS-Conducted or Supported Research Projects

1. In order to approve proposed research projects or proposed significant changes in ongoing research projects, the IACUC shall conduct a review of those components related to the care and use of animals and determine that the proposed research projects are in accordance with this Policy. In making this determination, the IACUC shall confirm that the research project will be conducted in accordance with the Animal Welfare Act insofar as it applies to the research project, and that the research project is consistent with the *Guide* unless acceptable justification for a departure is presented. Further, the IACUC shall determine that the research project conforms with the institution's Assurance and meets the following requirements:

    a. Procedures with animals will avoid or minimize discomfort, distress, and pain to the animal, consistent with sound research design.

b. Procedures that may cause more than momentary or slight pain or distress to the animals will be performed with appropriate sedation, analgesia, or anesthesia, unless the procedure is justified for scientific reasons in writing by the investigator.

c. Animals that would otherwise experience severe or chronic pain or distress that cannot be relieved will be painlessly sacrificed at the end of the procedure or, if appropriate, during the procedure.

d. The living conditions of animals will be appropriate for their species and contribute to their health and comfort. The housing, feeding, and nonmedical care of the animals will be directed by a veterinarian or other scientist trained and experienced in the proper care, handling, and use of the species being maintained or studied.

e. Medical care for animals will be available and provided as necessary by a qualified veterinarian.

f. Personnel conducting procedures on the species being maintained or studied will be appropriately qualified and trained in those procedures.

g. Methods of euthanasia used will be consistent with the recommendations of the American Veterinary Medical Association (AVMA) Panel on Euthanasia, unless a deviation is justified for scientific reasons in writing by the investigator.

2. Prior to the review, each IACUC member shall be provided with a list of proposed research projects to be reviewed. Written descriptions of research projects that involve the care and use of animals shall be available to all IACUC members, and any member of the IACUC may obtain, upon request, full committee review of those research projects. If full committee review is not requested, at least one member of the IACUC, designated by the chairperson and qualified to conduct the review, shall review those research projects and have the authority to approve, require modifications in (to secure approval), or request full committee review of those research projects. If full committee review is requested, approval of those research projects may be granted only after review at a convened meeting of a quorum of the IACUC and with the approval vote of a majority of the quorum present. No member may participate in the IACUC review or approval of a research project in which the member has a conflicting interest (e.g., is personally involved in the project) except to provide information requested by the IACUC; nor may a member who has a conflicting interest contribute to the constitution of a quorum.

3. The IACUC may invite consultants to assist in the review of complex issues. Consultants may not approve or withhold approval of an activity or vote with the IACUC unless they are also members of the IACUC.

4. The IACUC shall notify investigators and the institution in writing of its decision to approve or withhold approval of those activities related to the care and use of animals, or of modifications required to secure IACUC approval. If the IACUC decides to withhold approval of an activity, it shall include in its written notification a statement of the reasons for its decision and give the investigator an opportunity to respond in person or in writing.

5. The IACUC shall conduct continuing review of activities covered by this Policy at appropriate intervals as determined by the IACUC, including a complete review in accordance with IV.C.1.–4. at least once every three years.

6. The IACUC may suspend an activity that it previously approved if it determines that the activity is not being conducted in accordance with applicable provisions of the Animal Welfare Act, the *Guide*, the institution's Assurance, or IV.C.1.a–g. of this Policy. The IACUC may suspend an activity only after review of the matter at a convened meeting of a quorum of the IACUC and with the suspension vote of a majority of the quorum present.

7. If the IACUC suspends an activity involving animals, the Institutional Official in consultation with the IACUC shall review the reasons for suspension, take appropriate corrective action, and report that action with a full explanation to OLAW.

8. Applications and proposals that have been approved by the IACUC may be subject to further appropriate review and approval by officials of the institution. However, those officials may not approve an activity involving the care and use of animals if it has not been approved by the IACUC.

E. Recordkeeping Requirements

1. The awardee institution shall maintain:

    a. a copy of the Assurance which has been approved by the PHS;

    b. minutes of IACUC meetings, including records of attendance, activities of the committee, and committee deliberations;

    c. records of applications, proposals, and proposed significant changes in the care and use of animals and whether IACUC approval was given or withheld;

    d. records of semiannual IACUC reports and recommendations (including minority views) as forwarded to the Institutional Official; and

    e. records of accrediting body determinations.

2. All records shall be maintained for at least three years; records that relate directly to applications, proposals, and proposed significant changes in ongoing activities reviewed and approved by the IACUC shall be maintained for the duration of the activity and for an additional three years after completion of the activity. All records shall be accessible for inspection and copying by authorized OLAW or other PHS representatives at reasonable times and in a reasonable manner.

F. Reporting Requirements

1. At least once every 12 months, the IACUC, through the Institutional Official, shall report in writing to OLAW:

    a. any change in the institution's program or facilities which would place the institution in a different category than specified in its Assurance (see IV.A.2. of this Policy);

    b. any change in the description of the institution's program for animal care and use as required by IV.A.1.a.–i. of this Policy;

    c. any changes in the IACUC membership; and

    d. notice of the dates that the IACUC conducted its semi-annual evaluations of the institution's program and facilities and submitted the evaluations to the Institutional Official.

2. At least once every 12 months, the IACUC, at an institution which has no changes to report as specified in IV.F.1.a.–c. of this Policy, shall report to OLAW in writing, through the Institutional Official, that there are no changes and

inform OLAW of the dates of the required IACUC evaluations and submissions to the Institutional Official.

3. The IACUC, through the Institutional Official, shall promptly provide OLAW with a full explanation of the circumstances and actions taken with respect to:

    a. any serious or continuing noncompliance with this Policy;

    b. any serious deviations from the provisions of the *Guide*; or

    c. any suspension of an activity by the IACUC.

4. Reports filed under IV.F of this Policy shall include any minority views filed by members of the IACUC.

Appendix E

U.S. Government Principles for the Utilization and Care of Vertebrate Animals Used in Testing, Research, and Training

The development of knowledge necessary for the improvement of the health and well-being of humans as well as other animals requires *in vivo* experimentation with a wide variety of animal species. Whenever U.S. Government agencies develop requirements for testing, research, or training procedures involving the use of vertebrate animals, the following principles shall be considered; and whenever these agencies actually perform or sponsor such procedures, the responsible Institutional Official shall ensure that these principles are adhered to:

I. The transportation, care, and use of animals should be in accordance with the Animal Welfare Act (7 U.S.C. 2131 et seq.) and other applicable Federal laws, guidelines, and policies.*

II. Procedures involving animals should be designed and performed with due consideration of their relevance to human or animal health, the advancement of knowledge, or the good of society.

III. The animals selected for a procedure should be of an appropriate species and quality and the minimum number required to obtain valid results. Methods such as mathematical models, computer simulation, and *in vitro* biological systems should be considered.

IV. Proper use of animals, including the avoidance or minimization of discomfort, distress, and pain when consistent with sound scientific practices, is imperative. Unless the contrary is established, investigators should consider that procedures that cause pain or distress in human beings may cause pain or distress in other animals.

V. Procedures with animals that may cause more than momentary pain or slight pain or distress should be performed with appropriate sedation, analgesia, or anesthesia. Surgical or other painful procedures should not be performed on unanesthetized animals paralyzed by chemical agents.

VI. Animals that would otherwise suffer severe or chronic pain or distress that cannot be relieved should be painlessly killed at the end of the procedure or, if appropriate, during the procedure.

---

* For guidance throughout these principles, the reader is referred to the Guide for the Care and Use of Laboratory Animals prepared by the Institute of Laboratory Animal Resources, National Academy of Sciences.

VII. The living conditions of animals should be appropriate for their species and contribute to their health and comfort. Normally, the housing, feeding, and care of all animals used for biomedical purposes must be directed by a veterinarian or other scientist trained and experienced in the proper care, handling, and use of the species being maintained or studied. In any case, veterinary care shall be provided as indicated.

VIII. Investigators and other personnel shall be appropriately qualified and experienced for conducting procedures on living animals. Adequate arrangements shall be made for their in-service training, including the proper and humane care and use of laboratory animals.

IX. Where exceptions are required in relation to the provisions of these Principles, the decisions should not rest with the investigators directly concerned but should be made, with due regard to Principle II, by an appropriate review group such as an institutional animal care and use committee. Such exceptions should not be made solely for the purposes of teaching or demonstration.

Index

A

AAALAC, 2–3, 23:11
Abdominal surgery, 16:15, *see also* Surgery
Abnormal behavior, 27:22
Abnormalities, pain and distress, 16:53, 16:54
Academic freedom/privilege, 22:15
Academic institutions, *see also* Educational facilities; Institutions
    Institutions
  faculty rank of Chairperson, 5:18
  member appointment responsibility, 3:2
  reciprocal course admission, 5:32
  serving as institutional official, 4:5
Accessibility
  facility inspection and program review, 23:39
  IACUC protocols, 8:19
  veterinary care assessment, 27:19
Acclimation, 27:1
Accounting, number of animals
  breeding colonies, 14:16
  precise prediction, 14:15
  tracking, 14:13, 14:14
  in utero accounting, 14:19
Acepromazine, 17:11, 17:13
Acquisition and disposition, animals
  additional animal requests, 14:39
  adoption as pets, 14:47
  advance acquisition, 14:7, 14:8
  agricultural purposes, 14:3
  amphibians, 14:4
  animal control institution tissue and body parts, 14:29
  animal sources, 14:1
  antemortem manipulation, 14:32
  avian embryos, 14:23
  AWAR guidelines, 14:4, 14:5
  birds, 14:5, 14:23
  blood acquisition, 14:29
  body parts acquisition, 14:29, 14:30, 14:32
  breeding colonies, 14:16, 14:17, 14:18, 14:27
  commercial farms, 14:36
  compensation animals, 14:26
  defining acquired, 14:22
  duplication prevention, 14:39
  ectotherms, 14:4
  educational purposes, 14:42, 14:43
  embryonated bird eggs, 14:23
  endangered species, 14:40
  euthanasia, 14:18, 14:30, 14:31, 14:32
  excessive acquisition prevention, 14:27
  extra animals sent, 14:21

farm animal species, 14:48
field studies, 14:25
fish, 14:4, 14:24
foreign countries as tissue source, 14:33
hospital affiliations, 14:45
human hospitals, 14:45
information, 14:17
institutionally owned animals, 14:42
institutional programs, 14:41, 14:43
interstudy transfers, 14:37, 14:38
justification for number, 14:39
land grant college, 14:3
legal requirements, 14:2, 14:46
livestock, 14:48
maintained-only animals, 14:8
medical center affiliations, 14:45
nonsurvival training preparations, 14:26
number acquired and generated, 14:13, 14:14, 14:16
number justification, 14:39
ownership of animals, 14:42, 14:43
pets, 14:42, 14:44, 14:47
PHS Policy, 14:4, 14:5, 14:24
precise accounting, 14:15
preventing unnecessary duplication, 14:39
preweanling rodents, 14:19, 14:20
principal investigators, 14:21, 14:23, 14:29
privately owed animals, 14:42, 14:44, 14:47
procurement, 14:2, 14:7, 14:8, 14:9, 14:12, 14:13
programs related to veterinary care, 14:41
regulated species, 14:6
renewals of protocols, 14:39
reptiles, 14:4
requests for additional animals, 14:39
requirements for procurement, 14:2
retail pet stores, 14:35
rodents, 14:19
Seeing Eye dogs, 14:45
shared tissues, 14:30, 14:31
slaughterhouses, 14:28, 14:29, 14:48
sources of animals, 14:1
species' record keeping, 14:6
spontaneous mortality, 14:26
studies, transfers between, 14:37, 14:38
surgically modified animal acquisition, 14:34
technical colleges, 14:43
tissue acquisition, 14:2, 14:28, 14:30, 14:33
transfers between studies, 14:37, 14:38
unknown health status, 14:12
unnecessary duplication, 14:39
unplanned deaths, 14:26
unweaned animals, 14:19